Ecological Studies, Vol. 168

Analysis and Synthesis

Ecological Studies

Volumes published since 1997 are listed at the end of this book.

Springer-Verlag Berlin Heidelberg GmbH

H. de Kroon E.J.W. Visser (Eds.)

Root Ecology

With 72 Figures, 2 in Color, and 27 Tables

Springer

Prof. Dr. Hans de Kroon
Dr. Eric J. W. Visser

Section of Experimental Plant Ecology
Department of Ecology
University of Nijmegen
Toernooiveld
6525 ED Nijmegen
The Nederlands

ISSN 0070-8356

Library of Congress Cataloging-in-Publication Data

Root ecology / H. de Kroon, E.J.W. Visser (eds.)
p. cm. -- (Ecological studies, ISSN 0070-8356 ; vol. 168)
Includes bibliographical references and index.
ISBN 978-3-642-05520-1 ISBN 978-3-662-09784-7 (eBook)
DOI 10.1007/978-3-662-09784-7

1. Roots (Botany)--Ecology. I. Kroon, Hans de. II. Visser, E. J. W. (Eric J. W.), 1966-
III. Ecological studies ; v. 168.

QK644.R652 2003
575.5'4--dc21 2003041547

http://www.springer.de

Originally published by Springer-Verlag Berlin Heidelberg New York in 2003
Softcover reprint of the hardcover 1st edition 2003

Production: Friedmut Kröner, 69115 Heidelberg, Germany
Cover design: *design & production* GmbH, Heidelberg
Typesetting: Kröner, 69115 Heidelberg, Germany

31/31111 - 5 4 3 2 1 - Printed on acid free paper SPIN 11369011

Preface

The early vascular plants that invaded the land had a very simple morphology. Typically consisting of a rhizomatous axis with vertical aerial axes placed on top, they had a low degree of organ differentiation. Unlike the situation experienced by their aquatic ancestors, the source of water and mineral nutrients was located in the soil. To aid the uptake of these resources, these primitive plants possessed only rhizoids, root hair-like outgrowths from the rhizomatous axis providing some anchorage and increasing the surface area by which the plants had contact with the soil (Mogie and Hutchings 1990).

Much has happened since. In the course of evolution, a great variety of root systems developed that have overcome the many physical, biochemical and biological problems encountered in soil. It is the variety of advanced mechanisms by which roots have adapted to life in soil and the complex role of roots within the soil ecosystem that make roots a fascinating object of scientific study. This volume gives an overview of our current understanding of these mechanisms and roles, and suggestions for how to further deepen our insight into the ecology of roots.

We now know that roots are as extensive and important to plant growth and fitness as the plant's aboveground structures. However, roots have been rightfully coined the "the hidden half" (Waisel et al. 1996) because an appreciation of their significance has come rather late. The ignorance of the crucial role of roots for plant life has gradually disappeared as more information on the functioning of roots has seen the light of day. Recent scientific progress has depended strongly on sophisticated methodologies. Novel techniques continue to be developed (Smit et al. 2000) and so, in this volume, a number of chapters have sections on methods. This is an expression of an innovative field of research and much more is likely to be revealed in the future. Particular challenges are detailed in the "Summary and Prospects" sections that every chapter (except the opening chapter) concludes with.

The volume starts with an overview of the form and function of roots and the many problems that they encounter by life in soil (Chap. 1), introducing many of the topics that are discussed in more detail in the chapters that follow. Chapter 2 describes the spatial distribution of roots, including the responses to heterogeneous soils that are interpreted in terms of foraging for nutrients and water. Chapter 3 deals with root distribution in time, by reviewing the knowledge on turnover of roots in various ecosystems and their implications for ecosystem processes.

The following five chapters provide physiological background to the basic functions of roots, including carbon in- and output (Chap. 4) and water and solute movement (Chap. 5), and the physiological and morphological solutions that roots have developed to cope with three major abiotic stresses, i.e. hard soil structure (Chap. 6), drought (Chap. 7) and flooding (Chap. 8). The carbon balance in roots largely determines the growth of a root system, and is therefore crucial not only for our understanding of root proliferation, but also for the role of roots as carbon source for the soil ecosystem (Chap. 4). The allocation of such carbon compounds and the transport of nutrients taken up by the roots depend on a carefully controlled hydraulic balance (discussed in Chap. 5). Nutrient and water uptake, two prime tasks of a root system, may be severely hampered if a root cannot penetrate the bulk soil (Chap. 6), encounters low soil water potential (Chap. 7) or low oxygen concentrations (Chap. 8). The regulatory control of the anatomical and morphological changes that enable roots to overcome such adverse conditions is greatly similar among stresses, and based on key plant hormones such as ethylene and abscisic acid. Additionally, specific biochemical pathways add to the resistance of the roots to these extreme habitats (Chaps. 7 and 8).

The volume concludes with six chapters on biotic interactions emphasising the complex soil ecosystem that roots influence and, vice versa, influences the roots. Roots have evolved symbiotic interactions with mycorrhiza (Chap. 11), rhizobia (Chap. 12) and soil bacteria (Chaps. 12 and 13) that assist in the capture of soil resources such as nitrogen and phosphorus that are often in short supply. Roots compete for these resources with other roots (Chap. 9) and with soil microorganisms (Chap. 12) and are an important food source for a variety of soil herbivores (Chap. 14). Many of the biotic interactions involve the exudation of organic substances (Chap. 10) and release of gases, such as oxygen in flooded soils (Chap. 13). In this way, roots possess an array of intriguing mechanisms by which they manipulate the soil environment and its biota, facilitating the growth of soil bacteria that promote plant growth or suppress diseases (Chap. 12), stimulate microbial processes that accelerate soil nutrient cycling (Chap. 13), or provide a chemically hostile environment for competitor plants (Chap. 10).

We would like to end this preface with some words of thanks to the people who made this volume possible. First, and most importantly, we are grateful to the authors for their willingness to put their ideas into this volume, resulting in the creation of particularly challenging chapters. We also owe the many referees that have helped the authors to further improve their original contributions. Kees Blom, Professor in Experimental Plant Ecology at the University of

Nijmegen, initiated this project and selected and invited the various authors. His current position as Vice Chancellor has prevented him from completing his task as an editor, but we acknowledge his indispensable input into the early phases of the project. We finally thank Jose Broekmans for her assistance in the final stages of formatting and checking the final manuscript.

Hans de Kroon
Eric J.W. Visser Nijmegen, January 2003

References

Mogie M, Hutchings MJ (1990) Phylogeny, ontogeny and clonal growth in vascular plants. In: van Groenendael J, de Kroon H (eds) Clonal growth in plants – regulation and function. SPB Academic Publishing, The Hague, pp 3–22
Smit AL, Bengough AG, Engels C, van Noordwijk M, Pellerin S, van de Geijn SC (eds) (2000) Root methods. A handbook. Springer, Berlin Heidelberg New York
Waisel Y, Eshel A, Kafkafi U (eds) (1996) Plant roots. The hidden half, 2nd edn. Marcel Dekker, New York

Contents

Contributors

M.A. Bacon

The Lancaster Environment Centre, Lancaster University, Bailrigg,
Lancaster LA1 4YQ, UK

P.A.H.M. Bakker

Faculty of Biology, Section of Phytopathology, Utrecht University,
P.O. Box 800.84, 3508 TB Utrecht, The Netherlands

A.G. Bengough

Scottish Crop Research Institute,
Dundee DD2 5DA, UK

P.L.E. Bodelier

Netherlands Institute of Ecology, Centre for Limnology, Department of
Microbial Ecology, P.O. Box 1299, 3600 BG Maarssen, The Netherlands

W.J. Davies

The Lancaster Environment Centre, Lancaster University, Bailrigg,
Lancaster LA1 4YQ, UK

J.F. Farrar

School of Biological Sciences and School of Agricultural and Forest
Sciences, University of Wales Bangor, Bangor, Gwynedd LL57 2UW, UK

A. Fitter

Biology Department, University of York, P.O. Box 373, York YO10 5YW, UK

R. Gill

Department of Botany, Duke University, Durham, North Carolina 27705,
USA

A. Hodge

Biology Department, University of York, P.O. Box 373, York YO10 5YW, UK

M.J. Hutchings

School of Biological Sciences, University of Sussex, Falmer, Brighton, East Sussex BN1 9QG, UK

Inderjit

Department of Botany, University of Delhi, Delhi 110007, India

M.B. Jackson

School of Biological Sciences, University of Bristol, Woodland Road, Bristol BS8 1UG, UK

E.A. John

School of Biological Sciences, University of Sussex, Falmer, Brighton, East Sussex BN1 9QG, UK

D.L. Jones

School of Biological Sciences and School of Agricultural and Forest Sciences, University of Wales Bangor, Bangor, Gwynedd LL57 2UW, UK

H. de Kroon

Section of Experimental Plant Ecology, Department of Ecology, University of Nijmegen, Toernooiveld, 6525 ED Nijmegen, The Netherlands

W.K. Lauenroth

Department of Rangeland Ecosystem Science, Colorado State University, Fort Collins, Colorado 80523, USA

L.C. van Loon

Faculty of Biology, Section of Phytopathology, Utrecht University, P.O. Box 800.84, 3508 TB Utrecht, The Netherlands

L. Mommer

Section of Experimental Plant Ecology, Department of Ecology, University of Nijmegen, Toernooiveld, 6525 ED Nijmegen, The Netherlands

A. Nishiwaki

Department of Restoration Ecology, Faculty of Agriculture, Miyazaki University, 1–1 Gakuen Kibanadai Nishi, Miyazaki 889-2192, Japan

B. Ricard

INRA, Station de Physiologie Végétale,
BP 81, 33883 Villenave d'Ornon, Cedex, France

D. Robinson

Plant and Soil Science, School of Biological Sciences,
University of Aberdeen, Aberdeen AB24 3UU, UK

F.A. Smith

Department of Soil and Water, The University of Adelaide,
SA 5005, Australia

S.E. Smith

Department of Soil and Water, The University of Adelaide,
SA 5005, Australia

S. Timonen

Division of General Microbiology, Department of Biosciences,
P.O. Box 56, 00014, University of Helsinki, Finland

M.T. Tyree

Aiken Forestry Sciences Laboratory, USDA Forest Service,
705 Spear St, P.O. Box 968, Burlington, Vermont 05402, USA

L.A. Weston

Department of Horticulture, Cornell University, Plant Science Building,
Ithaca, New York 14853, USA

J.B. Whittaker

The Lancaster Environment Center, Lancaster University,
Lancaster LA1 4YQ, UK

1 Constraints on the Form and Function of Root Systems

D. Robinson, A. Hodge and A. Fitter

1.1 Introduction

This chapter sets the scene for many of the topics covered in detail later in this volume. We discuss first the basic problems that plants face when growing on land. These problems reflect the many physical, chemical and biological constraints that soil imposes on the functioning of roots in terms of growth and resource capture.

Second, we consider how these constraints are overcome or minimised by fundamental structural and physiological features of root systems. Such features include a capacity to penetrate soil pores, to branch hierarchically, to absorb and transport unpredictably available water and solute supplies, and to maintain and replace their constituent parts.

We then explore the ecologically important properties of root systems that emerge as a consequence of their 'primary' features. Some of the most important of these 'emergent properties' are the topology of the root system, its size and capacity for anchorage, and its relations with rhizosphere microbes, symbiotic or otherwise.

Little of what we discuss involves mechanisms. Much new information about the physiological and developmental control of root system form and function is being discovered for a few model species, e.g. *Arabidopsis thaliana* (Zhang et al. 1999; Forde and Lorenzo 2001; Williamson et al. 2001) and *Zea mays* (McCully 1999). Likewise, the molecular interactions between roots and soil microbes have been well documented for particular processes, e.g. rhizobia-induced nodulation of legume roots (Heidstra and Bisseling 1996) and the formation of arbuscular mycorrhizas (Harrison 1997), but our understanding of these processes in a wider range of plant taxa remains incomplete. Yet that comparative ignorance does not prevent significant advances being made at the larger, ecological scales of inquiry that we consider here.

Ecological Studies, Vol. 168
H. de Kroon, E.J.W. Visser (Eds.)
Root Ecology

1.2 Problems Associated with Life in Soil

When photoautotrophic plants colonised land (or, more accurately, invaded the air: Niklas 1997, p. 165), they faced some novel problems. They gained access to more light with a spectral composition better suited to efficient photosynthesis, but they lost the physical support provided by water, risking gravitational compression, and free access to nutrients and water, risking dehydration (Niklas 1997, p. 252). However, soil is obviously opaque: light cannot penetrate more than a few millimetres (Tester and Morris 1987), and belowground parts of plants therefore became dependent on the shoots for their carbon (C) supply. The invisibility of roots in soil also provides operational problems for root biologists: techniques such as computer-assisted tomography and magnetic resonance imaging (Asseng et al. 2000) or minirhizotrons (Chap. 3) must be used to visualise them with minimal disturbance.

Once plants had colonised land, the main water and nutrient reservoir became the soil. Soil provides plants with relatively predictable, long-term supplies of nutrients and water, and a secure anchorage. This is possible because of distinct physical, chemical and biological properties of soil. These same properties, however, also constrained how root systems evolved, and continue to constrain how they function.

1.2.1 Physical Problems

The diameter of soil pores (from <0.1 μm to >5 mm, depending on soil type and texture: Brady and Weil 1999, p. 147) allows water to be retained within them by polar forces, hydrogen bonding, cohesion, adhesion and surface tension. Water moves through pores, and drains from them, at rates dependent on pore diameter: large pores are more hydraulically conductive and correspondingly less retentive than narrow pores. The water-filled pores in soil are the pathways through which water and dissolved nutrients reach plant roots. Water moves convectively through soil pores towards the roots in response to the suction generated by transpiring plants. Solutes move in that mass flow of water and they also diffuse through the water down their concentration gradients (Tinker and Nye 2000). Most land plants acquire nutrients from the soil water via their roots, which can absorb ions directly or through their associated mycorrhiza-forming fungi (Chap. 11).

The tortuosity of the soil's pore network reduces the speed at which water and solutes can move from bulk soil to root surfaces compared with their rates of movement in free solution, potentially limiting their rates of uptake by roots and microbes (Tinker and Nye 2000).

Variations in bulk density, water content and particle size distribution influence soil strength, the extent to which soil resists deformation. A root penetrates soil by deforming it locally (Chap. 6). If a root cannot deform the soil around it, its penetration is opposed by friction and it risks being abraded, buckled and damaged. Soil strength also determines the ability of plants to anchor themselves: it declines with increasing wetness (and also in very dry soils) and this can result in the catastrophic windthrow of trees after heavy rain (Coutts 1986).

1.2.2 Chemical Reactivity

Ions dissolved in soil water exchange with those held electrostatically on the charged surfaces of soil solids including organic matter, and primary and secondary minerals (Brady and Weil 1999, p. 18). These surfaces act as reservoirs that can feed nutrients into, or remove ions from, the soil water in response to disturbances in electrochemical equilibria. Ion exchange between solid and solution phases can partly buffer the depletion of certain ions (e.g. phosphate, potassium) from the solution phase. The concentration of well-buffered solutes may be depleted in the soil around a root compared with their concentrations in bulk soil if the rate of uptake exceeds the rate of supply by diffusion and mass flow (Tinker and Nye 2000).

1.2.3 Biological Activity

Heterotrophic soil microbes decompose organic matter derived from plant, microbial and animal sources. This process liberates CO_2 into the atmosphere (a key link in the terrestrial C cycle) and releases plant-available nutrients (Chap. 13), especially nitrogen (N) and phosphorus (P). The living and dead organic fractions of soil are the major repositories of C and N in soil.

Many important ecological processes occur in soil, although they are often difficult to study. Predator–prey interactions among soil microbes and fauna such as amoebae, nematodes and earthworms and the resulting excretion of waste products drive much of the soil's biogeochemical processes. Plant–herbivore interactions occur when roots are grazed by arthropods and nematodes. Competition for limiting resources such as N can occur between and among microbes and plants and, as with most competitive interactions, it is hard to predict the eventual winner (Hodge et al. 2000). In addition, the soil harbours many potential plant pathogens (e.g. plant parasitic nematodes, fungal pathogens, soil-borne viruses).

All of this biological activity and diversity makes soil perhaps the most complex of all ecosystems. It also means that simple laboratory systems (e.g. Petri dishes, hydroponics) bear scant resemblance to a root system's normal environment. Clean, controlled laboratory conditions are needed to study particular physiological and molecular mechanisms, but they have limited ecological relevance because interactions among organisms and processes are of most interest.

1.2.4 Heterogeneity

Soil is not a well-mixed medium. Interactions between climate, parent rock, topography, vegetation, biological activity and the time over which these interactions occur result in soil formation and generate heterogeneity (Rowell 1994, p. 1). These interactions do not occur uniformly and so soil is a dynamic, three-dimensional mosaic whose properties vary from point to point.

Nutrients and water tend to be distributed patchily in soil, depending on the pore-size distribution and localised inputs of organic matter. Gases may also be distributed non-uniformly. In a strongly structured, biologically active soil, anaerobic microsites can persist at the centre of aggregates (Sexstone et al. 1985). Soil bulk density is also spatially variable.

A root system will never experience exactly the same solute concentrations, water potentials or penetration resistances simultaneously over its entire surface. Spatial variation in the resistance of soil to penetration can cause roots to clump in confined soil volumes. The resulting non-uniformity of root distribution can limit rates of water and nutrient capture by the vegetation (Passioura 1991). Consequently, each field site has a unique pattern of heterogeneity. Non-standard statistical techniques are often required to deal with the large spatial and temporal variability inherent to soil data (Ver Hoef and Cressie 2001).

1.3 Evolutionary Solutions

The constraints imposed by soil on photoautotrophic activity were major spurs to the evolutionary modifications to plant morphology and physiology that occurred following the colonisation of land (Raven and Edwards 2001). Nutrients and water are extracted from soil most effectively, and with greatest competitive advantage, if absorbing organs are placed in intimate contact with water-holding pores, ion-exchanging surfaces and regions of microbial activity, and if the positions and activities of those organs are not restricted in time and space. How could this be achieved? The basic 'design requirements', and their solutions, are summarised in Fig. 1.1 and described in detail below.

Figure 1.1 defines the *Bauplan* of a root system: an idealised, generalised, archetype (Niklas 2000). It is easy to argue retrospectively the general selective advantages of this *Bauplan*; after all, this *is* what evolved. Nevertheless, it is difficult to imagine any realistic alternatives to it (Harper et al. 1991).

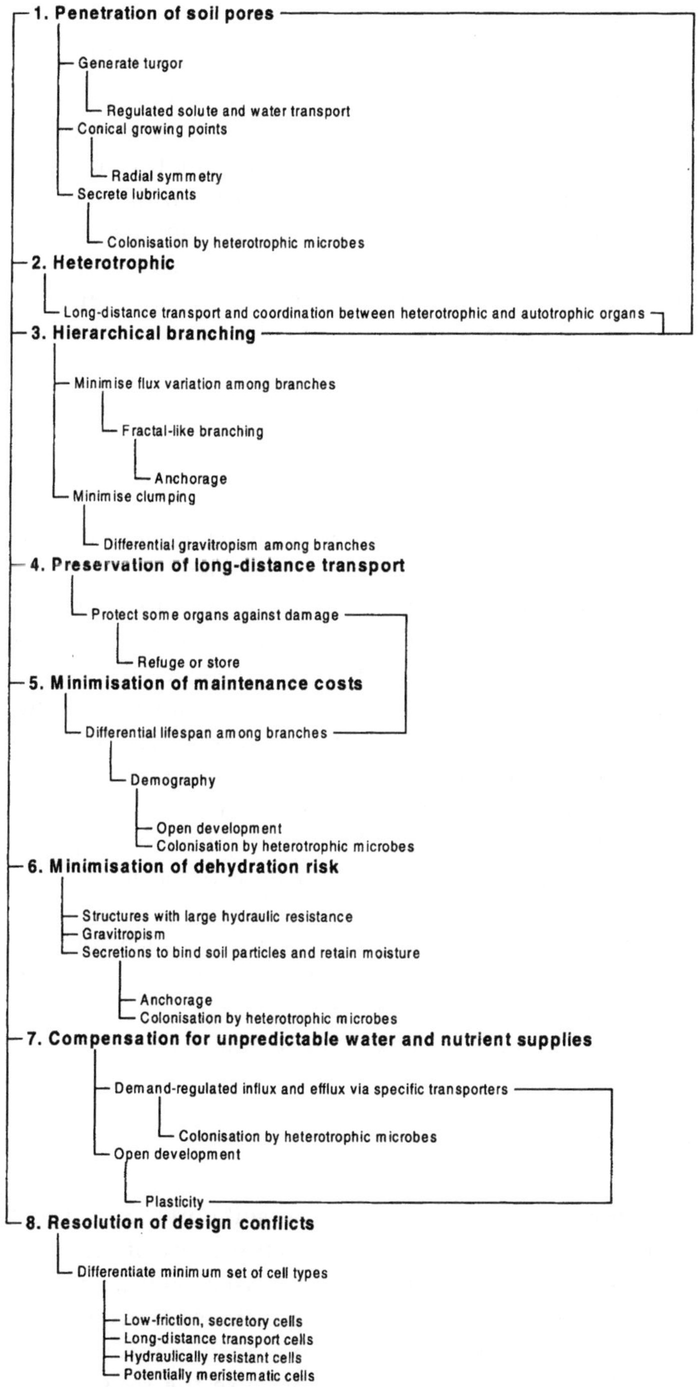

Fig. 1.1. Summary of the 'design requirements' for root systems. Primary requirements are in *bold*; also shown are some of the main secondary consequences arising from these. See Section 1.3 for details

1.3.1 Penetration of Soil Pores

Roots reduce root-soil friction by generating turgor behind a conical, apical growing point, and by secreting lubricants (Chap. 6). From these requirements, it follows that water and solute fluxes into the growing tip should be regulated to achieve and maintain sufficient turgor to deform the soil (see Sect. 1.3.7), root anatomy be radially symmetrical, and that root diameter be constrained to allow the penetration of narrow, water-filled pores.

Some of the organic lubricants secreted would inevitably be consumed by heterotrophic microbes in the soil, and lead to their colonisation of the root, the surrounding soil, or both (see Sects. 1.3.5–1.3.7).

1.3.2 Heterotrophy

Because they cannot fix CO_2 autotrophically, roots must import C from shoots. In return, roots provide shoots with water and nutrients. This mutual exchange of resources can occur only if root and shoot systems are connected by efficient long-distance mass-transport systems (Raven 1977; Chap. 5). 'Long-distance' in this context means more than the relatively few cell lengths beyond which molecular diffusion becomes an ineffective transport mechanism (Nobel 1991, p. 18). The most efficient transport vessels are the widest which, in turn, must be accommodated in wide roots. This requirement conflicts with that in Section 1.3.1 for narrow roots and is one reason why root systems branch hierarchically (see Sect. 1.3.3). Roots elongate at their tips, but expand radially once elongation is complete, allowing generation of large-diameter vessels in situ.

Farrar and Jones (Chap. 4) describe the mechanisms by which root and shoot systems regulate each other's activities, and how these interactions are manifested as coordinated changes in root and shoot growth.

The 'loss' of C from roots into soil is discussed in Section 1.4.5.

1.3.3 Hierarchical Branching

Thin roots are required for the penetration of narrow soil pores (Sect. 1.3.1), and fat roots are required for the efficient internal transport of resources (Sect. 1.3.2). Therefore, root systems must branch hierarchically, i.e. narrow branches arise from wider ones. Similarly, the long-distance transport systems contained within roots (and shoots) must also branch in this way. This has potential problems for resource distribution because the resistance to pressure-driven flow depends critically on vessel diameter and it increases with path length as the plant grows. A hierarchically branched network would tend to deliver resources preferentially to the ends of the shortest transport paths at the expense of more distant points in the system. West et al. (1999) suggested how a branched transport system should be constructed to overcome this difficulty. Their model is encapsulated by the equation:

$$Z_i = \left[\frac{1 - \left\{ \left(n^{1/3} - 1 \right) \left(l_T / l_N \right) \right\}^{(1-6b)}}{1 - n^{(1-6b)/3}} \right] Z_N \tag{1.1}$$

where Z_i is the resistance to flow through the entire system and Z_N the resistance through the terminal (i.e., apical) branches. n is the branching ratio, i.e. the number of daughter branches that emerge from a parental node; for root systems, typically n=2 (i.e. branching is trivalent). l_T is the total path length of the system and l_N the length of the apical branches. b is the rate at which vessel diameter changes at each level in the branching hierarchy, i.e.

$$b = -2 \frac{\ln\left(o_{k+1} / o_k \right)}{\ln(n)} \tag{1.2}$$

where o is the vessel radius at branch level k, with k=0 being the widest and k=N the narrowest, and N is the number of branching levels in the system. Equation (1.1) predicts that the total resistance, Z_i, is constant if $b \approx 1/6$, and is then independent of N and l_T, and the system is then able to transport materials equally rapidly to all of its parts irrespective of their distances from each other.

The hydraulic behaviour of *Acer saccharum* (sugar maple) shoots is described well by Eq. (1.1) with $b \approx 1/6$ (West et al. 1999). This has not yet been verified for the transport properties of root systems, but other evidence suggests that it will probably apply to them, too. Equation (1.1) predicts that root systems should have fractal dimensions of 1.5 in three-dimensional space, or of 1.33 in two dimensions. Fitter and Stickland (1992a) showed that the fractal dimension of *Trifolium pratense* (red clover) root systems grown between narrowly separated (≈2-D) glass sheets increased from 1.24 to 1.39 up to 27 days after germination, consistent with the model of West et al. (1999). Equation (1.1) offers a general biophysical explanation of why root systems branch as they do.

However, for any root system to achieve an efficient branching network, successive orders of roots should also be differentially gravitropic. If all roots grew vertically downwards, for example, severe root clumping would result with potentially disastrous consequences for resource capture and fitness.

Anchorage arises as a secondary property of the branching of a root system between soil particles (Sect. 1.4.4), as does the often-exponential decline in root abundance with soil depth (Sect. 1.4.3).

1.3.4 Long-Distance Transport

Because of the mutual interdependence of root and shoot systems, transport between them is vital for the continued survival of the plant. While damage to peripheral parts of branched root and shoot systems would not necessarily imperil the whole plant, the loss of main conducting roots or stems would. For this reason, main roots tend to be more durable than peripheral ones, as is obvious in woody species, but can also be deduced from the prolonged survival of individual grass roots in cohorts otherwise showing exponential decay (e.g. Fitter et al. 1997). This durability allows, in perennial plants, some roots and underground shoots to act as reservoirs of stored assimilates or of meristems that are relatively well protected from damage by aboveground herbivores and, in cold habitats, by frost.

If some organs are relatively well protected against, e.g. dehydration (Sect. 1.3.6) or grazing to prolong their functional life, this implies that others have less protection and will, on average, die sooner. Root demography (Sect. 1.3.7) is the inevitable result of this differential root lifespan.

1.3.5 Maintenance Costs

Once produced by a plant, a structure has to be maintained to keep it alive and functioning. Maintaining heterotrophic organs such as roots imposes a potential metabolic burden on the autotrophic parts, especially if the contribution of the former to water and nutrient capture diminishes as they deplete the soil around them of these resources. Economic models (Eissenstat and Yanai 1997) based on the investment (C) needed to achieve a given return (water and nutrient uptake) predict an optimum root lifespan beyond which further maintenance is a net drain on the plant's resources. Measurements show that roots do live for various lengths of time depending on species and environment (Chap. 3), although the extent to which this variation is explained by the need to minimise maintenance costs is unknown.

Whatever the physiological causes, the result is that, within a root system, roots are born, live for a certain period, and then die. Such patterns can be analysed demographically (Chap. 3). This is possible because roots have 'open and indeterminate' development (Niklas 2000) in which meristems can arise from non-meristematic tissue (Sect. 1.3.8) and perpetuate growth even if some meristems are lost. The branching structure of a root system at one time is, therefore, only a snapshot of the current balance between root births and deaths. No geometrical model of root system branching (cf. Sects. 1.3.3 and 1.4.1) yet includes a satisfactory description of turnover, and this is an area in which much remains to be done. Open and indeterminate development is responsible for the gradual accumulation of often massive amounts of root material in perennial vegetation (Sect. 1.4.2).

When roots grow, and more obviously when they die, they release some of their contents and structural material into the surrounding soil. This 'rhizodeposition' (Sect. 1.4.5) is a major source of plant-derived matter into the soil, and an important substrate for heterotrophic microbes.

1.3.6 Dehydration Risk

Roots in soil risk dehydration if the soil's water potential falls below that of the plant, as is common in arid areas and possible in dry topsoil even at wetter sites. The easiest way to minimise dehydration risk is for roots to grow into wetter (=deeper) soil, and this is another reason why roots are gravitropic and why the deepest roots are found in hot, dry habitats (Sect. 1.4.3).

A conspicuous anatomical feature of most mature roots is the presence of an endodermis between the stele and cortex (see Steudle and Peterson 1998). Endodermal cells are separated by hydrophobic layers of suberin between their radial walls. As well as directing water and solute flow through cells rather than between them (i.e. symplastically as opposed to apoplastically: Steudle and Peterson 1998; Chap. 5), the endodermis increases the resistance to water flow radially into and, crucially, out of the root. Because water efflux from roots can be minimised, the precious vascular and potentially meristematic pericycle tissues (cf. Sect. 1.3.4) can remain relatively hydrated even in dry soil. However, the cortical cells external to the endodermis would be vulnerable to water loss and shrinkage, reducing soil-root contact (Passioura 1988; Nye 1994). The cortex may be protected by a hypodermis lying immediately below the epidermis (Perumalla et al. 1990; Peterson and Perumalla 1990), and which performs the same function as the endodermis.

The endodermal barrier to water flow effectively separates the root into two coaxial, hydraulically differentiated regions. The water potential of the stele tends to track that of the shoot, whereas the water potential of the cortex follows that of the soil (Passioura 1988). Consequently, tissues internal to the endodermis are relatively well protected against dehydration at the expense of the cortex, which is potentially expendable as far as continued root functioning is concerned. In mature roots, the epidermis is almost always lost and the cortex may also disappear.

Dehydration risk is further minimised by root secretions that bind soil particles around the root, retaining moisture close to the root surface. This is especially evident around immature cereal roots (McCully 1999). Again, secondary functions arise from this primary adaptation: secretion of organic molecules supports heterotrophic microbes (Sect. 1.4.5) and the binding of soil particles aids anchorage (Sect. 1.4.4). Symbiotic mycorrhizal fungi have the same effect, binding the root to soil with their extra-radical mycelium (Chap. 11).

1.3.7 Compensation for Unpredictable Water and Nutrient Supplies

Plants must adjust their rates of C, water and nutrient capture to meet their genetically determined, metabolically feasible demands for these resources (Robinson 2003). To compensate for the spatial and temporal uncertainties in the supplies of water and nutrients (Sect. 1.2.4), rates of resource capture must be responsive to demand (Gutschick and Pushnik 2003). Plants can adjust the rates at which water and nutrients influx and efflux across root cell membranes by specific transport proteins, and can modify the amount of root that each volume of soil contains by means of their open and indeterminate developmental patterns.

These physiological and developmental responses allow the root system to grow and function with considerable plasticity. Perhaps the most renowned type of morphological plasticity is the proliferation of lateral roots in nutrient-rich patches (Robinson 1994a). Hutchings and John (Chap. 2) discuss the functional significance of this widespread, but not universal, response. Glass (2003) has reviewed the molecular regulation of membrane ion transporters in response to environmental stimuli and internal signals.

The solutes that are effluxed from roots include many organic molecules such as amino and organic acids (Jones and Darrah 1994; Jones 1998; Chap. 10). These contribute to the flow of plant-derived C into the soil that can be exploited by heterotrophic microbes.

Specific, regulated transport systems allow solute and water fluxes to be finely balanced to generate and regulate turgor, an essential requirement for soil penetration (Sect. 1.3.1).

1.3.8 Conflicting Design Requirements

These various design requirements would be impossible to meet without the differentiation of specialised cell types in root systems. At least four cell types are required:

1. low-friction secretory cells to aid soil penetration (Sect. 1.3.1);
2. cells to transport assimilates from autotrophic organs to the roots, and water and nutrients in the opposite direction (Sect. 1.3.2);
3. cells with a high hydraulic resistance able to prevent the backflow of water to the soil (Sect. 1.3.6);
4. cells capable of forming new meristems from which branches can arise, allowing open and indeterminate development (Sects. 1.3.5 and 1.3.7).

The need for this cellular diversity constrains certain fundamental aspects of root form. The minimum diameter of a root is set by that of each cell (c. 10 μm). Arranged radially (Sect. 1.3.1), this would give a minimum diameter of any root of ca. 60 μm. No root is built on quite such minimalist principles; even the thread-like roots of *Arabidopsis thaliana* are almost double this absolute minimum diameter (Ma et al. 2001). However, the constraints that determine root diameter remain poorly understood. Very fine roots are cheaper to construct (in proportion to the square of the radius) and may be able to penetrate finer soil pores. On the other hand, they are likely to be shorter lived than thicker roots, more vulnerable to biotic attack and slower growing. Guerrero-Campo and Fitter (2001) found that root characteristics were better correlated with seed size than were aboveground characteristics, and hypothesised that anatomical constraints mean that fat roots cannot be produced from small seeds. Root diameter may therefore have profound impacts on many areas of plant biology.

1.4 Emergent Properties

1.4.1 Topology

The overall form or architecture of root systems is as varied as is that of shoot systems. There are extensively branched systems and unbranched ones; deeply penetrating and shallow ones; wide-spreading systems and narrow ones (Fig. 1.2). Several authors have attempted to provide classifications of this variety, but none of those attempts have been successful, at least not to the extent that they have been adopted widely. All root systems start as an unbranched radicle or adventitious root. This primary root or axis generates lateral roots (primary laterals), which, in turn, generate secondary laterals, and so on. This developmental classification is easily applied to a root system as it grows. However, it is difficult to apply to a pre-existing root system, especially one that has been extracted from soil and been damaged in the process.

Root systems are modular, and each meristem can generate a potentially unlimited number of lateral meristems. This results in exceptionally flexible architectures that have a great ability to respond to changes in the local environment (Chap. 2). Because lateral meristems arise individually, the structure that develops is typically trivalent at each node (Box 1). An alternative classification scheme for roots that reflects this structure is based on graph theory, a branch of mathematics that describes branching structures. This topological classification (Box 1) can be applied in an unbiased fashion to any root system. It has been applied to plants grown in controlled (Fitter and Stickland 1991; Taub and Goldberg 1996; Arredondo and Johnson 1999; Bouma et al. 2001) and field (Fitter and Stickland 1992b; Farley and Fitter 1999) conditions. Because it reflects function (e.g. all terminal roots are classified in the same order, in contrast to the developmental model), the topological classifi-

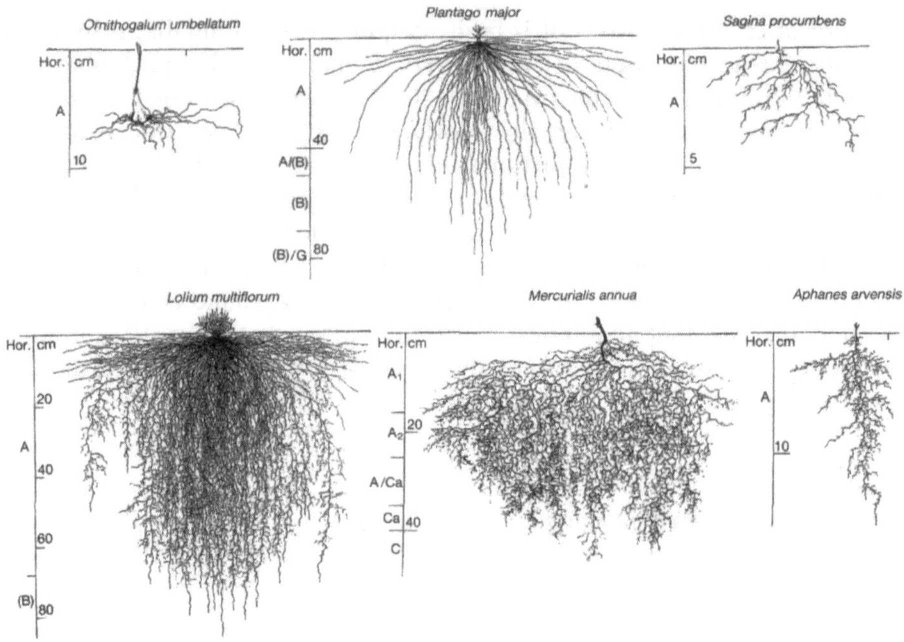

Fig. 1.2. Types of root system. The *top row* represents little-branched, the *bottom row* more strongly branched systems: in each case there is a trend to increasing dominance of a single main axis from left to right. Diagrams reproduced from Kutschera (1960) with kind permission of Axel Springer Verlag AG

cation can be used to predict the ecological significance of root system architecture.

Fitter et al. (1991) predicted that 'herringbone' root systems (i.e. those with a main axis and one or few developmental orders of laterals: Box 1) will be more expensive to construct than those with a dichotomous architecture. At the same time, herringbone systems are also more efficient at exploiting soil ('efficiency' being defined as volume of soil exploited per unit volume of root), especially for resources such as phosphate that are relatively immobile in soil (Hutchinson 2000). This prediction is perhaps counter-intuitive, and it depends partly on an assumption that root radius declines with increasing developmental order (Sects. 1.3.2 and 1.3.3).

Dicots from nutrient-poor habitats, for which acquisition of immobile soil resources would be expected to be most critical, conformed to the prediction that they would have more herringbone root systems, but this was not so for grasses (Fitter and Stickland 1991). Taub and Goldberg (1996) also found that the prediction held for dicots but not grasses among species from habitats ranged along a precipitation gradient from the Mediterranean coast to the Negev Desert. Along inundation gradients within Dutch salt marshes, root systems of members of the Chenopodiaceae became less herringbone-like and more dichotomous, but those of grasses did not show such a clear pattern (Bouma et al. 2001). The inundation gradients represented counter-variations in at least two limiting resources, O_2 and N. If herringbone root systems minimise O_2 leakage, that could explain why these topologies dominate on low, N-rich salt marshes that are inundated frequently. However, this does not explain why grasses can also thrive in those same sites even though their root systems do not share this topological response. Grasses apparently show little variation in root diameter across different levels of their branching hierarchy.

The topological model therefore explains some of the diversity of root system architecture in functional terms, specifically in relation to the ability of root systems to explore and exploit soil for resources. Other factors, including anchorage (Sect. 1.4.6), vegetative propagation, dispersal and herbivory, will also influence the overall form.

All species can express some plasticity in the morphology of their root systems (Chap. 2), but the extent to which each can adjust its phenotype to fill all the possible topological space between herringbone and dichotomous (Box 1) is unknown. Fitter (1987) showed that the root systems of *Trifolium pratense* tended towards dichotomy when water supply was ample, but became more herringbone-like as water availability decreased. Nutrients had relatively little effect on topology, but Farley and Fitter (1999) found that roots that proliferated in nutrient-rich patches were less herringbone-like than those in the nutrient-poor background soil.

Root system architecture can be strongly influenced by soil microbes. In ectomycorrhizas, for example, a fungal sheath encloses the root tip, root elongation is suppressed and often the roots undergo strictly dichotomous branching. These changes are directly induced by the formation of the symbiosis, although the responses can be seen as adaptive in relation to nutrient acquisition. In contrast, structural alterations in the root system are less pronounced in arbuscular mycorrhizas (AM) and may be related either to internal P concentrations in the shoots (Fitter 1985) or to an interaction with soil P status (Hetrick et al. 1991).

Some microbially induced plasticity in root system architecture is mediated by hormones, such as auxins and gibberellins released by rhizosphere *Azospirillum* (Marschner 1986, p. 192). Alternatively, microbes can genetically modify the plant by inserting plasmids into its genome; this is how *Agrobacterium rhizogenes* induces 'hairy roots' in the plants that it infects. This second mechanism is particularly interesting from an evolutionary perspective. Harper et al. (1991) discussed the likelihood that roots evolved after a soil microbe infected a rootless early land plant and induced it to produce root-like structures. Because the induced change involved genetic transformation, it was heritable. Advantages in resource capture enjoyed by progeny carrying 'rooty' genes would have favoured selection for the 'root' trait and its fixation in subsequent populations.

Box 1. Topological analysis of root system architecture

A root system comprises links (or edges, in mathematical terminology) and nodes (or vertices). The links can be divided into exterior (*e*, end in a meristem) or interior (*i*) and again subdivided as to the type of link they join (exterior-exterior, *ee*, and exterior-interior, *ei*):

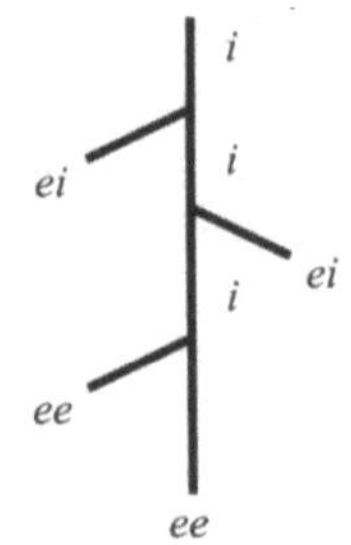

The branching pattern can be described in terms of the length of paths that traverse the system: these paths are longer in a herringbone system (below left), where branching is restricted to the main axis, and shorter in a dichotomous system (below right) where branching occurs equally:

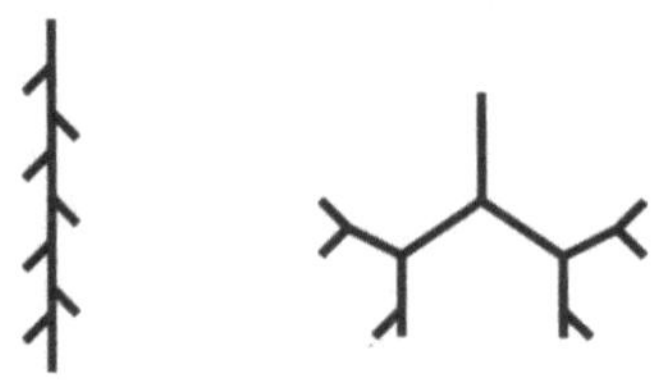

Since the topology of a root system changes as it grows, it is important to compare like with like. Root system size is measured as magnitude, the number of exterior links (μ); all root systems that follow these branching rules have μ exterior links and μ−1 interior links.

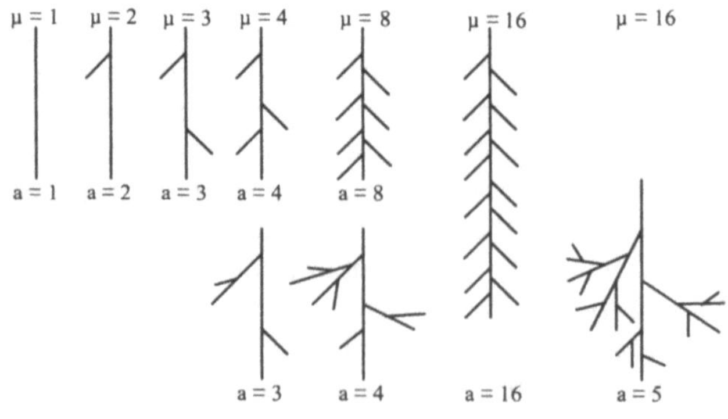

The altitude (a) is the number of links in the longest path that connects the base of the system to an exterior link. In a herringbone system, $a=\mu$; in a dichotomous system, $a=\log_2\mu$. A plot of log a against log μ reveals the branching rules followed by a root system:

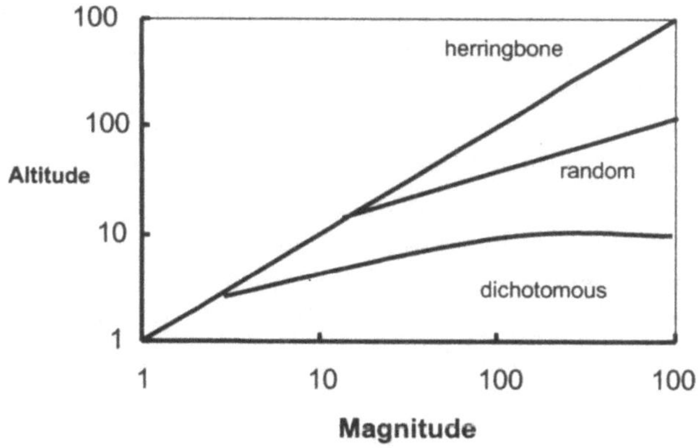

Herringbone and dichotomous systems are the extremes of a spectrum of topologies that branched systems can occupy. Real root systems fall somewhere between these extremes.

Topological analyses of root systems should take into account magnitude, as all the parameters are scale-dependent. Analysis of covariance (using magnitude as covariate) is often the best statistical approach for experimental data. Measurement of topological parameters is time-consuming unless automated. Several packages are available to do this, e.g., Win/MacRhizo (Regent Instruments, Quebec, Canada; www.regent.qc.ca).

1.4.2 Size

Quantifying the size of root systems has taxed the ingenuity and stamina of many investigators. The inaccessibility and branched structure of root systems prevents precise quantification of total root mass in all but the simplest circumstances (e.g. hydroponic or sand culture experiments). When information on root length, topology (Sect. 1.4.1) and turnover (Chap. 3) is required, and if more than one species is present, the problems are magnified several-fold. Nevertheless, information such as this is needed to estimate regional and global biogeochemical fluxes and interactions between vegetation and atmosphere (Schlesinger 1997, p. 136).

Jackson et al. (1996) reviewed the patterns of root mass and vertical distribution that had been reported for different terrestrial biomes. They used the following asymptotic relation (Gale and Grigal 1987) to describe the vertical distributions of roots:

$$Y=1-\beta^{d} \tag{1.3}$$

where Y is the cumulative fraction of root mass ($0<Y<1$) from the soil surface to a depth of d cm, and β is a fitted parameter. Larger values of β imply relatively deep-rooted vegetation.

The shallowest root systems occur in tundra, boreal forests and temperate grasslands (Table 1.1). In those biomes, up to 90 % of the root mass is confined to the top 30 cm of soil. In cold and warm deserts and temperate coniferous forests, this proportion is reduced to about 50 %. These patterns reflect interactions between temperature (deep tundra soil is permanently frozen, preventing root penetration) and water availability (desert soil is normally dry near the surface and water is available– if at all– only at depth for much of the year).

There are large differences among biomes in the mass of root material they contain (see also Chap. 3). Sclerophyllous shrubland and tropical evergreen forests contain, on average, about 5 kg m^{-2}, grasslands one-fifth of this amount, and cultivated soil one-twenty-fifth (Table 1.1). All of the roots in grassland may be classified as 'fine', but this fraction falls below one-fifth for shrublands and forests. Obviously, the taxonomic composition of the vegetation dictates the root diameter distributions in each biome: grass roots are finer than tree roots, because they lack secondary thickening.

Within the fine root fraction, less than two-thirds are, on average, estimated to be alive (however defined) in temperate coniferous forests and grasslands (Table 1.1). In boreal forests and tundra, only one-third of the fine roots are deemed to be alive. This reflects the huge accumulation and slow decomposition of belowground detritus at high latitudes, despite their small annual productivities (Swift et al. 1979). This contrasts with the greater productivities, but faster decomposition, in warmer regions.

The balance between above- and belowground mass also varies markedly across biomes (Table 1.1). Relatively little root mass is produced by crops which have been selected to maximise production of their harvestable parts, and these are usually aboveground. Aboveground production is also dominant in forests and all tropical biomes; only in temperate grasslands, cold deserts and tundra is root production significantly greater.

These data are crude in that they say nothing about annual productivities (for which information on turnover is required: Schlesinger 1997; Chap. 3), phenologies, or the mass of microbial (especially mycorrhizal) associations. When the latter are added, belowground production may be significantly greater, especially in forests (Fogel 1985).

When the data in Table 1.1 are scaled up according to the area of land occupied by each (or a similar) biome (Jackson et al. 1997), it can be seen that trop-

Table 1.1. Root distribution patterns in different biomes collated from published studies by Jackson et al. (1996, 1997) and Canadell et al. (1996). β is a parameter (Eq. 1.1) that increases in value as the vegetation becomes deeper rooted. n.a. = Data not available

Biome	β	Root mass fraction in upper 0.3 m	Maximum rooting depth (m)	Total root mass (kg m^{-2})	Fine root mass (kg m^{-2})	Live fine root mass (kg m^{-2})	Live fine root length (km m^{-2})	Fine/total root mass ratio	Live fine/total fine mass ratio	Root/shoot mass ratio
Boreal forest	0.943	0.83	2.0	2.9	0.60	0.23	2.6	0.25	0.36	0.32
Cultivated	0.961	0.70	2.1	0.2	n.a.	n.a.	n.a.	n.a.	n.a.	0.10
Cold desert	0.975	0.53	9.5	1.2	0.27	0.13	4.0	0.31	0.49	4.50
Warm desert				0.4				0.93		0.70
Sclerophyllous shrubs	0.964	0.67	5.2	4.8	0.52	0.28	8.4	0.14	0.59	1.20
Temperate coniferous forest	0.976	0.52	3.9	4.4	0.82	0.50	6.1	0.22	0.63	0.18
Temperate deciduous forest	0.966	0.65	2.9	4.2	0.78	0.44	5.4	0.21	0.56	0.23
Temperate grassland	0.943	0.83	2.6	1.5	1.5	0.95	112	1.0	0.64	3.70
Tropical deciduous forest	0.961	0.70	3.7	4.1	0.57	0.28	3.5	0.16	0.49	0.34
Tropical evergreen forest	0.962	0.69	7.3	4.9	0.57	0.33	4.1	0.13	0.59	0.19
Tropical grassland savanna	0.972	0.57	15.0	1.4	0.99	0.51	60.4	0.88	0.52	0.70
Tundra	0.914	0.93	0.5	1.2	0.96	0.34	7.4	0.98	0.36	6.60

Table 1.2. Global estimates of root mass and length, and of C and N contents of fine roots, in different biomes (Jackson et al. 1997). Biomes are those defined by Whittaker (1975) and do not coincide exactly with those listed in Table 1.1. Elemental contents were calculated assuming that total C, N and P concentrations of fine roots were 48.8, 1.17 and 0.11 % respectively. Note: 1 Pg=10^{15} g; 1 Tg=10^{12} g

Biome	Total root mass (Pg)	Fraction of total root mass	Fine root mass (Pg)	Fraction of fine root mass	C content of fine roots (Pg)	N content of fine roots (Tg)	Phosphorus content of fine roots (Tg)
Boreal forest	35	0.12	7.2	0.09	3.5	84.2	7.9
Cultivated	2.1	0.01	2.1	0.03	1.0	24.6	2.3
Desert	6.6	0.02	4.9	0.06	2.4	57.3	5.4
Savanna	21	0.07	14.9	0.19	7.3	174	16.4
Temperate deciduous forest	29	0.10	5.6	0.07	2.7	65.5	6.2
Temperate evergreen forest	22	0.07	4.1	0.05	2.0	48.0	4.5
Temperate grassland	14	0.05	13.6	0.17	6.6	159	15.0
Tropical rainforest	83	0.28	9.7	0.12	4.7	114	10.7
Tropical seasonal forest	31	0.11	4.3	0.05	2.1	50.3	4.7
Tundra/alpine	10	0.03	7.7	0.10	3.8	90.1	8.5
Woodland and shrubland	41	0.14	4.4	0.06	2.2	51.5	4.8
Total	295	1.00	78.5	1.00	38.3	919	86.4

ical rainforests contain by far the most root mass, 28 % of the total (Table 1.2). Other types of forest account for a further 40 %, leaving the remainder split between shrublands and grasslands. Cultivated land contains just 1 % of the world's root mass. Almost half of the 'fine' root mass is, however, in three biomes dominated by grasses and forbs: savanna, temperate grasslands and tundra regions; slightly less (40 %) is found in all forest types. Translated into C and N stocks, the 'fine' (and most labile) root fractions contain 5 % of the C present as atmospheric CO_2 (Jackson et al. 1997) and 1 % of the N fixed annually by diazotrophs in terrestrial ecosystems.

1.4.3 Depth

Perhaps surprisingly, the deepest roots occur in tropical savannas, reaching 15 m on average [although the deepest *recorded* roots, of *Boscia albitrunca* in the Kalahari, Botswana (Jennings 1974, cited in Canadell et al. 1996), were found at the astonishing depth of 68 m]. At the other extreme, the roots of tundra plants typically penetrate no more than 0.5 m into the soil. When classified according to plant functional group, maximum rooting depth decreases in the order: trees>shrubs>herbs>crops (Fig. 1.3).

These data dispel the impression that deep-rooted plants are rare. As Table 1.1 indicates, most roots are in the topsoil; the habit of root systems of branching from the base of stems makes this inevitable. However, the presence of *any* living roots at depth is important, especially in dry habitats. The phenomenon of 'hydraulic lift', by which deep water is transferred nocturnally into dry topsoil by deep-rooted species (Richards and Caldwell 1987), supplies water to shallow-rooted neighbours that have no direct access to it (Dawson 1993).

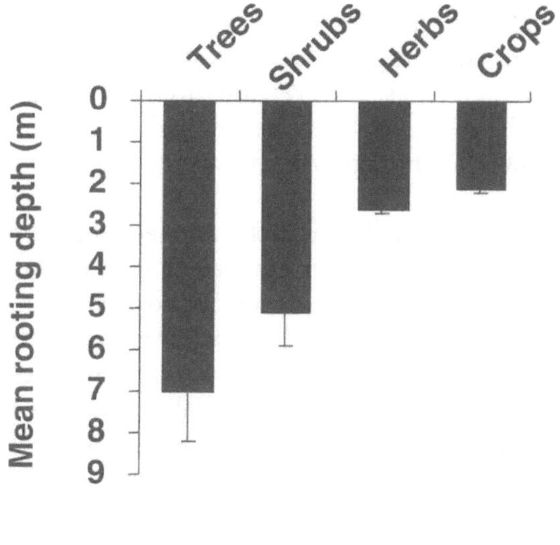

Fig. 1.3. Mean (+ 1 SE) maximum rooting depths of major plant functional groups (data from Canadell et al. 1996) averaged (with the exception of crops) across the biomes listed in Table 1.1

So little is known of microbial activity at extreme soil depths that it is not possible to assess the global impact of deep roots on C and nutrient turnover, for example. Locally, however, new information is appearing about the functional importance of deep roots. In an ingenious study, Jackson et al. (1999) solved the problem of how to identify the species to which a particular deep root belongs. At the soil surface, it is sometimes possible to trace a root back to its plant of origin. At depths of 25 (or 68) m, this is impossible. Jackson et al. (1999) compared DNA sequences [of the internal transcribed spacer (ITS) region of the 18S-26S ribosomal DNA repeat] from roots sampled from Texan limestone caverns 5–22 m below the soil surface with sequences collected from aboveground vegetation. They were able to unequivocally match roots to plants. Natural ^{18}O tracing of deep groundwater into stem water confirmed that the deep roots contributed to water capture. The ITS technique is, in principle, applicable to any mixture of roots and represents a major technical breakthrough in root ecology.

1.4.4 Anchorage

Anchorage arises both from the ability of roots to interact with the soil matrix, so dissipating into soil forces transmitted from root to shoot, and as an emergent property from the ramification of branched root systems among soil particles (Sect. 1.3.3). For small plants, therefore, anchorage will often be achieved incidentally. Early in the colonisation of land, however, competition for light led to the rapid development of woody stems (Niklas 1992). To support these required more massive belowground structures that played no part in resource acquisition.

The need for anchorage is in relation to three types of force. The plant must be able to resist the gravitational compression of its own mass, lateral forces due to wind, and vertical forces caused by grazing animals. The last of these is relevant only to small plants, whereas wind forces are important to large plants. The best-studied response of anchorage is that to wind, because of its economic significance in forestry. In all cases, however, the three key components of anchorage are the strengths of the roots, of the soil and of the bond between them. If any one of these fails, anchorage will fail causing the plant to fall.

Simple root systems, such as sunflower seedling radicles and leek seedlings with unbranched roots, generate anchorage forces equivalent to the root's breaking strength over short distances (20–30 mm) of root (Ennos 2000). This means that distal parts of the root system play no part in anchorage, since the root will break before forces can be transmitted to those parts. The distances over which this will be true will be greater for larger, especially woody, root systems, but the principles remain the same: only the basal parts of root systems are significant in anchorage. The finer, distal parts are involved solely in resource acquisition (Sect. 1.4.7), and, even though root hairs may increase the bond between root tips and soil, they play no part in whole-plant anchorage (Bailey et al. 2002).

When the large root systems of woody plants experience lateral forces in strong winds another important anchorage component comes into play. The roots bind the soil to make a plate whose weight acts as a massive anchor. Windthrown trees show that a large bowl-shaped mass of soil and roots has also been displaced. Failure occurs at the edges of this plate and in the roots that traverse those boundaries (Coutts 1986). Key determinants of anchorage strength here are the ability of the root system to bind the soil and the strength of the roots in tension on the windward side of the plate, as well as the strength of the soil itself. Wet soil is weaker and most windthrow occurs in winter when soils are wet; if anchorage is sufficient, trunks may break in very strong winds.

The architecture of the root system also determines anchorage strength. Lateral branches promote anchorage in both model root systems (Stokes et al. 1996) and in *Arabidopsis thaliana*, as shown by the reduced anchorage strength of a mutant with restricted branching (Bailey et al. 2002). The optimum branching angles appear to be 90° between an axis and a primary lateral, and <20° between primary and secondary laterals; the former angle corresponds well to that found in actual Sitka spruce (*Picea sitchensis*) root systems (Stokes et al. 1996). The importance of root architecture in resistance to windthrow is emphasised by the finding that Sitka spruce roots growing on the windward side of the root system of young seedlings are fatter, longer and more branched than those on the leeward side (Stokes et al. 1995). Such apparently adaptive responses imply strong selection pressures, although the mechanisms remain obscure.

1.4.5 Rhizosphere

Roots can lose considerable amounts of C to the surrounding soil by a variety of processes known collectively as 'rhizodeposition' (Chap. 4). Rhizodeposition includes exudations, secretions, lysates, sloughed-off cells, mucilage and gases, processes that are direct consequences of the need for root turnover, lubrication and solute exchange (Sect. 1.3). In addition, roots modify the surrounding soil by gas exchange and by removing nutrients and water (Chap. 13). Consequently, the biota of the soil around a root is markedly different from that in bulk soil, and this zone of influence is termed the rhizosphere.

Some of the C 'lost' by rhizodeposition may benefit the plant (Chap. 12). Root-derived exudates can improve the availability of nutrients around a root. For example, roots of white lupin (*Lupinus albus*) can acidify a P-deficient alkaline soil by exuding organic acids such as malate. The localised acidification is sufficient to solubilise phosphates and increase the availability of P in the rhizosphere (Dinkelaker et al. 1989). Specific exudates can be released in response to particular nutrient deficiencies. Root-derived non-proteinogenic amino acids ('phytosiderophores') can improve the supplies of iron, copper, zinc and manganese by forming plant-available complexes with these ions (Treeby et al. 1989; Jones et al. 1996).

Much of the C released into soil from roots is used as a substrate by heterotrophic microbes, which are often C-limited in soil (Bowen and Rovira 1991; Darrah 1998). Consequently, microbes are generally more abundant in the rhizosphere than in the bulk soil, although their apparent species diversity may be less. Many environmental factors influence the qualitative and quantitative composition of root exudates (Chap. 10).

Current C inputs from living, mycorrhizal root systems largely drive soil CO_2 fluxes. Högberg et al. (2001) demonstrated this in large-scale girdling experiments in a boreal pine forest. They removed the stem bark to the depth of the active xylem. This disrupted the supply of current assimilates into the root system without disturbing the functional relationships between roots and their associated microbes. Girdling reduced total CO_2 flux from the soil by about 50% within 1–2 months, but decreases of up to 37% were seen within 5 days.

An important function of rhizodeposition is the secretion of mucilage (McCully 1999). Mucilage binds soil particles to each other and to the root surface immediately behind the root cap, so retaining water close to the root, minimising dehydration even if water is lost from the bulk soil (Sect. 1.3.6). This also has profound effects on the structural integrity of soil.

The amounts of C involved in rhizodeposition should be set against those that root systems consume in total (see Chap. 4). Measurements of how much C a growing root system consumes include:

1. C used to construct roots – cellulose, lignin, protein, lipid, pectin, etc.;
2. C consumed during respiration associated with producing roots;
3. C consumed during respiration associated with maintaining roots;
4. C consumed during respiration associated with ion transport into roots; and
5. C lost from the root in rhizodeposition or transport into symbionts.

On average, about 5% of daily C fixation is deposited in the rhizosphere, and another 5% is used by root symbionts (Fig. 1.4). These figures will vary with species and environment. The lifetime 'costs' of rhizodeposition, and of other C-consuming processes listed in Fig. 1.4, should be offset against the 'benefits' that those roots provide in terms of water and nutrient capture (Eissenstat and Yanai 1992), and include a component to account for root turnover (Chap. 3).

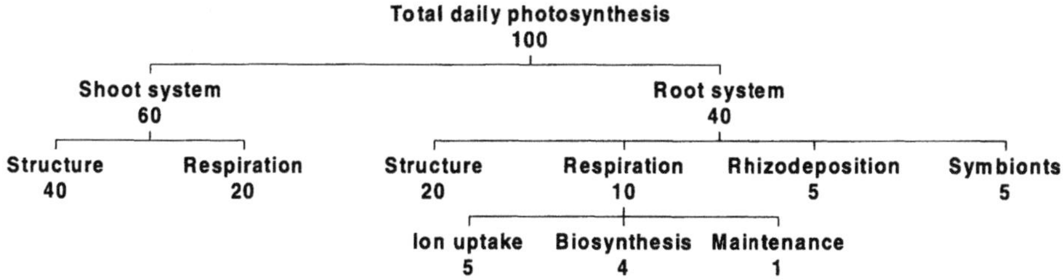

Fig. 1.4. Approximate C costs associated with shoot and root growth in a fast-growing herb with free access to nutrients. Numbers are percentages of daily C fixation, and are based on Lambers et al. (1998, p. 97)

1.4.6 Mycorrhizas

The most widespread (and, some would argue, the most important) root–microbe relation is the mycorrhiza. A mycorrhiza is probably best defined as a sustainable non-pathogenic biotrophic interaction between a fungus and a root (Fitter and Moyersoen 1996) with both intra- and extra-radical mycelium. Mycorrhizal associations were probably crucial in the colonisation of land. Fossil aquatic plants lacked structures analogous to root systems. Their terrestrial descendants probably depended on associated fungal hyphae to transfer nutrients (notably P) to them from the proto-soil into belowground stems (rhizomes). A likely sequence of evolutionary events is that sapro- or even necrotrophic fungi colonised plants as major C sources, but, because the plants were very P-deficient, the fungi were forced to import P from the soil to the internal hyphae. Some of this P would then have become available to the plant, beginning the evolution of a genuinely mutualistic association.

Read (1991) suggested that the dominant type of mycorrhizal symbiosis changes along an ecological gradient from ericoid (dominant in tundra), to ectomycorrhizas (boreal forests) to arbuscular mycorrhizas (in temperate and tropical biomes). Woodward and Kelly (1997) linked this sequence to the net primary productivities (NPP) of those biomes. Non-mycorrhizal plant species are more prevalent (but still in the minority) in nutrient-rich biomes, and most abundant in disturbed habitats (Peat and Fitter 1993), probably because of the repeated destruction of the fungal mycelium.

Enhanced nutrient uptake is the principal potential benefit that a plant obtains from being mycorrhizal (Chap. 11). The mycorrhizal hyphae radiate into the soil, effectively acting as fine extensions of the root system. Consequently, the rhizosphere becomes extended into the 'mycorrhizosphere' as the soil surrounding the root is influenced, not only by the root, but also the mycorrhizal hyphae. Many mycorrhizal fungi have very low host specificity so that their hyphae may link several root systems (Newman et al. 1994). The possibility that plants whose roots are connected by a common mycorrhizal network may gain extra ecological benefits is appealing (Simard et al. 1997), but the evidence for it is currently equivocal at best (Robinson and Fitter 1999).

1.4.7 Specialised Morphologies

Root systems vary greatly in morphology, and many types of specialised roots have evolved, including those that can contract, dragging bulbs deeper in soil; aerial roots that can anchor epiphytes as in strangler figs; storage roots; and the haustoria of root parasites such as *Lathraea* or *Rhinanthus*. Within the context of primary root functions, such as anchorage and resource acquisition, specialisation is less obvious, although buttress roots can be regarded as an anchorage speciality. Resource acquisition can largely be explained in terms of architecture and mycorrhizal symbioses. However, an apparent alternative to formation of a mycorrhizal symbiosis occurs in certain taxa: cluster roots. Along with N_2-fixing symbioses, cluster roots represent the third major strategy of nutrient acquisition to have evolved among land plants (Skene 1998).

Cluster roots are discrete bunches of closely spaced laterals that produce abundant root hairs when mature. Their development is accompanied by a short but marked exudative release of predominantly organic acids, notably citrate (Gardner et al. 1983; Dinkelaker et al. 1989; Watt and Evans 1999). This burst of acid exudation transiently increases the solubility of phosphates and micronutrients in the rhizosphere (Gardner et al. 1982; Grierson 1992; Dinkelaker et al. 1995) via the reducing, chelating and acidifying properties of the exudates (Dinkelaker et al. 1989; Lamont 1993). If the exudation of citrate per unit of root is sufficient to overwhelm the capacity of the soil to re-adsorb P released in this way, the cluster roots should, in theory, be capable of absorbing that P (Fitter 1999). The exudative burst can represent a considerable carbon cost for the plant. Dinkelaker et al. (1989) showed that the exudation of citric acid by *Lupinus albus* L. grown on a P-deficient calcareous soil was equivalent to 23 % of plant dry weight at 13 weeks (although any P solubilised by this exudation failed, in that experiment, to prevent *L. albus* suffering P deficiency as judged by the P concentration of its leaves).

Originally thought confined to the Proteaceae, cluster root formation has now been identified in other families including Fabaceae (notably lupins), Betulaceae and Myricaceae (Skene 1998). Although N_2-fixing nodules can form on species that produce cluster roots, mycorrhizas are usually absent (Lamont 1982; Skene 1998). Some mycorrhizal species can also produce cluster roots, but only when conditions for formation of the symbiosis are unfavourable (Reddell et al. 1997), and there is some evidence that exudates produced by cluster roots may directly inhibit mycorrhizal formation (Lamont 1982). As frequently observed in mycorrhizas, cluster root formation is inhibited by P fertiliser (Handreck 1991). Taken together, and as suggested by Lamont (1982), this suggests that cluster roots do provide an alternative to mycorrhizal symbiosis as a mechanism to boost nutrient (notably P) capture.

1.4.8 Global-Scale Processes

We calculated apparent inflows of N, P and water into roots for whole biomes. Inflow is the rate of nutrient uptake per unit root length; we use the term 'apparent' inflow to distinguish this parameter from measured fluxes for which root lengths, durations of uptake and other factors that affect the calculation are known (Robinson 1986, 1994b). The data of Jackson et al. (1996, 1997) for live fine root length (Table 1.1) were used, on the assumption that water and nutrients are absorbed via those roots and not by massive and/or seemingly dead roots. These were combined with NPP estimates (Schlesinger 1997), transpiration rates (Larcher 1980) and lengths of growing seasons (Swift et al. 1979) quoted for some of the biomes listed in Table 1.1. We assumed constant N and P concentrations in the vegetation (1.17 and 0.15% of dry matter for N and P respectively: Table 1.2).

Generally, as NPP increases, so do apparent inflows (Fig. 1.5). Exceptions to this trend occur in temperate grassland and savanna, biomes dominated by grasses. The relatively moderate water and N demands placed on the root systems of grasses are averaged over their enormous live root lengths per unit ground area, so decreasing their apparent inflows. The increase in apparent inflow with NPP parallels that of maximum photosynthetic rate per unit leaf area (Woodward and Kelly 1997) and it is well established that species diversity increases with NPP, at least in non-agricultural systems (Adams and Woodward 1989).

There is a three-order-of-magnitude range in apparent water inflow among biomes, and a slightly smaller range for N and P inflow (Fig. 1.5). This wide range is still, however, some two orders of magnitude smaller than actual water and N inflows that have been measured on single, intact roots under laboratory conditions (Robinson 1994b). In the more productive biomes, apparent P inflow approaches the theoretical maximum steady-state P inflow (~0.3 μmol m^{-1} day^{-1}) calculated by Sanders and Tinker (1973) for a sandy soil (in other soils this value would be smaller). This limit is set by the slow diffusivity of phosphate ions in soil compared with the rates at which roots are able to absorb P. This emphasises the strong reliance that most plants have on mycorrhizal symbioses (Sect. 1.4.5; Chap. 11) and other mechanisms (e.g., exudation: Sect. 1.4.4; Chap. 10) to increase the potential for P acquisition.

It would be unwise to conclude from Fig. 1.5 that the root systems of desert plants are in any way 'less effective' than those of tropical forest trees. Desert plants have access to meagre supplies of nutrients and, especially, water that

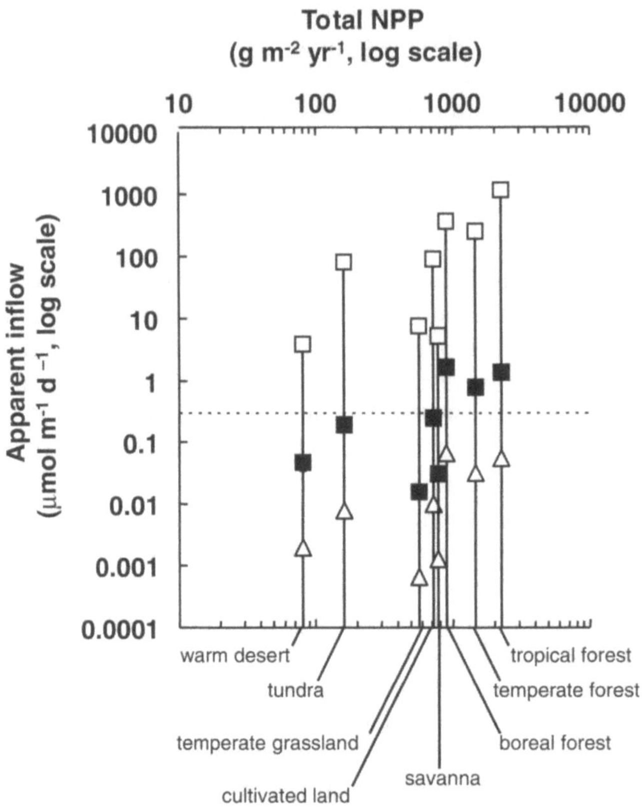

Fig. 1.5. Apparent inflows of water (*open squares*), N (*solid squares*) and P (*open triangles*) in relation to the net primary production (NPP) of eight biomes, calculated as described in Section 1.4.8. The *broken line* is the theoretical maximum steady-state P inflow calculated by Sanders and Tinker (1973)

occur discontinuously in time and space. Roots (and leaves) of such plants respond rapidly to rainfall. They do so by having a reservoir of meristems in relatively long-lived organs capable of forming new, physiologically active roots when required. For much of the year, the old roots do little (apart from keeping the plant well-anchored and providing a transport system, of course).

1.5 Concluding Remarks

Harper et al. (1991) described the evolution of roots as 'probably the most dramatic event in the evolution of the plant kingdom'. Without it, the bryophytes would still dominate the landscape. As we have tried to explain, once root systems *had* evolved, their basic form and functions were practically inevitable because the biophysical constraints imposed by life on the land would have ensured considerable evolutionary convergence (Niklas 2000).

The chapters that follow detail some of the most ecologically important and fascinating features of roots; no doubt, many more remain to be unearthed.

Acknowledgements. Angela Hodge is funded by the Biological and Biotechnological Research Council (BBSRC), UK. We appreciate the helpful comments of Rob Jackson, Mike Hutchings and the editors on an earlier draft.

References

Adams JM, Woodward FI (1989) Patterns in tree species richness as a test of the glacial extinction hypothesis. Nature 339:699–701

Arredondo JT, Johnson DA (1999) Root architecture and biomass allocation of three range grasses in response to nonuniform supply of nutrients and shoot defoliation. New Phytol 143:373–385

Asseng S, Aylmore LAG, MacFall JS, Hopmans JW, Gregory PJ (2000) Computer-assisted tomography and magnetic resonance imaging. In: Smit AL, Bengough AG, Engels C, van Noordwijk M, Pellerin S, van de Geijn SC (eds) Root methods. A handbook. Springer, Berlin Heidelberg New York, pp 343–363

Bailey PJH, Currey JD, Fitter AH (2002). The role of root system architecture and root hairs in promoting anchorage against uprooting forces in *Allium cepa* and root mutants of *Arabidopsis thaliana*. J Exp Bot 53:333–340

Bouma TJ, Nielsen KL, Van Hal J, Koutstaal B (2001) Root system topology and diameter distribution of species from habitats differing in inundation frequency. Funct Ecol 15:360–369

Bowen GD, Rovira AD (1991) The rhizosphere: the hidden half of the hidden half. In: Waisel Y, Eshel A, Kafkafi U (eds) Plant roots: the hidden half. Marcel Dekker, New York, pp 641–669

Brady NC, Weil RR (1999) The nature and properties of soil (12th edn). Prentice Hall, Upper Saddle River

Canadell J, Jackson RB, Ehleringer JR, Mooney HA, Sala OE, Schulze ED (1996) Maximum rooting depth of vegetation types at the global scale. Oecologia 108:583–595

Coutts MP (1986) Components of tree stability in Sitka spruce on peaty gley soil. Forestry 59:173–197

Darrah PR (1998) Interactions between root exudates, mineral nutrition and plant growth. In: Lambers H, Poorter H, van Vuuren MMI (eds) Inherent variation in plant growth. Physiological mechanisms and ecological consequences. Backhuys, Leiden, pp 159–181

Dawson TE (1993) Hydraulic lift and water-use by plants – implications for water-balance, performance and plant-plant interactions. Oecologia 95:565–574

Dinkelaker B, Römheld V, Marschner H (1989) Citric acid excretion and precipitation of calcium citrate in the rhizosphere of white lupin (*Lupinus albus* L.). Plant Cell Environ 12:285–292

Dinkelaker B, Hengeler C, Marschner H (1995) Distribution and function of proteoid roots and other root clusters. Bot Acta 108: 183–200

Eissenstat DM, Yanai RD (1997) The ecology of root lifespan. Adv Ecol Res 27:1–60

Ennos AR (2000) The mechanics of root anchorage. Adv Bot Res 33:133–157

Farley RA, Fitter AH (1999) The response of seven co-occurring woodland herbaceous perennials to localized nutrient-rich patches. J Ecol 87:849–859

Fitter AH (1985) Functional significance of root morphology and root system architecture. In: Fitter AH, Atkinson D, Read DJ, Usher MB (eds) Ecological interactions in soil. Blackwell, Oxford, pp 87–106

Fitter AH (1987) An architectural approach to the comparative ecology of plant root systems. New Phytol 106 (Suppl):61–77

Fitter AH (1999). Roots as dynamic systems: the developmental ecology of roots and root systems. In: Press M (ed) Plant physiological ecology, British Ecological Society Symposium No. 39. Blackwell Scientific Publications, Oxford, pp 115–131
Fitter AH, Moyersoen B (1996) Evolutionary trends in root-microbe symbioses. Philos Trans R Soc Lond B 351:1367–1375
Fitter AH, Stickland TR (1991) Architectural analysis of plant root systems. 2. Influence of nutrient supply on architecture in contrasting plant species. New Phytol 118:383–389
Fitter AH, Stickland TR (1992a) Fractal characterization of root-system architecture. Funct Ecol 6:632–635
Fitter AH, Stickland TR (1992b) Architectural analysis of plant root systems. III. Studies on plants under field conditions. New Phytol 121:243–248
Fitter AH, Stickland TR, Harvey ML, Wilson GW (1991) Architectural analysis of plant root systems. 1. Architectural correlates of exploitation efficiency. New Phytol 118:375–382
Fitter AH, Graves JD, Wolfenden J, Self GK, Brown TK, Bogie D, Mansfield TA (1997) Root production and turnover and carbon budgets of two contrasting grasslands under ambient and elevated atmospheric carbon dioxide concentrations. New Phytol 137:247–255
Fogel R (1985) Roots as primary producers in below-ground ecosystems. In: Fitter AH, Atkinson D, Read DJ, Usher MB (eds) Ecological interactions in soil: plants, microbes and animals. Blackwell, Oxford, pp 23–36
Forde BG, Lorenzo H (2001) The nutritional control of root development. Plant Soil 232:51–68
Gale MR Grigal DK (1987) Vertical root distributions of northern tree species in relation to successional status. Can J For Res 17:929–834
Gardner WK, Barber DA, Parbery DG (1982) The acquisition of phosphorus by *Lupinus albus* L. I. Some characteristics of the soil/root interface. Plant Soil 68:19–32
Gardner WK, Barber DA, Parbery DG (1983) The acquisition of phosphorus by *Lupinus albus* L. III. The probable mechanism by which phosphorus movement in the soil/root interface is enhanced. Plant Soil 70:107–124
Glass ADM (2003) Homeostatic processes for the optimization of nutrient absorption: physiology and molecular biology. In: BassiriRad H (ed) Root ecophysiology. Springer, Berlin Heidelberg New York (in press)
Grierson PF (1992) Organic acids in the rhizosphere of *Banksia integrifolia* L. Plant Soil 144:259–265
Guerrero-Campo J, Fitter AH (2001) Relationships between root characteristics and seed size in two contrasting floras. Acta Oecol 22:77–85
Gutschick VP, Pushnik JC (2003) Internal regulation of nutrient uptake by relative growth rate and nutrient use efficiency. In: BassiriRad H (ed) Root ecophysiology. Springer, Berlin Heidelberg New York (in press)
Handreck KA (1991) Interactions between iron and phosphorus in the nutrition of *Banksia ericifolia* L. f. var. *ericifolia* (Proteaceae) in soil-less potting media. Aust J Bot 39:373–384
Harper JL, Jones M, Sackville Hamilton NR (1991) The evolution of roots and the problems of analysing their behaviour. In: Atkinson D (ed) Plant root growth, an ecological perspective. Blackwell, London, pp 3–22
Harrison MJ (1997) The arbuscular mycorrhizal symbiosis: an underground association. Trends Plant Sci 2:54–60
Heidstra R, Bisseling T (1996) Nod factor-induced host responses and mechanisms of Nod factor perception. New Phytol 133:25–43
Hetrick BAD, Wilson GWT, Leslie JF (1991) Root architecture of warm- and cool-season grasses: relationship to mycorrhizal dependence. Can J Bot 69:112–118

Hodge A, Robinson D, Fitter AH (2000) Are microbes more effective than plants at competing for nitrogen? Trends Plant Sci 5:304–308

Högberg P, Nordgren A, Buchmann N, Taylor AFS, Ekblad A, Högberg MN, Nyberg G, Ottosson-Lofvenius M, Read DJ (2001) Large-scale forest girdling shows that current photosynthesis drives soil respiration. Nature 411:789–792

Hutchinson JMC (2000) Three into two doesn't go: two-dimensional models of bird eggs, snail shells and plant roots. Biol J Linn Soc 70:161–187

Jackson RB, Canadell J, Ehleringer JR, Mooney HA, Sala OE, Schulze ED (1996) A global analysis of root distributions for terrestrial biomes. Oecologia 108:389–411

Jackson RB, Mooney HA, Schulze ED (1997) A global budget for fine root biomass, surface area, and nutrient budgets. Proc Natl Acad Sci USA 94:7362–7366

Jackson RB, Moore LA, Hoffman WA, Pockman WT, Linder CR (1999) Ecosystem rooting depth determined with caves and DNA. Proc Natl Acad Sci USA 96:11387–11392

Jones DL (1998) Organic acids in the rhizosphere: a critical review. Plant Soil 205:25–44

Jones DL, Darrah PR (1994) Amino-acid influx at the soil-root interface of *Zea mays* L. and its implications in the rhizosphere. Plant Soil 163:1–12

Jones DL, Darrah PR, Kochian LV (1996) Critical evaluation of organic acid mediated iron dissolution in the rhizosphere and its potential role in root iron uptake. Plant Soil 180:57–66

Kutschera L (1960) Wurzelatlas mitteleuropäischer Ackerunkräuter und Kulturpflanzen. DLG Verlag, Frankfurt

Lambers H, Chapin FS III, Pons TL (1998) Plant physiological ecology. Springer, Berlin Heidelberg New York

Lamont BB (1982) Mechanisms for enhancing nutrient uptake in plants with particular reference to Mediterranean South Africa and western Australia. Bot Rev 48:597–689

Lamont BB (1993) Why are hairy root clusters so abundant in the most nutrient-impoverished soils of Australia? Plant Soil 156:269–272

Larcher W (1980) Physiological plant ecology, 2nd edn. Springer, Berlin Heidelberg New York

Ma Z, Bielenberg DG, Brown KM, Lynch JP (2001) Regulation of root hair density by phosphorus availability in *Arabidopsis thaliana*. Plant Cell Environ 24:459–467

Marschner H (1986) Mineral nutrition of higher plants. Academic Press, London

McCully ME (1999) Roots in soil: unearthing the complexities of roots and their rhizospheres. Annu Rev Plant Physiol Mol Biol 50:695–718

Newman EI, Devoy CLN, Easen NJ, Fowles KJ (1994) Plant-species that can be linked by VA mycorrhizal fungi. New Phytol 126:691–693

Niklas KJ (1992) Plant biomechanics. University of Chicago Press, Chicago

Niklas KJ (1997) The evolutionary biology of plants. University of Chicago Press, Chicago

Niklas KJ (2000) The evolution of plant body plans – a biomechanical perspective. Ann Bot 85:411–438

Nobel PS (1991) Physicochemical and environmental plant physiology. Academic Press, San Diego

Nye PH (1994) The effect of root shrinkage on soil water inflow. Philos Trans R Soc Lond B 345:395–402

Passioura JB (1988) Water transport in and to roots. Ann Rev Plant Physiol Plant Mol Biol 39:245–265

Passioura JB (1991) Soil structure and plant growth. Aust J Soil Res 29:717–728

Peat HJ, Fitter AH (1993) The distribution of arbuscular mycorrhizas in the British flora. New Phytol 125:845–854

Perumalla CJ, Peterson CA, Enstone DE (1990) A survey of angiosperm species to detect hypodermal Casparian bands. 1. Roots with a uniseriate hypodermis and epidermis. Bot J Linn Soc 103:93–112

Peterson CA, Perumalla CJ (1990) A survey of angiosperm species to detect hypodermal Casparian bands. 2. Roots with a multiseriate hypodermis or epidermis. Bot J Linn Soc 103:113–125

Raven JA (1977) The evolution of vascular land plants in relation to supracellular transport processes. Adv Bot Res 5:153–219

Raven JA, Edwards D (2001) Roots: evolutionary origins and biogeochemical significance. J Exp Bot 52:381–401

Read DJ (1991) Mycorrhizas in ecosystems. Experientia 47:376–391

Reddell P, Yun Y, Shipton WA (1997) Cluster roots and mycorrhizae in *Casuarina cunninghamiana*: their occurrence and formation in relation to phosphorus supply. Aust J Bot 45:41–51

Richards JH, Caldwell MM (1987) Hydraulic lift: substantial nocturnal water transport between soil layers by *Artemisia tridentata* roots. Oecologia 73:486–489

Robinson D (1986) Limits to nutrient inflow rates in roots and root systems. Physiol Plant 30:491–494

Robinson D (1994a) The response of plants to non-uniform supplies of nutrients. New Phytol 127:635–674

Robinson D (1994b) Resource capture by single roots. In: Monteith, JL, Scott RK, Unsworth MH (eds) Resource capture by crops. Nottingham University Press, Nottingham, pp 53–76

Robinson D (2003) Integrated root responses to variations in nutrient supply. In: Bassiri-Rad H (ed) Root ecophysiology. Springer, Berlin Heidelberg New York (in press)

Robinson D, Fitter A (1999) The magnitude and control of carbon transfer between plants linked by a common mycorrhizal network. J Exp Bot 50:9–13

Rowell DL (1994) Soil science: methods and applications. Addison Wesley Longman, Harlow

Sanders FE, Tinker PB (1973) Phosphate flow into mycorrhizal roots. Pestic Sci 4:385–395

Schlesinger WH (1997) Biogeochemistry, 2nd edn. Academic Press, San Diego

Sexstone AJ, Revsbech NP, Parkin TB, Tiedje JM (1985) Direct measurement of oxygen profiles and denitrification rates in soil aggregates. Soil Sci Soc Am J 49:645–651

Simard SW, Perry DA, Jones MD, Myrold DD, Durall DM, Molina R (1997) Net transfer of carbon between ectomycorrhizal tree species in the field. Nature 388:579–582

Skene KR (1998) Cluster roots: some ecological considerations. J Ecol 86:1060–1064

Steudle E, Peterson CA (1998) How does water get through roots? J Exp Bot 49:775–788

Stokes A, Fitter AH, Coutts MP (1995) Responses of young trees to wind and shading: effects on root architecture. J Exp Bot 46:1139–1146

Stokes A, Ball J, Fitter AH, Brain P, Coutts MP (1996) An experimental investigation of the resistance of model roots systems to uprooting. Ann Bot 78:415–421

Swift MJ, Heal OW, Anderson JM (1979) Decomposition in terrestrial ecosystems. Blackwell, Oxford

Taub DR, Goldberg D (1996) Root system topology of plants from habitats differing in soil resource availability. Funct Ecol 10:258–264

Tester M, Morris C (1987) The penetration of light through soil. Plant Cell Environ 10:281–286

Tinker PB, Nye PH (2000) Solute movement in the rhizosphere. Oxford University Press, Oxford

Treeby M, Marschner H, Römheld V (1989) Mobilisation of iron and other micronutrients from a calcareous soil by plant-borne microbial and synthetic metal chelators. Plant Soil 114:217–226

Ver Hoef JM, Cressie N (2001) Spatial statistics: analysis of field experiments. In: Scheiner SM, Gurevitch J (eds) Design and analysis of ecological experiments, 2nd edn. Oxford University Press, Oxford, pp 289–307

Watt M, Evans JR (1999) Linking development and determinacy with organic acid efflux from proteoid roots of white lupin grown with low phosphorus and ambient or elevated atmospheric CO_2 concentration. Plant Physiol 120:705–716

West GB, Brown JH, Enquist BJ (1999) A general model for the structure and allometry of plant vascular systems. Nature 400:664–667

Whittaker RH (1975) Communities and ecosystems, 2nd edn. MacMillan, London

Williamson LC, Ribrioux SPCP, Fitter AH, Leyser HMO (2001) Phosphate availability regulates root system architecture in *Arabidopsis*. Plant Physiol 126:875–882

Woodward FI, Kelly CK (1997) Plant functional types: towards a definition by environmental constraints. In: Smith TH, Shugart HH, Woodward FI (eds) Plant functional types. Cambridge University Press, Cambridge, pp 47–65

Zhang H, Jennings A, Barlow PW, Forde BG (1999) Dual pathways for regulation of root branching by nitrate. Proc Natl Acad Sci USA 96:6529–6534

2 Distribution of Roots in Soil, and Root Foraging Activity

M.J. Hutchings and E.A. John

2.1 Introduction

A knowledge of the pattern of root distribution in soil is critical to a number of areas of ecology. For example, our ability to model the interactions between climate and vegetation depends in part on our knowledge of the global pattern of distribution of belowground biomass at various soil depths, and how it will change as one vegetation type replaces another (Jackson et al. 1996). Similarly, our understanding of ecosystem processes is currently limited by poor understanding of the distribution, quantity and productivity of fine roots within a variety of ecosystems, even though, for example, the annual production of fine roots may be twice that of leaves in Northern American hardwood forests (Fahey and Hughes 1994), and up to 80 % of the biomass of some ecosystems is underground (Jackson et al. 1996). At a more local scale, we are limited in our ability to accurately model the processes involved in plant competition by our inadequate knowledge of the fine-scale distributions of roots of individual plants and of the distribution of their associated symbionts (Mou et al. 1995; Casper and Jackson 1997; Casper et al. 2000).

That we know less about root distributions than we would like reflects the inherent difficulties in studying roots in the natural environment. Not only are most roots hidden within an opaque medium, but they are also so enmeshed in the soil and with each other that they are very hard to extract without causing them damage. A further complication is the difficulty of distinguishing roots belonging to different species and plants. Studies of roots in situ either involve painstaking removal of soil (e.g. Reinartz and Popp 1987; Pecháčková et al. 1999) followed by detailed recording of the arrangement of roots revealed, or the insertion into the soil of transparent-sided viewing devices against which roots grow and in which they can be recorded (e.g. the minirhizotrons of Farley and Fitter 1999; the Biotron of Pregitzer et al. 1993; Chap. 3). While the latter techniques allow researchers to view roots nondestructively and to monitor their birth and death, the behaviour of roots in

Ecological Studies, Vol. 168
H. de Kroon, E.J.W. Visser (Eds.)
Root Ecology

such situations is unlikely to be entirely natural. A further technique is to remove cores or sections of soil and their roots from known positions and to wash, weigh and measure the roots in the laboratory (e.g. Milchunas and Lauenroth 1989; Mou et al. 1995). These techniques allow accurate investigation of root distributions but only at the scale of the soil sample. A core freezing technique (Caldwell et al. 1996) allows extremely detailed analysis of root distributions with respect to each other.

While there is no doubt that our current understanding of underground patterns and processes is far from complete, advances are being made. This chapter reviews the distribution of roots in soil with particular reference to individual plants, and discusses some of the consequences of root distribution for plant growth. We discuss the vertical and horizontal distribution of roots, segregation of root systems, and foraging for resources by roots.

Globally, root distributions reflect patterns of climate and soil development (Chap. 1). Locally, heterogeneity in the supply of soil-based plant resources is a major cause of variation in root distribution in both the vertical and horizontal dimensions. Heterogeneity in the distribution of water and nutrients is ubiquitous. Its causes include the low mobility of many nutrient ions, the tendency for much of the supply of essential resources to arise at or near the soil surface (e.g. from litterfall and from animal-derived inputs), and the influence of plants, which may both add and remove resources. Roots tend to concentrate in parts of the soil in which resources are abundant (Fitter 1994).

The concentration of essential resources often declines with depth in the soil. For most plants, root development starts close to the soil surface. For these reasons, root biomass and length per unit volume of soil usually decline with depth. There is less predictability in resource availability and patterns of root distribution in the horizontal dimension, although biomass and density of the coarser roots often decline with distance from the aerial shoots from which root systems originate.

Fine roots and root hairs are probably the most significant parts of root systems for resource acquisition (Dumortier 1991), and their distribution within the soil is therefore more relevant to resource acquisition than that of coarser roots. The rates of production and turnover of fine roots and root hairs are high, and often vary significantly, even within the same species (Hendrick and Pregitzer 1993). For example, Wells and Eissenstat (2001) have shown that the survivorship of fine roots of apple is negatively correlated with their diameter. However, quantification of this fraction of root systems is extremely difficult (Fahey and Hughes 1994), and therefore data on root distribution patterns are usually more reliable when based on the coarser root fraction.

2.2 Plant Rooting Patterns in the Vertical and Horizontal Dimensions

Root structure and architecture can be greatly modified by growing conditions, but there is considerable variation between species in the effects of abiotic conditions on the distribution of roots with depth. For example, when water is abundant, root systems might be expected to be shallower and smaller than they are under drought conditions, because sufficient moisture to satisfy the transpiration requirements of the plant can be obtained from a smaller volume of soil. However, a range of responses was seen in an investigation of the maximum rooting depth of seedlings of 42 vascular species in response to regular watering or lengthy drought (Reader et al. 1993). Rooting depth increased significantly in seven species when watering was withheld, decreased significantly in five, and did not respond significantly in the remainder. Rooting depth in unwatered soil tended to increase in species that had deep root systems in watered soil, whereas many species that had shallow root systems in watered soil decreased their rooting depth when water was withheld (Fig. 2.1).

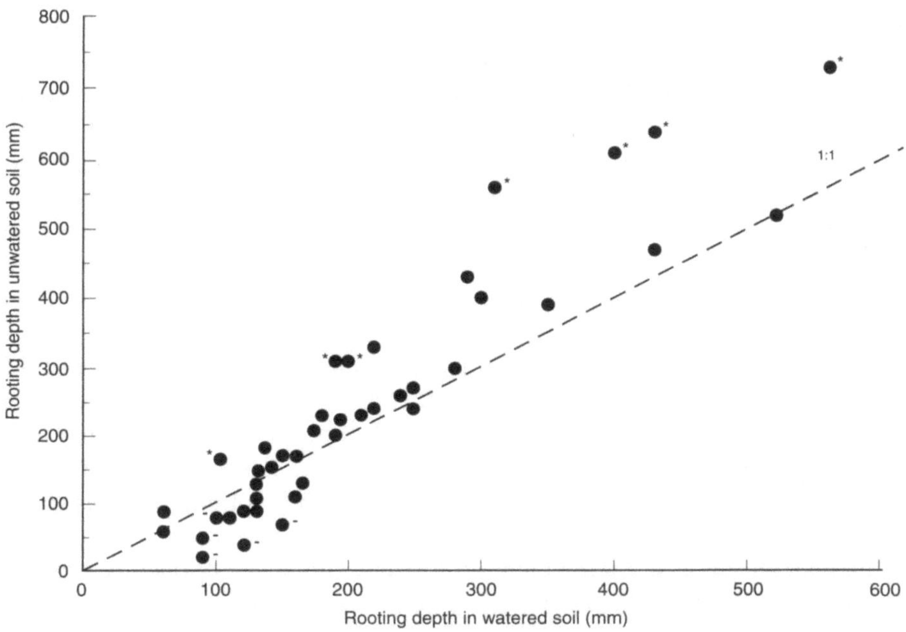

Fig. 2.1. Mean rooting depth of seedlings of 42 vascular plant species in watered vs. unwatered soil. Rooting depth in unwatered soil was either significantly (P <0.05) greater (*) or significantly less (–) or no different (blank) from rooting depth in watered soil. Data for species showing no change in rooting depth would lie on the *dashed line*. (Adapted from Reader et al. 1993)

The location within the soil profile of roots that are actively growing and acquiring resources may also change as the growing season proceeds. Such changes have been recorded both within species and between species that are physiologically active within communities at different times during the growing season. For example, Fernandez and Caldwell (1975) studied root growth during the growing season within the soil of a semi-desert habitat in northern Utah. Root growth of three shrub species shifted to progressively greater depths as time passed. The upper soil froze in winter, and subsequent soil warming progressed from the soil surface downwards. The increasing depth of root growth as time passed was thought to be associated with this warming. Cessation of root growth in the upper part of the soil profile in mid-summer appeared to be related to declining photosynthetic activity rather than to low soil water potential. At the driest period of the year, *Atriplex confertifolia* could explore deeper soil than other species, enabling it to continue to transpire and achieve positive net assimilation at times when other species could not.

Many authors have published profile drawings of the excavated root systems of a range of species (Yeaton et al. 1977; Fitter 1994; references in Schenk et al. 1999). In contrast, quantitative analyses of rooting profiles of individual species are not common, although some detailed and informative studies have been undertaken. For example, Mou et al. (1995) analysed root distribution with depth in dense monospecific stands of 3-year-old *Liquidambar styraciflua* (sweetgum) and *Pinus taeda* (loblolly pine) by washing roots from different soil strata. Both species had similar coarse root and total root masses per unit volume of soil, and these measures declined steeply with depth (Fig. 2.2a,b). In contrast, loblolly pine had approximately twice as much fine (<2 mm diameter) root mass as sweetgum (Fig. 2.2 c), but the fine roots of sweetgum were thinner, resulting in both species having similar fine root length densities at intermediate depths in the soil (Fig. 2.2d). As in other studies (e.g. Fahey and Hughes 1994), fine roots from the upper soil were thinner than those at greater depth. This could be because thicker roots do not buckle so easily under given levels of soil impedance (Chap. 6) or because more of the nutrients are in the top strata of the soil. 51 % of the fine roots of sweetgum and 71 % of those of loblolly pine were in the top 10 cm of soil, where much nutrient acquisition takes place. In their northern hardwood forest study, Fahey and Hughes (1994) found that 43 % of the fine root biomass was in the organic horizon.

Plants can increase their potential nutrient and water uptake by producing a greater length of roots from a given mass of roots (i.e. by increasing the specific root length, SRL), although the effectiveness of this is nutrient-dependent. For example, nitrate uptake is, in theory, affected very little, whereas phosphate uptake is strongly affected (Robinson et al. 1999). In the study by Mou et al. (1995), sweetgum developed considerably more root length per unit weight of root than loblolly pine at all soil depths, thus pos-

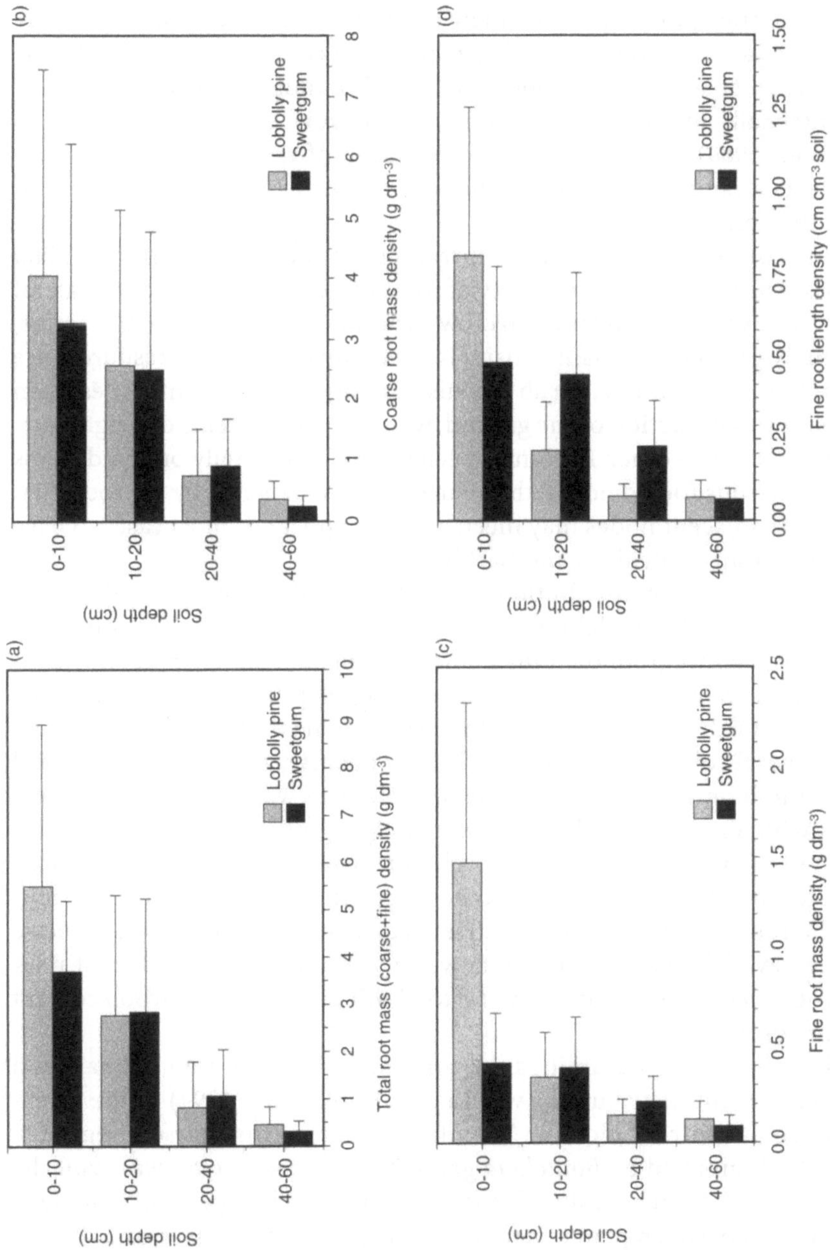

Fig. 2.2a–d. Vertical distributions in the soil of mean root mass densities (g dm^{-3}) and their standard deviations for **a** total roots, **b** coarse roots only, **c** fine roots only, and **d** the vertical distribution of mean fine-root-length densities (cm root length cm^{-3}), and their standard deviations, for 3-year-old sweetgum and loblolly pine monocultures. (Mou et al. 1995)

sibly compensating for a lower root biomass. Whereas most data on the distribution of roots with depth are based on analyses of root weight, the distribution of root length with depth may be a better indicator of the relationship between soil quality and the likely benefits obtained by the plant from exploring the soil volume. However, thicker roots can penetrate soil more easily and anchor plants more effectively. Moreover, because the diameter of the stele increases with root diameter, thicker roots have higher water and nutrient transport capacities than thinner roots. Finally, under waterlogged and anaerobic conditions, which are more common at greater depths in many types of soil, thicker roots may permit enough oxygen to be transported internally for the root to function and grow, even if aerenchyma is absent (Boot 1989). Thus, the depth distribution of roots of differing diameters reflects a range of factors that may influence the success of the plant in utilising the soil environment.

In many communities, the lateral spread of the root systems of individual plants is constrained by the proximity of roots of neighbouring plants. Lateral spread of root systems in relation to crown width, and root/shoot (R/S) ratios, are normally greater, and plant density is generally lower, when resources are less abundant. For example, shrub crowns in many desert communities often cover a very small fraction of the ground, whereas root systems of neighbouring shrubs abut each other. In clonal species with horizontally oriented stems bearing numerous rooted nodes, the volumes of soil depleted by the roots that emerge from adjacent nodes may meet, but they appear in most cases to have very little overlap (Hutchings and Barkham 1976).

In their study of root distribution in sweetgum and loblolly pine, Mou et al. (1995) also analysed the relationships between the distributions of root mass and both local aboveground biomass and soil nutrient status in the horizontal plane. Roots were excavated from different soil depths in blocks with a horizontal area of 0.25 m^2. At this scale, total root biomass of both species, measured at several soil depths, was significantly positively correlated with local aboveground biomass. When only fine root biomass was considered, the correlations with aboveground biomass were only significant at greater depths. Fine root density, however, was positively correlated with available phosphorus and, for sweetgum, with potassium, suggesting that the spatial distribution of the roots most involved in nutrient acquisition was locally responsive to nutrient availability. In contrast, however, there were no significant relationships between fine root density and total soil nitrogen and carbon content at the time of the study.

Another study which examined the correlation between root biomass and aboveground plant distributions was that of Hook et al. (1994) in the shortgrass steppe of central Colorado. They found that in this sparsely vegetated community dominated by *Bouteloua gracilis* (plant basal cover was typically around 20–40 %), there was significantly less root biomass per unit volume of soil in the gaps between plants than under the plants themselves, and the difference increased with increasing gap size. However, some roots were present even in the centre of the largest gaps (50–85 cm diameter). Both root biomass and its horizontal variability decreased with depth.

In intensively grazed ecosystems, herbivory can determine the spatial heterogeneity of root (and shoot) biomass distributions. For example, in a study carried out in Colorado shortgrass steppe, Milchunas and Lauenroth (1989) showed that 47 years of intensive grazing had hardly affected belowground biomass, but that the root biomass from under the grazed areas was significantly more homogeneous in its distribution than that from otherwise similar ungrazed areas. This pattern was reflected above ground in the development of a grazing lawn.

An extremely detailed study of the distribution of belowground plant parts in the horizontal and vertical dimensions was carried out by Pecháèková et al. (1999) in a montane grassland in the Czech Republic. The exact positions of roots and rhizomes were exposed by carefully removing the soil from successive strata down to 17 cm depth. All excavated parts were identified to species and weighed. Root biomass, excluding rhizomes, ranged from 1.2–1.5 kg m^{-2}, and from 1.7–2.0 kg m^{-2} when rhizomes were included. The ratio of roots to aboveground biomass ranged from 3.4–4.8, and the ratio of below- to aboveground biomass ranged from 4–7. Spatial autocorrelation analyses showed significant correlations between a variety of belowground biomass variables over distances up to 10 cm, and at all depths analysed. The numbers of such correlations were far higher than expected by chance. It was suggested that the pattern of belowground biomass reflected spatial heterogeneity in the availability of resources for plant growth in the soil. The presence of roots and rhizomes of individual species in the surface soil (0–3 cm depth) was closely correlated with the presence directly above ground of the same species. There were also several negative spatial associations between the roots and rhizomes of different species within this soil layer, and these mirrored negative associations between the same species above ground. This pattern indicates a high level of interdependence in the spatial arrangement of species in the upper soil layer and the aboveground vegetation. Correlations between species above ground and in deeper soil (3–6 cm) were strongest when there was a horizontal distance of approximately 2 cm between the sites being compared, probably because the roots and rhizomes of many species grow diagonally down through the soil. Most of the negative spatial associations between parts of different species disappeared at this depth, suggesting that there was less interdependence between vegetation structure above ground and at this depth in the soil, and that the underground parts of different species were more intermixed at this depth. One of the authors' conclusions was that "whereas the shallow roots are … constrained by the spatial distribution of the aboveground parts and plant rhizomes, the deep roots are more free to follow soil heterogeneities". The results of Pecháèková et al. (1999) show that horizontal root distribution patterns are at least partly decoupled between different soil strata, suggesting that root systems respond very locally to the factors affecting their development and that these factors differ between strata.

2.3 Segregation of Root Systems

Although root systems of different plants might be expected to aggregate in resource-rich soil patches (Caldwell et al. 1996; Casper et al. 2000), inter- and intraspecific segregation of root systems in time and space is frequently recorded. This could be important in reducing competition and facilitating species coexistence (e.g. Yeaton et al. 1977; Sydes and Grime 1984; see Chap. 9). In most cases, segregation is believed to be a consequence of either species-specific differences in root architecture or resource depletion (Schenk et al. 1999).

2.3.1 Segregation of Root Systems in the Vertical Dimension

In a species-rich grassland habitat, Fitter (1986) showed that species that were active in terms of nutrient uptake early in the growing season were shallower-rooted than those that were active later. The earlier-growing species were also less productive. He proposed that being active earlier in the growing season enabled less productive species to avoid competition with more aggressive species, and to exploit the flush of resources that became available near the soil surface early in the growing season. The greater depth of root activity later in the growing season was probably also related to drying out of the surface soil at this time of year.

The previous example, and that of Fernandez and Caldwell (1975) described earlier, show that the depth of root activity can change during a growing season. In other cases, rooting depth may depend on the presence or absence of particular neighbouring species. In a study of competition between the invasive *Carpobrotus edulis* and two native Californian coastal scrub species, D'Antonio and Mahall (1991) showed that at sites where *C. edulis* was present, the root systems of the native species were displaced downwards in the soil profile. This reduced, but did not eliminate, overlap with *C. edulis*. Removal experiments showed that the presence of *C. edulis* deprived the native species of water, resulting in smaller shoot sizes and other morphological changes (Fig. 2.3). A further example suggests that segregation of rooting depth reduces competition between species. In a Sonoran Desert community where all species together produced less than 10 % ground cover, the sizes of nearest neighbours suggested that there was significant competition between neighbouring plants of the same species (Yeaton et al. 1977).

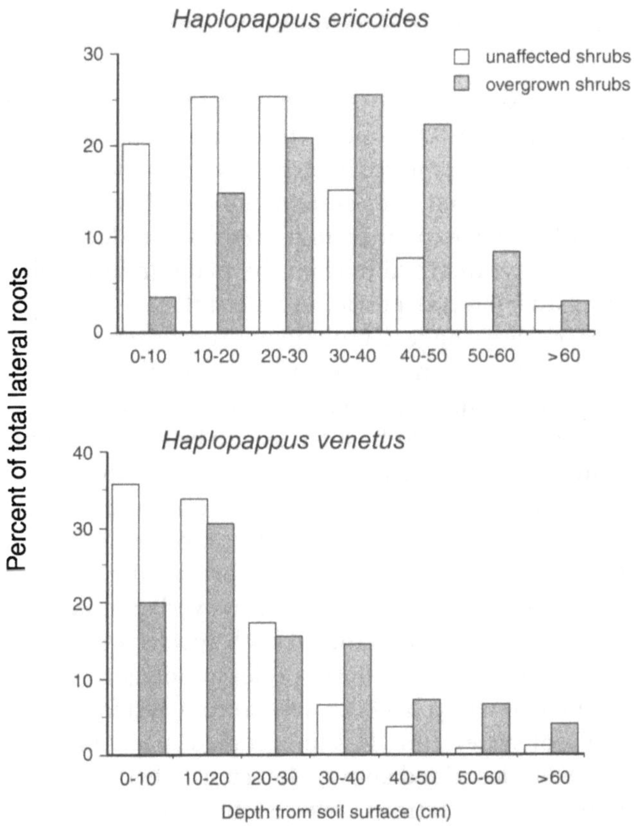

Fig. 2.3. Percentage of all lateral roots of *Haplopappus ericoides* (top) and *H. venetus* (*below*) produced at different soil depths in the presence (*hatched columns*) and absence (*open columns*) of the invasive alien *Carpobrotus edulis*. (D'Antonio and Mahall 1991)

Although aboveground parts of neighbouring plants were distant from each other, root systems met, and this could have resulted in competition for limited water. There was also evidence of interspecific competition, but only between species with root systems occupying similar depths in the soil.

Berendse (1981, 1983) demonstrated that two common grassland species, *Anthoxanthum odoratum* and *Plantago lanceolata*, exhibit niche differentiation, and thereby partially avoid competition, by rooting at different depths in the soil. *A. odoratum* roots are concentrated in the top 10 cm of soil, whereas *P. lanceolata* produces roots down to a depth of 50 cm. The root systems of the species segregate to a greater extent in fertilised soil, because *A. odoratum* develops a dense mat of roots even closer to the soil surface. This rooting pattern enables it to acquire more nutrients than the unfertilised rooting pattern. This in turn leads to *A. odoratum* developing more foliage, which results in greater suppression of *P. lanceolata* through competition for light.

2.3.2 Segregation of Root Systems in the Horizontal Dimension

As noted above, decreasing the overlap between root systems of different plants is likely to reduce competition for limiting resources. This has been well documented in environments where water is scarce. For example, Brisson and Reynolds (1994) demonstrated that root systems in neighbouring creosotebush (*Larrea tridentata*) plants expanded in a way that produced less overlap than if they had grown symmetrically from the plant and occupied the same area (i.e. if growth had been concentric, with no influence of neighbours; Fig. 2.4A–D). These results suggest both that little water was left unused by the plants in this desert environment and that this efficient resource usage was achieved with minimum inter-plant competition.

Soil nutrients, particularly nitrogen and phosphorus, can be at least as limiting in mesic forests as water is in desert environments. In the study of Mou et al. (1995) of dense monospecific populations of sweetgum and loblolly pine, roots of sweetgum were more densely packed (in terms of cm roots/cm^3 soil) than those of loblolly pine, which extended further horizontally from the tree trunks, causing root systems of neighbouring pines to overlap more than those of sweetgum. Root system overlap of plants of both species was less at greater depths, and root territories drawn for both species as polygons, rather than circles, again produced less overlap between several root systems, more space occupied by only one root system, and less unoccupied space. As in the study by Pecháèková et al. (1999) described above, Mou et al. (1995) observed depth-related differences in the interactions between the root systems of neighbouring plants, implying again that the factors determining the structure of the community below ground changed with depth.

Some studies suggest that active mechanisms cause segregation of root systems. For example, different clones of northern prickly ash (*Xanthoxylum americanum*) exhibit almost complete absence of root system overlap (Reinartz and Popp 1987). When roots of different clones meet they either turn away or stop growing. Caldwell et al. (1996) demonstrated that when nutrient patches were created between the shrub *Artemisia tridentata* and one of two tussock grasses (*Pseudoroegneria spicata* or *Agropyron desertorum*), the roots within the patch were aggregated at the scale of the nutrient patch, but there was significant spatial segregation between shrub and grass roots at the smaller scale of the fine roots. Competition for soil-based resources did not cause this segregation, because there was no relationship between the proportions of shrub and grass roots in different soil microsites and nutrient concentration. The most studied examples of root system segregation involve the desert shrubs *Larrea tridentata* and *Ambrosia dumosa* (Mahall and Callaway 1992). Proximity to *Larrea* roots inhibits root elongation in both *Larrea* and *Ambrosia*. In contrast, contact between roots is necessary for *Ambrosia* to inhibit root growth, and only roots of other *Ambrosia* individuals are affected. Addition to the soil of activated carbon, which

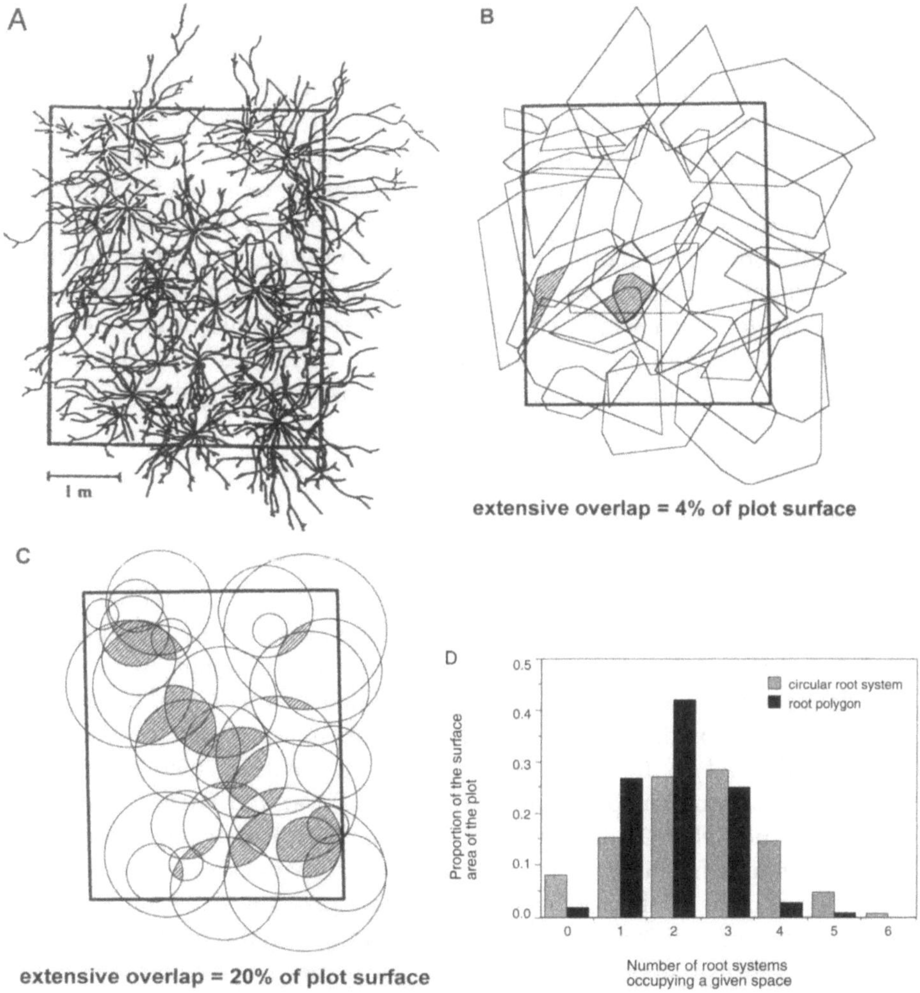

Fig. 2.4A–D. **A** Map of the lateral extension of root systems of 32 creosotebush (*Larrea tridentata*) shrubs excavated from a 4 × 5 m rectangular plot (*solid lines*) in the Chihuahuan Desert, New Mexico. **B** Map of areas occupied by each root system expressed as the smallest closed-angle polygon that encompassed all the roots of each shrub. **C** Map of the area occupied by hypothetical root systems of each shrub. These hypothetical root systems are circular in shape, centred on the location of each shrub, and have a surface area equal to that of the corresponding polygon. In **B** and **C** the *shaded area* represents the parts of the plot where there is overlap between at least four root systems. **D** Root system overlap expressed as the proportion of the plot area covered by *n* root systems, using both the polygonal root areas and the hypothetical circular root areas. Note that the proportion of fully mapped area that was occupied by roots from only one plant (i.e. where there was no interplant competition for resources) was almost twice as much using polygons as when circular territories were drawn. Polygons produced more overlap between pairs of root systems than circular root systems with the same area, but far less overlap of three or more root systems. Moreover, the polygonal territories occupied almost all of the mapped area, unlike the circular territories. (Brisson and Reynolds 1994)

adsorbs organic compounds, reduces the inhibition caused by *Larrea*, but does not alter the inhibitory effect of *Ambrosia*. This suggests that *Larrea* inhibits by diffusion of a non-specific chemical from its roots, while *Ambrosia* inhibition requires direct contact, and a self/non-self recognition mechanism. The inhibiting chemicals or signals have not been identified in any of these examples. Several other examples of root segregation are listed by Schenk et al. (1999). Studies of the response of perennial grass roots to inter- or intraspecific contact with other roots also suggest the existence of recognition systems. Plants of *Pseudoroegneria spicata* grown with roots in interspecific contact change their specific root length (Huber-Sannwald et al. 1996) and root growth rate (Krannitz and Caldwell 1995) compared with plants with roots growing alone or in contact only with intraspecific neighbours.

2.4 Foraging by Roots

Edaphic heterogeneity can take several forms. Patches of soil with different nutrient or water contents can contrast slightly or substantially in quality. The spatial and temporal scales at which different nutrient ions and toxins vary in abundance in the soil are widely different, and depend on rates of diffusion in soil solution and on the texture, organic, mineral and water content of the soil. They also depend on the dominant vegetation (e.g. Wilson 2000). All of these properties also vary in space and in time (e.g. Kovar and Barber 1988,1989; Lechowicz and Bell 1991; Jackson and Caldwell 1993; Farley and Fitter 1999). There is often sufficient variation at small enough spatial and temporal scales for different parts of the root systems of single plants to experience significantly different conditions. For example, Jackson and Caldwell (1993) reported that, in a sagebrush-steppe community, roots less than a metre apart belonging to the same individuals of both shrub and tussock grass species can experience as much variation in soil properties as they might if separated by much larger distances. They suggest that, because some soil patches are more favourable than others, "root plasticity and active foraging in a heterogeneous soil environment are likely to be important to the nutrient balance of many plants".

Foraging has been defined by Hutchings and de Kroon (1994) as "the processes whereby an organism searches, or ramifies within its habitat, which enhance its acquisition of essential resources" (modified after Slade and Hutchings 1987). Foraging involves the use of morphological plasticity, in response to heterogeneous conditions, to selectively place resource-acquiring structures, such as roots or leaves, in more favourable patches of habitat. Changes in the size of the plant (i.e. growth in response to habitat quality) that do not also involve changes in form are not manifestations of foraging (Hutchings and de Kroon 1994; de Kroon and Hutchings 1995; Fransen et al. 1999).

2.4.1 Root Foraging Responses to Spatial Heterogeneity in Availability of Soil-Based Resources

The most common morphological responses to higher nutrient availability include increased root growth rate and root branching rate. Early observations (Weaver 1919) and experiments (Drew et al. 1973) showed that such responses could be displayed at a local level, rather than throughout the plant, thus enabling the plant to invest resources at scales equivalent to that of nutrient patches, or individual roots. Drew (1975) grew seminal roots of barley (*Hordeum vulgare*) downwards through a sequence of three vertically stacked layers of sand, in which the concentration of selected nutrients could be independently controlled. When the sand in only one of the compartments contained a high nitrate, ammonium or phosphate concentration, the section of seminal root within that compartment produced significantly more primary and secondary laterals, with significantly higher extension rates, than equivalent parts of seminal roots in other compartments and in control treatments. These effects were independent of level of nitrate, ammonium or phosphate in the other compartments. In contrast, there was no effect of potassium on the growth of lateral roots. The rate of extension of the seminal root was not affected by the concentration of single nutrient ions (Drew 1975). Recent work has demonstrated that there is an important genetic component to the lateral root proliferation response to nitrate (Zhang and Forde 1998). Experimental enrichment of specific compartments with different nutrients showed that the branching and extension of lateral axes depended on the nutrient involved (Drew et al. 1973; Drew 1975; Drew and Saker 1978). One important outcome of this experiment in the context of the present chapter is that exposure of only a few percent of the seminal root to nutrient-rich sand resulted, after a short lag, in whole-plant relative growth rates similar to those achieved by plants with all of their seminal root in nutrient-rich sand (Drew and Saker 1975). This was presumably because localised proliferation of laterals in the heterogeneous treatments allowed nutrient acquisition to match that achieved in homogeneously nutrient-rich conditions.

Many other studies of root behaviour under heterogeneous conditions corroborate the results of Drew and his co-workers for a range of species. However, a review of these studies (Robinson 1994) showed that such responses are not ubiquitous; although they occurred in the majority of cases, about 30 % failed to show significant responses. In summarising the studies he reviewed, Robinson reported that the scale of response depended on the nutrient that was in patchy supply, the nutrient status of the plant, and the stage of growth being considered.

Localised proliferation of parts of root systems in the nutrient-rich patches of heterogeneous habitats is often associated with reduced growth of parts of the same root system located where nutrients are scarce. For example, Gersani and Sachs (1992) divided the root systems of pea plants equally between two containers of nutrient solution. If the solutions in the containers differed in concentration, root development was greater in the more nutrient-rich container. A greater difference in concentration caused a greater proportion of the roots to develop in the nutrient-rich container, but, overall, the same total number of root primordia was produced in all situations. These results led Gersani and Sachs to suggest that the whole root system developed in a coordinated manner involving trade-offs between growth in locations of different quality.

An important study by Gersani et al. (1998) also commenced by examining the relationship between local nutrient concentration and biomass of roots developed by pea plants in different containers. They began their experiments by dividing the roots of plants referred to as "fence-sitters" equally between two containers. When fence-sitters were grown alone, their root growth in the two containers was proportional to the concentration of nutrients in each container, as in the study by Gersani and Sachs (1992). Next, the effect of introducing competitors on root growth and fitness was examined. Competitors ranged in number from 0 to 5 in different treatments, and their root systems were all confined to one of the two containers. The root systems of the fence-sitters were again divided equally between the two containers (Fig. 2.5A). Experiments were undertaken with nutrient solution of equal strength in both containers, or with the container in which competitors were rooted having solution twice as concentrated as that in the container with no competitors. In the first case, the total root weight of the fence-sitter was unaffected by competitors. Strikingly, however, as competitor number increased, progressively more of the fence-sitter root biomass was produced in the container without competitors (Fig. 2.5B). In the second experiment, the weight of fence-sitter roots in the container with no competitors matched that in the container with double the nutrients *when this pot also contained one competitor.* This suggests that doubling the number of plants with roots accessing the nutrients in this container also halved the availability of nutrients to each plant. The fence-sitter compensated for this by producing a greater fraction of its root system in the container without competitors (Fig. 2.5C). These results indicate strong coordination of root system development in pea, as Gersani and Sachs (1992) suggested, and a fine level of control over root placement patterns.

These experiments illustrate important plant responses to edaphic heterogeneity, whether it is abiotic or biotically induced. Root systems of competitors were confined to a single (homogeneous) container in which the nutrients available to each plant fell as competitor number increased. The fitness of these plants (measured as weight of pea fruits produced) declined as their number increased. In contrast, the root systems of the fence-sitters occupied two containers of differing quality. As availability of nutrients in one container declined because of the presence of competitors, the fence-sitters acquired more of their nutrients from the other container. This presumably

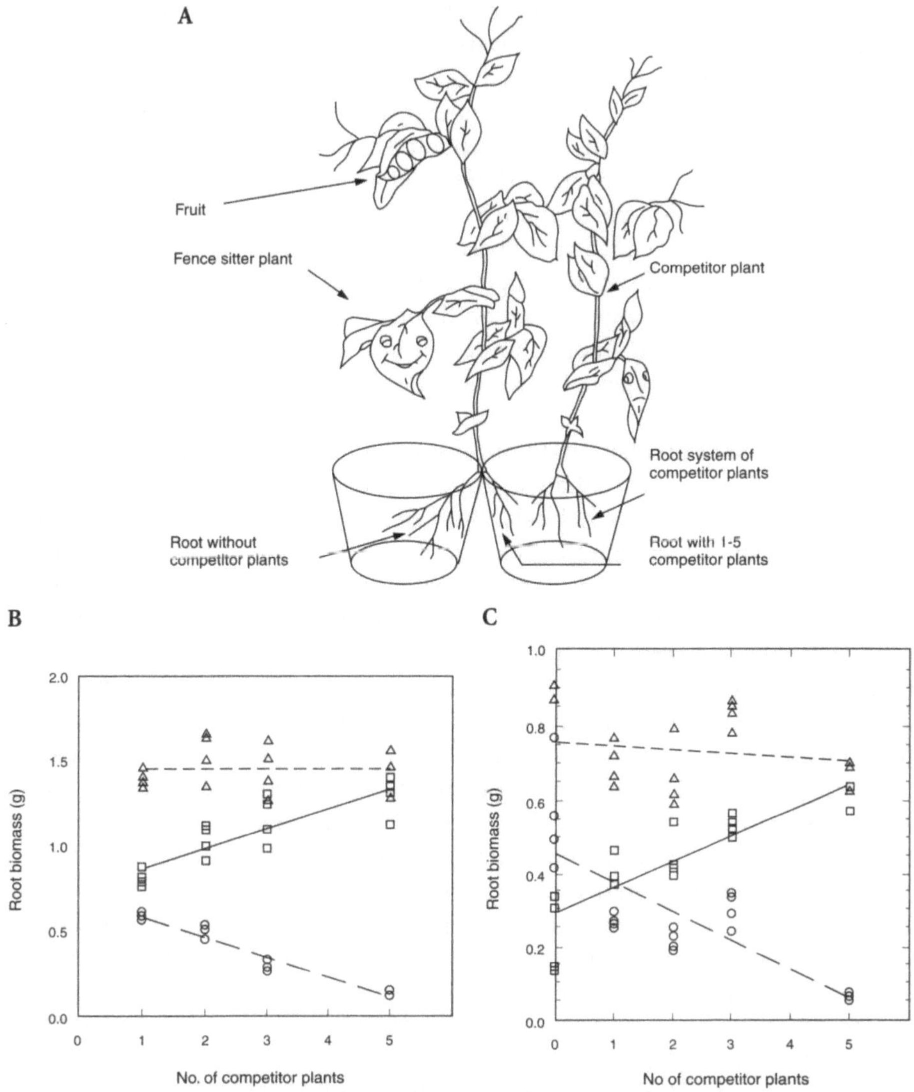

Fig. 2.5A–C. **A** Schematic illustration of the experimental system used by Gersani et al. (1998) to examine root growth and root distribution patterns in fence-sitter and competitor pea plants. **B** Relationship between the total root biomass of the fence-sitter plant and number of competitors (*open triangles*), and the biomass of fence-sitter roots in the container with (*open circles*) and without (*open squares*) the competitors. The substrate in both containers was watered with nutrient solution of the same concentration. **C** The same as in **B**, but the nutrient solution used to water the substrate in the container with competitors was twice as concentrated as the solution used to water the substrate in the container without competitors

enabled them to maintain their nutrient uptake despite the presence of competitors in part of the habitat they were occupying. They also maintained the same fitness across the treatments, unlike the competitors. Thus there were very different consequences for plants that occupied, and foraged for nutrients in, homogeneous and heterogeneous habitats. In the first case, the rewards from utilising all parts of the habitat are equal. However, in heterogeneous conditions, plant performance depended on the quality of the patches selected for resource acquisition. These results should serve as a warning that experiments on the effects of competition on plant performance and fitness conducted under homogeneous conditions may give a very misleading impression of what may happen under natural conditions, which are always characterised by heterogeneity (see also Robinson et al. 1999; Fransen et al. 2001; Chap. 9).

Development of greater length or mass of roots, particularly fine roots, per unit of substrate, may enhance nutrient acquisition from nutrient-rich patches in heterogeneous habitats. Both density and biomass of fine roots were significantly correlated with available soil P and K in the study by Mou et al. (1995). This may have promoted greater uptake of these nutrients, but no information was given in this study about nutrient uptake. There was no correlation between quantity of fine roots and local availability of total N or carbon. In an experiment with five grass species, Fransen et al. (1998) also showed that root length density (RLD) was higher for parts of plant root systems located in a nutrient-rich patch in heterogeneous soil than for parts in nutrient-poor soil. The difference was significant for the three fastest growing species. The total root biomass produced by the five species was unaffected by the spatial configuration of nutrients in the soil. Only one species (*Anthoxanthum odoratum*) acquired more nitrogen altogether in the heterogeneous treatment than in the homogeneous treatment, and phosphate uptake did not differ significantly between treatments for any species. This was also the only species for which total plant biomass differed significantly between treatments. Its biomass was higher under heterogeneous conditions, perhaps because of greater nitrogen acquisition.

Finally, a method allowing quantification and comparison, between and within species, of the architecture of entire root systems has been described by Fitter (1987) and Robinson et al. (Chap. 1). Different root branching patterns appear to offer differences in efficiency for exploration of soil and exploitation of soil-based resources. However, the more efficient branching patterns appear to involve greater production costs.

2.4.2 Morphological vs. Physiological Plasticity: Responses to Total Resource Supply and to the Spatial and Temporal Patterns of Resource Provision

Many of the morphological responses to heterogeneity of soil-based resources described above require investment of energy and biomass in the production of roots or other structures for exploration of new volumes of soil or exploitation of resource-rich patches of soil. As the overall level of resources in the environment, and the frequency of resource-rich patches, decline in space and time, so does the likelihood of acquiring sufficient resources to balance the cost of investment in exploratory structures. Consequently, it has been predicted (Grime et al. 1986) that species characteristic of resource-poor habitats will exhibit lower morphological plasticity in response to a patchy resource supply than species from resource-rich habitats. Instead, much of their resource supply will be obtained by changing *rates* of acquisition on the rare occasions when resource availability increases. Thus, such species are predicted to exhibit physiological plasticity more than investment in exploratory structures. Either way of expressing plasticity can be confined to the parts of the root system that experience nutrient enrichment, rather than being displayed throughout the root system. Thus, these plastic responses are deployed economically.

There have been some revealing experiments on this topic. Crick and Grime (1987) compared the responses of *Agrostis stolonifera*, a species of nutrient-rich habitats, and *Scirpus sylvaticus*, a species from nutrient-poor habitats, to a range of treatments in which nutrients were supplied in different spatial and temporal arrangements to all or part of the root systems of single plants. In all treatments *S. sylvaticus* allocated a higher proportion of its biomass to roots than *A. stolonifera*. When parts of the root system of *A. stolonifera* received nutrients for a long period, these parts proliferated, but no such response was found when parts of the root system were given brief nutrient pulses. The root system of *S. sylvaticus* did not respond morphologically to either pattern of nutrient enrichment. However, despite clear evidence of morphological responses to predictable patterns of nutrient enrichment in *A. stolonifera*, and of their absence in *S. sylvaticus*, detailed analysis of nutrient uptake did not convincingly indicate greater physiological response in the latter species (Hutchings and de Kroon 1994).

Campbell and Grime (1989) compared nutrient acquisition and growth in *Arrhenatherum elatius*, a species from fertile habitats, and *Festuca ovina*, a species from nutrient-poor habitats, subjected to low nutrient supply interrupted by pulses of nutrients provided at the mid-point of 6-day intervals. Pulse duration ranged from 80 s to 6 days in various treatments. *A. elatius* exhibited more variation in growth and more plasticity in allocation of biomass to roots and shoots in different treatments than *F. ovina*, providing at least some support for the proposal that species from more fertile habitats are more morphologically plastic.

Different species from the same habitat also differ in the way they respond to episodes of nutrient enrichment. For example, Jackson et al. (1990) studied the responses of three desert grasses to localised phosphate addition to the soil. Within a few days, roots of all three species taken from the enriched patches of soil showed significant enhancement in phosphate uptake, but the degree of physiological plasticity differed between species, as did the extent to which roots proliferated in the enriched soil patches.

In conclusion, different species may be more striking in their capacity to exhibit either morphological or physiological plasticity in response to spatially and temporally patchy nutrient supply. However, it is likely that most species utilise a combination of both types of plastic response, and adjust the extent to which each is used to enhance resource acquisition in different circumstances.

2.4.3 Patterns of Root Placement in Heterogeneous Environments and Their Consequences

Campbell et al. (1991) predicted that there would be a negative relationship between the scale of a species' root system and its precision in placing its roots in the more nutrient-rich parts of a heterogeneous soil. Experimental evidence appeared to support this prediction by showing that the subordinate species in an artificial community had smaller root systems than the dominant species, but that, when they were grown alone in heterogeneous substrate, they produced a higher proportion of their new roots in high quality soil patches. This may make subordinate species more efficient at obtaining nutrients and enable them to persist despite strong competition from larger species. Subsequent studies have not supported this prediction, however. Einsmann et al. (1999) obtained a positive correlation between scale of root system (measured as root mass density and root length density) and precision of root placement for ten species from different vegetation types along a successional sequence. Differences in experimental procedures and in the methods used to measure scale and precision may have contributed to the different conclusions from these studies, but it is also likely that the prediction is too simplistic. For example, the ability of a species to forage precisely for nutrients also depends on there being a match between the size of its root system and the scale of environmental heterogeneity, and on its ability to perceive and respond to the heterogeneity (Wijesinghe et al. 2001). In addition, the precision with which roots can be located in nutrient-rich patches depends on the positions of those patches with respect to the place at which the plant starts growing, and on the size of the species. Thus, a species with a small root system growing in a nutrient-rich patch may produce all of its roots within the patch. It will have high precision, and this will be beneficial to its growth. However, if the same species is in a nutrient-poor patch, it may be unable to extend much of its root system into even very close nutrient-rich patches (Hutchings et al. 2003). Thus, precision depends on both the species and its environmental context. Finally, there is evidence that roots are less branched when colonised by mycorrhizal fungi. Plants also exhibit less lateral root extension and less root hair development, than when they lack mycorrhizas (Steeves and Sussex 1989; Hetrick et al. 1991; Chap. 11). The hyphae of many mycorrhizal fungi become more branched when provided with soil of higher nutrient status (St. John et al. 1983a,b; Cui and Caldwell 1996a,b). Clearly, association with mycorrhizal fungi enables plants to invest less effort to obtain nutrients efficiently from a given volume of soil. Thus, it would be predicted that plants with mycorrhizal associates would exhibit lower precision in root placement. Evidence to date does not support this prediction, however (Wijesinghe et al. 2001).

Ideally, in a heterogeneous environment, root distribution between soil patches of different quality should be in proportion to the rewards that may be obtained from each patch (Fretwell 1972; Pulliam and Caraco 1984). That plants can closely approach this state is illustrated in the studies by Gersani et al. (1998) on pea discussed above. A similar result was also reported by Wijesinghe and Hutchings (1999) from a more complex experiment on the clonal species *Glechoma hederacea*. This species grows by the extension of horizontally oriented branched stems (stolons) that produce ramets that root at intervals as they grow across the substrate. Under heterogeneous conditions, this means that different ramets may root in sites that differ in quality. This pattern of growth was utilised in an experiment in which genetically identical clones were grown in environments consisting of good and poor patches, each of the same size, arranged in a chequer-board pattern. Different treatments provided six different degrees of contrast between good and poor patches. The experiment was also performed at two patch scales (Fig. 2.6A).

The proportion of clone root biomass in good and poor patches closely matched the proportion of nutrients in the patches in all but the highest contrast treatment (100:0, Fig. 2.6B). The poorer match between patch nutrient quality and root distribution in the 100:0 treatment was caused by the need for clones to root in patches consisting entirely of sand in order to cross them. However, they probably gained benefits other than nutrients – particularly water – from doing so. While the proportion of roots located in patches of different quality was independent of patch size (Fig. 2.6C), root biomass was not (Fig. 2.6D). The total area and volume of soil in patches of a given quality were the same between different patch size treatments, but more root biomass was produced per unit volume of substrate in good patches of a given quality when patch scale was larger. Thus, the potential of the species to acquire nutrients was greater when nutrients at a given concentration were available in larger patches. An earlier study specifically on

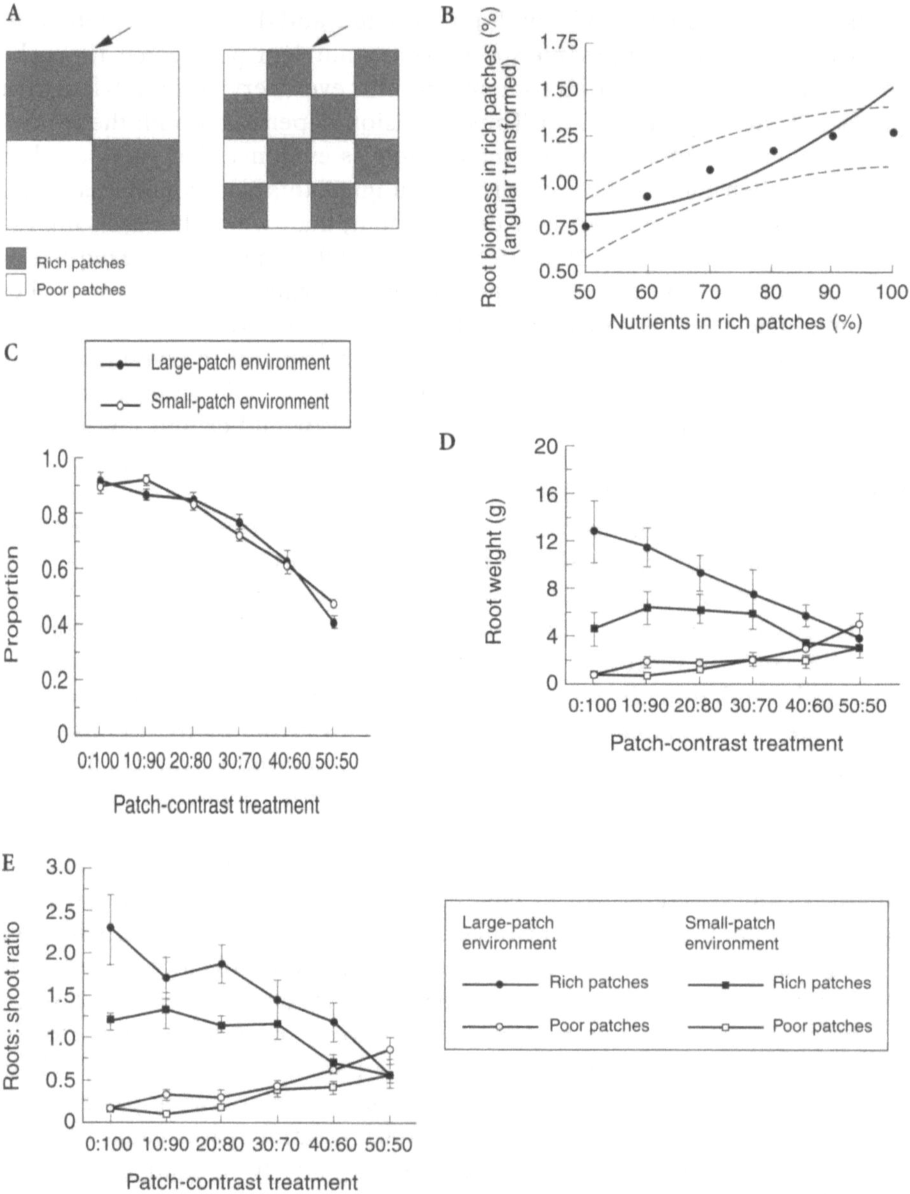

Fig. 2.6A–E. **A** Clones of *Glechoma hederacea* were grown in heterogeneous environments with two patch scales and six different degrees of contrast between good and poor patches. Patches of different quality were produced by mixing compost and sand in different ratios: 50:50 (homogeneous environments), 60:40, 70:30, 80:20, 90:10 and 100:0. All treatments provided the same total amount of nutrients, but in different spatial arrangements. A single ramet of *G. hederacea* was planted in each box at the position indicated by *the arrow*. The two emerging stolons that grew from this ramet were directed towards the opposite sides of the box, and new ramets were allowed to root freely. **B** Relationship between the percentage (angular transformed) of clone root biomass in rich patches and the percentage of the total nutrient supply in rich patches. Results from the two patch

the effect of patch scale on clone behaviour showed that root biomass per unit volume of substrate, *and* the ability of the clone to locate roots in higher quality patches, were dependent on the scale of the patches (Wijesinghe and Hutchings 1997). The nutrient-foraging ability of *G. hederacea* thus depends on both scale of environmental heterogeneity and on contrast between good and poor patches.

Parts of clones located in the higher quality patches allocated a higher proportion of their biomass to roots (i.e. they had a higher local R/S ratio than parts located in poor quality patches), and the difference in local R/S ratio increased as the contrast between patches increased. R/S ratios were also higher in good patches of a given quality when these patches were larger (Fig. 2.6E). Thus, clones also responded to soil quality by exhibiting morphological specialisation at a local level, and this response was graded according to local habitat quality and scale of patches. Several studies of clonal species have demonstrated similar localised adjustments of R/S ratios in response to nutrient availability (reviewed in Alpert and Stuefer 1997). This feature of species with multiple root systems entering the substrate at several different localities contrasts with that usually reported for species with only one root system, where a higher R/S ratio is normally associated with nutrient scarcity. However, it makes sense from an economic perspective, because the most efficient way to acquire patchily distributed resources is to concentrate effort where the resources are abundant rather than scarce. This maximises the return per unit of investment. Similar behaviour may also occur in plants with single root systems when placed in heterogeneous substrates (e.g. the barley studied by Drew and coworkers, described above). Drew did not directly demonstrate that proliferation of roots in nutrient-rich patches increased nutrient acquisition. Nevertheless, barley plants with only a few percent of their roots exposed to nutrient-rich conditions were able to achieve similar whole-plant relative growth rates to plants with all of their roots exposed to nutrient-rich conditions. It is a striking fact that experimental (van Vuuren et al. 1996; Fransen et al. 1998) and modelling studies (Robinson 1996) of the relationship between proliferation of roots in nitrate-rich patches and nitrate uptake tend to show very weak correlations. Studies of such relationships for other nutrients would be valuable.

scales were pooled because there was no significant difference in proportions between them. The *dots* indicate a quadratic relationship between the plotted variables based on the observed data ($Y=-0.74+0.039X-0.0002X^2$, $F_{2,57}$, $P<0.0001$) and the *broken lines* are the 95 % confidence limits for this regression. The *solid curve* shows the expected percentage of root biomass in the rich patches if there were an exact match between root placement and nutrient availability. **C** Proportion and **D** biomass of roots of *G. hederacea* clones located in rich and poor patches of the experimental treatments. **E** Mean (±SE) root/shoot ratio of clone parts located in rich and poor patches of the experimental treatments. The values in C–E are means (±SE). (Wijesinghe and Hutchings 1999)

Another response of *Glechoma hederacea* to soil patchiness was reported by Birch and Hutchings (1994). Performance was compared when clones were given the same quantity of nutrients in either a spatially homogeneous or heterogeneous pattern. A standard stage in the development of each ramet was chosen as the point at which it was deemed to have started growth, and its age was determined from that point onwards. Careful records were then made of the age at which each ramet established roots. In clones subjected to homogeneous conditions, this age was the same for all ramets throughout the clone. In comparison, in the heterogeneous treatment, ramets located on nutrient-rich substrate rooted earlier, and those on nutrient-poor substrate rooted later in their ontogeny. Thus, developmental processes were locally responsive to external conditions. There was on average 5 days difference in the age of ramet rooting on and off nutrient-rich substrate. Subsequent root extension was exponential. Consequently, ramets in nutrient-rich patches were able to begin acquiring nutrients far earlier than those in nutrient-poor patches, and their capacity to acquire nutrients at a given age was also greatly increased.

These studies reveal plants as having highly developed abilities to perceive and respond to variation in external conditions occurring at scales below that of the whole plant. Environmental heterogeneity at such scales is therefore potentially important to plant growth. Indeed, several recent studies have shown that plant performance is significantly different in environments providing the same quantity of nutrients in either homogeneous or spatially heterogeneous patterns. Moreover, different forms of heterogeneity produce plant yields that can either exceed or fall short of that achieved under homogeneous conditions. When the type of heterogeneity can be perceived by the plant, and suitable local responses can be made, considerable whole-plant benefits can ensue (Birch and Hutchings 1994; Wijesinghe and Hutchings 1997, 1999; Einsmann et al. 1999). These responses include active foraging for better quality sites from which to acquire nutrients, localised morphological specialisation, and localised changes in developmental processes. These responses result in greater root growth in higher quality sites, and this can promote greater nutrient acquisition, plant yield, and ultimately greater fitness, even when the supply of nutrients does not change. In the most extreme case of heterogeneity affecting plant biomass reported to date, yield of *Glechoma hederacea* clones varied by a factor of four when a given quantity of nutrients was provided in different heterogeneous patterns. These responses of plants and their roots to soil quality can be expected to have important implications for both pure ecology and its applications. The effects of environmental heterogeneity on single plants are becoming better understood, but its effects on competition between plants, and on plant populations and plant communities, are still largely unknown (see Chap. 9). These topics deserve greater attention from ecologists in the future.

2.5 Summary and Prospects

If we are to understand the processes that determine the structure of plant communities we need a clearer picture of the belowground distributions of plant parts than is currently available. Detailed descriptive studies like those of Pecháčková et al. (1999) and Mou et al. (1995) are informative, but we lack similar studies from most ecosystems and vegetation types. We also need manipulative studies that will help us to understand the processes that determine the belowground responses of plants to their biotic and abiotic environments. The development of methods that would ease the identification of roots in soil would greatly help in such work. The extent to which our current lack of knowledge may handicap understanding of ecology at the levels of individual plants, and populations and communities of plants, should be apparent from the fact that belowground plant parts in most situations account for more current biomass, and often more turnover of biomass, than aboveground parts.

In particular, we need to know more about the effects of soil heterogeneity on root distribution. When soils exhibit heterogeneity over scales smaller than the root systems of individual plants, localised responses can be invoked, and these can have important consequences for plant growth both at a local and plant-wide level. Although it is not a universal response, proliferation of roots in nutrient-rich soil patches is common under heterogeneous conditions. The extent to which this enhances nutrient uptake is dependent on the nutrient under consideration, but for mobile nutrients like nitrate there often seems to be little benefit unless plants are competing. Under these circumstances, plants that proliferate faster acquire more nitrate at the expense of their competitors (Robinson et al. 1999). Nevertheless, over-proliferation of roots in ephemeral nutrient patches appears to be a distinct possibility, and this would incur costs either in maintaining roots produced in the patch, or due to their death once they have ceased to acquire nutrients. Such costs might not be offset by the benefit gained from exploitation of the patch. Analysis of the costs and benefits involved in exploiting ephemeral patches for different nutrients would shed much light on this important phenomenon. Quantification is also needed of the costs and benefits associated with the growth of isolated and competing plants, in homogeneous and heterogeneous environments containing the same quantity of nutrients. Currencies in such analyses might include nutrient acquisition, biomass and direct measures of plant fitness. While the reactions of isolated plants in heterogeneous conditions are beginning to be understood, we clearly have much to learn about the effects of heterogeneity on competitive interactions and competitive outcomes, and about its effects on plant fitness.

There are still very few comparative studies of the effects on different species of different types of heterogeneity in nutrient supply. The few studies that have been carried out show that heterogeneity itself, and its various facets, including the scale of patches, and the contrast between patches of different quality, can significantly affect root growth and placement patterns, the proportion of plant biomass allocated to roots, and the timing of root growth. Many of the effects seen are specific to local patches, although it is also apparent that high spatial or temporal unpredictability in growing conditions can significantly affect the capacity of plants to achieve suitable local adaptations, and can therefore have an adverse impact on whole-plant growth (Oborny 1994; Wijesinghe and Hutchings 1997). Further comparative studies are needed to determine the links between the age, size and morphological scale of plant species and their responses to scale of heterogeneity. It would also be of great interest to know whether species that inhabit natural environments with different levels of nutrient supply exhibit similar patch selectivity in heterogeneous conditions.

Finally, new programmes of research are now needed to extend our understanding of the effects of heterogeneity at the levels of individual plants, plant populations and plant communities. Our understanding of underground patterns and processes is expanding rapidly as new techniques for the study of roots are developed, and a wide range of plant responses to heterogeneity is gradually being uncovered. This growing field of study in pure and applied plant ecology is a challenging area for research, in which further exciting insights and developments can be expected in the near future.

Acknowledgements. We gratefully acknowledge support (Grant GR3/11069) from the Natural Environment Research Council, UK, during the preparation of this manuscript, and David Robinson, Hans de Kroon, Eric Visser and an anonymous referee for helpful and challenging comments on an earlier draft.

References

Alpert P, Stuefer JF (1997) Division of labour in clonal plants. In: De Kroon H, Van Groenendael J (eds) The ecology and evolution of clonal plants. Backhuys, Leiden, pp 137–154

Berendse F (1981) Competition between plant populations with different rooting depths. Oecologia 48:334–341

Berendse F (1983) Interspecific competition and niche differentiation between *Plantago lanceolata* and *Anthoxanthum odoratum* in a natural hayfield. J Ecol 71:379–390

Birch CPD, Hutchings MJ (1994) Exploitation of patchily distributed soil resources by the clonal herb *Glechoma hederacea*. J Ecol 82:653–664

Boot RGA (1989) The significance of size and morphology of root systems for nutrient acquisition and competition. In: Lambers H, Cambridge ML, Konings H, Pons TL (eds) Causes and consequences of variation in growth rate and productivity of higher plants. SPB Academic Publishing, The Hague, pp 299–311

Brisson J, Reynolds JF (1994) The effect of neighbours on root distribution in a creosotebush (*Larrea tridentata*) population. Ecology 75:1693–1702

Caldwell MM, Manwaring JH, Durham SL (1996) Species interactions at the level of fine roots in the field: influence of soil nutrient heterogeneity and plant size. Oecologia 106:440–447

Campbell BD, Grime JP (1989) A comparative study of plant responsiveness to the duration of episodes of mineral nutrient enrichment. New Phytol 112:261–267

Campbell BD, Grime JP, Mackey JML, Jalili A (1991) The quest for a mechanistic understanding of resource competition in plant communities and the role of experiments. Funct Ecol 5:765–772

Casper BB, Jackson RB (1997) Plant competition underground. Ann Rev Ecol Syst 28:545–570

Casper BB, Cahill JF, Jackson RB (2000) Plant competition in spatially heterogeneous environments. In: Hutchings MJ, John EA, Stewart AJA (eds) The ecological consequences of environmental heterogeneity. Blackwell, Oxford, pp 111–130

Crick JC, Grime JP (1987) Morphological plasticity and mineral nutrient capture in two herbaceous species of contrasted ecology. New Phytol 122:618–625

Cui M, Caldwell MM (1996a) Facilitation of plant phosphate acquisition by arbuscular mycorrhizas from enriched soil patches. I. Roots and hyphae exploiting the same soil volume. New Phytol 133:453–460

Cui M, Caldwell MM (1996b) Facilitation of plant phosphate acquisition by arbuscular mycorrhizas from enriched soil patches. II. Hyphae exploiting root-free soil. New Phytol 133:461–467

D'Antonio CM, Mahall BE (1991) Root profiles and competition between the invasive, exotic perennial, *Carpobrotus edulis*, and two native shrub species in California coastal scrub. Am J Bot 78:885–894

De Kroon H, Hutchings MJ (1995) Morphological plasticity in clonal plants: the foraging concept reconsidered. J Ecol 83:143–152

Drew MC (1975) Comparison of the effects of a localised supply of phosphate, nitrate, ammonium and potassium on the growth of the seminal root system, and the shoot, in barley. New Phytol 75:479–490

Drew MC, Saker LR (1975) Nutrient supply and the growth of the seminal root system in barley. II. Localized, compensatory increases in lateral root growth and rates of nitrate uptake when nitrate supply is restricted to only part of the root system. J Exp Bot 26:79–90

Drew MC, Saker LR (1978) Nutrient supply and the growth of the seminal root system in barley. III. Compensatory increases in growth of lateral roots, and in rates of phosphate uptake, in response to a localised supply of phosphate. J Exp Bot 29:435–451

Drew MC, Saker LR, Ashley TW (1973) Nutrient supply and the growth of the seminal root system in barley. I. The effect of nitrate concentration on the growth of axes and laterals. J Exp Bot 24:1189–1202

Dumortier, M. (1991). Below-ground dynamics in a wet grassland ecosystem. In: Atkinson D (ed) Plant root growth – an ecological perspective. British Ecological Society Special Symposium 10, Blackwell, Oxford, pp 301–309

Einsmann JC, Jones RH, Pu M, Mitchell RJ (1999) Nutrient foraging traits in 10 co-occurring plant species of contrasting life forms. J Ecol 87:609–619

Fahey TJ, Hughes JW (1994) Fine root dynamics in a northern hardwood forest ecosystem, Hubbard Brook Experimental Forest, NH. J Ecol 82:533–548

Farley RA, Fitter AH (1999) Temporal and spatial variation in soil resources in a deciduous woodland. J Ecol 87:688–696

Fernandez OA, Caldwell MM (1975) Phenology and dynamics of root growth of three cool semi-desert shrubs under field conditions. J Ecol 63:703–714

Fitter AH (1986) Spatial and temporal patterns of root activity in a species-rich alluvial grassland. Oecologia 69:594–599

Fitter AH (1987) An architectural approach to the comparative ecology of plant root systems. New Phytol 106(Suppl):61–77
Fitter AH (1994) Architecture and biomass allocation as components of the plastic response of root systems to soil heterogeneity. In: Caldwell MM, Pearcy RW (eds) Exploitation of environmental heterogeneity by plants: ecophysiological processes above- and belowground. Academic Press, San Diego, pp 305–323
Fransen B, de Kroon H, Berendse F (1998) Root morphological plasticity and nutrient acquisition of perennial grass species from habitats of different nutrient availability. Oecologia 115:351–358
Fransen B, de Kroon H, de Kovel CGF, van den Bosch F (1999) Disentangling the effects of root foraging and inherent growth rate on plant biomass accumulation in heterogeneous environments: a modelling study. Ann Bot 84:305–311
Fransen B, de Kroon H, Berendse F (2001) Soil nutrient heterogeneity alters competition between two perennial grass species. Ecology 82:2534–2546
Fretwell SD (1972) Populations in a seasonal environment. Princeton University Press, Princeton, New Jersey
Gersani M, Sachs T (1992) Developmental correlations between roots in heterogeneous environments. Plant Cell Environ 15:463–469
Gersani M, Abramsky Z, Falik O (1998) Density-dependent habitat selection in plants. Evol Ecol 12:223–234
Grime JP, Crick JC, Rincon JE (1986) The ecological significance of plasticity. In: Jennings DH, Trewavas AJ (eds) Plasticity in plants. Biologists Limited, Cambridge, pp 5–29
Hendrick RL, Pregitzer KS (1993) Patterns of fine root mortality in two sugar maple forests. Nature 361:59–61
Hetrick BAD, Wilson GWT, Leslie JF (1991) Root architecture of warm- and cool-season grasses: relationship to mycorrhizal dependence. Can J Bot 69:112–118
Hook PB, Lauenroth WK, Burke IC (1994) Spatial patterns of roots in a semiarid grassland: abundance of canopy opening and regeneration gaps. J Ecol 82:485–494
Huber-Sannwald E, Pyke DA, Caldwell MM (1996) Morphological plasticity following species-specific recognition and competition in two perennial grasses. Am J Bot 83:919–931
Hutchings MJ, Barkham JP (1976) An investigation of shoot interactions in *Mercurialis perennis* L., a rhizomatous perennial herb. J Ecol 64:723–744
Hutchings MJ, de Kroon H (1994) Foraging in plants: the role of morphological plasticity in resource acquisition. Adv Ecol Res 25:159–238
Hutchings MJ, John EA, Wijesinghe DK (2003) Towards an understanding of the consequences of patchy distribution of soil-based resources for populations and communities of plants. Ecology (in press)
Jackson RB, Caldwell MM (1993) Geostatistical patterns of soil heterogeneity around individual perennial plants. J Ecol 81:683–692
Jackson RB, Manwaring JH, Caldwell MM (1990) Rapid physiological adjustment of roots to localised soil enrichment. Nature 344:58–60
Jackson RB, Canadell J, Ehleringer JR, Mooney HA, Sala OE, Schulze ED (1996) Global analysis of root distributions for terrestrial biomes. Oecologia 108:389–411
Kovar JL, Barber SA (1988) Phosphorus supply characteristics of 33 soils as influenced by seven rates of phosphorus addition. Soil Sci Soc. Am J 52:160–165
Kovar JL, Barber SA (1989) Reasons for differences among soils in placement of phosphorus for maximum predicted uptake. Soil Sci Soc Am J 53:1733–1736
Krannitz PG, Caldwell MM (1995) Root growth responses of three Great Basin perennials to intra- and interspecific contact with other roots. Flora 190:161–167
Lechowicz MJ, Bell G (1991) The ecology and genetics of fitness in forest plants. II. Microspatial heterogeneity of the edaphic environment. J Ecol 79:687–696

Mahall BE, Callaway RM (1992) Root communication mechanisms and intracommunity distributions of two Mojave Desert shrubs. Ecology 73:2145–2151

Milchunas DG, Lauenroth WK (1989) Three-dimensional distribution of plant biomass in relation to grazing and topography in the shortgrass steppe. Oikos 55:82–86

Mou P, Jones RH, Mitchell RJ, Zutter B (1995) Spatial distribution of roots in sweetgum and loblolly pine monocultures and relations with above-ground biomass and soil nutrients. Funct Ecol 9:689–699

Oborny B (1994) Growth rules in clonal plants and environmental predictability – a simulation study. J Ecol 82:341–351

Pecháčková S, During HJ, Rydlová V, Herben T (1999) Species-specific spatial pattern of below-ground plant parts in a montane grassland community. J Ecol 87:569-582

Pregitzer KS, Hendrick RL, Fogel R (1993) The demography of fine roots in response to patches of water and nitrogen. New Phytol 125:575–580

Pulliam HR, Caraco T (1984) Living in groups: is there an optimal group size? In: Krebs JR, Davies NB (eds) Behavioural ecology, 2nd edn. Blackwell, Oxford, pp 214–244

Reader RJ, Jalili A, Grime JP, Spencer RE, Matthews N (1993) A comparative study of plasticity in seedling rooting depth in drying soil. J Ecol 81:543–550

Reinartz JA, Popp JW (1987) Structure of clones of northern prickly ash (*Xanthoxylum americanum*). Am J Bot 74:415–428

Robinson D (1994) Tansley review No. 73. The responses of plants to non-uniform supplies of nutrients. New Phytol 127:635–674

Robinson D (1996) Resource capture by localised root proliferation: why do plants bother? Ann Bot 77:179–185

Robinson D, Hodge A, Griffiths BS, Fitter AH (1999) Plant root proliferation in nitrogen-rich patches confers competitive advantage. Proc R Soc Lond B 266:431–435

Schenk HJ, Callaway RM, Mahall BE (1999) Spatial root segregation: are plants territorial? Adv Ecol Res 28:145–180

Slade AJ, Hutchings MJ (1987) The effects of nutrient availability on foraging in the clonal herb *Glechoma hederacea*. J Ecol 75:95–112

St. John TV, Coleman DC, Reid CPP (1983a) Association of vesicular-arbuscular mycorrhizal hyphae with soil organic particles. Ecology 64:957–959

St. John TV, Coleman DC, Reid CPP (1983b) Growth and spatial distribution of nutrient-absorbing organs: selective placement of soil heterogeneity. Plant Soil 71:487–493

Steeves TA, Sussex IM (1989) Patterns in plant development, 2nd edn. Cambridge University Press, Cambridge

Sydes CL, Grime JP (1984) A comparative study of root development using a simulated rock crevice. J Ecol 72:937–946

Van Vuuren MMI, Robinson D, Griffiths BS (1996) Nutrient inflow and root proliferation during the exploitation of a temporally and spatially discrete source of nitrogen in soil. Plant Soil 178:185–192

Weaver JE (1919) The ecological relations of roots. Carnegie Institute, Washington

Wells CE, Eissenstat DM (2001) Marked differences in survivorship among apple roots of different diameters. Ecology 82:882–893

Wijesinghe DK, Hutchings MJ (1997) The effects of spatial scale of environmental heterogeneity on the growth of a clonal plant: an experimental study with *Glechoma hederacea*. J Ecol 85:17–28

Wijesinghe DK, Hutchings MJ (1999) The effects of environmental heterogeneity on the performance of *Glechoma hederacea*: the interactions between patch contrast and patch scale. J Ecol 87:860–872

Wijesinghe DK, John EA, Beurskens S, Hutchings MJ (2001) Root system size and precision in nutrient foraging: responses to spatial pattern of nutrient supply in six herbaceous species. J Ecol 89:972–983

Wilson SD (2000). Heterogeneity, diversity and scale in plant communities. In: Hutchings MJ, John EA, Stewart AJA (eds) The ecological consequences of environmental heterogeneity. Blackwell, Oxford, pp 53–69
Yeaton RI, Travis J, Gilinsky E (1977) Competition and spacing in plant communities: the Arizona upland association. J Ecol 65:587–595
Zhang H, Forde BG (1998) An *Arabidopsis* MADS box gene that controls nutrient-induced changes in root architecture. Science 279:407–409

3 Turnover of Root Systems

W.K. Lauenroth and R. Gill

3.1 Introduction

Turnover of tissues is a fundamental process that operates continuously in all vascular plants. In some, especially those in seasonal environments, entire organs may be shed at the end of the growing season. Winter or drought deciduous trees are a clear example of this process. Herbaceous plants may also episodically lose all or most of their aboveground organs at the onset of a cold or dry season, but most of them also undergo continuous replacement of leaves throughout a growing season (Larcher 1995; Lambers et al. 1998). Many grasses are specifically adapted to withstand high turnover of their aboveground organs be it by grazing, fire, or drought because of their evolutionary histories of exposure to such selection pressures (Mack and Thompson 1982; Coughenour 1985; Milchunas et al. 1988). In the case of deciduous trees, turnover of the canopy is 100 % per year. Similarly, grasses in seasonal environments often have essentially 100 % canopy turnover per year.

The concept of turnover, i.e. the replacement of a particular standing stock, is equally applicable to belowground plant organs as to those aboveground. While assessment of canopy turnover is relatively straightforward, evaluation of the turnover of belowground organs (e.g. roots, rhizomes, corms, tubers, etc.) is very difficult and continues to be one of the most challenging issues in root system ecology. Our knowledge of the anatomy, morphology, and physiology of individual belowground organs is quite good even though much of our knowledge has come from studying roots removed from the field environment (Esau 1965; Davis and Haissig 1994; Altman and Waisel 1997). We know considerably less about belowground organ systems (we will refer to these as root systems), but we have relatively good information about the static aspects of root systems. Our gaps in knowledge about root systems are related to their dynamic aspects of which root system turnover is of central importance.

The objectives of this chapter are to evaluate the state of our knowledge about root system turnover under field conditions. Specifically, we will dis-

Ecological Studies, Vol. 168
H. de Kroon, E.J.W. Visser (Eds.)
Root Ecology

cuss the structure of root systems and characteristics of individual roots, methods used to estimate root turnover in the field, published results for root turnover and net primary production in forest, grassland and shrubland ecosystems, relationships between root turnover and environmental variables, and end with an assessment of future needs in research on root system turnover.

3.2 Overview of the Structure of Root Systems

The major benefits to a plant of investing in a root system are the connections it provides with the substrate and the ability it confers to absorb water and mineral nutrients. While not all members of the plant kingdom have true roots, our current analysis is limited to those that do (vascular plants). Plants have "solved" the problems of anchorage and absorption in a number of ways (see Chap. 1). In our analyses of root systems, we will simplify the diversity of root system types by dividing the plant kingdom into woody and herbaceous types and further subdividing the herbaceous plants into those in which the root system of the established individual is organized around the primary root and those in which that is not the case. This leads to a division of vascular plants into three groups: conifers and woody dicots, herbaceous dicots, and monocots. The first two groups share the unifying characteristic that the primary root or taproot "...commonly produces, by elongation and branching, the root system of the plant." (Esau 1965). By contrast, the root systems of most monocots, and particularly grasses (Poaceae) and grass-like plants (Cyperaceae and Juncaceae), consist entirely of stem-borne adventitious roots (Esau 1965). Our treatment of monocots will be heavily weighted to the grasses (Poaceae), the most ubiquitous family in the group and the dominant in grasslands worldwide.

The reason for dealing with root system structure in a chapter on root turnover is that we think our three simplified groups provide both an important way to organize our thinking about the variety of root systems that exists in the natural environment and understanding of how this variety will influence our ability to design sampling methods to estimate root system turnover. For instance, conifers and woody dicots always produce secondary root tissue (woody roots) and monocots do not. This results in a large portion of the root systems of conifers and woody dicots being composed of coarse roots (>2–3 cm in diameter). Harris et al. (1978) reported for a conifer dominated forest and a woody dicot dominated forest that elements of the root systems less than 2 cm in diameter accounted for 36 and 23 % of the total root biomass, respectively. The largest diameter and woodiest portions of the root systems (central root and laterals to a radius of 60 cm) made up 50 % or more of the root system biomass. By contrast, Liang et al. (1989) reported for five shortgrass steppe sites that roots greater than 2 mm composed an average of 7 % of root biomass to a depth of 90 cm. Understanding how root biomass is distributed among different sized elements of a root system and with depth in the soil is important to understanding how best to sample to estimate root turnover and to understanding how to interpret root turnover data.

3.2.1 Conifers and Woody Dicots

The key attribute of this group that differentiates it from the others is secondary growth of elements of the root system (Esau 1965). While some herbaceous dicots produce secondary root tissue, all conifers and woody dicots exhibit secondary growth in their taproot and main branch roots. Fine roots are the most dynamic components of the root systems of conifers and woody dicots and are the only portion of the root system that is commonly considered in evaluation of turnover. Although there is considerable variability in how they are defined, Gill and Jackson (2000) found that the majority of studies used either 0–3 or 0–5 mm as the diameter size range for fine roots.

One of the consequences of the production of secondary tissues is the ability of the taproot and main branch roots to achieve large diameters. Large diameter roots can develop large cross-sectional areas of conducting tissues and therefore effectively transport water and nutrients over long distances. This is consistent with the observation that many woody plants have deep or widely spreading root systems (Canadell et al. 1996). Such deep and widely spreading root systems are likely to be difficult to sample to obtain good estimates of root biomass and or turnover using the commonly applied methods (Lauenroth 2000; Sect. 3.3). Furthermore, the portions of the root system of conifers and woody dicots that account for the majority of the biomass, and likely an important part of belowground net primary production (taproot and main branches), are usually ignored in studies of root system turnover.

3.2.2 Herbaceous Dicots

Root systems of herbaceous dicots are characterized by the presence of a taproot and limited or no secondary tissue. Despite the lack of secondary growth, the relatively large diameter taproots of herbaceous dicots provide this group with the ability to penetrate deeply into the soil. This ability, which is not always expressed, leads to enormous diversity in the vertical and horizontal distribution of roots (Kutschera and Lichtenegger 1992). Most herbaceous dicots have the bulk of their roots in the top 100 cm of the soil, although some may penetrate much deeper (Kutschera and Lichtenegger 1992; Chap. 1). These deep-growing species have woody or semi-woody taproots and main branches although none achieve the massiveness of conifers and woody dicots. Except for the deepest portions of the root systems, the additional sampling problems posed by these species is minimal. The true herbaceous dicots, those with most of their roots in the top 100 cm of the soil, are effectively sampled by most of the commonly used techniques for root biomass and turnover (Lauenroth 2000; Sect. 3.3). Studies of root turnover in ecosystems in which individuals of this group occur usually include all of the elements of the root system and therefore include them in estimates of turnover.

3.2.3 Monocots

Stem-borne adventitious roots are the most common element of the root systems of mature individuals of monocots. In the Poaceae, which is the most widely distributed family in the group, seedlings begin life with a root system organized around the primary root (taproot). Long-term survival depends on the individual making a transition from dependence on the primary root to a root system consisting of stem-borne adventitious roots. In contrast to the herbaceous dicots and their taproots, the adventitious root system consists of many main laterals none of which is very large in diameter. A large main lateral root may be ≤5 mm in diameter (Weaver 1968). A commonly described characteristic of monocots especially those in the family Poaceae is a very dense fibrous root system. This is particularly true of the upper layers of the soil and is one of the most striking features evident in root system drawings (Kutschera and Lichtenegger 1992). In addition to a concentration of roots in the upper layers of the soil, monocots tend to have a large fraction of their root system concentrated near their tiller(s) (Kutschera and Lichtennegger 1992; Hook et al. 1994). The lack of woody roots and their proximity to the plant and the soil surface make root systems in this group relatively easy to sample. A large body of root turnover literature is available for grasslands; with respect to turnover they may be the single best-understood ecosystem type (Gill and Jackson 2000).

3.3 Methods of Assessing Root Turnover

Root system turnover can be calculated in terms of a ratio of the quantity of material added to a system with a particular standing stock. The most common representation is to use belowground net primary production (BNPP) as the quantity added and belowground biomass as the standing stock. In this case, root system turnover is:

$$TC = \frac{BNPP}{Belowground \quad Biomass} \tag{3.1}$$

where *TC* is the root system turnover coefficient with units of per unit time (t^{-1}), *BNPP* is belowground net primary production with units of g m^{-2} t^{-1} and *Belowground Biomass* is the standing stock of belowground organs at a particular point in time or an average over a particular time period with units of g/m^2. Gill and Jackson (2000), following the lead of Dahlman and Kucera (1965), recommended that maximum root biomass be used in the denominator. *TC* can also be calculated from root length or root area data.

Root system turnover coefficients can be estimated directly or indirectly. Direct methods involve using a label and calculating the amount of time it remains in belowground organs. Indirect methods rely upon estimates of BNPP and belowground biomass or their root length or area analogs. We will briefly describe the key features of the most commonly used methods for both direct and indirect estimates of TC. A complete description and analysis of the methods can be found in Lauenroth (2000).

3.3.1 Direct Estimates of Root System Turnover Coefficients Based on ^{14}C Turnover

The only methods available to directly estimate turnover coefficients (TC) utilize a carbon label (Dahlman and Kucera 1967; Caldwell and Camp 1974; Milchunas et al. 1985; Milchunas and Lauenroth 1992). ^{14}C is the isotope that has been used most frequently, although conceptually ^{13}C should also work. Recent attempts to use ^{13}C to estimate TC have reported difficulties introducing enough label into the plants to distinguish subsequent samples from background (Stewart and Metherell 1999; D.G. Milchunas, pers. comm.). This method relies on introducing the label via photosynthesis and assuming that fixed carbon containing the label will be incorporated into the structural portion of the root system.

Caldwell and Camp (1974) used this method to estimate TC in a shrub steppe ecosystem. They pulse-labeled the vegetation to establish a $^{14}C/^{12}C$ ratio and subsequently sampled to estimate changes in the ratio. TC was calculated as:

$$TC = \frac{^{14}C_{t1}/^{12}C_{t1}}{^{14}C_{t2}/^{12}C_{t2}} - 1 \tag{3.2}$$

The simplest method of estimating TC using a carbon label is to introduce the label and follow it until it is no longer detectable in the root system (Dahlman and Kucera 1967; Milchunas and Lauenroth 1992). Dahlman and Kucera (1967) fit a linear equation to 2 years of loss data and predicted that the TC for their tallgrass prairie site was 0.25 per year which was very similar to TC calculated from biomass. Milchunas and Lauenroth (1992) did the same calculation with their loss data after 3 years and estimated that TC for the shortgrass steppe ranged from 0.2 to 0.14 per year.

3.3.2 Indirect Estimates of Root System Turnover Coefficients

3.3.2.1 Biomass

The most common method to estimate TC from field data is to use Eq. (3.1). In fact, in most cases estimating TC is not the objective. Most often the objective is to estimate BNPP (Milchunas and Lauenroth 1992; Nadelhoffer and Raich 1992; Publicover and Vogt 1993). Estimating TC from biomass data requires two sets of decisions. The first set of decisions is related to how to sample biomass (Böhm 1979) and the second set how to calculate BNPP from the biomass samples (Lauenroth 2000; Sala 2000).

3.3.2.2 Ingrowth Cores

This method involves removing all of the roots from a volume of soil and measuring the regrowth (Jordan and Escalante 1980; Persson 1983; Neill 1992). Soil samples are extracted using a metal core and the roots are removed. This first sample is an estimate of root biomass. The soil without roots is returned, in a mesh bag, to the hole from which it came. At a subsequent time, the bags are removed from the soil and the roots are again separated from the soil. This second sample of root biomass is an estimate of BNPP. A TC for the root system can be calculated by combining the root biomass information with the estimate of BNPP using Eq. (3.1).

3.3.2.3 Nitrogen Balance

Here, the problem of estimating the biomass increment of the root system is exchanged for one of estimating the amount of nitrogen allocated to roots and the average nitrogen concentration of the root system (Aber et al. 1985; Nadelhoffer et al. 1985). BNPP is calculated as:

$$BNPP = \frac{N_r}{N_c} \tag{3.3}$$

where N_r is the amount of nitrogen allocated to roots (g m^{-2} t^{-1}) and N_c is the concentration of nitrogen in fine roots (g/g). Combining this information with an estimate of root biomass allows one to estimate the TC of a root system.

3.3.2.4 Minirhizotrons

All of the previous methods provide information about root turnover on the entire root system or more accurately the entire mass of roots in the portion of the soil that is sampled. The minirhizotron method is very different in that it provides an estimate of turnover using information about individual roots. This information at the individual root level can be combined into an estimate of turnover for the area sampled.

Methods involving direct observations of roots and rhizomes were among the first methods developed to study root systems (Böhm 1979). However, the effective implementation of these methods for quantitative studies of root systems and root turnover has had to await the development of small video cameras (Upchurch and Ritchie 1983; Ferguson and Smucker 1989; Cheng et al. 1990; Hendrick and Pregitzer 1992). These small video cameras have revolutionized the use of rhizotrons in the form of transparent tubes placed into the soil which, after a period of equilibration, allow repeated nondestructive observations of root systems (Brown and Upchurch 1987). The transparent tubes are referred to as minirhizotrons.

Root observations are recorded on videotape (Brown and Upchurch 1987) and the images converted to digital form for analysis. The current state of the technology requires that converting the images to digital form be done by hand digitizing the root outlines (Hendrick and Pregitzer 1992) or counting the number of intersections of roots with the tube (Merrill and Upchurch 1994). This is the limiting step in using this technology. Hand digitization of each image can take an enormous amount of time.

Root length is recorded at each sample date and the change in root length is often expressed as a proportion of the initial root length (Hendrick and Pregitzer 1992, 1993). The ratio of root length added to the initial root length is a turnover coefficient.

Aerts et al. (1989) compared minirhizotrons with an estimate of turnover using biomass and found that minirhizotron estimates were more reliable. They used Eq. (3.4) to calculate root length turnover (RLT; year^{-1}) for each tube.

$$RLT = \frac{Rlp}{Arl} \tag{3.4}$$

Rlp is annual root length production (cm/year) and *Arl* is average living root length (cm). Hendrick and Pregitzer (1992) found that growth and death of fine roots in a northern hardwood forest occurred simultaneously and that the ability of minirhizotrons to separate growth and mortality was a major strength of the technique. Other methods that could not handle simultaneous growth and death would underestimate production and turnover.

3.4 The Growth, Life Span, and Death of Roots

Root turnover is caused by the growth and death of roots. Theoretically, anything that increases root growth (BNPP) will increase root turnover (Eq. 3.1). By the same reasoning, anything that decreases BNPP will decrease root turnover. Root death affects turnover by decreasing root biomass. If BNPP remains constant, a decrease in root biomass will cause an increase in root turnover (Eq. 3.1).

Factors that influence growth and death operate at the individual root, the whole plant, and the ecosystem levels. Understanding how these factors influence growth and death of roots is an important step in understanding root turnover. Unlike leaves, which have well-understood patterns of growth and death, root growth and death are poorly understood in almost all plants and are sensitive to short-term flushes or deficits of resources.

While root growth is observable using minirhizotrons, one of the most challenging aspects of root studies is determining an individual root's life span, or even when a root has died or is no longer functional. Unlike leaves, which have a clear physiological process of senescence, root death is not cued by the formation of an abscission layer and may occur gradually. Even after cortical cells begin sloughing, roots are capable of hydraulic conductance and apoplastic uptake of nutrients (Larcher 1995). In addition, while some studies that use sequential sampling techniques indicate that live root biomass may follow a unimodal or bimodal seasonal pattern (Vogt et al. 1986), direct observation and the plasticity of root longevity indicate that many roots have indeterminate life spans (Pregitzer et al. 1993; Bloomfield et al. 1996). Direct observations and C isotopic analysis of roots have shown that, in many species, a portion of the root system is incredibly dynamic, with life spans shorter than weeks (Ares 1976), while other individual small diameter roots may live years, or perhaps decades (Gaudinski et al. 2000).

3.4.1 Effects at the Individual Root Level

3.4.1.1 Water and Nutrients

The availability of water and mineral nutrients is a key control over individual root growth, longevity, and death (Caldwell 1976; Lauenroth et al. 1987; Jackson and Caldwell 1989, 1992; Gross et al. 1993; Pregitzer et al. 1993; Hook et al. 1994; Zhang et al. 1999; Chap. 7). The need to maintain contact with wet soil is an important driver of root growth and the negative effect of drought is an important control on root death (Caldwell 1976; Kramer and Boyer 1995). The influence of root proliferation on the ability of a plant to capitalize on transient water and/or nutrient availability is likely an important mechanism of competition in water and nutrient-limited environments (Chap. 2). Pregitzer et al. (1993) found that fine roots in a second-growth hardwood forest grew more rapidly in sites enriched with nitrogen and water than in control microsites, and that the roots produced in the fertile microsites lived longer than control roots. Jackson and Caldwell (1989) found great plasticity in the response of roots to nutrient additions among cold-desert perennial plants. *Agropyron desertorum* roots had three to four times more rapid mean relative growth rate in fertile microsites compared with those in control microsites. In contrast, *Agropyron spicatum* showed no tendency to proliferate roots into fertile patches during the 2-week experiment. Hook et al. (1994) reported that *Bouteloua gracilis*, the dominant perennial grass in the North American shortgrass steppe, produced significant root proliferation in response to patches of increased water availability.

3.4.1.2 Soil Temperature

Soil temperature may partially control individual root longevity directly through altering root respiration or, more likely, indirectly through controlling rates of nitrogen mineralization or pathogen activity. Hendrick and Pregitzer (1993) contrasted root life span at two sites dominated by *Acer saccharum* and found that roots persisted 75 days longer at the cooler of the two sites. According to a root efficiency model developed by Eissenstat and Yanai (1997), the longer life span of roots grown at the cooler site can only be partially explained by the direct effect of temperature on maintenance respiration. They suggested that the indirect effects of temperature may drive observed temperature effects on root life span. This conclusion is supported by the work of Fitter et al. (1998, 1999) who found that artificial soil warming decreased root life span. However, they concluded that temperature was not the proximate driving force for higher root production in warmed soils, but rather that temperature indirectly altered root longevity by changing soil nutrient status.

3.4.1.3 Root Diameter

In studies of woody plants, root diameter is another commonly cited factor influencing root life span and death. Coarse and fine roots differ in their ability to acquire water and mineral nutrients and fine roots are more dynamic than coarse roots (Pregitzer et al. 1993, 1997). Since length and surface area are the primary determinants of a root's ability to acquire water and nutrients, small diameter roots are a more efficient use of carbon to collect belowground resources than large diameter roots (Nye and Tinker 1977; Eissenstat 1992, 1997). In addition, small diameter roots have a higher relative extension rate into nutrient patches than larger diameter roots (Fitter 1994; Eissenstat 1997). However, small diameter roots come at a cost. The smaller the root diameter, the less likely they are to be well defended against herbivores or pathogens due to lack of suberized or lignified cell walls in the cortex or epidermis (Eissenstat 1997). In addition, fine roots may be less efficient at water transport than large diameter roots because of their smaller cross-sectional area of xylem.

3.4.1.4 Root Symbionts

Symbiotic associations may increase root life span. Root longevity has been reported to be increased by symbiotic mycorrhizal fungi infection in woody plants (Harley 1969; Vogt et al. 1982; Harley and Smith 1983), although Hooker et al. (1995) found that only 16 % of poplar roots colonized by mycorrhizae lived longer than 49 days, compared with the 49 % of uncolonized roots that lived beyond 49 days. Bloomfield et al. (1996) suggested that the increase in root life span for mychorrhizal roots is primarily the consequence of protection from pathogens. Mychorrizal fungi produce antifungal and antibacterial secondary compounds, as well as the physical protection of root tips (Pfleger and Linderman 1994; Bloomfield et al. 1996).

3.4.1.5 Herbivory

Herbivory must be an important source of root death, although there are very few data available about it in the literature. What information we have is speculative and based on correlative information. Stanton (1988) suggested root herbivory in grasslands might decrease net primary productivity by as much as 28 %. Direct observation of root cohorts provides indirect evidence for the prevalence of root herbivory. Based on cost-benefit analyses of root longevity, the majority of root deaths should occur at an optimal life span, with high survivorship among young roots (Eissenstat and Yanai 1997). However, cohort analyses have shown that root mortality is either highest in young roots (Hendrick and Pregitzer 1993; Ruess et al. 1996) or constant throughout the life span of a root (Kosola et al. 1995). Young roots are especially vulnerable to herbivory or parasitism because they lack the thick secondary walls of established roots (Eissenstat and Yanai 1997).

3.4.2 Effects at the Whole-Plant Level

Understanding root growth, maintenance, and death within an entire plant requires that we determine what controls the allocation of carbon between shoots to roots. Patterns of carbon allocation above- and belowground appear to be regulated by a network of competing carbon sinks (Friend et al. 1994; Chap. 4). The source-sink relationships between above- and belowground organs are clearly demonstrated in woody plants with determinate and semi-determinate growth patterns. The determinate shoot growth of the genus *Picea* occurs in a single flush of shoot growth (Kozlowski 1992; Friend et al. 1994). Following this early-season shoot growth, roots receive almost all assimilates, with sustained root growth throughout the remaining growing season. In contrast, the semi-determinate growth of the genus *Quercus* occurs in a cyclic pattern of allocation to either shoots or roots. In this recurring flushing growth pattern, carbohydrates are transported up the stem to new leaves during flush periods, followed by lag periods where nearly all carbon is allocated to root growth (Kozlowski 1992; Friend et al. 1994). Understanding how and when carbon is allocated belowground will help us understand root turnover.

3.4.2.1 Elevated CO_2

Manipulations of plant carbon status have shown the sensitivity of root system growth and longevity to carbon availability. Experiments that manipulate atmospheric CO_2 concentrations have shown that an increase in photosynthesis often increases root production and standing crop (Rogers et al. 1994; Pregitzer et al. 1995; Berntson and Bazzaz 1997), although this response is not universal (Rogers et al. 1996; Arnone et al. 2000). For annual crops, there may be changes in demography and temporal shifts in growth patterns, but Pritchard and Rogers (2000) concluded that changes in annual plant turnover rates are unlikely with increased CO_2. Pregitzer et al. (1995) used minirhizotrons to observe root dynamics of *Populus* trees under ambient and twice-ambient atmospheric CO_2. They consistently found higher root production in elevated CO_2 compared with ambient conditions, independent of soil fertility. Berntson and Bazzaz (1997) demonstrated that elevating CO_2 increased both gross root production and sped-up development of the root system in *Betula papyrifera* seedlings. However, elevating CO_2 decreased root life span, resulting in higher root turnover, but similar levels of root standing stock between CO_2 treatments. Tingey et al. (2000) reported that mycorrhizal infection and colonization increase under elevated CO_2, which may contribute to an increased ability to exploit soil resources without necessarily altering root growth dynamics.

3.4 2.2 Pathogens and Herbivores

Carbohydrate limitation caused by environmental stresses may cause root death, either directly through decreased carbon allocation to roots or through increased susceptibility to pathogens or herbivores (Wargo 1972; Waring and Schlesinger 1985). Wargo et al. (1993) observed that as carbohydrate reserves decreased in large structural roots there was increased mortality in secondary roots in red spruce trees. Plants already stressed due to some carbohydrate limitation may be more susceptible to pathogen invasion than healthy plants, and stressed plants may lose a larger portion of their root system following pathogen inoculation (Wargo and Houston 1974). By growing seedlings in a low light environment, Matson and Waring (1984) showed an increased susceptibility of mountain hemlock (*Tsuga mertensiana*) to infection by laminated root rot. The low light decreased plant carbon reserves and presumably the ability of roots to resist the infection of pathogens.

The influence of grazing on root growth and turnover is problematic. The old adage "stress the shoot, reduce the root" implies that consuming leaves should result in lower root production (Crawley 1993). However, a meta-analysis by Milchunas and Lauenroth (1993) showed no consistent relationship between grazing and root mass. Of the 29 studies Milchunas and Lauenroth (1993) reviewed, 11 reported negative effects of grazing on roots and 18 reported positive effects. Unfortunately, there are few studies that adequately document root turnover in paired grazed and ungrazed systems, making it difficult to determine how reducing aboveground biomass influences root production and longevity.

3.5 Field Estimates of Root Turnover and Net Primary Production

Our approach to discussing the currently available data for turnover and belowground net primary production will be to separate it by ecosystem types. We have chosen three broad types of ecosystems, forests, grasslands and shrublands, and within each of these categories we will deal with climatically defined subgroups (i.e. tropical forests, temperate grasslands, etc.). This approach is similar to the one taken by Gill and Jackson (2000) in a meta-analysis of root production and turnover for ecosystems globally. Our analysis builds on their work by including additional studies. The details of the methods used in collecting and analyzing these data can be found in Gill and Jackson (2000).

3.5.1 Forests

Forests occur in the most productive environments on earth and therefore include some of the most productive ecosystems (Ajtay et al. 1979; Waring and Running 1998). These environments receive more annual precipitation than can be fully exploited by grasses and occur over a broad range of annual temperatures (Bailey 1998).

Several factors complicate any effort to synthesize information on root turnover in forest ecosystems. First, the complexity of tree root systems has forced researchers to create categories of roots that confound the interpretation of root growth and turnover. Fine roots are considered the most dynamic component of the root system; therefore, they are often assumed to constitute the majority of root turnover. However, many studies report total root turnover, using allometry to estimate coarse and fine root production (Gill and Jackson 2000). We chose to only include studies that report fine root production. A second complicating factor is that root studies are very unevenly distributed over forest types. Belowground processes have been extensively characterized in temperate forests, while very few studies report root dynamics in tropical forests. Finally, methods for determining root growth have changed over time as researchers have become dissatisfied with traditional strategies, forcing redefinition of both sampling and calculation methods (Lauenroth 2000; Sect. 3.3). As a result, there are clear methodological biases and non-independence in the data (Singh et al. 1984; Lauenroth 2000).

3.5.1.1 Temperate

Forests in the temperate zone consist of either conifers, angiosperms, or mixtures of the two tree types. While there are slight differences in fine root turnover rates between conifers and angiosperm trees, the variability within conifers and angiosperms is much greater than between groups. Estimates of fine root turnover for temperate conifers range from 0.08 $year^{-1}$ in subalpine forests in the US Rocky Mountains (Arthur and Fahey 1992) to 2.57 $year^{-1}$ in a US *Tsuga heterophylla* forest (Santantonio and Hermann 1985). Fine root turnover in broadleaf dominated forests likewise range over nearly an order of magnitude, from 0.18 $year^{-1}$ in *Quercus borealis* dominated forest (Aber et al. 1985) to 1.42 $year^{-1}$ in an *Acer rubrum* forest in the mid-Atlantic US (Fredericksen and Zedaker 1995).

Perhaps the most important factor in determining fine root turnover in temperate forests is how "fine" roots are defined. Nearly all of the studies we reviewed used a root diameter threshold to define the fine root portion of the root system. As the mean diameter of fine roots increases, root turnover decreases, supporting the presumption that tree roots have a large variability

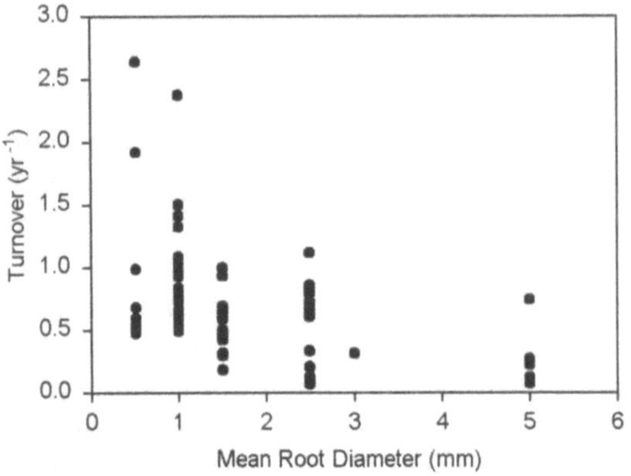

Fig. 3.1. Relationship between mean diameter of fine roots and annual turnover rate

in life span even with small changes in root diameter (Wells and Eissenstat 2001). We found only two studies that determined both standing crop and production for roots of different diameter increments, and the results from these studies are somewhat contradictory. Arunachalam et al. (1996) showed that relative production rates were constant for five different diameter increments, from 0–1 to 10–15 mm. Gholz et al. (1986), however, showed a decrease in root turnover with increasing root diameter, consistent with our analysis (Fig. 3.1).

3.5.1.2 Boreal

Root turnover in boreal forests is typically quite low, averaging 0.34 $year^{-1}$ for conifer dominated forests and 0.26 $year^{-1}$ for broadleaf forests (Gill and Jackson 2000). However, Ruess et al. (1996) reported that, on average, more than two-thirds of the root system turned over annually in central Alaska, US, so it is possible for boreal forest root systems to have high turnover rates. The initiation of the BOREAS study, in addition to the work being conducted through the US National Science Foundation Long-Term Ecological Research Program, has greatly enhanced our understanding of boreal forest root dynamics. In particular, Steele et al. (1997), as part of the BOREAS program, provided an excellent example of the use of multiple sampling techniques to contrast root turnover within and among boreal forest sites. They concluded that ingrowth cores underestimated root production because they failed to account for simultaneous production and mortality of roots. Because of the difference in methods, turnover estimates for *Pinus banksiana* and *Picea mariana* were higher using minirhizotrons than using the ingrowth core method.

3.5.1.3 Tropical

Tropical forests are among the most poorly understood biomes regarding root dynamics. Very few studies have reported production or turnover in tropical forests, possibly due to the difficulty of sampling roots in tropical ecosystems (Clark et al. 2001). The length of the tropical growing season, high spatial variability of tropical forest roots (Carvalheiro and Nepstad 1996; Ostertag 1998), along with the depth of root exploration in tropical systems (Nepstad et al. 1994; Canadell et al. 1996), likely contribute to highly variable estimates of root production in tropical forest ecosystems. Estimates of fine root production range from approximately 100 g m^{-2} $year^{-1}$ in the Ivory Coast to 845 g m^{-2} $year^{-1}$ in Puerto Rico. Fine root turnover is higher in tropical forests than in temperate or boreal forests, with almost all estimates of turnover exceeding 1.0 $year^{-1}$; two studies reported turnover in excess of 2.0 $year^{-1}$. The high turnover rates in the tropics are understandable given that maintenance respiration rates are higher with increasing temperature, the virulence of plant pathogens in tropical zones, and the length of the growing season in tropical forests.

3.5.2 Grasslands

The grassland literature contains the most extensive data on belowground biomass dynamics covering BNPP and root turnover from the high arctic to tropical savannas and from the early twentieth century to the present (Gill and Jackson 2000). Our understanding of root turnover in grasslands has been heavily influenced by the crucial role that belowground processes play in grasslands where estimates of the amount of net primary production allocated belowground ranges from 20 % to more than 80 % (Coleman 1976; Fogel 1985; Milchunas and Lauenroth 1992).

3.5.2.1 Temperate

There is an extensive literature on root dynamics in temperate grasslands that began in the 1930s with the pioneering work of John Weaver in the tallgrass prairies of the central US (Weaver et al. 1935; Weaver and Zink 1946). Alastair Fitter and his colleagues at the University of York, UK, have recently contributed a great deal to our understanding of environmental controls over root turnover in temperate grasslands (Fitter et al. 1996–1999). In addition, the research that began with the International Biological Program and continues today as part of the shortgrass steppe Long-term Ecological Research project in Colorado, USA, reveals how variable estimates of root dynamics can be among years and methods (Sims and Singh 1978; Milchunas and Lauenroth 1992, 2001).

Estimates of root turnover for temperate grasslands are highly variable, ranging from a high of 0.83 in the central Netherlands (Aerts et al. 1992) to no significant turnover in the US shortgrass steppe (Milchunas and Lauenroth 1992). Mean root turnover for temperate grasslands is 0.47 $year^{-1}$ (Gill and Jackson 2000). The long-term research on root dynamics in the shortgrass steppe demonstrates that variation in root turnover at a single site, associated with different methods used at different times, may be almost as large as inter-site variation across temperate grasslands (Sims and Singh 1978; Milchunas and Lauenroth 1992, 2001; Gill 1998).

3.5.2.2 High Latitude

Root turnover for graminoids at high latitudes are among the lowest reported for any ecosystem. Mean turnover for the 16 studies reported by Gill and Jackson (2000) was 0.29 $year^{-1}$. The work conducted in tundra ecosystems as part of the International Biological Program contributed over half of the reported values for root turnover in high-latitude grasslands, and many of these studies occurred in some of the most remote ecosystems in the world. The lowest values come from the most extreme environments, including turnover rates of less that 0.15 $year^{-1}$ in northern Alaska, US (Shaver and Billings, 1975, Miller et al. 1980), above the Arctic Circle in central Canada (Bliss 1975), and in northwestern Poland (Szanser 1997). The highest reported root turnover rate for high latitude grasslands comes from Macquarie Island, in the southern Pacific Ocean. Hnatiuk (1993) reported that the *Poa foliosa* grasslands on the island had a turnover of 0.76 $year^{-1}$, although Jenkin (1975) reported turnover in the same grasslands as 0.49 $year^{-1}$.

3.5.2.3 Tropical

Root turnover in tropical grass-dominated systems is typically very rapid. The highest estimate of root turnover comes from a modeling exercise, with some field validation in a *Panicum maximum* grassland in the Ivory Coast (Picard 1979). Picard estimated that the mean root life span for *P. maximum* roots was between 2 and 4 months, corresponding to a turnover of 2.5–4.6 $year^{-1}$. Several other studies found that roots turned over more than once annually in tropical grasslands, including studies from Central America (Garcia-Moya and Castro 1992), Africa (Kinyamario and Imbamba 1992; Scholes and Walker 1993), India (Tiwari 1986), and southeast Asia (Kamnalrut and Evenson 1992). While site-specific estimates of turnover for tropical grasslands typically exceed those of temperate and high latitude grasslands, several tropical studies have reported relatively low turnover rates. Shackleton et al. (1988) found that root turnover was between 0.24 and 0.30 $year^{-1}$for a *Tristachya leucothrix* site in Transkei, South Africa. In a similar finding, Saxena et al. (1996) reported root turnover of 0.35 $year^{-1}$ for *Brachiara mutica* (para grass) in a savanna in northern India. Overall, mean root turnover in tropical grasslands is 0.87 $year^{-1}$.

3.5.3 Shrublands

Shrublands are one of the least studied ecosystems on earth particularly with respect to root dynamics and turnover. The most heavily studied Mediterranean-type shrublands have received only a fraction of the attention of forests or grasslands.

3.5.3.1 Temperate

Caldwell et al. (1977) used ^{14}C pool-dilution to estimate belowground production and turnover in cold-desert shrublands dominated by either *Ceratoides lanata* or *Atriplex confertifolia*. Between these two species, *Atriplex* roots turned over more quickly than *Ceratoides* roots; however, turnover was slow ranging between 0.12 and 0.21 $year^{-1}$. Oechel and Lawrence (1981) reported turnover for a chaparral site in California of 0.71 $year^{-1}$, substantially greater than the cold desert and comparable to values reported for grasslands. Berendse et al. (1987), working in wet heathlands in The Netherlands, found turnover rates approaching 0.96 $year^{-1}$. Turnover for the other three temperate shrublands for which we found data in the literature ranged from 0.348 to 0.362 for shrublands in North Carolina and Michigan, USA, as well as southwestern Spain (Wentz and Chamie 1980; Saterson and Vitousek 1984; Martinez et al. 1998). Average turnover for the temperate shrublands we examined was 0.44 $year^{-1}$.

3.5.3.2 High Latitude

Similar to high latitude grassland roots, shrub roots at high latitudes have long life spans and low turnover rates. Nearly all of the shrubland root-dynamics studies conducted in the Arctic and other high-latitude sites found that shrub roots had turnover rates less than 0.2 $year^{-1}$, and many of these were less than 0.1 $year^{-1}$ (Gill and Jackson 2000). Mean root turnover was 0.17 $year^{-1}$, although this estimate should be considered tentative given the small number of studies available from high-latitude shrub-dominated ecosystems.

3.5.3.3 Tropical

We found only three studies from tropical, shrub-dominated ecosystems. Kumar and Joshi (1972) characterized a site near Pilani, India, that was dominated by thin scrub of *Capparis decidua* and *Prosopis cineraria*, as well as some intermixed grasses and forbs. They reported root turnover rates of 0.74 $year^{-1}$. Rawat et al. (1994) found that root turnover in a shrub-dominated system in central Himalaya was 0.73 $year^{-1}$. Schroth and Zech (1995) reported root turnover of 1.23 $year^{-1}$ for a hedgerow in an alley cropping system in West Africa.

3.6 Relationship of Root Turnover to Environmental Factors

Gill and Jackson (2000) reported significant exponential relationships between root turnover and temperature, for grasslands, shrublands, and forests. The shrubland relationship explained 55 % of the variability in root turnover (Fig. 3.2a, r^2=0.55). The relationship between mean annual temperature and root turnover for grasslands explained 48 % of the variability (Fig. 3.3a, r^2=0.48). Within the forest studies, there was a weak relationship between mean annual temperature and root turnover that explained only a small proportion of the variability (Fig. 3.4a, r^2=0.19), although Gill and Jackson (2000) found a slightly better relationship between annual thermal amplitude and root turnover for forest fine roots (Fig. 3.4 c, r^2=0.24). It appears that for all three vegetation types, the increase in root turnover with temperature was driven by increases in belowground net primary production with higher temperature, while there is no trend in maximum biomass with increasing temperature (Figs. 3.2b,d, 3.3b,d, 3.4b,d).

The weak relationships between root turnover and precipitation or potential evapotranspiration (PET) surprised us (Figs. 3.2e, 3.3e, 3.4e). Given the importance of moisture in controlling plant production in arid and semiarid regions (Lauenroth and Sala 1992), we expected that precipitation would be a strong correlate of root turnover patterns, particularly in shrublands and grasslands. Inter-annual variation in precipitation locally is important in controlling root production and longevity (Milchunas and Lauenroth 1992). However, we found only weak, non-significant relationships between root turnover and mean annual precipitation (grasslands: r^2=0.02; forests: r^2=0.02; shrublands: r^2=0.04). The relationship between turnover and potential evapotranspiration was slightly better, although, in all cases, it was not as strong as with mean annual temperature (grasslands: r^2=0.14; forests: r^2=0.08; shrublands: r^2=0.22).

It is unlikely that mean annual temperature is the direct cause for the observed differences in root turnover (Fitter et al. 1999, Gill and Jackson

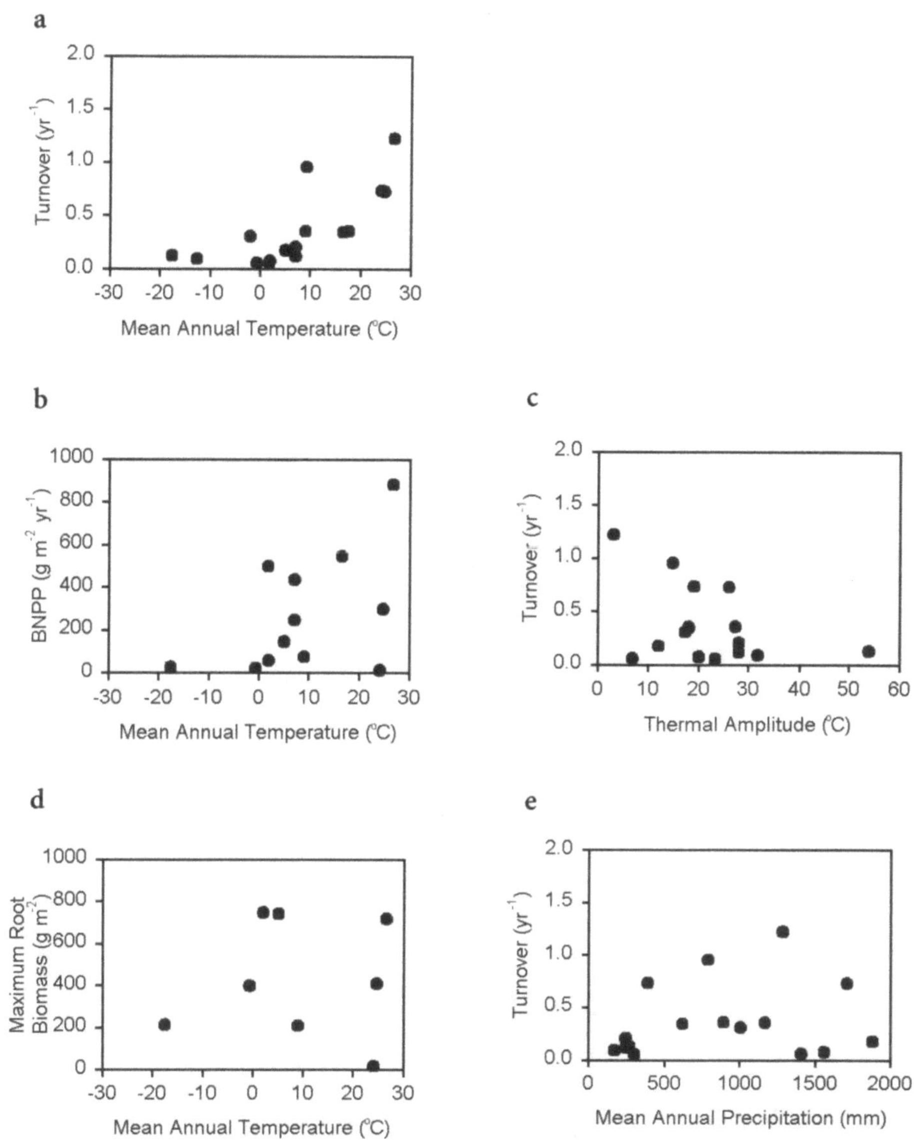

Fig. 3.2a–e. Root turnover, production and biomass relationships for shrublands: **a** turnover versus mean annual temperature; **b** belowground net primary production versus mean annual temperature; **c** turnover versus annual thermal amplitude; **d** maximum root biomass versus mean annual temperature; and **e** turnover versus mean annual precipitation

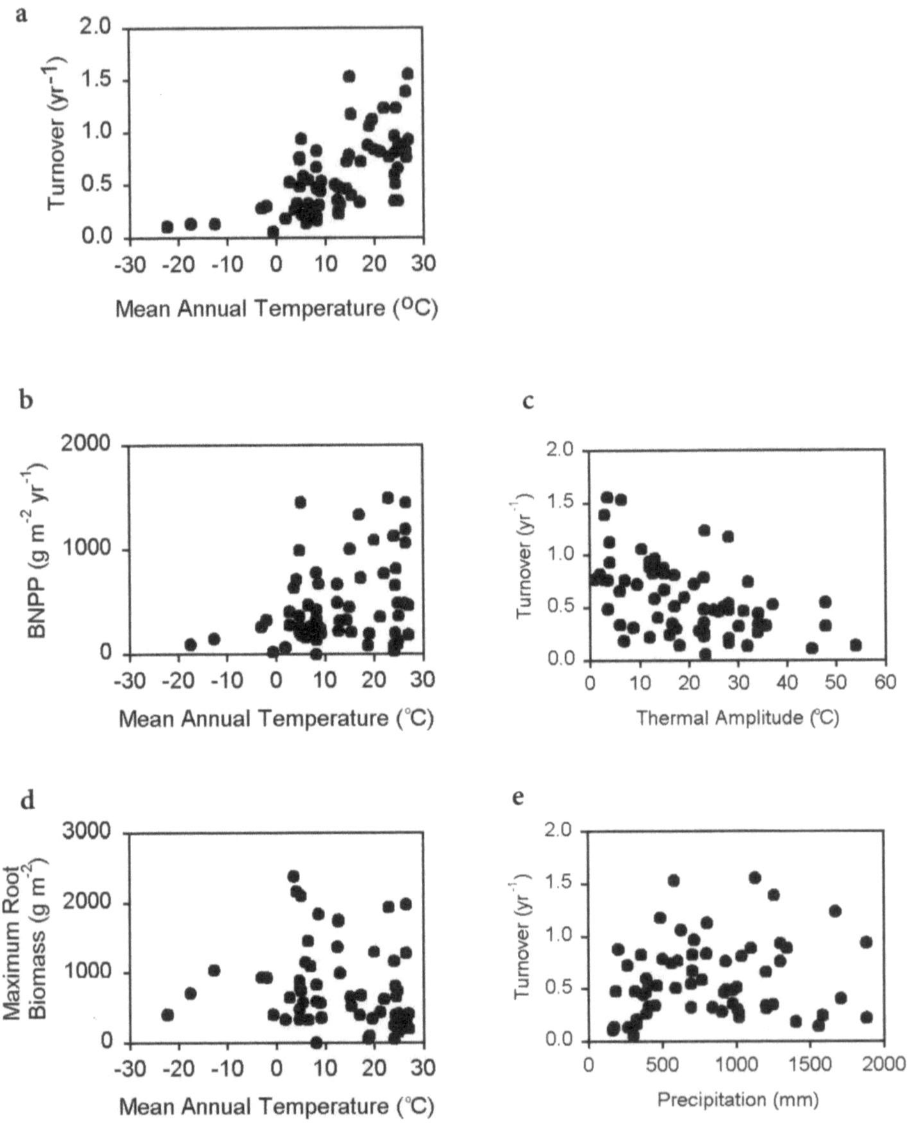

Fig. 3.3a–e. Root turnover, production and biomass relationships for grasslands: **a** turnover versus mean annual temperature; **b** belowground net primary production versus mean annual temperature; **c** turnover versus annual thermal amplitude; **d** maximum root biomass versus mean annual temperature; and **e** turnover versus mean annual precipitation

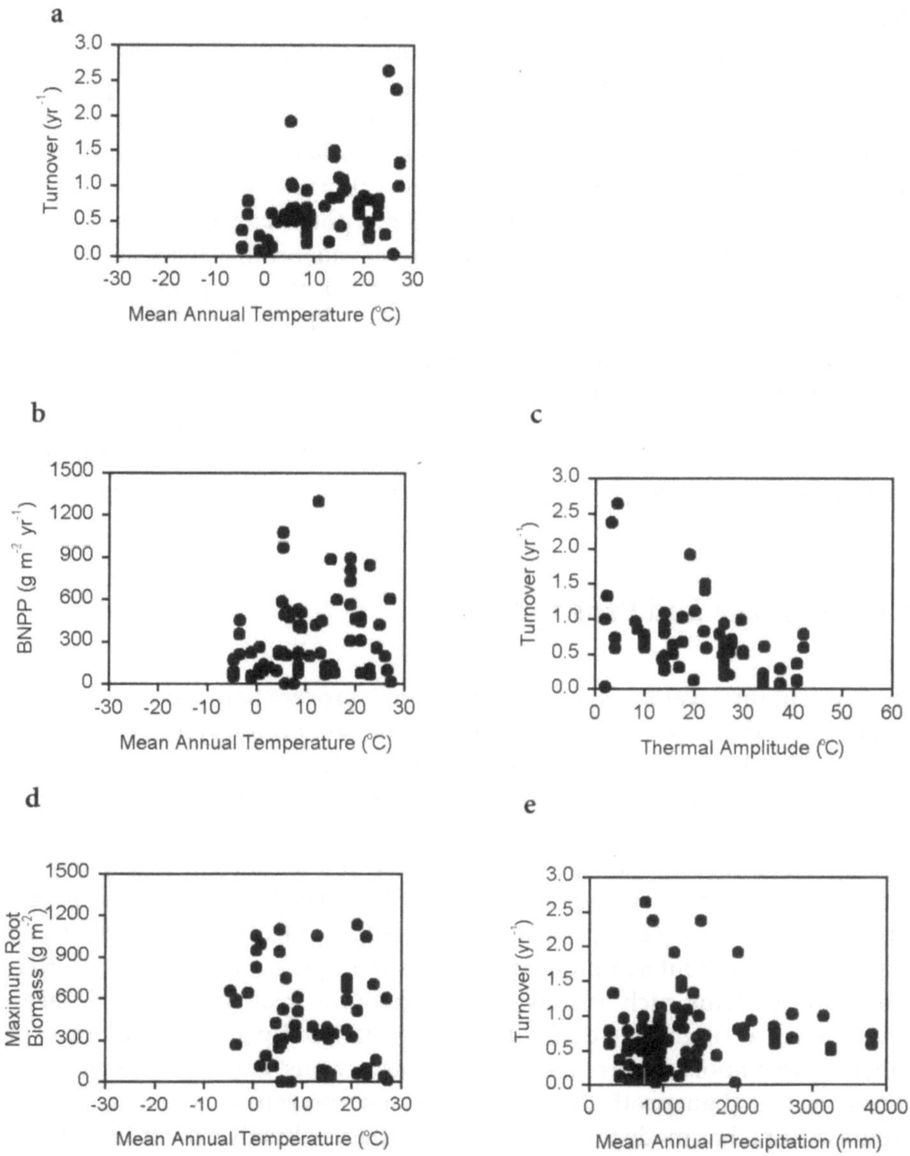

Fig. 3.4a–e. Root turnover, production and biomass relationships for forests: **a** turnover versus mean annual temperature; **b** belowground net primary production versus mean annual temperature; **c** turnover versus annual thermal amplitude; **d** maximum root biomass versus mean annual temperature; and **e** turnover versus mean annual precipitation

2000). Rather, temperature is most likely a covariate for other, process-driving factors. Other factors that influence root turnover that could increase with higher temperatures include (1) root pathogens, (2) maintenance respiration, (3) solar radiation, and (4) nutrient mineralization rates. Further research will be necessary before we can definitively address what are the most important factors controlling root turnover across broad climactic gradients.

Root sampling and calculation methods are a potentially overlooked factor controlling observed patterns of root turnover (Lauenroth 2000; Sect. 3.3). For example, in order for a site to have a higher BNPP than maximum root biomass in a given year, and therefore a turnover >1, the calculation method must account for concurrent root production and death. We found that a higher proportion of tropical or warm-temperate studies than cool climate studies used budget methods that consider changes in live biomass, necromass, and estimates of decomposition to calculate BNPP. This may partially account for the observed increases in root turnover with higher temperatures.

3.7 Summary and Prospects

Root system turnover is a specific instance of a process that operates on all of the tissues and organs of vascular plants. In the case of root systems, it is commonly defined as the ratio of belowground net primary production to the amount of biomass in the root system. Because of the enormous variety in the structure of root systems across the range of vascular plants, collecting the data required to estimate turnover has been one of the most challenging activities in community and ecosystem ecology. For instance, plants with a large diameter woody component to their root systems such as woody dicots require very different sampling approaches than plants with entirely herbaceous root systems such as grasses. A large number of methods have been developed in an attempt to deal with the huge variability in root systems, all of which have important strengths as well as serious and well-documented shortcomings (Lauenroth 2000). This has not prevented us from making progress, but it has made the rate of progress painstakingly slow. It also means that this is an area that holds great promise and great rewards for future innovators.

This chapter is an attempt to summarize currently available data on root system turnover for ecosystems. While an expectation that there should be an inverse relationship between turnover rate and mean root diameter seems reasonable, available data do not unequivocally support this idea and even those data that do show substantial variability especially in the smallest diameter classes (Fig. 3.1). Comparisons of turnover for only fine roots indicate that maximum and minimum values are similar for grasslands, shrublands and forests ranging from less than 0.1 to approximately 1.5 year^{-1}. Several values from forested sites were greater than 1.5 year^{-1} but they were very much in the minority. Analysis of relationships between environmental factors and root turnover indicated that temperature variables were most strongly related to turnover. Root turnover increased with mean annual temperature and decreased with thermal amplitude in grasslands, shrublands and forests.

Our limited knowledge about root system turnover is one of the key gaps in our understanding of ecosystem structure and function. While we have long suspected that the biodiversity residing belowground in ecosystems far exceeds that occurring aboveground, the data to confirm this has been very slow in accumulating (Adams and Wall 2000; Wolters et al. 2000). Understanding the dynamics of belowground plant parts will be critical to understanding belowground food webs. Turnover of belowground plant biomass will be an essential part of this understanding. Compared with aboveground plant processes, increasing our knowledge about belowground processes is not simply a matter of allocating more effort to the topic. We are fundamentally limited by methods of assessing belowground plant processes and especially by methods that will allow us to assess turnover at the individual plant or larger scale and methods that can be used across ecosystem types to provide comparable data. To a very large extent the next major advances in our understanding of root turnover at the stand and ecosystem scales must await the invention of technologies to do for belowground research what eddy flux methods and aircraft and satellite remote sensing have done for aboveground research.

References

Aber JD, Melillo JM, Nadelhoffer JK, McClaugherty CA, Pastor J (1985) Fine root turnover in forest ecosystems in relation to quantity and form of nitrogen availability: a comparison of two methods. Oecologia 66:317–321

Adams GA, Wall DH (2000) Biodiversity above and below the surface of soils and sediments: linkages and implications for global change. BioScience 50:1043–1048

Aerts R, Berendse F, Klerk NM, Bakker C (1989) Root production and root turnover in two dominant species of wet heathlands. Oecologia 81:374–378

Aerts R, Bakker C, De Caluwe H (1992) Root turnover as determinant of the cycling of C, N, and P in a dry heathland ecosystem. Biogeochemistry 15:175–190

Ajtay GL, Ketner P, Duvigneaud P (1979) Terrestrial primary production and phytomass. In: Bolin B, Degens ET, Kempe S, Ketner P (eds) The global carbon cycle. Wiley, Chichester, pp 129–181

Altman A, Waisel Y (1997) Biology of root formation and development. Plenum Press, New York

Ares J (1976) Dynamics of the root system of blue grama. J Range Manage 29:208–213

Arnone JA III, Zaller JG, Sphen EM, Niklaus PA, Wells CE, Körner C (2000) Dynamics of root systems in native grasslands: effects of elevated atmospheric CO_2. New Phytol 147: 73–85

Arthur MA, Fahey TJ (1992) Biomass and nutrients in an Englemann spruce-subalpine fir forest in north central Colorado: pools, annual production, and internal cycling. Can J For Res 22:315–325

Arunachalam A, Pandey HN, Tripathi RS, Maithani K (1996) Biomass and production of fine and coarse roots during regrowth of a disturbed subtropical humid forest in north-east India. Vegetatio 123: 73–80

Bailey RG (1998) Ecoregions. Springer, Berlin Heidelberg New York

Berendse F, Beltman B, Bobbink R, Kwant R, Schmitz M (1987) Primary production and nutrient availability in wet heathland ecosystems. Acta Oecol 8:265–279

Berntson GM, Bazzaz FA (1997) Elevated CO_2 and the magnitude and seasonal dynamics of root production and loss in *Betula papyrifera*. Plant Soil 190:211–216

Bliss LC (1975) Devon Island, Canada. In: Rosswall T, Heal OW (eds) Structure and function of tundra ecosystems: papers presented at the IBP Tundra Biome V. International Meeting on Biological Productivity of Tundra, Abisko, Sweden, April 1974. Swedish Natural Science Research Council, Stockholm Sweden, pp 17–60

Bloomfield J, Vogt K, Wargo PM (1996) Tree root turnover and senescence. In: Waisel Y, Eshel A, Kafkafi U (eds) Plant roots: The hidden half. Marcel Decker, New York, pp 363–381

Böhm, W (1979) Methods of studying root systems. Springer, Berlin Heidelberg New York

Brown DA, Upchurch DR (1987) Minirhizotrons: a summary of methods and instruments in current use. In: Taylor HM (ed) Minirhizotron observation tubes: methods and applications for measuring rhizosphere dynamics. Am Soc Agron, Madison, Wisconsin, pp 15–30

Caldwell MM (1976) Root extension and water absorption. In: Lange OL, Kappen L, Shulze E-D (eds) Water and plant life. Springer, Berlin Heidelberg New York, pp 63–85

Caldwell MM, Camp LB (1974) Belowground productivity of two cool desert communities. Oecologia 17:123–130

Caldwell MM, White RS, Moore RT, Camp LB (1977) Carbon balance, productivity, and water use of cold-winter desert shrub communities dominated by C_3 and C_4 species. Oecologia 29:275–300

Canadell J, Jackson RB, Ehleringer JR, Mooney HA, Sala OE, Shulze ED (1996) Maximum rooting depth for vegetation types at the global scale. Oecologia 108:583–595

Carvalheiro KD, Nepstad DC (1996) Deep soil heterogeneity and fine root distribution in forests and pastures of eastern Amazonia. Plant Soil 182:279–285

Cheng W, Coleman DC, Box JE Jr (1990) Root dynamics, production and distribution in agroecosystems on the Georgia Piedmont using minirhizotrons. J Appl Ecol 27:592–604

Clark DA, Brown S, Kicklighter DW, Chambers JQ, Thomlinson JR, Ni J, Holland EA (2001) Net primary production in tropical forests: an evaluation and synthesis of existing field data. Ecol Appl 11:371–384

Coleman, DC (1976) A review of root production processes and their influence on soil biota in terrestrial ecosystems. Anderson JM, Macfadyen A (eds) The role of terrestrial and aquatic organisms in decomposition processes: Proc Br Ecol Soc Symp April 15–18, Blackwell, Oxford, pp 417–434

Coughenour MB (1985) Graminoid responses to grazing by large herbivores: adaptations, exaptations, and interacting processes. Ann MO Bot Gar 72:852–863

Crawley MJ (1993) Herbivory. Studies in Ecology 10. University of California Press, Berkeley

Dahlman RC, Kucera CL (1965) Root productivity and turnover in native prairie. Ecology 46:84–89

Dahlman RC, Kucera CL (1967) Carbon-14 cycling in the root and soil components of a prairie ecosystem. Proc Natl Symp Radioecol 1967:652–660

Davis TD, Haissig BE (1994) Biology of adventitious root formation. Plenum Press, New York
Eissenstat DM (1992) Costs and benefits of constructing roots of small diameter. J Plant Nutr 15:763–782
Eissenstat DM (1997) Trade-offs in root form and function. In: LE Jackson (ed) Ecology in agriculture. Academic Press, San Diego, pp 173–199
Eissenstat DM, Yanai RD (1997) The ecology of root lifespan. Adv Ecol Res 27:1–60
Esau K (1965) Plant anatomy. Wiley, New York
Farrar JF, Jones DL (2000) The control of carbon acquisition by roots. New Phytol 147:43–53
Ferguson JC, Smucker AJM (1989) Modifications of the minirhizotron video camera system for measuring spatial and temporal root dynamics. Soil Sci Soc Am J 53:1601–1605
Fitter AH (1994) Architecture and biomass allocation as components of the plastic response of root systems to soil heterogeneity. In: Caldwell MM, Pearcy RW (eds) Exploitation of environmental heterogeneity of plants: ecophysiological processes above and belowground. Academic Press, New York, pp 305–323
Fitter AH, Self GK, Wolfenden J, van Vuuren MMI, Brown TK, Williamson L, Graves JD, Robinson D (1996) Root production and mortality under elevated atmospheric carbon dioxide. Plant Soil 187:299–306
Fitter AH, Graves JD, Wolfenden J, Self GK, Brown TK, Bogie D, Mansfield TA (1997) Root production and turnover and carbon budgets of two contrasting grasslands under ambient and elevated atmospheric carbon dioxide concentrations. New Phytol 137:247–255
Fitter AH, Graves JD, Self GK, Brown TK, Bogie DS, Taylor K (1998) Root production, turnover and respiration under two grassland types along an altitudinal gradient: influence of temperature and solar radiation. Oecologia 114:20–30
Fitter AH, Self GK, Brown TK, Bogie DS, Graves JD, Benham D, Ineson P (1999) Root production and turnover in an upland grassland subjected to artificial soil warming respond to radiation flux and nutrients, not temperature. Oecologia 120:575–581
Fogel R (1985) Roots as primary producers in below-ground ecosystems. In: Fitter AH, Atkinson D, Read DJ, Usher MB (eds) Ecological interactions in soil. British Ecological Society Special Publication 4. Blackwell, Oxford, pp 23–35
Fredericksen TS, Zedaker SM (1995) Fine root biomass, distribution, and production in young pine-hardwood stands. N For 10:99–110
Friend AL, Coleman MD, Isebrands JG (1994) Carbon allocation to root and shoots systems of woody plants. In: Davis TD, Hassig BE (eds) Biology of adventitious root formation. Plenum Press, New York, pp 245–273
Garcia-Moya E, Castro PM (1992) Saline grassland near Mexico City. In: Long SP, Jones MB, Roberts MJ (eds) Primary productivity of grass ecosystems of the tropics and sub-tropics. Chapman and Hall, London, pp 70–99
Gaudinski JB, Trumbore SE, Davidson EA, Zheng SH (2000) Soil carbon cycling in a temperate forest: radiocarbon-based estimates of residence times, sequestration rates and partitioning of fluxes. Biogeochemistry 51:33–69
Gill RA (1998) Biotic controls over the depth distribution of soil organic matter. PhD Diss, Colorado State University, Fort Collins, CO, USA
Gill RA, Jackson RB (2000) Global pattern of root turnover for terrestrial ecosystems. New Phytol 147:13–31
Gholz HL, Hendry LC, Cropper WP Jr (1986) Organic matter dynamics of fine roots in plantations of slash pine (*Pinus elliottii*) in north Florida. Can J For Res 16:529–538
Gross KL, Peters A, Pregitzer KS (1993) Fine root growth and demographic responses to nutrient patches in four old-field plant species. Oecologia 95:61–64

Harley JL (1969) The biology of mycorrhiza. Leonard Hill, Glasgow
Harley JL, Smith SE (1983) Mycorhhizal symbiosis. Academic Press, London
Harris WF, Kinerson RS, Edwards NT (1978) Comparisons of belowground biomass of natural deciduous forest and loblolly pine plantations. Pedobiologia 17:369–381
Hendrick RL, Pregitzer KS (1992) The demography of fine roots in a northern hardwood forest. Ecology 73:1094–1104
Hendrick RL, Pregitzer KS (1993) The dynamics of fine root length, biomass and nitrogen content in two northern hardwood ecosystems. Can J For Res 23:2507–2520
Hnatiuk RJ (1993) Grasslands of the sub-antarctic islands. In: Coupland RT (ed) Natural grasslands. Elsevier, Amsterdam, pp 411–482
Hook PB, Lauenroth WK, Burke IC (1994) Spatial patterns of roots in a semiarid grassland: abundance of canopy openings and regeneration gaps. J Ecol 82:485–494
Hooker JE, Black KE, Perry RL, Atkinson D (1995) Arbuscular mycorrhizal fungi induced alteration to root longevity of poplar. Plant Soil 172: 327–329
Jackson RB, Caldwell MM (1989) The timing and degree of root proliferation in fertile-soil microsites for three cold-desert perennials. Oecologia 81:149–153
Jackson RB, Caldwell MM (1992) Shading and the capture of localized soil nutrients: nutrient contents, carbohydrates, and root uptake kinetics of a perennial tussock grass. Oecologia 91:457–462
Jenkin JF (1975) Macquarie Island, Subantarctic. In: Rosswall T, Heal OW, (eds) Structure and function of tundra ecosystems: papers presented at the IBP Tundra Biome V. International Meeting on Biological Productivity of Tundra, Abisko, Sweden, April 1974. Swedish Natural Science Research Council, Stockholm, Sweden, pp 375–397
Jordan CF, Escalante G (1980) Root productivity in an Amazonian rain forest. Ecology 61:14–18
Kamnalrut A, Evenson JP (1992) Monsoon grassland in Thailand. In: Long SP, Jones MB, Roberts MJ (eds) Primary productivity of grass ecosystems of the tropics and sub-tropics. Chapman and Hall, London, pp 100–126
Kinyamario JI, Imbamba SK (1992) Savanna at Nairobi National Park, Nairobi. In: Long SP, Jones MB, Roberts MJ (eds) Primary productivity of grass ecosystems of the tropics and sub-tropics. Chapman and Hall, London, pp 25–69
Kosola KR, Eissenstat DM, Graham James H (1995) Root demography of mature citrus trees: The influence of *Phytophthora nicotianae*. Plant Soil 171:283–288
Kozlowski TT (1992) Carbohydrate sources and sinks in woody plants. Bot Rev 58:107–222
Kramer PJ, Boyer JS (1995) Water relations of plants and soils. Academic Press, San Diego
Kumar A, Joshi MC (1972) The effects of grazing on the structure and productivity of the vegetation near Pilani, Rajasthan, India. J Ecol 60:665–674
Kutschera L, Lichtenegger E (1992) Wurzelatlas mitteleuropäischer Grünlandpflanzen. Gustav Fischer, Stuttgart
Lambers H, Chapin FS III, Pons TL (1998) Plant physiological ecology. Springer, Berlin Heidelberg New York
Larcher W (1995) Physiological plant ecology, 3rd edn. Springer, Berlin Heidelberg New York
Lauenroth WK (2000) Methods of estimating belowground net primary production. In: Sala OE, Jackson RB, Mooney H, Howarth RW (eds) Methods in ecosystem science. Springer, Berlin Heidelberg New York, pp 58–71
Lauenroth WK, Sala OE (1992) Long-term forage production of North American shortgrass steppe. Ecol Appl 2:397–403
Lauenroth WK, Sala OE, Milchunas DG, Lathrop RW (1987) Root dynamics of *Bouteloua gracilis* during short-term recovery from drought. Funct Ecol 1:117–124
Liang Y, Hazlett DL, Lauenroth WK (1989) Water use efficiency of five plant communities in the shortgrass steppe. Oecologia 80:148–153

Mack RN, Thompson JN (1982) Evolution in steppe with few large, hooved animals. Am Nat 119:757–773

Martínez F, Merino O, Martín A, Martín DG, Merino J (1998) Belowground structure and production in a Mediterranean sand dune shrub community. Plant Soil 201:209–216

Matson PA, Waring RH (1984) Effects of nutrient and light limitation on mountain hemlock: susceptibility to laminated root rot. Ecology 65:1517–1524

Merrill SD, Upchurch DR (1994) Converting root numbers observed at minirhizotrons to equivalent root length density. Soil Sci Soc Am J 58:1061–1067

Milchunas DG, Lauenroth WK (1992) Carbon dynamics and estimates of primary production by harvest, ^{14}C dilution and ^{14}C turnover. Ecology 73:1593–1607

Milchunas DG, Lauenroth WK (1993) A quantitative assessment of the effects of grazing on vegetation and soils over a global range of environments. Ecol Monogr 63:327–366

Milchunas DG, Lauenroth WK (2001). Belowground primary production by carbon isotope decay and long-term root biomass dynamics. Ecosystems 4:139–150

Milchunas DG, Lauenroth WK, Singh JS, Cole CV, Hunt HW (1985) Root turnover and production by C-14 dilution: implications of carbon partitioning in plants. Plant Soil 88:353–65

Milchunas DG, Sala OE, Lauenroth WK (1988) A generalized model of the effects of grazing by large herbivores on grassland community structure. Am Nat 132:87–106

Miller PC, Webber PJ, Oechel WC, Tieszen LL (1980) Biophysical processes and primary production. In: Brown J, Miller PC, Tiezen LL Bunnell FL (eds) An Arctic ecosystem: the coastal tundra at Barrow, Alaska. Dowden, Hutchinson and Ross, Stroudsburg, PA, pp 66–101

Nadelhoffer KJ, Raich JW(1992) Fine root production estimates and belowground carbon allocation in forest ecosystems. Ecology 73:1139–1147

Nadelhoffer KJ, Aber JD, Melillo JM (1985) Fine roots, net primary production, and soil nitrogen availability: a new hypothesis. Ecology 66:1377–1390

Neill C (1992) Comparison of soil coring and ingrowth methods for measuring belowground production. Ecology 73:1918–1921

Nepstad DC, de Carvalho CR, Davidson EA, Jipp PH, Lefebvre PA, Negreiros GH, da Silva ED, Stone TA, Trumbore SE, Vieira S (1994) The role of deep roots in the hydrological and carbon cycles of Amazonian forests and pastures. Nature 372:666–669

Nye PH, Tinker PB (1977) Solute movement in the soil-root system. Blackwell, Oxford

Oechel WC, Lawrence W (1981) Carbon allocation and utilization. In: Miller PC (ed) Resource use by chaparral and matorral. Springer, Berlin Heidelberg New York, pp 185–236

Ostertag R (1998) Belowground effects of canopy gaps in a tropical wet forest. Ecology 79:1294–1304

Persson HA (1983) The distribution and production of fine roots in boreal forests. Plant Soil 71:87–101

Pfleger FL, Linderman RG (1994) Mycorrhizae and plant health. APS Press, St. Paul, MN

Picard D (1979) Evaluation of the organic matter supplied to the soil by the decay of the roots of an intensively managed *Panicum maximum* sward. Plant Soil 51:491–501

Pregitzer KS, Hendrick RL, Fogel R (1993) The demography of fine roots in response to patches of water and nitrogen. New Phytol 125:575–580

Pregitzer KS, Zak DR, Curtis PS, Kubiske ME, Teeri JA, Vogel CS (1995) Atmospheric CO_2, soil nitrogen and turnover of fine roots. New Phytol 129:579–585

Pregitzer KS, Kubiske ME, Yu CK, Hendrick RL (1997) Relationships among root branch order, carbon, and nitrogen in four temperate species. Oecologia 111:302–308

Pritchard SG, Rogers HH (2000) Spatial and temporal deployment of crop roots in CO_2-enriched environments. New Phytol 147:55–71

Publicover DA, Vogt KA (1993) A comparison of methods for estimating forest fine root production with respect to sources of error. Can J For Res 23:1179–1186

Rawat YS, Bhatt YD, Pande P, Singh SP (1994) Production and nutrient cycling in *Arundinaria falcata* and *Lantana camara*: the two converted ecosystems in central Himalaya. Trop Ecol 35:53–67
Rogers HH, Runion GB, Krupta SV (1994) Plant responses to atmospheric CO_2 enrichment with emphasis on roots and the rhizosphere. Environ Pollut 83:155–189
Rogers HH, Prior SA, Runion GB, Mitchell RJ (1996) Root to shoot ratio of crops as influenced by CO_2. Plant Soil 187:229–248
Ruess RW, van Cleve K, Yarie J, Viereck LA (1996) Contributions of fine root production and turnover to the carbon and nitrogen cycling in taiga forests of the Alaskan interior. Can J For Res 26:1326–1336
Sala OE (2000) Methods of estimating aboveground net primary production. In: Sala OE, Jackson RB, Mooney H, Howarth RW (eds) Methods in ecosystem science. Springer, Berlin Heidelberg New York, pp 31–43
Santantonio D, Hermann RK (1985) Standing crop, production, and turnover of fine roots on dry, moderate, and wet sites of mature douglas-fir in western Oregon. Ann Sci For 42:113–142
Saterson KA, Vitousek PM (1984) Fine-root biomass and nutrient cycling in *Aristida stricta* in a North Carolina coastal plain savanna. Can J Bot 62:823–829
Saxena AK, Rana BS, Rao OP, Singh BP (1996) Seasonal variation in biomass and primary productivity of para grass (*Brachiaria mutica*) under a mixed tree stand and in an adjacent open area in northern India. Agrofor Syst 33:75–85
Scholes RJ, Walker BH (1993) Primary production. In: Scholes RJ, Walker BH (eds) An African savanna: synthesis of the Nylsveley Study. Cambridge University Press, Cambridge, pp 144–167
Schroth G, Zech W (1995) Above- and below-ground biomass dynamics in a sole cropping and an alley cropping system with *Gliricidia sepium* in the semi-deciduous rainforest zone of West Africa. Agrofor Syst 31:181–198
Shackleton CM, McKenzie B, Granger JE (1988) Seasonal changes in root biomass, root/shoot ratios and turnover in two coastal grassland communities in Transkei. S Afr J Bot 54:465–471
Shaver GR, Billings WD (1975) Root production and root turnover in a wet tundra ecosystem, Barrow, Alaska. Ecology 56:401–409
Sims PL, Singh JS (1978) The structure and function of ten western North American grasslands: III. Net primary production, turnover and efficiencies of energy capture and water use. J Ecol 66:573–97
Singh JS, Lauenroth WK, Hunt HW, Swift DM (1984) Bias and random errors in estimators of net root production: a simulation approach. Ecology 65:1760–1764
Stanton NL (1988) The underground in grasslands. Annu Rev Ecol Syst 19:573–589
Steele SJ, Gower ST, Vogel JG, Norman JM (1997) Root mass, net primary production and turnover in aspen, jack pine, and black spruce forests in Saskatchewan and Manitoba, Canada. Tree Physiol 17:577–587
Stewart DPC, Metherell AK (1999) Carbon (C-13) uptake and allocation in pasture plants following field pulse-labelling. Plant Soil 210:61–73
Szanser M (1997) Root production and biomass of *Arrhenatheretalia* meadows of different age. Ekol Pol 45:633–646
Tiwari SC (1986) Variations in net primary production of Garhwal Himalayan grasslands. Trop Ecol 27:166–173
Tingey DT, Phillips DL, Johnson Mark G (2000) Elevated CO_2 and conifer roots: effects on growth, life span and turnover. New Phytol 147:87–103
Upchurch DR, Richie JT (1983) Root observations using a video recording system in minirhizotrons. Agron J 75:1009–1015

Vogt KA, Grier CC, Meier CE, Edmonds RL (1982) Mycorrhizal role in net primary production and nutrient cycling in *Abies amabilis* ecosystems in western Washington. Ecology 63:370–380

Vogt KA, Grier CC, Vogt DJ (1986) Production, turnover, and nutrient dynamics of above- and belowground detritus of world forest. Adv Ecol Res 15:303–377

Wargo PM (1972) Defoliation-induced chemical changes in sugar maple roots stimulate growth of *Armillaria mellea*. Phytopathology 62:1278–1283

Wargo PM, Houston DR (1974) Infection of defoliated sugar maple trees by *Armillaria mellea*. Phytopathology 64:817–822

Wargo PM, Bergdahl DR, Tobi DR, Olson CW (1993) Root vitality and decline of red spruce. In: Fuhrer E, Schutt P (eds) Contributiones Biologiae Arborum. Ecomed, Landsberg/Lech, Germany

Waring RH, Schlesinger WH (1985) Forest ecosystems, concepts and management. Academic Press, San Diego

Waring RH, Running SW (1998) Forest ecosystems. Academic Press, San Diego

Weaver JE, Houghen VH, Weldon MD (1935) Relation of root distribution to organic matter in prairie soil. Bot Gaz 96:389–420

Weaver JE (1968) Prairie plants and their environment. University of Nebraska Press, Lincoln

Weaver JE, Zink E (1946) Annual increase of underground materials in three range grasses. Ecology 27:115–127

Wells CE, Eissenstat DM (2001) Marked differences in survivorship among apple roots of different diameters. Ecology 82:882–892

Wentz WA, Chamie JPM (1980) Determining the belowground productivity of *Chamaedaphne calyculata*, a peatland shrub. Int J Ecol Environ Sci 6:1–4

Wolters V, Silver WL, Bignell DE, Coleman DC, Lavelle P, Van Der Putten WH, De Ruiter P, Rusek J, Wall DH, Wardle DA, Brussard L, Dangerfield JM, Brown VK, Giller KE, Hooper DU, Sala O, Tiedje J, Van Veen JA (2000) Effects of global changes on above- and belowground biodiversity in terrestrial ecosystems: implications for ecosystem functioning. BioScience 50:1089–1098

Zhang H, Jennings A, Barlow PW, Forde BG (1999) Dual pathways for regulation of root branching by nitrate. Proc Natl Acad Sci 96:6529–6534

4 The Control of Carbon Acquisition by and Growth of Roots

J.F. FARRAR and D.L. JONES

4.1 Introduction

What controls the rate of growth of roots? Behind this deceptively simple question lie a very complex set of processes within the plant and a wide range of environmental variables that affect root growth. To begin to answer it, we will simplify by making the assumption that the question is nearly the same as this: what controls the rate of net acquisition of carbon by roots? A consideration of the gross fluxes of carbon (C) that together constitute the net flux into a root (Table 4.1) is thus central to our argument.

We have recently provided a brief review of the main hypotheses for the control of C acquisition by roots (Farrar and Jones 2000). We concluded that each of the fluxes that contributes to the net acquisition of C exerts some degree of control over the process; we called this the 'shared control hypothesis'. Whilst many of these fluxes occur in the root, others are in the leaf or stem. We rejected two hypotheses which are much simpler. These were the 'push hypothesis', which suggests that net C acquisition is controlled by the supply of C from the shoot; and the 'pull hypothesis', where C acquisition is controlled by demand from within the root itself. Here we will concentrate on processes within roots, but these alone cannot give a complete understanding of control of fluxes within roots.

Roots grow very quickly; young fibrous root systems can increase their dry weight by 25 % per day; root apices commonly elongate at 2 cm/day. To grow this quickly, the fluxes of assimilates and other compounds into, and metabolic rates in, the root must be correspondingly high. A root can gain C by import of carbohydrate and other compounds from the shoot, and by uptake of a variety of organic molecules from the soil (Table 4.1). It can lose carbon by export to the shoot, by respiration, and by rhizodeposition involving loss of dead cells or exudation. The challenge is to determine just what controls each of the individual fluxes, and to evaluate just how important each is in the control of net C flux to the root.

Ecological Studies, Vol. 168
H. de Kroon, E.J.W. Visser (Eds.)
Root Ecology

Table 4.1. Main fluxes of carbon involved in the net acqui5sition of carbon by roots

Gains of C	
	Import in the phloem
	Uptake from the soil
	CO_2 fixation by PEPcarboxylase
Losses of C	
	Respiration
	Exudation of organics
	Exchange of bicarbonate
	Export in the xylem
	Loss to symbionts
	Death of cortex or whole roots
Fluxes of C not resulting directly in gain or loss	
	Phloem unloading
	Short Short-distance transport
	Cellular compartmentation
	Storage
	Synthesis and growth

These considerations provide proximate explanations for the rate at which roots acquire C. The question of how fast roots grow has another dimension: what rate has been selected for? We argue that the rate of root growth is proportional to that of shoot growth, in order that plants acquire above- and belowground resources in the balance needed for continued growth.

We begin with the production of carbohydrate in leaves and its transport in the phloem. The importance and quantification of C fluxes within roots is then discussed, with an emphasis on exudation. We then discuss the concept of shared control, and the quantifying of where control lies, before proceeding to a brief review of the mechanisms of control. Lastly, we consider the outcome of C allocation: what is the ecological significance of the relative allocation to shoot and root?

4.2 Production of Carbohydrate in Source Leaves

The carbon that is imported by roots originates from photosynthesis in source leaves. The rate of carbon fixation is a function of leaf area, amount of photosynthetic machinery per unit area, and leaf environment; the first two are subject to internal regulation in part due to the carbon status and demand for carbon by the root system. Reviews of the mechanisms involved are given by Farrar et al. (2000), Stitt (1996) and Moore et al. (1999). Here, we need only note that both sugars and nitrogenous compounds, particularly amino acids, have been invoked as taking messages about status and demand from sinks such as roots to source leaves, where the density of photosynthetic machinery is increased or decreased as appropriate, probably by changes in the expression of the relevant genes.

What proportion of the C fixed by photosynthesis is partitioned to roots? This question should be refined by considering mature source leaves and asking two questions: what proportion of the C fixed is exported from the source leaf (the export fraction) and what proportion of that exported is then allocated to the roots (the partitioning fraction)? The latter is our main concern here.

4.3 Import of Carbohydrates by Roots: the Phloem Path

Carbohydrates enter the roots in the phloem. Sucrose seems to be transported in all species, and is the dominant sugar in many; however, in a significant number of taxa, it is joined by a range of sugars and polyhydric alcohols (Lewis 1984; Milburn and Baker 1989). The mechanism of phloem transport is widely believed to be pressure-driven mass flow in accordance with the Munch hypothesis (Farrar 1992; Farrar et al. 1995; Minchin et al. 2002). The evidence for pressure flow is cumulative rather than decisive; nearly all experiments can be interpreted as compatible with it, whilst there are none that definitively rule it out, and yet some doubts must remain. These doubts are mainly quantitative, and are most serious in large plants where phloem is operating over long distances. Can sufficient pressure be generated to overcome the resistance to flow down a phloem pathway many metres long? The relay hypothesis (Lang 1979) proposes a series of short sections of phloem in relay, each of which operates according to the Munch hypothesis with its own active loading of solute, but there is no direct evidence to support it. Qualitatively, the Munch hypothesis does not explain the presence of sieve plates, unless these prevent dissipation of the high pressure in expanding sieve element radius, nor has it been fully reconciled with symplastic loading of phloem (van Bel 1993). Overall, whilst the Munch hypothesis is still the best we have, there must remain questions over just how assimilate is moved from source leaves to roots especially over the long distances typical of trees and lianas. Once C has been loaded into the phloem in a source leaf, its passage into the root is rapid, with the time-course reflecting that of the loading process (Fig. 4.1).

A consequence of phloem operating by pressure flow is that the rate of transport must be a function of several factors in source leaf, phloem conduits and in sink which jointly determine phloem turgor gradients and the result-

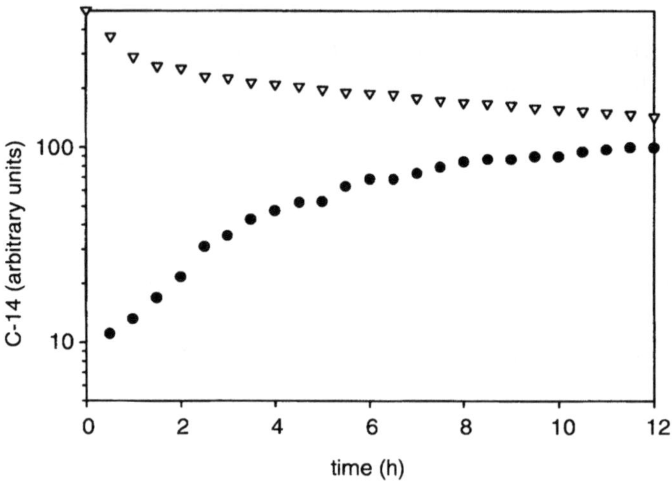

Fig. 4.1. Export of C-14 from a source leaf of barley (*upper line*) and its import into the root (*lower line*) after supplying $^{14}CO_2$ to the leaf for 15 min. The loss of C-14 from the source leaf in the first 2 h follows the loading of sucrose that has not passed through mesophyll cell vacuoles (Farrar 1989). The time-course of import into the root mirrors leaf export. Note that the y-axis is arbitrarily scaled so the C-14 contents of leaf and root should not be compared directly. Counting by Geiger-Muller tubes attached to the leaf and root throughout the experiment

ing solute flux. Import into roots must therefore be a whole-plant property (Farrar 1992, 1996; Minchin et al. 1993; Farrar and Jones 2000). When there is more than one source or sink the system behaves in a complex way best revealed by modelling. For example, when one source is supplying two dissimilar sinks the proportioning of assimilate between the two sinks changes with the flux out of the source (Minchin et al. 1993). A second consequence is the lack of a simple link between the turgor established in the phloem and metabolism in the tissue containing it; nor need there be a simple link between sugar loaded into the phloem and turgor generation since many other solutes contribute to turgor.

4.4 Import of Carbohydrates by Roots: Phloem Unloading and Short-Distance Transport

Once carbohydrate has entered a root in phloem, it is unloaded and transported to receiver cells. How this happens depends on the type of root. Unloading can occur apoplastically – across the plasmalemma bounding the sieve tubes into the surrounding apoplast, possibly via sugar carriers – or symplastically, via plasmodesmata directly into the cytosol of surrounding cells.

4.4.1 Fibrous Roots

In fibrous roots, phloem is unloaded symplastically via plasmodesmata into phloem parenchyma within the stele; there is then symplastic connection out to the cortex (Dick and Rees 1975; Warmbrodt 1987; Farrar 1992; Oparka et al 1994; Patrick 1997; Pritchard 1998a,b, Pritchard et al. 2000). Unloading is localised to near the root tip (Fig. 4.2), which may be due to the constriction of plasmodesmata linking sieve tubes to parenchyma as roots age (A. Schulz, per. comm.). The regulation of unloading may therefore centre on the operation of plasmodesmata, which are still poorly understood (van Bel and Knoblauch 2000). If ATP is involved in their regulation, there is a potential link with metabolism and demand for sugars.

The journey after unloading, from phloem parenchyma cells to the receiver cells which will use the sugars, will presumably be diffusive and driven by concentration gradients. If phloem unloads more than a very short distance from the receiver cells, it is doubtful whether those gradients are sufficient to

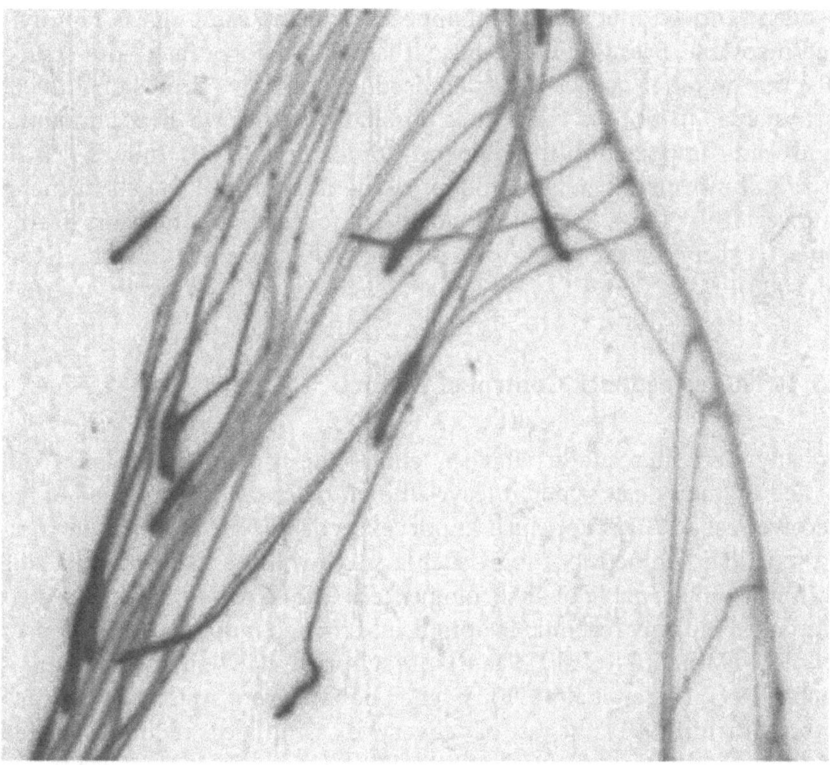

Fig. 4.2. New photoassimilate is localised mainly in root tips (cap, meristem and elongation zone) and lateral initials. An autoradiograph of a portion of barley root 1 h after supplying $^{14}CO_2$ to the shoot

drive the necessary flux (Bret-Harte and Silk 1994). Alternatively, the phloem may unload directly into expanding cells, where the dry weight increment is greatest (Farrar et al. 1995; Pritchard et al. 2000) and the flux of phloem contents will be driven by turgor pressure. Critical experiments are needed to clarify the route of short-distance transport, its relation to cell expansion, and how it may be controlled.

4.4.2 Storage Roots

In storage roots, such as sugar beet and carrot, unloading may be apoplastic, symplastic, or a mixture of both (Patrick 1997; Tomos et al. 2000). Apoplastic unloading will involve membrane-bound sugar transporters, which may be turgor-regulated in sugar beet storage roots (Wyse et al. 1986; Bell et al. 1996). In carrot taproots, rapid longitudinal import of sucrose in the phloem is followed by slow, diffusive, radial transport into xylem and phloem tissues (Korolev et al. 2000; Tomos et al. 2000). The transport is faster into the phloem, as the cambium apparently offers a barrier to movement to the xylem. Indeed much material appearing in the cambium is rapidly rendered insoluble, presumably because it is being incorporated into structure, whilst that in other tissues is metabolised less quickly (Tomos et al. 2000). In this tissue, as in fibrous roots, it is possible that net water accumulation is due to water imported in the phloem (Tomos et al. 2000). Indeed it is likely that radial movement in the phloem tissue is from cell to cell in the symplast; certainly, there is limited potential for uptake of sugars from the apoplast, except for the cambium and a ring of xylem just inside it (Korolev et al. 2000).

4.4.3 Is There Feedback Control of Import?

In steady state, the rate at which C enters the root in the phloem will be equalled by the rate at which it leaves the phloem and is consumed or stored in receiver cells. What happens if import exceeds the rate at which the root can use the C? The phloem occupies such a small volume that potential storage within it is only capable of absorbing excess C for a few minutes. A reduction in import is the only reasonable option, and indeed it does occur. When a root is cooled (Minchin et al.1994a,b) or supplied with sucrose (Farrar and Minchin 1991; Farrar et al. 1995), or when half of a root system is excised (Farrar and Minchin 1991), the rate of delivery of assimilate is reduced. The reduction can occur within minutes of treatment (Minchin et al. 1994a,b) and is rapidly followed by reduced phloem loading in the source leaf and a reduced partitioning to the root (Minchin et al. 2002). (These effects occur in minutes or hours; longer-term responses are discussed below.)

Each of these treatments will increase the sucrose content of roots. We have no idea how such metabolic events in receiver cells are transmitted to the phloem. Patrick and Offler (1995) have pointed out that it cannot be the carbohydrate status of bulk tissue that directly modulates import, since the concentrations of sugars are too low to appreciably alter the turgor of phloem. Turgor in expanding cells will be in part a consequence of the rate of use of assimilate versus its rate of import, and thus could act as a direct mediator of metabolic activity in the key region of the growing root. The question of whether import into roots is regulated by roots alone, or is shared around the plant, is considered in Section 4.7.

4.4.4 Are There Plant Growth Substances That Control Import?

As the regulation of transport to the root is essentially a problem in coordinating the growth of the whole plant, there have been many suggestions that plant growth substances are centrally involved in it. Thomas (1986) and Patrick (1987) summarise the early literature. Morris (1996) provides a thoughtful update which stresses the importance of the target cell in determining plant hormone action.

Can plant growth substances directly affect the loading of phloem? Indoleacetic acid (IAA) and giberellic acid (GA) supplied exogenously increase the uptake of sugars by vascular bundles of celery (Daie et al. 1986), but IAA has no effect on loading in wheat (Bauermeister et al. 1980). Can they alter phloem unloading? IAA does not alter phloem unloading in wheat (Bauermeister et al. 1980). Can they alter the direction of phloem transport to favour one sink over another? For example, do they increase import to roots? There are strong correlations between supply of absciscic acid (ABA) from roots to leaves, and increased partitioning to roots, in droughted plants; ABA-deficient mutants have relatively small roots (Munns and Cramer 1996). There are also correlations between cytokinin status of shoots and roots, export of cytokinins from roots, and reduced root growth (Beck 1996). Low GA mutants of tomato have relatively much more root (Nagel et al. 2001). It is not at all clear if any of these processes are of importance in wild-type plants, and, although the evidence from low-hormone mutants is particularly interesting, there is no obvious logic to a range of plant growth substances having an effect on root growth.

Can they alter the metabolism of specific sinks such that import to that sink is increased? Here the evidence is perhaps the most interesting, since a number of reports suggest that invertases and other hydrolases may be induced by IAA (sometimes GA); although it is difficult to separate these effects from those on cell elongation, the net effect for partitioning would be an increase in flux to the sink concerned (Morris 1996). This area, where there is a clear logical connection between hormone, carbohydrate metabolism, and mechanisms of partitioning, has attracted less attention than it merits.

What is the role of plant growth substances? Note that most evidence comes from applying growth substances externally, not from establishing that those occurring naturally are important. A further barrier to accepting that they are involved is the absence of a convincing mechanism. How are they produced to match the supply or consumption of assimilate? Once they have been detected at their target organ, how do they alter partitioning? Currently, it seems best to accept that there are effects of adding plant growth substances exogenously, but no convincing evidence they are naturally involved.

4.4.5 Are There Genes That Control Import?

This brief survey of how carbohydrates enter roots has touched on issues of what might control this import. Is import into roots regulated by genes? There are three possibilities. Where unloading is symplastic, partial control could be exercised by gene-level regulation of plasmodesmatal frequency and size exclusion limits (see above). Where unloading is apoplastic, sucrose transporters may partially control import, and they are becoming increasingly understood at the gene level (Hellmann et al. 2000). And, in both cases, genes that regulate the metabolism that uses the compounds delivered in the phloem will indirectly regulate transport (Heineke et al. 1999). Perhaps, surprisingly, we know rather less about this possibility, perhaps because it is harder to define precisely than transporters; it will be considered further in Sections 4.5 and 4.7. Ultimately, these genes must be of more importance than transporters in regulating root growth, since they set the demand for C within the root.

4.5 Carbon Fluxes Within Roots and Their Role in Growth and Import

The growth of roots is a measure of the net flux of carbon through them, which in its turn is a product of root metabolism. We next consider some major metabolic carbon fluxes; some have been reviewed recently (Farrar 1999a). There are five main fates for the carbon newly imported into roots: incorporation into structure, respiratory loss, storage, exudation, and transfer to symbionts. Herbivory, senescence and rhizodeposition of carbon by sloughing of old cells are alternative fates for C in the longer term. Indeed the first two of these may be of major ecological importance, yet are relatively poorly understood, at least quantitatively. We need to consider gross, not net, fluxes, and move beyond description to ask what controls these fluxes.

4.5.1 Fluxes That Increase C Content

The dominant root C fluxes are listed in Table 4.1; attempts at quantification are given by Farrar and Williams (1990) and Lambers (1987). The dominant source of gross C gain by roots is import in the phloem from the shoot, which nearly completely describes gross C acquisition.

The second means by which roots can acquire carbon is by fixation of CO_2 by PEP carboxylase, but since the activity of this enzyme is typically low, and the reaction requires ATP, the amount of C fixed in this way is small. Much more significant, and usually underestimated, is uptake of low molecular weight organic compounds from the soil solution, which may be in part a retrieval mechanism for compounds lost by root exudation. Many reports in the early biochemical literature support the idea that roots are avid accumulators of organic compounds, including sugars and amino acids, from their environment. Exudation causes a proliferation of microorganisms in the endo- and ectorhizosphere (Chaps. 12 and 13). Whilst symbiotic microorganisms may improve plant fitness (e.g. mycorrhizae, N_2-fixing bacteria), the rhizosphere is also a primary point of pathogen entry. There is little competitive advantage to the loss of simple sugars and proteinaceous amino acids to the rhizosphere as they have little nutrient complexing power and they rarely act as specific microbial signals (Jones and Darrah 1994). The plant reduces the net C loss by active uptake of these low molecular weight compounds from the soil via H^+-ATPase-driven proton co-transporters that are typically constitutively expressed in all regions of the root and appear to be capable of recapturing large amounts of lost C in hydroponics and to soil (Jones and Darrah 1994, 1996). Although the competitive influence of the soil microbial biomass on this C-retrieval mechanism has yet to be quantified, based on independent kinetic studies it appears that competition from microbes will be strong (Jones and Hodge 1999). In addition to recovery of its own C, this C-retrieval mechanism may also significantly contribute to N uptake (Chapin et al. 1993).

4.5.2 Fluxes That Cause Loss of C

The major loss of root C – about 40 % – is by respiration (Lambers 1987; Farrar and Williams 1990), the loss being due to the generation of energy for synthetic metabolism, maintenance of osmotic and concentration gradients across membranes, and net uptake of ions from the soil solution. Typically, a growing root might use one-third of its respiratory energy for each process (Lambers 1987; Farrar and Williams 1990). There is a tight stoichiometry between respiration and the processes for which it provides energy and carbon skeletons (Amthor 1989; Farrar and Williams 1990; Atkin et al. 2000).

What controls the rate of respiration? Respiration is not an independent process, but is tightly coupled to growth and other energy-consuming processes for which it provides C skeletons, ATP and NADPH. In the short term it is regulated largely by the availability of ADP and NADP, products of reactions consuming energy and substrates for respiration itself; thus it is said that respiration is regulated by adenylate turnover. Thus fibrous roots do not immediately respire faster after supplying them with extra sugars, although they are taken up (Bingham and Farrar 1988; Farrar and Williams 1990); increasing import of sucrose in the phloem is similarly ineffective. (Formally, we should say that the control of respiration is shared, but the majority of control is exercised by adenylate turnover and only a small amount by substrate supply.) Thus the question of what controls the rate of respiration is equivalent to asking what controls the demand for respiratory energy and carbon skeletons. It is not at all clear how the demand for respiratory energy is determined. Sustained increase in sugar supply results in an increased rate of growth by roots, and an increased rate of respiration and amount of respiratory machinery (Farrar and Williams 1990; Farrar et al. 2000). The amount of sucrose-metabolising and respiratory machinery rises as a result of sugar-induced enhancement of expression of appropriate genes including those for sucrose synthase, cytochrome c oxidase, acid invertase, and fumarase (Koch 1997; B. Collis, C. Pollock and J.F. Farrar, unpubl.). This cannot be the cause of the increases in growth and respiration rate, since respiration is under adenylate control and is not determined by the amount of machinery or substrate supply. Rather, sustained sugar supply in excess of that currently used in growth and supporting respiration must trigger an increased capacity to grow. We discuss this later in Section 4.7.

Fluxes to symbionts are appreciable (perhaps 15 % to legume nodules and 10 % to mycorrhizas). How these fluxes are determined and the nature of the compounds moving between the symbionts have been discussed elsewhere (Farrar and Lewis 1987; Hampp and Wingler 1997; Ourry et al. 1997; Chap. 11).

Losses by exudation are considered in Section 4.6. Export of C to the shoot in the xylem is very low but certainly not negligible; both amino acids and sugars are involved (Canny 1990).

4.5.3 Turnover and Metabolism Within Roots

4.5.3.1 Localisation and Compartmentation

Even the simplest root contains perhaps 12 different types of cells, so carbon metabolism and physiology within roots is compartmented both between cell types, and between organelles within cells. Strangely, intracellular compartmentation is easier to study, since the tonoplast has such low permeability to most metabolites that it vacuolar contents are readily distinguished from the cytosol, and, in some cases (e.g. taproot of beet), physical isolation of vacuoles has been possible. Both isotope washout experiments and incubation with dimethyl sulfoxide (DMSO; which increases the permeability of the plasmalemma more than that of the tonoplast) suggest that about half of the sugars in fibrous roots are in the vacuole (60 % in barley, Farrar 1985; Chapleo and Hall 1989; Farrar and Williams 1990; Williams and Farrar 1990). Much of the storage is of hexose, not sucrose; concentrations approximating 25 mM in the cytosol and 4 mM in the vacuole, in barley roots (Farrar 1985). The concentration difference, together with the low flux outwards across the tonoplast, suggests that the tonoplast has a major role in control of carbohydrate fluxes within roots, but we know little of how it is exercised or what factors regulate it. Certainly, fluxes across the tonoplast are larger when overall C fluxes in the root are larger (Farrar and Jones 1986; Williams and Farrar 1990).

4.5.3.2 Size of Pools Relative to Fluxes

A striking feature of the pools of stored soluble and structural carbohydrate in fibrous roots is how small they are relative to the fluxes through them. In consequence, it is unwise to use even an apparently large carbohydrate pool as evidence for there being an adequate supply of carbon from source leaves. Typically, the pools may represent only a few hours worth of import from the shoot – for example, the cytosolic pool of barley roots is turned over every 0.4 h, and vacuolar sugars every 9 h (Farrar 1985; Farrar and Jones 1986). Swollen roots which store sugar, such as carrot and sugar beet, accumulate the sugar in their vacuoles to very high concentrations. Apoplastic concentrations of sugars can be high in these roots, with the consequence of keeping the turgor of phloem low, promoting import (Bell et al. 1996). Other roots store polymers; fructan-storing roots include those of *Bellis perennis*, *Iris pseudacorus* and *Tussilago farfara* (Hendry 1987), whilst others accumulate starch. The turnover of such pools is very low, as typically they accumulate during warm seasons and are mobilised to drive growth after winter.

4.5.3.3 Flux to Structure (Including Maintenance)

A large proportion – typically 40 % – of carbon entering a growing root is used in the synthesis of new structure. The proportion of C entering the root which is allocated to structure remains relatively constant when flux alters (Farrar and Jones 1986; J.F. Farrar and D.G. Jones, unpubl.), which might imply that it is use of C which determines import. However, Farrar and Jones (2000) argue that demand for and supply of sucrose both determine import, consistent with the concept of shared control. In Section 4.7 we discuss how demand might be set in response to the integrated history of supply of sucrose to the root. C-14 has been a powerful tool in establishing the flux of C-14 to structure within roots (Farrar 1985) so it is important to note that the C-14 allocated to structure may come in part from a different pool from that going to respiration (J.F. Farrar, unpubl.).

4.5.3.4 Localisation of Metabolism to Different Cell Types

The root tip was a classical model for problems of growth and development, based on the serial changes in cell number, size and metabolism from genesis in the apical meristem to mature differentiated cells. More recently, the water and solute relations of cell expansion in root tips have been examined at the level of the single cell (Tomos and Pritchard 1994; Pritchard 1998b, Pritchard et al. 2000).

The difference between cell types is well illustrated by recent work on the taproot of carrot which involved sampling of the contents of individual cells from distinct tissues (Korolev et al. 1999, 2000; Tomos et al. 2000). In mature taproot, in which appreciable accumulation of sugars had occurred, fructose, glucose and sucrose all are at appreciable concentrations in phloem and xylem tissue (sampling would favour parenchyma cells), but at very low concentrations in periderm and cambium. Potassium and malate are abundant in the periderm, and at lower concentrations where sugars accumulate (Korolev et al. 1999). Time-series images of the distribution of photosynthetically acquired C-14 demonstrate a highly heterogeneous distribution of newly imported assimilate, which is partitioned preferentially to the phloem zone (Korolev et al. 2000; Tomos et al. 2000). Together with magnetic resonance images of water distribution, a plausible model of how assimilate moves between cells when in transit between sieve tubes and receiver cells, and the problem of crossing the cambium to reach the xylem, have been described (Tomos et al. 2000). Features of this model include: phloem supplying the bulk of the water for cell expansion, and probable diffusive movement of assimilate across the cambium to the xylem (Tomos et al. 2000).

An intriguing feature of cell-level compartmentation is the localisation of different isozymes of the same enzyme to different regions of the root (Koch 1997). Thus one isozyme of sucrose synthase is generally distributed, and one phloem-localised (Koch 1997). The same is true for acid invertase and some N-metabolising genes; these isoforms are also differentially induced by sucrose (Koch 1997). Whilst there have been attempts to see a logic in this arrangement (Hellmann et al. 2000), it is safer to await more information before trying to interpret it. Clearly, it is unlikely that we will fully understand root carbon fluxes without a full appreciation of how they are compartmentalised between different cell types.

4.6 Exudation

The release of C compounds from roots into the surrounding soil (exudation) is a ubiquitous phenomenon as a result of the inherently leakiness of root cells. Further, the concentration of solutes in the cytoplasm is often manifold greater than present in the soil solution due to the continual removal of released C by rhizosphere microorganisms. Therefore, roots lose many C compounds by passive diffusion down a concentration gradient over which they have little control. However, under a variety of biotic and abiotic stress conditions, C efflux can be regulated to achieve: (1) the prevention of rhizotoxic compounds entering the root, (2) the removal of potentially rhizotoxic chemicals from entering the roots (C dumping), and (3) the enhanced mobilisation and uptake of nutrients under deficiency.

4.6.1 How Large Is the Root Exudation C Flux?

Measurement of C loss from roots can be determined by two experimental approaches; (1) $^{14}CO_2$ feeding of shoots of soil-grown plants followed by separation of ^{14}C-root, ^{14}C-shoot, belowground $^{14}CO_2$ (root+microbial) and ^{14}C-soil components after a suitable chase period, or (2) collecting specific exudate compounds in the culture medium of hydroponically grown plants. Whilst both techniques are fraught with potential problems (Jones and Darrah 1993; Meharg 1994), many studies using both approaches have indicated large amounts of C loss from roots. A summary of 95 whole-plant ^{14}C-labelling studies performed on a broad range of plant species indicates that approximately 10 % of the net fixed C can be recovered in soil (Table 4.2).

In contrast to these ^{14}C budget studies, experiments performed in hydroponics have demonstrated that typically only 0.5 to 1.5 % of a plant's net fixed

Table 4.2. Carbon partitioning in long-term whole-plant ^{14}C-labelling experiments performed in soil under non-sterile conditions. Values are mean ±SEM (n=95). The mean duration of the experiment was 46 days while the average shoot dry weight was 1.3 g and root dry weight was 1.0 g

	Percentage partitioning of C
Shoot	58 ±2
Root	17 ±1
Root respiration	14 ±1
Soil	10 ±1

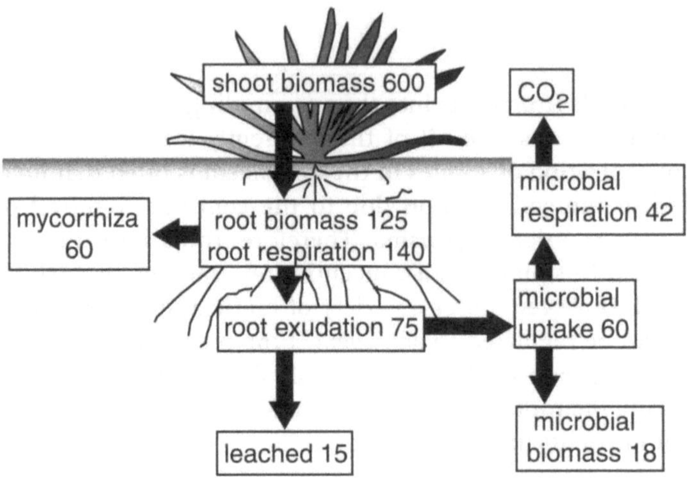

Fig. 4.3. Partitioning of C within the plant and the distribution of root exudates within the soil after the net fixation of 1000 C units. (Adapted from Heulin et al. 1987; Martin 1977; Lambers 1987; Whipps 1990; Haller and Stolp 1985)

C is lost as rhizodeposition. The disparity between the results of the two experimental approaches may reflect the different plant growth conditions; however, it is more likely that the soil pool recovered in the whole-plant ^{14}C-labelling studies includes dead root material and symbiont cells (e.g. mycorrhizas) thereby overestimating rhizodeposition. In contrast, in the sterile hydroponic cultures there is no microbial removal of C from the medium leading to unnatural diffusion gradients that may cause an underestimation of exudation fluxes. Based on a conservative estimate, it is likely that the amount of net fixed C lost by roots is within the range 0.5–5 % (Fig. 4.3).

4.6.2 What Are the Dominant Exudate Components?

The composition of and amount of rhizodeposition varies widely with plant species, age, genetics, distance along root, environmental conditions, presence of microbes/pathogens, etc. (Curl and Trueglove 1986). Plants at different developmental stages appear to be able to modify the characteristics of their exudates, especially in response to nutrient availability (Klein et al. 1989), suggesting that, through their physiological processes, plants exert close control of some exudates. From sterile hydroponic studies it has been shown that most of the chemical compounds present in roots can be recovered in the surrounding solution (Table 4.3).

Much of the material lost from a root is thought to be of a soluble low molecular weight (diffusible) nature with the principal components reflecting the root's cytoplasmic content; sugars, amino acids, organic acids and phenolic compounds (Tables 4.3 and 4.4; Chap. 10). In addition, however, a significant quantity of C is also lost as insoluble mucilage which is defined as a gelatinous material ensheathing roots and which is composed of negatively charged

Table 4.3. Estimates of the loss of various components of rhizodeposition

Exudate	Plant	Rate of C loss (μg C g^{-1} root DW day^{-1})	Rate of N loss (μg N g^{-1} root DW day^{-1})	Reference
Carbon dioxide	*Zea mays*	86,400	0	Saglio and Pradet (1980)
Lactic acid	*Zea mays*	19,458	0	Xia and Saglio (1992)
Mucilage	*Vigna unguiculata*	34,000	952	Horst et al. (1982)
Sugars	*Zea mays*	7880	0	Schonwitz and Ziegler (1982)
Phytosiderophores	*Hordeum vulgare*	2880	560	Römheld (1991)
Ethanol	*Pisum sativum*	130	0	Smucker and Erickson (1987)
Amino acids	*Gossypium hirsutum*	123	51	Cakmak and Marschner (1988)
Phenolics	*Triticum aestivum*	334	–	Cakmak and Marschner (1988)
Organic acids	*Zea mays*	104	0	Kraffczyk et al. (1984)
Sloughed cells	*Arachis hypogaea*	72	9	Griffin et al. (1976)
Fatty acids	*Brassica napus*	139	0	Svenningsson et al. (1990)
Sterols	*Brassica napus*	12	0	Svenningsson et al. (1990)
Vitamins	*Zea mays*	6	1	Schonwitz and Ziegler (1982)
Flavenoids	*Phaseolus vulgaris*	3	0	Hungria et al. (1991)
Growth hormones	*Arachis hypogaea*	0.07	0.04	Reddy et al. (1989)

Table 4.4. Quantity of root exudates (μg C plant^{-1} day^{-1}) detected in the culture medium of plants grown under sterile or non-sterile hydroponic culture conditions. (Jones and Darrah 1993)

	Sterile	Non-sterile
Sugars	0.95	0.12
Amino acids	0.03	0.01
Proteins	0.01	0.01
Phenolics	0.02	0.01
Others (inc. organic acids)	1.94	0.18

polysaccharide (e.g. polygalacturonic acid, 93%), protein (7%), phenolic acids (0.2%) and polyuronic acids (pectin, 0.2%). The amount of mucilage observed on root tips is highly dependent on environment conditions and is closely regulated by the root. Speculation on its function includes lubrication for root growth, maintenance of soil–root contact, chelation of toxic metals, nutrients storage via its exchange capacity, sites for microbial colonisation, prevention against desiccation, sites for extracellular enzyme activity (Bacic et al. 1986; McCully 1999; Iijima et al. 2000). Unlike the soluble and insoluble exudates of roots, there are few quantitative and mechanistic studies on the loss of gaseous components other than CO_2 from roots. However, the loss of a broad range of volatile compounds has been detected, although not in large quantities (Vancura and Stotzky 1976).

4.6.3 Localisation of Root Exudation

The exact site of exudate release along the root will dictate its fate in soil as a significant chemical and biological gradient exists in the soil along the root length. Generally, it is conventionally thought that microbial populations increase and nutrient levels decrease with distance away from the root apex towards the root base (Marschner 1995). Whole-plant ^{14}C-labelling studies indicate that root exudation is maximal at the root apex and at lateral branching points where the main C sinks are located (McDougall and Rovira 1970), entirely consistent with the localisation of newly imported assimilate shown in Fig. 4.2. Later studies have subsequently confirmed these findings showing distinctly spatially localised excretion at the root apex in response to environmental stress (e.g. organic acids under Al toxicity, mucilage under mechanical stress, phosphatases under P deficiency; Hoffland et al. 1989; Marschner 1995; Ryan et al. 1995; McCully 1999). In contrast, flavenoid compounds that are involved in the chemotaxis of symbiotic N_2 fixing *Rhizobium* towards the root are only exuded from the root hair region where nodulation occurs (Peters and Long 1988).

4.6.4 Mechanistic Basis of Root Exudation

It has commonly been assumed that rhizosphere C flow is a unidirectional flux whereby C is lost from roots in response to the large concentration gradient that exists between the root cytoplasm and the soil solution (Curl and Trueglove 1986; Marschner 1995). Whilst the traditional unidirectional flux approach may adequately describe the loss of high molecular weight exudate components (e.g. mucilage), there is a significant amount of literature that suggests that the amount of C loss from roots is actually a net loss (i.e. efflux–influx). From mechanistic studies of transport processes, it is now clear that roots can control the levels of low molecular weight components accumulating in the rhizosphere by at least two distinct mechanisms, those regulated by C efflux and those regulated by C influx.

4.6.4.1 Root Exudation Regulated by C Influx

Uncharged or net-neutral C compounds such as sugars and amino acids move across the plasma membrane into the soil purely as a result of the concentration gradient that exists between the cytoplasm and apoplast/soil solution. In the case of sugars and amino acids this gradient is typically 100- to 1000-fold. The root can be expected to exert little direct control over the loss of sugars and amino acids from the root as cytoplasmic solute concentrations remain fairly constant and membrane permeability is optimised for other metabolic processes (e.g. nutrient uptake, cell elongation, etc.). However, there is increasing evidence to suggest that the degree of C accumulation in the rhizosphere is regulated by the root's attempt to recapture these compounds from the soil. Evidence suggests that uptake is mediated via plasma membrane H^+-ATPase-driven, proton co-transporters which are capable of moving these compounds back into the root greatly reducing net exudation (Fig. 4.4; Jones and Darrah 1993, 1994, 1996). This recapture of exudates benefits the roots by lowering rhizosphere C concentrations thereby reducing the potential for microbial (and pathogen) growth and competition for nutrient resources. Exudate recapture can occur along the whole root length and the

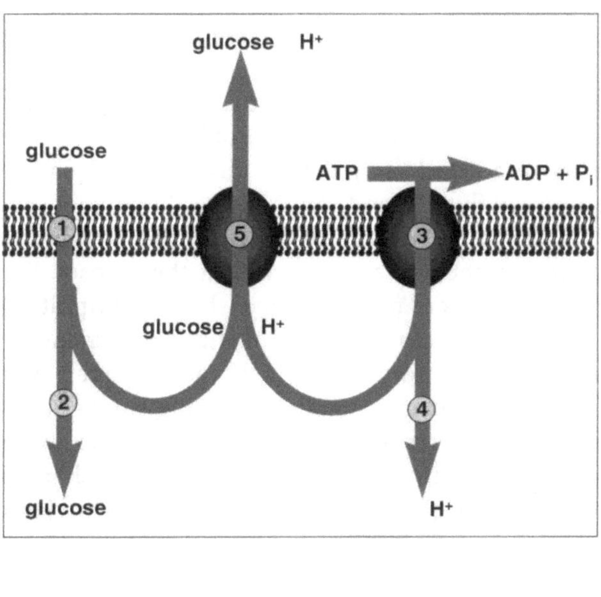

Fig. 4.4. Schematic representation of the efflux and influx of low molecular weight C compounds at the soil–root interface. Although glucose is shown here, the process is also characteristic of most sugars and amino acids. *1* Passive diffusion of solutes across the plasma membrane; *2* diffusion of solutes away into the soil; *3* release of H^+ by the plasma membrane H^+-ATPase; *4* diffusion of H^+ into the soil; and *5* recapture of C compounds by a plasma membrane H^+-cotransporter

Table 4.5. Comparison of the Michaelis-Menten root uptake kinetic parameters for C-containing compounds and for mineral nutrients

	Plant species	Km (μM)	Vmax (μmol g root DW^{-1} h^{-1})	Reference
Carbon compounds				
Lysine	Barley	3.9	0.6	Soldal and Nissen (1978)
Proline	Barley	1.6	2.2	Soldal and Nissen (1978)
Putrescine	Maize	120	4.0	DiTomaso et al. (1992)
Glucose	Maize	800	62	Xia and Saglio (1988)
Fructose	Maize	5900	75	Xia and Saglio (1988)
Inorganic nutrients				
Ca	Barley	5	0.3	Barber (1984)
K	Maize	16	7.2	Barber (1984)
P	Maize	3	0.7	Barber (1984)
NO_3^-	Maize	10	1.8	Barber (1984)
SO_4^{2-}	Maize	115	1.1	Ferrari and Renosto (1972)

kinetics of the transporters are similar to those for inorganic nutrients (Jones and Darrah 1994, 1996; Table 4.5). In certain situations this mechanism can also contribute to a plant's N economy through the capturing of low molecular weight dissolved organic N present in the soil solution that is transported into the root via the amino acid transport system (Jones and Darrah 1994).

4.6.4.2 Root Exudation Regulated by C Efflux

The efflux of negatively charged solutes such as organic acids can be regulated directly. Here, the plasma membrane electrochemical potential gradient is the driving force for transport. Due to the charge gradient, influx is extremely small and C loss can be described by unidirectional efflux. From a knowledge of membrane biophysics and the cellular concentrations of organic acids, the net exudation of solutes can be predicted using the net flux density equation (Nobel 1991). Here the net outward flux depends on the inward flux, root surface area, membrane permeability, the charge of the ion, the membrane potential, and its concentration in the soil solution and in the cytosol. For example, we have calculated the net flux of malate from a wheat root tip, at 3.3×10^{-2} pmol mg^{-1} root FW s^{-1}. This is in close agreement with experimentally derived malate efflux rates for wheat root tips which range from 1.4×10^{-2} pmol mg^{-1} root FW s^{-1} under normal conditions rising up to 3.3×10^{-1} pmol mg^{-1} root FW s^{-1} in the presence of toxic levels of Al and when membrane permeability is enhanced through the opening of anion channel proteins (Fig. 4.5; Ryan et al. 1995). The gating of these anion channels, at least

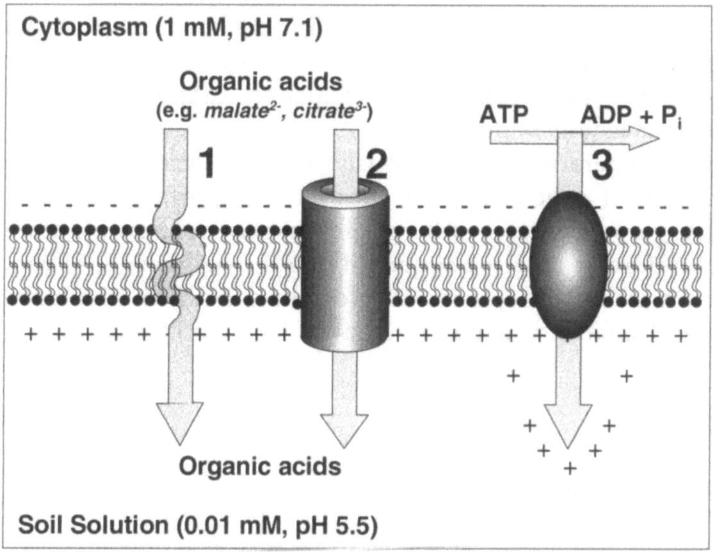

Fig. 4.5. Schematic representation of the two main routes of organic acid efflux from the cytoplasm of root cells into the soil solution. *1* represents the slow passive diffusion across the lipid bilayer, whereas *2* represents efflux through a plasma membrane channel protein. The direction of transport by both mechanisms is controlled by the electrochemical potential gradient (charge and concentration) across the membrane which is largely generated by *3* the H^+-ATPase

for the Al-associated malate efflux in wheat, appears to be under the control of a single gene locus that at present remains unidentified. From kinetic analysis of organic acid release in wheat root tips, channel-mediated malate release appears to be almost instantaneous and thus cannot involve *de novo* synthesis of proteins. From these results, some researchers have speculated that the genes controlling malate release (e.g. *Alt1*) therefore encode not the organic acid transport proteins but rather signalling elements necessary to trigger the release of malate and citrate. The nature of this signalling protein, however, remains unknown but may constitute an Al^{3+} receptor located in the plasma membrane (Kochian 1995). A similar channel-enhanced efflux of organic acids may also occur in some plants under P deficiency where citrate is released to enhance P mineral dissolution in the rhizosphere; however, more work is required to confirm this (e.g. lupins; Johnson et al. 1996).

4.6.5 Exudation: Conclusion

It is clear that root exudation represents a small but significant C flux in plants involving a wide variety of compounds. Further, exudation is a spatially and temporally complex process that is highly dependent on the root's physiological status and the nature of its surrounding environment. It can be expected that the spectrum and concentration of exudates at any one time will represent a unique fingerprint. Root exudation cannot be simply explained by a single mechanism but is moreover a combination of complex multidirectional fluxes operating simultaneously. While we currently possess a basic understanding of root exudation, its overall importance in plant nutrition, pathogen responses, etc., still remains largely unknown. Future research should therefore be directed at quantifying the significance of root exudates in realistic plant–soil systems. Current knowledge of the agronomic and ecological effects of exudates is given by Inderjit and Weston (Chap. 10).

4.7 Integration of Fluxes

4.7.1 Shared Control of Carbon Flux

We have presented elsewhere the idea that control of import into roots is shared around the plant, being partly localised in the root, in the transport system and in the shoot (Farrar and Jones 2000). There is both a theoretical argument, and evidence, to support the idea of shared control, which also applies to leaf properties being determined partly from outside of the leaf (Gunn et al. 1999b), for sinks in general (Farrar 1996) and for whole plants (Farrar 1999b). The theory is that of metabolic control analysis (Fell 1997). Its central tenet is that in complex, multistep linear or branched systems, every step contributes to the control of flux and so control is distributed throughout the system. Roughly, the control due to a single step is the fractional change in flux through that step in response to a fractional change in the activity of machinery catalysing it. It is a central theme of this review that shared control applies to all features of root physiology – from metabolism within single cells to fluxes into and out of whole roots.

Top-down metabolic control analysis can be applied to carbon flux in whole plants in two ways. Sweetlove et al. (1998), using transgenic potato and ^{14}C to estimate flux, conclude that about 80 % of control resides in source leaves and the remainder within sinks. We have used a single source of unchanging photosynthetic area supplying assimilate to a single sink – a root system growing at a constant absolute rate (a true single-source/single-sink system) by rooting a soybean leaf from the base of the lamina. Sucrose produced in the source lamina has only two significant fates: storage in the lamina, or export to the root. Our initial data (Table 4.6) ascribe about 80 % of control of C flux to the source leaf, and about 20 % to the root; our agreement with the result of Sweetlove et al. (1998) is probably fortuitous.

Table 4.6. Top-down metabolic control analysis of the carbon flux in single-rooted soybean leaves. The table gives control coefficients for a two-block system where the source leaf produces, and the root consumes, the common intermediate sucrose (S. Gunn and J.F. Farrar, unpubl.)

Age of single-rooted leaves	14 days	16 days
Leaf/root dry weight	2.02	1.42
Control in source leaf	0.64	0.79
Control in root	0.36	0.21

There is an urgent need to extend this type of analysis to wild plants, since the implications for ecosystem processes are obviously considerable.

4.7.2 Additional Evidence for Shared Control of Import into Roots

The transport system has a role in control, since length of the transport pathway affects flux to sinks (Canny, 1973; Cook and Evans 1978; Minchin et al. 1993). Models of phloem transport also show clearly that there is the potential for control by transport itself (Minchin et al. 1993), and indeed shared control is implicit in the Munch hypothesis (see above).

Frequent suggestions that carbohydrate flux into sinks is linearly dependent on photosynthesis are valid descriptively, but the relationship between photosynthesis and export from source leaves is easily broken, so import itself must also be regulated by other factors. For example, Ho (1978) altered both light intensity and carbon dioxide concentration to vary the photosynthetic rate of tomato leaves, but found that the rate of export from those leaves was to an appreciable degree independent of photosynthesis. We have similar results from barley (Table 4.7): whilst leaves on plants grown at 700 ppm CO_2 photosynthesise and export faster than those on plants at 350 ppm, just 24 h after a transfer between CO_2 concentrations the rate of export does not reflect photosynthetic rate. We conclude that downstream events in sinks such as roots partly regulate flux from source to sink.

Experiments designed to address the problem of whether growth is limited by sources or by sinks have often been unsatisfactory. If control is partly in source leaves then an experiment designed to detect, qualitatively, the presence of some source limitation will find it, but will fail to quantify it or even to detect any sink limitation. The literature is therefore littered with opposing qualitative claims about whether source or sink limits growth. This can be seen as a false argument, since both source and sink will always offer some control – the critical question is how much.

Table 4.7. Export from second leaf of barley at elevated CO_2: effect of sinks. Barley plants were grown at either 350 or 700 ppm CO_2 and some switched (>) to the other CO_2 concentration 2 d after expansion of the second leaf. Carbohydrate content, and rates of photosynthesis and export, were measured 24 h later. All differences in export are significant (B. Collis, C. Pollock, J.F. Farrar, unpubl.)

CO_2 concentration	350	700	350>700	700>350
Net photosynthesis (μmol m^{-2} s^{-1})	11.2	19.0	18.2	11.8
Soluble carbohydrate (g m^{-2})	6.1	10.4	9.6	5.3
Export of sucrose (g m^{-2} h^{-1})	0.93	1.73	1.58	1.22

4.7.3 Mechanisms Underlying Shared Control of Carbon Flux

We have argued that control of carbon flux into roots is shared. We now need to discuss how this is achieved, and how the distribution of control can be altered. We propose two suitable mechanisms: phloem transport, and gene regulation by sugars and other resource compounds such as amino acids. Phloem has been discussed in Sections 4.3 and 4.4.

Just how resource compounds, particularly sugars and amino acids, regulate gene expression is discussed by Koch (1996) and Farrar et al. (2000). In essence, the expression of genes encoding proteins key to source leaf and sink metabolism is regulated by sugars, nitrate or amino acids. Typically, photosynthetic genes are down-, and sink metabolism genes up-, regulated by high sugar concentrations. It is an appealing idea that plentiful carbohydrate regulates whole-plant metabolism to both reduce its production and to increase its consumption, whilst a shortage does the reverse (Koch 1997). The signal molecules include sucrose and glutamine, or their close metabolites; they are xylem and phloem mobile, and so can take messages of sugar and N status from leaf to root, and vice versa. Clearly, such a system means that control of C influx into roots is shared since the capacity of a root to metabolise sucrose and thus sustain its import will be set by the history of sugar supply from the shoot. A critical feature of such a system is that the pools of resource compounds such as sucrose are small enough in relation to fluxes through them such that they respond to internal or external changes. This is certainly so for sucrose, which is thus a time-integrated sensor of the balance between production and consumption of sugars.

Many details of this signalling system are still to be worked out (Farrar et al. 2000; Smeekens 2000). We have probably not yet identified the key genes that regulate the rate of root growth and carbon import in response to sugar and amino acid status; they may, for example, include genes regulating the cell cycle. The growth of a root tip can be divided arbitrarily into the processes of cell division, cell expansion, and addition of dry weight. There is no obvious mechanism whereby the biophysics of expansion can be altered directly (osmotically) by sugar supply. Simple mass action cannot provide a mechanism for increase of dry weight increment by increased sugar supply, since respiration is mainly under adenylate, not substrate, control (Bingham and Farrar 1988). The extra sugar arriving in the expanding cell may enhance the activity of genes, as yet unidentified, which are centrally involved with determining metabolic rate, growth, and demand for respiratory energy. It is striking how cell division and root growth rate are synchronised; there is never a build-up of unexpanded cells. Either there is communication between the zones of expansion and division, with the former controlling the latter, or control of root growth is exercised at the level of cell division, presumably the cell cycle. Good evidence is badly needed.

4.7.4 What Is Root 'Demand'?

Roots thus partly (but by no means completely) regulate their import of carbon, and they signal their status to the shoot via phloem in the short term, and in the long term via their ability to metabolise sugars. Does this mean that a measurement can be made which defines root demand for carbon? Certainly there is much acceptance of the idea that demand is involved in regulating partitioning (Marcelis 1996). Although the word 'demand' is frequent in the source-sink literature, it is rarely defined precisely. There are three ways of defining demand: (1) the current flux (the sum of C partitioned to growth plus the fluxes resulting in loss of C); (2) the current capacity – the flux using the current amount of enzymes and transporters but with no limitation from substrate; and (3) inducible capacity: the flux unconstrained by substrate supply when the genes coding for all machinery have been maximally induced. We argue that the most useful is the third, inducible capacity, since it can be used to compare with the current flux to determine whether the shoot is supplying roots at a rate substantially below their potential. There is no simple or agreed method of measuring inducible capacity, although the flux after 24 h on extra sucrose (Bingham and Farrar 1988; Gunn and Farrar 1999) or rate of sucrose uptake from solution (Farrar and Jones 1985; Ciereszko et al. 1999) could be candidates.

Species vary in the extent to which their root respiration – a useful and non-destructive measure of current flux – is constrained by an inadequate assimilate supply. In one experiment, *Poa annua* was so constrained but *Dactylis glomerata* and *Bellis perennis* were not (Gunn and Farrar 1999). Indeed, given the mechanisms we have just described, we would predict that the amount of enzymes and transporters in roots is adjusted to match the history of assimilate supply, rendering uncommon a short-term substrate limitation of respiration and growth. Rather, the disparity will be between current flux and capacity.

4.7.5 The Remarkable Consequences of Darkening

Finally, we give a striking example of how import into roots is indeed a whole-plant property. When plants are kept in the dark for prolonged periods, the relative effects on shoot and root are unexpected. Typically, a root has only enough stored carbohydrate to sustain metabolism for a few hours at the current rate, whereas the source leaves are carbohydrate-rich – they have enough carbohydrate to sustain their current respiration for about 2 days. The roots would appear to be potentially vulnerable to carbohydrate starvation if the shoot is subject to prolonged shading. Remarkably, the converse is true. In barley darkened for 8 days, root respiration and growth rate decline only after several days during which sugar, starch and protein contents, and activities of key respiratory enzymes, are little affected (Table 4.8; Farrar 1999a). Roots only lose metabolic capacity after darkening for 6 days. Source leaves are quite different: carbohydrate content (initially very high), protein content and respiration rate all fell within 1–2 days of darkening, and there was loss of photosynthetic and respiratory enzymes. The unexpected reason is that the shoot continues to export carbohydrate to the root for many days, so root metabolism is supported. Surprisingly, it is the shoot, far richer than the root in carbohydrate when the plant is first darkened, which first shows the effects of darkening. This is a striking example of root demand having a large effect on import. More importantly, if this result for a crop plant applies to wild species, it suggests that root systems will survive far better than might have been predicted when the photosynthesis of source leaves is impaired.

Table 4.8. Respiration in leaves and roots of barley in extended darkness. Values are % value before darkening. At the time of darkening, roots contain enough carbohydrate to sustain respiration for 6 h at control rates (P. Dwivedi and J.F. Farrar, unpubl.)

	Leaf			Root		
Days in dark	2	4	8	2	4	8
Rate of respiration	58	–	–	100	92	46
CHO content	11	5	9	53	53	65
Protein content	61	23	5	67	72	32

4.8 Allocation of C and Dry Weights to Roots Relative to Shoots

We have demonstrated that the net flux of carbon into roots is controlled by all parts of the plant to some degree. Indeed, flux to the root is part of the whole-plant process of allocating assimilates between the various sinks, which include the shoot apex and young leaves, secondary thickening in the stem, inflorescences and fruits. How does allocation to the roots compare with these? Presumably there has been strong natural selection for the appropriate allocation of resources.

4.8.1 The Conservation of Shoot/Root Ratio

Remarkably, net allocation, described as either the shoot/root ratio or root weight ratio, is rather conserved. Within a single species, single-plant S/R is much less variable than weight (Table 4.9). Although it may change three-fold during growth in controlled environments (Farrar and Gunn 1996) other features such as relative growth rate change far more. Even when measured in the field, the range of values of S/R is modest, typically being between 0.1 and 5. This 50-fold range is far smaller than the range of absolute plant size, or of relative growth rate. Thus Körner and Renhardt (1987) describe species from alpine grassland with a 34-fold range of biomass but only a 15-fold range of S/R, and Larcher (1995) describes communities with a 254-fold range of biomass but only a 63-fold range of S/R. Whilst we would not argue that differences in S/R are not of considerable interest, its relative constancy needs explaining.

Why should it be relatively conserved? Firstly, it is regulated, in the sense that removal of either part of the root or of the shoot of single plants results in regrowth to restore the S/R ratio (Brouwer 1981; Farrar and Gunn 1998; J.F. Farrar, unpubl.). Secondly, the value to which it is restored is not solely a function of the species and its stage of development: it also depends closely on the

Table 4.9. Variability of S/R in ryegrass vars. Perma (low carbohydrate) and Aurora (high carbohydrate); data expressed as coefficient of variation

	Weight		S/R	
	Fresh	Dry	Fresh	Dry
Perma	28	27	16	13
Aurora	29	28	16	14

environment. Classically, the S/R ratio reflects the relative abundance of different resources, being relatively larger when a resource utilised by the shoot (e.g. light) is in low supply relative to those utilised by the root, and being smaller when a resource utilised by the root – say N or P – is in low supply relative to light. Whilst this generalisation is not universally valid – atmospheric CO_2 and soil K are relatively ineffective at altering S/R, for example – it applies in many cases (Wilson 1989). The simplest explanation for these observations is that S/R is adjusted to maximise resource capture, and so represents a balance between the amount of root to acquire water and nutrients, and leaf to capture light and CO_2. It is therefore not surprising that S/R is normally conserved, since the relative requirement for these resources is rather constant. Adjustment to differential supply of resources from the environment can explain phenotypic changes in S/R.

Are whole shoot and root weights the best measures of allocation? If resource acquisition is indeed the main criterion for partitioning between root and shoot, then leaf area and fine root length may be the features that partitioning is aiming to balance. Suggestions to this effect (Körner 1994; Farrar and Gunn 1998) are supported by the growth of plants such as carrot, where allometry between shoot and root weights becomes non-linear once the taproot starts to form; constant allometry between leaf weight and fine root weight is maintained throughout the life of the plant (A. Korolev and J.F. Farrar, unpubl.).

If allocation to root relative to shoot is indeed an exercise in balanced resource acquisition, the mechanisms by which it is achieved are unlikely to be confined to those we have discussed so far. Rather, some measure of resource status within the plant needs to be monitored and partitioning between shoot and root adjusted accordingly. Sucrose may play this role (Farrar 1996) but it cannot do so alone; at least one sensor of resource acquired by the root is also needed. Nitrate can alter growth of roots, either generally (Wilson 1989) or locally (Drew 1975), and there are suggestions that it is sensed internally to alter expression of a MADS box gene when altering root growth (Zhang and Forde 1998).

4.8.2 The Case of Phosphate

The supply of phosphate is a determinant of S/R. Here we use it as an example of some of the issues involved in controlling the relative growth of the root by an environmental variable. When phosphate supply in the root environment is low, partitioning alters to favour the root at the expense of the shoot, and S/R falls; further the architecture of the root is altered (Drew 1975; Wilson 1989; Ciereszko et al. 1999). The phosphate is not detected on the outside of the root. Ramets of *Agrostis stolonifera*, grown in pairs joined by a stolon so that they can exchange P and with one of each pair in high and one in low phosphate, have

Table 4.10. S/R ratio and root architecture. Correlation coefficients with measures of plant P status of individual ramets of *Agrostis capillaris* grown in ramet pairs in all combinations of 1330 and 7 mmol P m^{-3} (R. Solbe, C. Marshall, J.F. Farrar, unpubl.)

	Shoot P content	Shoot P conc.	cyt P in young lf	cyt P in old lf	Root P conc.
S/R ratio (weight)	0.524	0.539	0.510	0.330	0.338
Root architecture (length/weight)	0.885	0.909	0.852	0.726	0.881

S/R ratios intermediate between ramets grown singly in high or low P (R. Solbe, C. Marshall and J.F. Farrar, unpubl.). Therefore, some measure of the phosphorus status of plant tissue must determine S/R. We know neither exactly where the P is detected, nor the chemical form, but in the same system of *Agrostis* with paired ramets and heterogeneous ramets and heterogeneous P supply, S/R and root architecture are better correlated with P concentration in the shoot than that in the root (Table 4.10). Further, these measures of response to P correlate well with cytosolic rather than total P, and with cytosolic P in young leaves. In young leaves of *Agrostis stolonifera* of greatly differing P status, once the P content of the cytosol is above c. 10 nmol g^{-1} leaf DW, extra P accumulates as Pi in vacuoles (R. Solbe, C. Marshall and Farrar, unpubl.); below that concentration, presumably P deficiency can be sensed. Thus we suggest that cytosolic P in young leaves is a strong candidate for the regulator of S/R in response to altered P supply. This makes biological sense: it is likely to be young leaves that are most sensitive to a reduced supply of P, since, although P is recycled internally, the fluxes involved are modest.

How might reduced P status in the cytosol of a young leaf be translated into an increased fraction of assimilate being partitioned to the root? We have no idea. The problem is this: in the long term, C is directed preferentially to the root when both major sinks – shoot apex and roots – are of lower P status. Thus it cannot be just that low tissue P is a trigger for more C import, but we do know that P status of both leaf (Zulu et al. 1991) and root (Ciereszko et al. 1999) is reflected in altered amount, fluxes and metabolism of carbohydrates. Thus the phenomenon could be sugar-mediated, but we badly need critical experiments.

4.8.3 Functional Equilibrium

The conventional view of how partitioning between root and shoot is a function of resource acquisition is summarised by the functional equilibrium hypothesis (Thornley 1977; Brouwer 1983; Hunt et al. 1990). It states that both shoot and root acquire different resources but that each needs both resources in a relatively stable ratio. Therefore, growth of the root depends on the provision of C from the shoot as well as N, and so, as it gets larger, a larger shoot is needed to provide sufficient C. Growth of shoot and root are thus inseparably linked by this functional equilibrium. This hypothesis is successful at the level of description (Brouwer 1962, 1983; Wilson 1989) when light, water, nitrogen, and phosphorus supplies vary. It is not wholly satisfactory even at the descriptive level – for example, it predicts that a shortage of K should favour root growth, but this is not a universal finding (Wilson 1989), and that increased atmospheric CO_2 will favour root growth, although in carefully conducted experiments this ratio remains unaltered (Farrar and Gunn 1996; Gunn et al. 1999a). The functional equilibrium hypothesis has not been tested in swards or communities, although it might be expected to apply (Tilman 1988; Grime 1994). However, its major limitation is that it is not truly mechanistic. It describes the allocation of net weight, not gross carbon, so differential losses by respiration are ignored. It proposes that carbon is retained in the shoot unless in excess, when it is exported to the root; but most carbon is fixed in mature leaves and exported from them, and only then partitioned between shoot and root – so its focus on export rather than partitioning is misleading. We believe it is time to abandon this hypothesis in favour of ideas based on a better mechanistic understanding of partitioning.

4.9 Summary and Prospects

The growth of roots is dominated by the import of carbohydrates from the shoot, and the subsequent metabolism and associated respiration of those carbohydrates. A small but significant proportion of C entering the root is lost by exudation, a process that may have more adaptive significance than has sometimes been assumed. Uptake of organic molecules – particularly sugars and amino acids – from the soil solution may make a significant contribution to root C and plant N balance in some soils.

We have shown that the import of carbon by, and thus growth of, roots is a complex property. The component processes that determine import and growth are not confined to the root: carbon metabolism and transport in source leaves and stem are also important. The concepts of metabolic control analysis provide the underlying theory for the idea that control of carbon flux is shared by many processes in all parts of the plant. Two levels of mechanism, one short-term regulation of phloem transport and the other control of gene expression by compounds such as sugars, underlie shared control. Many facets of our current knowledge of root physiology can be re-interpreted in these terms.

The partitioning of carbon and dry weight between root and shoot is rather tightly regulated and conserved. Although flexible in response to environmental variables, S/R ratio can take only a limited range of values, in order that the plant acquires resources from the aerial and soil environments in the balance required for continued growth. The future of understanding this partitioning, and of root growth and C import, lies in a full mechanistic understanding of the component processes.

So what are the prospects? We need to understand the control of fluxes much better, and to understand the significance of exudation in a natural setting. We need to explore the ecological consequences of a conserved S/R and the mechanisms of its regulation in response to key environmental variables such as nitrogen. The key to advances will be the adoption of a broad approach, with questions derived from roots functioning in their natural environment, rigorously controlled experiments aimed at a reductionist understanding of mechanism, and the use of functional genomics where it is appropriate.

Acknowledgements. We would like to acknowledge the financial support of the BBSRC, the NERC, and DFID.

References

Amthor JS (1989) Respiration and Crop Productivity. Springer, Berlin Heidelberg New York

Atkin OK, Edwards EJ, Loveys BR (2000) Response of root respiration to changes in temperature and its relevance to global warming. New Phytol 147:141–53

Atkinson CJ and Farrar JF 1983 Allocation of photosynthetically fixed carbon in *Festuca ovina* and *Nardus stricta*. New Phytol 95:5519-5531Bacic A, Moody S, Clarke AE (1986) Structural analysis of secreted root slime from maize (*Zea mays* L.). Plant Physiol 80:771–777

Barber SA (1984) Soil nutrient bioavailability: a mechanistic approach. Wiley Interscience, New York

Bauermeister A, Dale JE, Williams EJ, Scobie J (1980) Movement of ^{14}C- and ^{11}C-labelled assimilate in wheat leaves: the effect of IAA. J Exp Bot 31:1199–1209

Beck E (1996) Regulation of shoot/root ratio by cytokinins from roots in *Urtica dioica*. Plant Soil 185:3–12

Bell CI, Milford GFJ, Leigh RA (1996) Sugar beet. In: Zamski E, Schaffer AA (eds) Photoassimilate distribution in plants and crops. Dekker, New York, pp 691–707

Bingham IJ, Farrar JF (1988) Regulation of respiration in roots of barley. Physiol Plant 70:491–498

Bret-Harte MS, Silk WK (1994) Nonvascular symplastic diffusion of sucrose cannot satisfy the carbon demands of growth in the primary root tip of *Zea mays*. Plant Physiol 105:19–33

Brouwer R (1962) Nutritive influences on the distribution of dry matter in the plant. Neth J Agric Sci 10:399–408

Brouwer H (1981) Functional equilibrium: sense or nonsense? Neth J Agric Sci 31:335–348

Cakmak I, Marschner H (1988) Increase in membrane permeability and exudation in roots of zinc deficient plants. J Plant Physiol 132:356–361

Canny MJ (1973) Phloem translocation. Cambridge University Press, Cambridge

Canny MJ (1990) What becomes of the transpiration stream? New Phytol 114:341–368
Chapin FS, Moilanen L, Kielland K (1993) Preferential use of organic nitrogen for growth by a nonmycorrhizal arctic sedge. Nature 361:150-153
Chapleo S, Hall JL (1989) Sugar unloading in roots of *Ricinus communis*. New Phytol 111:381–390
Ciereszko I, Farrar JF, Rychter A (1999) Compartmentation and fluxes of sugars in roots of *Phaseolus vulgaris* under phosphate deficiency. Biol Plant 42:223–231
Cook MG, Evans LT (1978) Effect of relative size and distance of competing sinks on the distribution of photosynthetic assimilate in wheat. Aust J Plant Physiol 5:495–509
Curl EA, Trueglove B (1986) The rhizosphere. Advanced series in agricultural science 15. Springer, Berlin Heidelberg New York
Daie J, Watts M, Aloni B, Wyse RE (1986) In vitro and in vivo modification of sugar transport and translocation in celery by phytohormones. Plant Sci 46:35–41
Dick PS, ap Rees T (1975) The pathway of sugar transport in roots of *Pisum sativum*. J Exp Bot 26:305–314
DiTomaso JM, Hart JJ, Kochian LV (1992) Transport kinetics and metabolism of exogenously applied putrescine in roots of intact maize seedlings. Plant Physiol 98:611–620
Drew MC (1975) Comparison of the effects of a localised supply of phosphate, nitrate, ammonium and potassium on the growth of the seminal root system and the shoot of barley. New Phytol 75:479–489
Farrar JF (1985) Fluxes of carbon in roots of barley plants. New Phytol 99:57–69
Farrar JF (1989) Fluxes and turnover of sucrose and fructans in healthy and diseased plants. J Plant Physiol 134:137–140
Farrar JF (1992) The whole plant: carbon partitioning during development. In: Pollock CJ, Farrar JF, Gordon AJ (eds) Carbon partitioning within and between organisms. Bios Scientific Publishers, Oxford, pp 163–179
Farrar JF (1996) Regulation of root weight ratio is mediated by sucrose: opinion. Plant Soil 185:13–19
Farrar JF (1999a) Carbohydrate: where does it come from, where does it go? In: Bryant JA, Burrell MM, Kruger NJ (eds) Plant carbohydrate biochemistry. Bios Scientific Publishers, Oxford, pp 29–46
Farrar JF (1999b) Acquisition, partitioning and loss of carbon. In: Press MC, Scholes JD (eds) Advances in plant physiological ecology. Blackwell, Oxford, pp 25–43
Farrar JF, Gunn S (1996) Effects of temperature and atmospheric carbon dioxide on source-sink relations in the context of climate change. In: Zamski E, Scheffer AA (eds) Photoassimilate distribution in plants and crops. Dekker, New York, pp 389–406
Farrar JF, Gunn S (1998) Allocation: allometry, acclimation – and alchemy? In Lambers H, Poorter H, Van Vuuren MMI (eds) Inherent variation in plant growth. Backhuys, Leiden, pp 183–198
Farrar JF, Jones CL (1985) Modification of respiration and carbohydrate status of barley roots by selective pruning. New Phytol 102:513–521
Farrar JF, Jones D (2000) The control of carbon acquisition by roots. New Phytol 147:43–53
Farrar JF, Lewis DH (1987) Nutrient relations in pathogenic and mutualistic infections. In: Ayres PG, Pegg GF (eds) Plant infecting fungi. Cambridge University Press, Cambridge, pp 92–132
Farrar JF, Minchin PEH (1991) Carbon partitioning in split root systems of barley: relation to metabolism. J Exp Bot 42:1261–1269
Farrar JF, Williams JHH (1990) Control of the rate of respiration in roots. In: Emes MJ (ed) Compartmentation of plant metabolism in non-photosynthetic tissues. Cambridge University Press, Cambridge, pp 167–188

Farrar JF, Minchin PEH, Thorpe MR (1995) Carbon import into barley roots: effects of sugars and relation to cell expansion. J Exp Bot 46:1859–1865
Farrar JF, Pollock CJ, Gallagher J (2000) Sucrose and the integration of metabolism in vascular plants. Plant Sci 154:1–11
Fell D (1997) Understanding the control of metabolism. Portland Press, London
Ferrari G, Renosto F (1972) Comparative studies on the active transport by excised roots of inbred and hybrid maize. J Agric Sci 79:105–108
Griffin GH, Hale MG, Shay FJ (1976) Nature and quantity of sloughed organic matter produced by roots of axenic peanut plants. Soil Biol Biochem 8:29–32
Grime JP (1994) The role of plasticity in exploiting environmental heterogeneity. In: Caldwell M, Pearcy R (eds) Exploitation of environmental heterogeneity in plants. Academic Press, New York, pp 1–19
Gunn S, Farrar JF (1999) Effects of a 4 °C increase in temperature on partitioning of leaf area and dry mass, root respiration and carbohydrates. Funct Ecol 13:12–20
Gunn S, Bailey SJ, Farrar JF (1999a) Partitioning of dry mass and leaf area within plants of three grown at elevated CO_2. Funct Ecol 13:3–11
Gunn S, Farrar JF, Collis BE, Nason M (1999b) Specific leaf area in barley: individual leaves versus whole plants. New Phytol 145:45–51
Haller T, Stolp H (1985) Quantitative estimation of root exudation of maize plants. Plant Soil 86:207–216
Hampp R, Wingler A (1997) The role of mycorrhiza. In: Foyer CH, Quick WP (eds) A molecular approach to primary metabolism in higher plants. Taylor and Francis, London, pp 275–291
Heineke D, Kauder F, Frommer W (1999) Application of transgenic plants in understanding responses to atmospheric change. Plant Cell Environ 22:623–628
Hellmann H, Barker L, Funch D, Frommer WB (2000) The regulation of assimilate allocation and transport. Aust J Plant Physiol 27:583–594
Hendry G (1987) The ecological significance of fructan in a native flora. New Phytol 106:201–216
Heulin T, Guckert A, Balandreau J (1987) Stimulation of root exudation of rice seedlings by Azospirillum strains: Carbon budget under gnotobiotic conditions. Biol Fert Soils 4:9–14
Ho LC (1978) The regulation of carbon transport and the carbon balance of mature tomato leaves. Ann Bot 42:155–164
Hoffland E, Findenegg GR, Nelemans JA (1989) Solubilization of rock phosphate by rape. 2. Local root exudation of organic acids as a response to P starvation. Plant Soil 113:161–165
Horst WJ, Wagner A, Marschner H (1982) Mucilage protects root meristems form aluminium injury. Z Pflanzenphysiol 105:435–444
Hungria M, Joseph CM, Phillips DA (1991) Rhizobium nod gene inducers exuded naturally from roots of common bean (*Phaseolus vulgaris* L.). Plant Physiol 97:759–764
Hunt R, Warren Wilson J, Hand DW (1990) Integrated analysis of resource capture and utilisation. Ann Bot 65:643–648
Iijima M, Griffiths B, Bengough AG (2000) Sloughing of cap cells and carbon exudation from maize seedling roots in compacted sand. New Phytol 145:477–482
Johnson JF, Allan DL, Vance CP, Weiblen G (1996) Root carbon dioxide fixation by phosphorus-deficient *Lupinus albus*: contribution to organic-acid exudation by proteoid roots. Plant Physiol 112:19–30
Jones DL, Darrah PR (1993) Re-sorption of organic compounds by roots of *Zea mays* L. and its consequences in the rhizosphere. II. Experimental and model evidence for simultaneous exudation and re-sorption of compounds. Plant Soil 153:47–59

Jones DL, Darrah PR (1994) Influx and efflux of amino acids from *Zea mays* L. roots and its implications in the rhizosphere and N nutrition. Plant Soil 163:1–12
Jones DL, Darrah PR (1996) Re-sorption of organic compounds by roots of *Zea mays* L. and its consequences in the rhizosphere. III. Spatial, kinetic and selectivity characteristics of sugar influx and the factors controlling efflux. Plant Soil 178:153–160
Jones DL, Hodge A (1999) Biodegradation kinetics and sorption reactions of three differently charged amino acids in soil and their effects on plant organic nitrogen availability. Soil Biol Biochem 31:1331-1342
Klein DA, Frederick BA, Redente EF (1989) Fertilizer effects on soil microbial communities and organic matter in the rhizosphere of *Sitanion hystrix* and *Agropyron* smithii. Arid Soil Res Rehabil 3:397–404
Koch KE (1996) Carbohydrate-modulated gene expression in plants. Annu Rev Plant Physiol Plant Mol Biol 47:509–540
Koch KE (1997) Molecular crosstalk and the regulation of C- and N-responsive genes. In: Foyer CH, Quick WP (eds) A molecular approach to primary metabolism in higher plants. Taylor and Francis, London, pp 105–124
Kochian LV (1995) Cellular mechanisms of aluminum toxicity and resistance in plants. Annu Rev Plant Physiol Plant Mol Biol 46:237–260
Körner C (1994) Biomass fractionations in plants: a reconsideration of definitions based on plant functions. In: Roy J, Garnier E (eds) A whole plant perspective on carbon-nitrogen interactions. SPB, The Hague, pp 173–185
Körner C, Renhardt U (1987) Dry matter partitioning and root length/leaf area ratios in herbaceous perennial plants with diverse altitudinal distribution. Oecologia 74:411–418
Korolev A, Tomos AD, Bowtell R, Farrar JF (1999) Spatial and temporal distribution of solutes in the carrot taproot measured at single-cell resolution. J Exp Bot 51:567–577
Korolev A, Tomos AD, Farrar JF (2000) The trans-tissue pathway and chemical fate of ^{14}C photoassimilate in carrot taproot. New Phytol 147:299–306
Kraffczyk I, Trolldenier G, Beringer H (1984) Soluble root exudates of maize: Influence of potassium supply and rhizosphere microorganisms. Soil Biol Biochem 16:315–322
Lambers H (1987) Growth, respiration, exudation and symbiotic association: the fate of carbon translocated to roots. In: Gregory PJ, Lake JV, Rose DA (eds) Root development and function. Cambridge University Press, Cambridge, pp 125–146
Lang A (1979) A relay mechanism for phloem translocation. Ann Bot 44:141–145
Larcher W (1995) Physiological plant ecology. Springer, Berlin Heidelberg New York
Lewis DH (1984) Storage carbohydrates in vascular plants. Cambridge University Press, Cambridge
Marcelis LFM (1996) Sink strength as a determinant of dry matter partitioning in the whole plant. J Exp Bot 47:1281–1292
Marschner H (1995) Mineral nutrition of higher pants. Academic Press, London
Martin JK (1977) Effect of moisture on the release of organic carbon from wheat roots. Soil Biol Biochem 9:303–304
McCully ME (1999) Roots in soil: unearthing the complexities of roots and their rhizospheres. Annu Rev Plant Physiol Plant Mol Biol 50:695–718
McDougall BM, Rovira AD (1970) Sites of exudation of ^{14}C-labelled compounds from wheat roots. New Phytol 69:999–1003
Meharg AA (1994) A critical review of labeling techniques used to quantify rhizosphere carbon-flow. Plant Soil 166:55–62
Milburn JA, Baker DA (1989) Transport of photoassimilates. Longman, London, pp 306–343
Minchin PEH, Thorpe MR, Farrar JF (1993) A simple mechanistic model of phloem transport which explains sink priority. J Exp Bot 44:947–955

Minchin PEH, Farrar JF, Thorpe MR (1994a) Partitioning of carbon in split root systems of barley: effect of temperature of the root. J Exp Bot 45:1103–1109

Minchin PEH, Thorpe MR, Farrar JF (1994b) Short-term control of root: shoot partitioning. J Exp Bot 45:615–622

Minchin PEH, Thorpe MR, Farrar JF (2002) Source-sink coupling in young barley plants and control of phloem loading. J Exp Bot 53:1671–1676

Moore BD, Cheng SH, Sims D, Seeman JR (1999) The biochemical and molecular basis for photosynthetic acclimation to elevated atmospheric CO_2 Plant Cell Environ 22:567–582

Morris DA (1996) Hormonal regulation of source-sink relationships. In : Zamski, E, Scahher AA (eds) Photoassimilate distribution in plants and crops. Dekker, New York, pp 441–466

Munns R, Cramer GR (1996) Is co-ordination of leaf and root growth mediated by abscisic acid? Plant Soil 185:33–49

Nagel OW, Konings H, Lambers H (2001) Growth rate and biomass partitioning of wild-type and low gibberellin tomato plants. Physiol Plant 111:33–39

Nobel PS (1991) Physiochemical and environmental plant physiology. Academic Press, London

Oparka KJ, Duckett CM, Prior DAM, Fisher DB(1994) Real-time imaging of phloem unloading in the root tips of *Arabidopsis*. Plant J 6:759–766

Ourry A, Gordon AJ, Macduff JH (1997) Nitrogen uptake and assimilation in roots and root nodules. In: Foyer CH, Quick WP (eds) A molecular approach to primary metabolism in higher plants. Taylor and Francis, London, pp 237–253

Papernik LA, Kochian LV (1997) Possible involvement of Al-induced electrical signals in Al tolerance in wheat. Plant Physiol 115:657–667

Patrick JW (1987) Are hormones involved in assimilate transport? In: Hoad GV (ed) Hormone action in plant development. Butterworth, London, pp 175–188

Patrick JW (1997) Phloem unloading: sieve element unloading and post-sieve element transport. Annu Rev Plant Physiol Plant Mol Biol 48:191–222

Patrick JW, Offler CE (1995) Post-sieve element transport of sucrose in developing seeds. Aust J Plant Physiol 22:681–702

Peters NK, Long SR (1988) Alfalfa root exudates and compounds which promote or inhibit induction of *Rhizobium meliloti* nodulation genes. Plant Physiol 88:396–400

Pritchard J (1998a) Aphid stylectomy reveals an osmotic step between sieve tube and cortical cells in barley roots. J Exp Bot 47:1519–1524

Pritchard J (1998b) Control of root growth: cell walls and turgor. In Lambers H, Poorter H, Van Vuuren MMI, (eds) Inherent variation in plant growth. Backhuys, Leiden, pp 21–39

Pritchard J, Winch S, Gould N (2000) Phloem water relations and root growth. Aust J Plant Physiol 27:539–548

Reddy MN, Pokojska A, Kampert M, Strzelczyk E (1989) Auxin, gibberelin-like substances and cytokinins in the seed and root exudates of groundnut. Plant Soil 113:283–286

Römheld V (1991) The role of phytosiderophores in the acquisition of iron and other micronutrients in graminaceous species. An ecological approach. Plant Soil 130:127–134

Ryan PR, Delhaize E, Randall PJ (1995) Characterization of Al-stimulated efflux of malate from the apices of Al-tolerant wheat roots. Planta 196:103–110

Saglio PH, Pradet A (1980) Soluble sugars, respiration, and energy charge during ageing of excised maize root tips. Plant Physiol 66:516–519

Schonwitz R, Ziegler H (1982) Exudation of water soluble vitamins and some carbohydrates by intact roots of maize seedlings (*Zea mays* L.) into a mineral nutrient solution. Z Pflanzenphysiol 107:7–14

Smeekens S (2000) Sugar-induced signal transduction in plants. Annu Rev Plant Physiol Pantl Mol Biol 51:49–82

Smucker AJM, Erickson AE (1987) Anaerobic stimulation of root exudates and disease in peas. Plant Soil 99:423–433

Soldal T, Nissen P (1978) Multiphasic uptake of amino acids by barley roots. Physiol Plant 43:181–188

Stitt M (1996) Metabolic regulation of photosynthesis In Baker NR (ed) Photosynthesis and the environment. Kluwer, Dordrecht, pp 151–190

Svenningsson H, Sundin P, Lljenberg C (1990) Lipids, carbohydrates and amino acids exuded from the axenic roots of rape seedlings exposed to water deficit stress. Plant Cell Environ 13:155–162

Sweetlove LJ, Kossmann J, Riesmeier JW, Trethewey RN, Hill SA (1998) The control of source to sink carbon flux during tuber development in potato. Plant J 15:697–706

Thomas TH. (1986) Hormonal control of assimilate movement and compartmentation. In: Bopp MP (ed) Plant growth substances 1985. Springer, Berlin Heidelberg New York, pp 350–359

Thornley JHM (1977) Root:shoot interactions. In: Jennings DH (ed) Integration of activity in the higher plant. University of Cambridge Press, Cambridge, pp 367–389

Tilman DT (1988) Plant strategies and the dynamics and structure of plant communities. Princeton, New Jersey

Tomos AD, Pritchard J (1994) Biophysical and biochemical control of cell expansion in roots and leaves. J Exp Bot 45:1721–1731

Tomos AD, Korolev A, Farrar J, Nicolay K, Bowtell R, Kockenberger W (2000) Water and solute relations of the carrot cambium studied at single-cell resolution. In: Savage R, Barnett J, Napier R (eds) Cell and molecular biology of wood formation. Bios Scientific Publishers, Oxford, pp 101–112

van Bel AJE (1993) Strategies of phloem loading. Annu Rev Plant Physiol Plant Mol Biol 44:253–281

van Bel AJE, Knoblauch M (2000) Sieve element and companion cell. Aust J Plant Physiol 27:477–487

Vancura V, Stotsky G (1976) Gaseous and volatile exudates from germinating seeds and seedlings. Can J Bot 54:518–532

Warmbrodt RD (1987) Solute concentration in the phloem and associated vascular and ground tissue of the roots of *Hordeum vulgare.* In: Cronshaw J, Lucas WJ, Giaquinta RT (eds) Phloem transport. Liss, New York, pp 435–444

Whipps JM (1990) Carbon economy. In: Lynch JM (ed) The rhizosphere. Wiley Interscience, New York, pp 59–98

Williams JHH, Farrar JF (1990) Control of barley root respiration. Physiol Plant 79:259–266

Wilson JB (1989) A review of evidence on the control of shoot:root ratio, in relation to models. Ann Bot 61:433–449

Wyse RE, Zamski E, Tomos AD (1986) Turgor regulation of sucrose transport in sugar beet taproot tissue. Plant Physiol 81:478–481

Xia JH, Saglio PH (1988) Characterisation of the hexose transport system in maize root tips. Plant Physiol 88:1015–1020

Xia JH, Saglio PH (1992) Lactic acid efflux as a mechanism of hypoxic acclimation of maize root tips to anoxia. Plant Physiol 100:40–46

Zhang H, Forde BG (1998) An *Arabidopsis* MADS box gene that controls nutrient-induced changes in root architecture. Science 279:407–409

Zulu JN, Farrar JF, Whitbread R (1991) Effects of phosphate supply on wheat seedlings infected with powdery mildew: carbohydrate metabolism of first leaves. New Phytol 118:553–558

5 Hydraulic Properties of Roots

M.T. Tyree

Symbols

A_r	Surface area of root (m^{-2})
C	Concentration of solutes (osmolal)
$\overline{C}_s$	Average solute concentration between inside and outside of root
d	Diameter
D	Diffusion coefficient ($m^2\ s^{-1}$)
F_m	Mass flow of solution ($kg\ s^{-1}$)
H_r	Classical root hydraulic conductance ($kg\ s^{-1}\ MPa^{-1}$)
J_r	Mass flux density or F_m per unit root surface area ($kg\ s^{-1}\ m^{-2}$)
J_s	Solute flux density ($mol\ s^{-1}\ m^{-2}$)
K_r	Thermodynamic root hydraulic conductance ($kg\ s^{-1}\ MPa^{-1}$)
ln(2)	natural log of 2
L	Length (m)
L_r	Thermodynamic root hydraulic conductance per unit root surface area ($kg\ s^{-1}\ m^{-2}\ MPa^{-1}$)
P	Pressure (MPa)
P_s	Solute permeability
RT	Gas constant times Kelvin temperature ($Mpa\ osmolal^{-1}$)
T_p	Time constant for pressure relaxation (s)
V	Volume
T_{s1} and T_{s2}	Time constants for pressure relaxation after changing outside solute concentration

Ecological Studies, Vol. 168
H. de Kroon, E.J.W. Visser (Eds.)
Root Ecology

α	Fraction of water root surface area occupied by water in apoplast
δ	Path length (m)
Δ	Difference
π	Osmotic potential (MPa) except in Eq. (5.8) where it denotes pi=3.14159...
Ψ	Water potential (MPa)
η	Viscosity of water
σ	Reflection coefficient

Subscripts: *i* inside; *o* outside; *r* root; *s* solute; *L* leaf; *sh* shoot

5.1 Introduction

The primary function of roots is to provide water and solute (primarily mineral ion) transport from the soil to the shoots of plants. Water uptake is ecologically important because it is the universal solvent for all biochemical reactions. Also, water uptake is constantly needed to replace water loss by transpiration. Transpiration is a necessary ecological cost of gas exchange for carbon gain. Rather little is known about the pathway of water movement in roots and the mechanism of water uptake. Even less is known about the comparative ecophysiology of water uptake between species. A few instances are discussed in Section 5.6 of this chapter and rather more can be learned in a recent review of the ecological aspects of root permeability to water (Nardini et al. 2002). Hence, this chapter will concentrate primarily on the mechanism and pathway of water movement in roots.

Water transport across roots is generally described as being purely passive, i.e., it involves no input of metabolic energy. Water is not taken up actively, but instead moves passively through the root in response to a 'force' set up by transpiration. Flow is assumed to be proportional to the force. The force involved is usually equated with the difference in water potential ($\Delta\Psi$) across the root, which is approximately true for high flow rates, but becomes increasingly inaccurate at low flows. This is in contrast to ion uptake by cells, which often involves an active process requiring the input of metabolic energy and the mediation of ion pumps located in membranes.

In this chapter, we will review what is known about the biophysics of water transport across roots, i.e., the equations that describe transport and how the equations relate to root anatomy. Along the way we will review the current methods used to measure root conductance, how solute and water transport are coupled and the experimental approaches used so far to quantify the distribution of hydraulic resistances along the water transport pathway, i.e., radially from the root surface to the stele and then axially from the stele to the shoot. The chapter will conclude with a discussion on how to scale root conductance measurements to plant size illustrated with some results and ecological interpretations.

5.2 Root Structure and Possible Pathways of Water Movement

The 'typical' structure of monocot and dicot roots is illustrated in Fig. 5.1. The transport properties of roots cannot be interpreted without a clear understanding of their structure. However, root anatomy is highly variable because major differences are seen between species and between roots of the same species grown in different habitats. There are also differences along the length of an individual root. Common examples for such differences are the formation of aerenchyma and the development of endo- and exodermis, usually with Casparian bands, suberin lamellae, and thickened, modified walls. Roots are also altered by the death of the epidermis or even the entire central cortex or by the development of bark and lateral roots. This means that the knowledge gained on one root is not easily generalized to all roots.

The only good example of a study that correlates structure with the physiology of water and solute uptake is the work done on maize roots grown in hydroponic culture (Peterson and Steudle 1993; Peterson et al. 1993; Steudle et al. 1993). Maize roots are characterized by having a living epidermis and central cortex, an immature exodermis, mature proto- and early metaxylem,

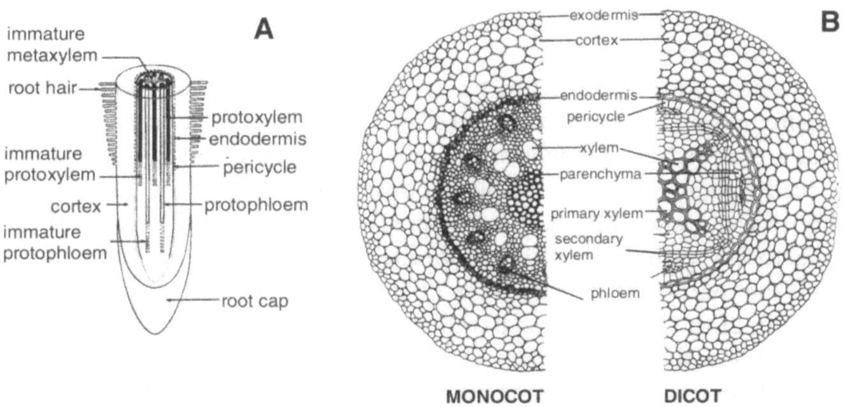

Fig. 5.1A,B. **A** Enlargement of a dicot root tip about 0.6 mm basal diameter. **B** Cross section of monocot and dicot roots. (Adapted from Tyree 1999)

immature late metaxylem (i.e., the vessels still have end walls and living cytoplasm), and an endodermis with Casparian bands, but no suberin lamellae or thickened walls (Fig. 5.2). Towards the tip of the roots there is a hydraulically isolated zone with immature metaxylem and a mature but non-functioning protoxylem.

Many people presume that the Casparian bands or suberin lamellae prevent water moving from epidermis to xylem entirely through cell walls, i.e., forcing water and/or solutes at some point to traverse at least one cell layer by a transcellular pathway en route to the stele. The stele is defined as the morphologic unit of the root consisting of the vascular system and the associated ground tissue – pericycle, interfascicular tissue, and pith (Esau 1960). However, even in the relatively simple and unthickened cortex, it is unclear what the pathway is for water and solute flow. There are three possible pathways: apoplastic, symplastic and transcellular (Fig. 5.3). The part of any plant tissue outside the plasma membrane of the living cells is termed apoplast. It includes cell walls, intercellular spaces, and the lumina of dead cells (e.g., mature vessels and tracheids or dead fiber cells). The symplast is the continuum of cytoplasm interconnected by plasmodesmata and excluding the vacuoles. The terms apoplastic and symplastic transport refer to movements within the two compartments previously defined. While this may be a reasonable definition for ion transport it may not be sufficient for water transport because water permeability across membranes is several orders of magnitude more than for ions. So a third path for water flow (the transcellular path) can be defined as one in which water moves across membranes to get from cell to cell. Two plasma membranes would have to be crossed per cell layer as well as the short distance of wall space between adjacent cells, which is normally presumed not to be rate-limiting. Although we define three pathways, there could be a combination of paths. For example, water flow might be 30 % apoplastic and 70 % transcellular for the whole root radius or the pathways may vary depending on radial position so flow may start out in the symplast for some distance then might pass through a plasma membrane and move within the cell wall for the rest of the path.

Hence, water and solute transport in roots must be viewed as going though a composite membrane consisting of cell walls, membranes and plasmodesmata that are spatially arranged into parallel and serial pathways. Each component (cell wall, membrane and plasmodesmata) will have a different permeability to water and solutes. If we define the whole root annulus from the epidermis to the vessels to be the 'membrane' limiting water and solute transport, then the permeability properties of the root membrane will be a complex function of the sum of these parallel and serial components. A complete discussion of how the transport parameters of a composite membrane relates quantitatively to its components is beyond the scope of this chapter, but such literature can be accessed through Kedem et al. (1962). At this point we do not have enough information about component properties to use existing theory.

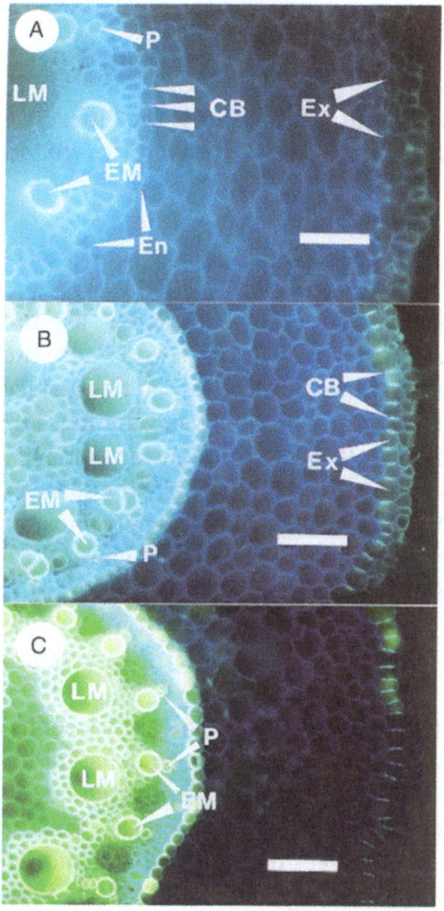

Fig. 5.2A–C. Cross sections of maize primary roots at various stages of development. Specimens were photographed under UV/violet light (wavelength 390–420 nm) following staining with berberine and either aniline blue or toluidine blue O. *Scale bars* 100 μm. **A** Immature exodermis (*Ex*), endodermis (*En*) with Casparian bands (*CB*), mature protoxylem (*P*), mature early metaxylem (*EM*), immature late metaxylem (*LM*). **B** Mature exodermis, endodermis with asymmetrically thickened walls, mature protoxylem and early metaxylem, immature late metaxylem. **C** As in **B**, but late metaxylem (*LM*) is mature. Walls of cells surrounding the late metaxylem are thickened and lignified. (Steudle and Peterson 1998)

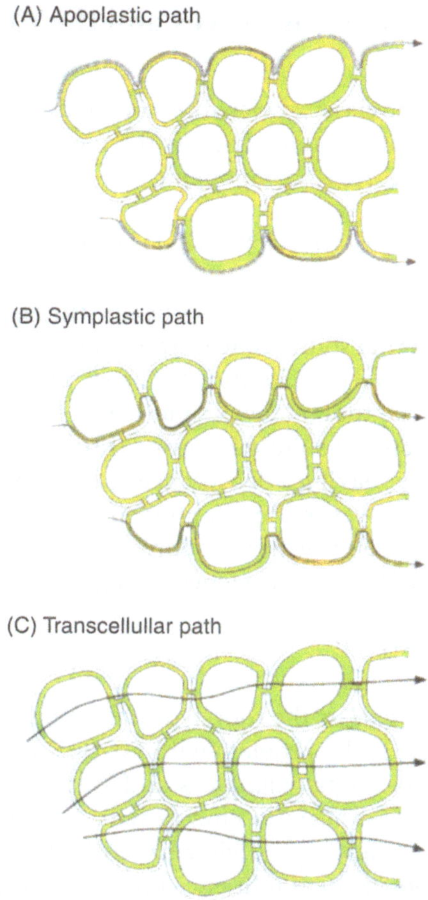

Fig. 5.3A–C. Routes of water flow in plant tissue. The tissue is represented by four cell layers arranged in series. **A** Denotes the apoplastic path (cell walls, *grey*) around protoplasts. **B** The symplastic path is mediated by plasmodesmata, which bridge the cell walls between adjacent cells so that a cytoplasmic continuum is formed (*green*). During the passage of the apoplast and symplast, no membranes have to be crossed. **C** In the transcellular path, two plasma membranes have to be crossed per cell layer. Due to the rapid water exchange between protoplasts and adjacent apoplast, there should be local water flow equilibrium between the two compartments at any time. In the root, the apoplastic flow component is modified by the existence of apoplastic barriers (Casparian bands, suberin lamellae). (Steudle and Peterson 1998)

5.3 Driving Forces and the 'Composite Membrane'

In the classical view, the force acting on water entry into roots is the difference in water potential across the root, $\Delta\Psi$, which is the sum of its osmotic (π) and pressure (P) components:

$$\Delta\Psi=\Delta\pi+\Delta P\cong -RT(C_i-C_o)+(P_i-P_o) \tag{5.1}$$

where RT is the gas constant times absolute temperature, C_i and P_i are the internal solute concentration and pressure, respectively, and C_o and P_o are the external (soil solution) concentration and pressure, respectively, at the surface of the root. Many people seek a functional relationship between the mass or volume of solution passing through roots and $\Delta\Psi$. Since the solution is mostly water having a density near 1 kg/l, mass flow rate in kg/s is nearly the same as l/s. However, the relationship between $\Delta\Psi$ and mass flow (F_m, where most of the mass is water) is more complicated than given by the classical view.

In the classical view, F_m (kg s^{-1} or m^3 s^{-1}) is proportional to $\Delta\Psi$, i.e.,

$$F_m=H_r\Delta\Psi \tag{5.2}$$

where H_r is the root hydraulic conductance (a root water permeability coefficient). The problem with Eq. (5.2) is that H_r was found not to be constant for some roots (e.g., Mees and Weatherley 1957; see Sect. 5.5.) In particular, H_r often appears to change with F_m. Irreversible thermodynamics also tells us that the classical view has to be modified to reflect the fact that $\Delta\pi$ and ΔP have unequal influence on F_m in many situations, e.g., a $\Delta\pi$ of 0.1 MPa might not cause the same flow as a ΔP of 0.1 MPa. One approach to deal with the problem is to introduce another transport constant called the reflection coefficient (σ) which is a number generally between 0 and 1 that measures the relative impact of $\Delta\pi$ versus ΔP on F_m and indicates the impermeability of a membrane to solutes. This permits us to improve the classical equation:

$$F_m=K_r(\sigma\Delta\pi+\Delta P)=K_r[-\sigma RT(C_o-C_i)+(P_o-P_i)] \tag{5.3}$$

where K_r is a different kind of root conductance, which unlike H_r is more likely to be a constant, i.e., independent of the driving forces of $\Delta\pi$ and ΔP. Strictly speaking σ has a different value for each solute interaction with any given 'membrane', but in many experiments only one solute is varied at a time so Eq. (5.3) can be used as a first approximation. Equation (5.3) is more frequently written in terms of flux density, i.e., flow per unit surface area, $J_v=F_m/A_r$, where A_r is the root surface area over which flow is presumed to occur.

$$J_v=L_r(\sigma\Delta\pi+\Delta P)=L_r[-\sigma RT(C_o-C_i)+(P_o-P_i)] \tag{5.4}$$

where $L_r=K_r/A_r$ is a root conductance per unit root-surface area.

The classical equation for solute flux (J_s) in a root can be written as:

$$J_s \cong P_s(C_o-C_i)+J^*_s \tag{5.5a}$$

where J^*_s is the active solute uptake flux density and P_s is the passive permeability of the root to the solute. In many situations Eq. (5.5a) is sufficiently accurate, but in some cases we need to take into account the coupling between flux of water (J_v) and solute so the equation from irreversible thermodynamics is:

$$J_s = \overline{C}_s\,(1-\sigma)J_v + P_s\,(C_o - C_i) + J_s^* \tag{5.5b}$$

and $\overline{C}_s$ is the average concentration of the solute in the root 'membrane'. The first term in Eq. (5.5b) accounts for the influence of water flux density (J_v) on solute flux, the second term accounts for passive diffusion, and the last for active transport.

However, mass flow (F_m) can also be viewed as 'quasi-active' because water flow is coupled to solute flow in Eq. (5.3). This follows because the rate of active solute uptake into a root can change the value of C_i. The coupling between water and solute flux is most when ΔP is small, i.e., usually at times of low transpiration. The situation is even more complex when we view the anatomical details of the pathway of water and solute movement in roots. The different anatomical components of roots provide different pathways for solute and water flow with each pathway having potentially different transport permeabilities. Hence, Steudle has argued that we must view roots as composite membranes (Steudle et al. 1993; Steudle and Frensch 1996; Steudle and Peterson 1998).

Some information can be gained about the role of cell types in the transport of water and solutes in roots by measuring transport of water and solutes in roots before and after damaging different cells or regions of roots. However, before entering into a detailed discussion of what has been learned about maize roots, we need to digress to discuss methods used for measuring water and/or solute transport in root systems.

5.4 Methods of Measuring Hydraulic Conductances

A number of different methods have been used in the past for measuring solute and water transport in excised root systems. In most instances, roots were measured while immersed in aqueous solution so that the experimenter had better control of the solute composition and concentration. Roots were grown either in hydroponic culture or in soils and extracted to the aqueous solutions in the measuring apparatus.

5.4.1 Root Chamber Methods

Fiscus (1977) was one of the major users of the root chamber method for measuring hydraulic properties of roots. This method is generally used for measuring steady-state fluxes of water and solute on large root systems. A root system is enclosed in a metal chamber. The root medium is generally an aqueous solution of known composition, but it can also be a soil. The pressure of the root medium can be adjusted by connecting the root chamber to a compressed air source (Fig. 5.4A). The root system is excised together with a length of stem, which passes out of the root chamber via a rubber seal. The stem is connected to water-filled pipes and valves, which in turn are connected to pressure sensors or flow sensors. Many roots will exude solution and the rate of exudation can be measured by opening the valve to the flow sensor. Exudation occurs because the root accumulates solutes increasing C_i above C_o in Eq. (5.3) causing a flow even when $P_i=P_o$. By measuring the flow rate of exudation (F_m) and the concentration of a given solute in the exudates (C_{si}, mol kg^{-1}) the rate of solute uptake (F_s, mol s^{-1}) can be estimated under steady-state conditions from $F_s=F_m C_{si}$. F_m is usually measured by directing flow to a balance and measuring the weight of root exudates at fixed time intervals. The root pressure under zero flow can also be measured by closing the valve to the flow sensor. Flow rate can also be measured as a function of (P_o-P_i) because the value of P_o can be changed by admitting gas into the chamber from the compressed air source (Fig. 5.5A).

5.4.2 Nobel Method

The Nobel method is a very simple technique that has been used on small root systems extracted from soil; the method is used to estimate steady-state solution flow through roots resulting from an imposed pressure drop (P_o-P_i). A root is sealed to a capillary tube and immersed in solution. A partial vacuum of 10–50 kPa is drawn from the end of the capillary tube, which induces water uptake. Solution flow through the root is computed by noting the rate of advance of the air/water interface in the capillary tube. A plot of flow versus pressure drop is similar to that in Fig. 5.5A. The same method can be used to estimate the vascular resistance to water flow by mounting a root segment in the capillary tube rather than a whole root. A possible disadvantage of the Nobel method is that flow rates and applied pressure drop are smaller than

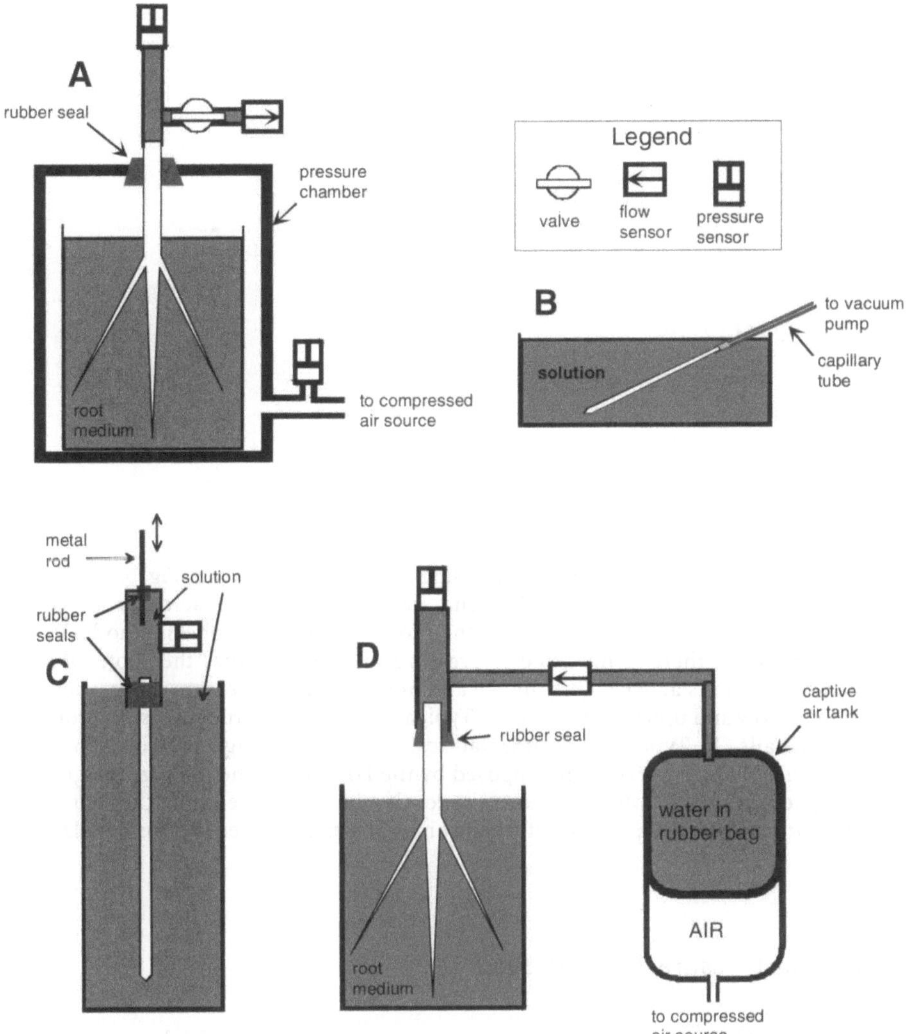

Fig. 5.4A–D. Methods for measuring root transport properties. **A** Root chamber method. **B** Nobel method. **C** Root pressure probe (RPP) method. **D** High-pressure flowmeter (HPFM) method. See text for details

normal physiological ranges, hence the data could be influenced by osmotic contributions to flow in plants with high rates of active uptake of solutes (see Eq. 5.4).

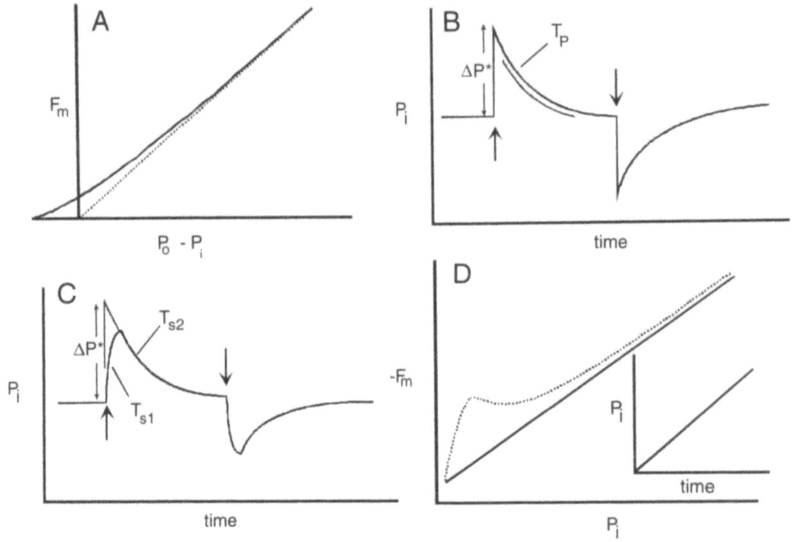

Fig. 5.5A–D. Examples of experimental results from the methods in Fig. 5.4. **A** Typical relationship between mass flow (F_m) and pressure difference across root (P_o–P_i) obtained for steady-state experiments by the root chamber and Nobel methods. *Dashed line* Results when there is no osmotic pressure difference across the root; *solid line* results when there is active solute uptake and hence osmotic pressure differences are a function of flow and uptake rate. **B** and **C** Typical results for the root pressure probe. See text for details. **D** Typical experimental results with the high-pressure flowmeter (HPFM). *Inset* Flow P_i versus time imposed by the HPFM and the main graph gives the resulting flow versus P_i. *Solid line* shows the result with no air present in the root system and the *dotted line* is a typical result when some air is present. See text for details

5.4.3 Root Pressure Probe Method

Steudle (1993) developed the root pressure probe (RPP), which is a variation on a cell pressure probe (Fig. 5.4C). The RPP uses a pressure relaxation method (dynamic method) for measuring solute and water transport parameters of root systems. The RPP has the advantage over all other methods because it can be used to measure all the important parameters for solute and water transport, i.e., L_r, σ, and P_s (see Eqs. 5.4 and 5.5). The RPP works best with small root systems but has also been used with larger, branched root systems of relatively small seedlings. A root is excised and the basal end is sealed into a plastic chamber filled with solution. The pressure of the fluid is measured with a pressure sensor and the pressure can be dynamically altered either by moving a metal rod, also sealed into the chamber, or by changing the solution concentration in the external solution (C_o).

The root reaches a stable internal pressure (P_i) after it has been mounted in the pressure probe for several hours. A typical pressure relaxation experiment

is illustrated in Fig. 5.5B. At the up-arrow the metal rod is rapidly advanced into the RPP displacing a known volume of solution, ΔV, which causes an initial pressure increase, ΔP^*. The ratio $\Delta P^*/\Delta V$ is a measure of the absolute elasticity of the root plus pressure probe. The increase in pressure immediately causes an efflux of water and a gradual relaxation of the pressure. By analysis of the relaxation curve, the value of L_r can be determined provided the absolute elasticity is a constant, i.e., provided ΔP^* is a linear function of ΔV during the pressure relaxation. If air bubbles are present in the root (either in vessels or intercellular spaces) this requirement of constant elasticity is not met because some of the volume displacement of the rod goes to compress the bubbles and pressure of the bubbles is inversely proportional to the volume (based on the ideal gas law). Hence the root system has to remain under pressure for many hours to dissolve all the air bubbles prior to measurement. If the metal rod is withdrawn rapidly from the RPP (down-arrow, Fig. 5.5B), the pressure change and relaxation is in the opposite direction.

The relaxation curve has a half-time, T_p, which describes the rate at which ΔP approaches zero, and the root conductance, L_r, is calculated from:

$$L_r = \frac{\Delta V}{\Delta P^*} \frac{T_P}{A_r \ln(2)} \tag{5.6}$$

where A_r is the root surface area. Equation (5.6) is valid only for an exponential decay process. Generally the shape of the relaxation curve is not a true exponential decay of ΔP, but the middle portion of the curve (highlighted by dots in Fig. 5.5B) is approximately exponential. The computation of L_r of Eq. (5.6) has been validated by independent measurements of L_r.

The pressure changes in Fig. 5.5C can also be induced by rapid changes in C_{so} of a solute in the solution bathing the root. If C_{so} is changed from an initially high to low value (up-arrow, Fig. 5.5C) the pressure increases. The time delay for the increase in pressure, T_{s1}, is presumed (by the author, see Sect. 5.5) to be due to the time for the solute to diffuse through unstirred layers and root tissue to reach the solute permeability 'membrane', which is presumed to be the endodermis. The second time delay (half-time, T_{s2}) is governed by the time it takes the solute to permeate from outside the root to the xylem conduits. The permeability of the root to the solute is computed from:

$$P_{sr} = \frac{V_x}{A_r} \frac{T_{s2}}{\ln(2)} \tag{5.7}$$

where V_x is the volume of the xylem conduits.

The reflection coefficient can be calculated from the initial change in pressure and concentration, $\sigma = \Delta P^*/RT\Delta C_{so}$; when $T_{s1} << T_{s2}$ the value of ΔP^* is easily evaluated, but when T_{s1} and T_{s2} are more nearly equal then a small correction has to be made for the probable ΔP^* by extrapolating the T_{s2} curve back to 'time zero' when the solution concentration was changed (as shown in Fig. 5.5C).

When C_{so} is increased from a low to a high value the initial pressure change and relaxation will be in the opposite direction (down-arrow, Fig. 5.5C).

5.4.4 The High-Pressure Flowmeter Method

The high-pressure flowmeter (HPFM) method was developed by Tyree et al. (1995) to provide dynamic and steady-state measurements of root and shoot hydraulic conductance. In roots it is best to use the dynamic method (Tyree et al. 1994). The HPFM method can be used on much larger root systems than the RPP method. There is some overlap in measurable root sizes by both methods, i.e., the smaller root sizes measurable by the HPFM overlap with the larger root sizes measurable by the RPP. The HPFM method has been calibrated versus the pressure chamber method and both methods produce comparable values of K_r (Eq. 5.3).

In the HPFM method a root system in solution or soil is excised from the shoot and connected to the HPFM via a rubber seal (Fig. 5.4D). The pressure (P_i) is measured above the excised root as well as the flow into the root. The value of P_i is controlled by adjusting the air pressure in a captive air tank (CAT) where water is held in a rubber bag inside the CAT. As P_i increases water flows into the base of the root and exits through the root surfaces, so flow is opposite to the direction during transpiration. In a typical dynamic measurement, air is admitted into the CAT at a constant rate causing a linear increase in P_i versus time (Fig. 5.5D, insert). Typically, the pressure is increased from 0 to 0.5 MPa. If no air is present in the root, a plot of F_m vs. P_i is linear (Fig. 5.5D, solid line) with a non-zero y-intercept caused by the elasticity of the root plus HPFM. The slope of the line equals K_r.

Generally, there is some air present in the root being measured, but the accuracy of K_r determined by the HPFM is not as seriously affected by air as in the RPP method so a long period of pre-pressurization is not required. When a moderate amount of air is present a curve like the dashed line (Fig. 5.5D) is found, but the limiting slope at high P_i is the same as for a root without air. However the HPFM does overestimate K_r in large woody root systems when a lot of air is evenly distributed throughout the wood.

The advantage of the HPFM over the RPP is that many roots can be measured per day; the time required to mount and measure a root is typically 10–20 min per root. Also the HPFM can be used on plants growing in soil in pots or growing in the field.

5.5 Distribution of Hydraulic Resistances in Roots

Generally, the axial resistance to water flow in roots is much less than the radial resistance. This is because axial flow is carried by vessels or tracheids whereas radial flow involves passage of water through non-vascular tissue. It is often assumed that the main barriers for radial flow of solutes and water are both in the endodermis, but some experiments have shown otherwise.

5.5.1 Axial Water Flow – Poiseuille's Law

Axial hydraulic conductivity in roots tends to increase from apex to base, i.e., with root age, because as roots get older the number and/or diameter of vascular conduits (vessels or tracheids) tend to increase. Poiseuille's law states that the hydraulic conductance of a cylindrical pipe of uniform diameter increases with the fourth power of the diameter. Poiseuille's law does not strictly apply to vascular conduits in roots, because they are not cylindrical nor of uniform diameter along their lengths, but Poiseuille's law does provide a useful first approximation, so for a root segment with *n* vessels of diameter *d*:

$$K_{axial} = \sum_{i=1}^{n} \frac{\pi\ d_i^4}{128\eta L} \tag{5.8}$$

where η is the viscosity of the solution in the vessels and L is the length. Steudle and Peterson (1998) report that young maize roots have protoxylem vessels near the tip that are 5–10 µm in diameter, then 25 mm from the tip early metaxylem vessels are 23 µm, and, finally, at distances of 250 mm from the tip, the late metaxylem elements are about 100 µm. So Poiseuille's law would predict that one old metaxylem vessel would be as conductive as 357 early metaxylem vessels. There are typically 14 early metaxylem vessels versus 7 late metaxylem vessels. If we compare the hydraulic conductance of 100 mm of root with early metaxylem versus 100 mm of root with late metaxylem, the old metaxylem part of the root would have an axial conductance that is 179 times that of the early metaxylem part. The theoretical axial conductance of maize roots is much more than the total root conductance and this has been confirmed experimentally as indicated below. Hence the radial (non-vascular) path limits the rate of water flow through roots.

5.5.2 Radial Water Flow and Role of Endodermis and Exodermis

Root conductivities per unit surface area, L_r, have been measured on maize roots about 1 mm diameter and 200–500 mm long by the RPP method and values tend to be around 2.3×10^{-5} kg s^{-1} m^{-2} MPa^{-1} (Steudle and Peterson 1998). These values agree with those measured by more traditional methods (Newman 1973; Miller 1985). The 225 mm length of maize root containing early metaxylem has a surface area of $A=7 \times 10^{-4}$ m^2, hence K_r for that region is $AL_r=1.6 \times 10^{-7}$ kg s^{-1} MPa^{-1} or a resistance of 6.2×10^6 MPa s^{-1} kg^{-1} ($=1/1.6 \times 10^{-7}$). The 14 early metaxylem vessels in this same length of root would have an axial conductance of 4.4×10^{-7} (from Eq. 5.6) or a resistance of 2.25×10^6. Hence, in young roots, the radial resistance is two to three times the axial. In slightly older maize roots with old metaxylem we have already shown that the axial resistance of the same 225 mm would be 179 times less, hence the radial resistance would be 300–500 times more than the axial resistance. Clearly, the ratio of resistance depends rather strongly on the diameter and number of xylem conduits in the roots so the theoretical calculations are rather approximate. However, these theoretical calculations are consistent with experiments on young roots in which hydraulic resistance is measured before and after root tips are excised, which decreases the resistance by a factor of 3–20.

5.5.3 Experiments to Locate Major Barriers to Water and Solute Flow

From the above considerations we can conclude that the main barrier to water flow may be non-vascular, i.e., somewhere in the radial pathway. But where in the radial pathway is the barrier located and is it at the same location for water and solutes? Some useful insights result from wounding experiments in which the effect of mechanical damage to the cortex and/or endodermis is measured in terms of the effect of such damage of root pressure, reflection coefficient, hydraulic conductivity and solute permeability.

The Casparian band is thought to reduce the water permeability of the cell walls in the endodermis forcing the pathway of water movement to be transcellular in the vicinity of the endodermis. Hence water would have to pass through at least two plasmalemma membranes. The surface area of this membrane pathway would be at least as great as the surface area of the endodermis. The area could be more if a significant fraction of the water transport is symplastic via plasmodesmata, hence water could enter through several layers of cortical cells, pass through the symplast, then exit through several layers of cells in the stele before reaching the xylem conduits. This notion is approximately consistent with observed values of root and membrane conductance. Maize root conductance (Steudle et al. 1993) measured on roots 90–180 mm long and 1 mm diameter is about 2.7×10^{-8} m s^{-1} MPa^{-1} (normalized to the epidermal surface area) and, since the endodermis is about half the

diameter of the epidermis, the conductance normalized to the endodermis surface would be about double (5.4×10^{-8}). In contrast, maize-root, cortical-cell membrane conductance is 2.4×10^{-8} m s^{-1} MPa^{-1} (as measured by a cell pressure probe) so the conductance of two membranes in series would be 1.2×10^{-8} or about one-quarter that of the root conductance normalized to the endodermal area. From this we can conclude that the membrane (transcellular) pathway accounts for at least one-quarter of the root conductance and the rest is via the apoplast. Alternatively, a larger fraction of the root conductance could be transcellular if water enters several cells, and thus through a much larger area of membrane, and then travels through the symplast. In an earlier study where 90 measurements were done on shorter maize roots (70–110 mm), L_r=$1.1\pm0.1 \times 10^{-8}$ (mean and SEM, Steudle et al. 1987) and cortical-cell membrane conductance was again $2.4\pm0.3 \times 10^{-8}$, so from these data we might conclude that about half the hydraulic resistance to water transport was transcellular across two membranes in the endodermis.

Wounding experiments provide more insights. Maize roots (90–180 mm long and 1 mm diameter) were mounted in a RPP until a stable root pressure (P_i) of 0.15±0.03 MPa was reached (Peterson et al. 1993; Steudle et al. 1993). A fine glass rod was inserted radially 400 μm into the roots causing a 20 μm diameter wound in the endodermis ($<10^{-4}$ or 10^{-5} of the endodermal surface area). The wound caused a rapid (100–200 s) drop in root pressure to 0.05±0.01 MPa, but no significant change in L_r. When P_i is constant in a RPP there is no water flow [F_m=0 in Eq. (5.3)]. Setting F_m=0 in Eq. (5.3) and solving for P_i (the root pressure) it follows that:

$$P_i=P_o-\sigma RT(C_o-C_i) \tag{5.9}$$

Since P_o=0 (atmospheric pressure) and C_o is unchanged, the rapid drop in root pressure could only be due to a rapid drop in C_i along the whole length of the xylem vessels. Diffusion would be too slow to account for the drop of C_i in <200 s, hence there must have been a mass flow leak of xylem sap through the wound. From this Steudle et al. (1993) conclude that the endodermis is the principle barrier to solute diffusion but not to water flow (otherwise L_r would also have increased). I am not convinced by the argument regarding water flow, because a wound to $<10^{-4}$ of the endodermis root surface would be effective only if all the water could freely exit through a very small surface area of cells in the stele and this resistance is unknown but probably high. In other experiments, Peterson et al. (1993) removed rather larger fractions of the cortical tissue of maize roots (15–38%) and this caused an approximately proportional increase in L_r, from which they conclude (rather more convincingly) that the barrier to water flow is diffusely distributed over the entire cortex.

Another approach to the location of the water barrier is to measure the pressure drop across the radius of a root while water is flowing; the principle hydraulic resistance would correspond to the region with the biggest drop in

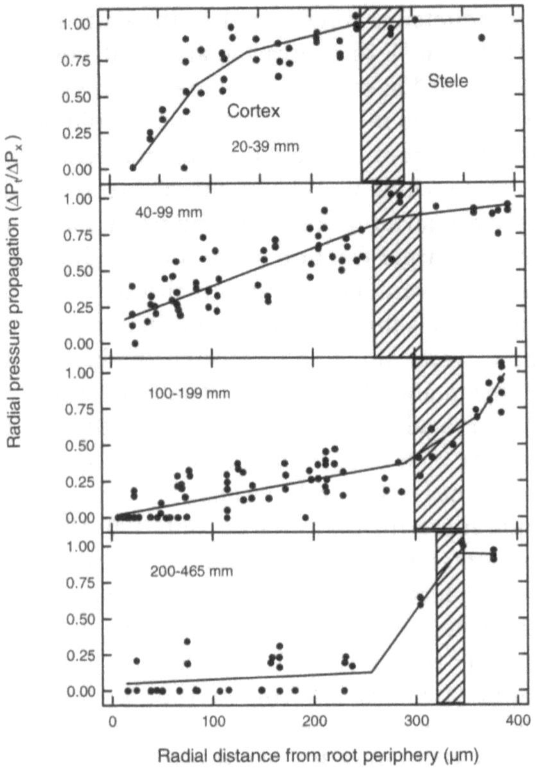

Fig. 5.6. Radial propagation of pressure across the cortex in four different zones of a maize root. Data points represent turgor responses (ΔP_t) following either an increase or a decrease in xylem pressure (ΔP_x). Locations of the endodermis within each zone are marked by *cross-hatched areas*, as determined by microscopic observations. Pressure gradient trend lines are drawn in *by hand*; steeper slopes indicate where the hydraulic resistance is most. (Adapted from French et al. 1996)

pressure. This has been accomplished more or less directly in experiments done on 500 mm long maize roots by Frensch et al. (1996). They used a cell pressure probe to measure changes in cortical and endodermal cell turgor pressure resulting from changes in xylem pressure. Although we do not know if the main pathway is transcellular or apoplastic, the changes in cell turgor should reflect changes in apoplastic pressure in adjacent cell walls. Results were obtained for root regions of different ages, i.e., different distances from the root tip (Fig. 5.6). Over the first 100 mm of the root (measured from the tip) there is a more or less uniform pressure gradient from the stele to the epidermis, which is consistent with the earlier finding. In the next section (100–200 mm from the tip) about two-thirds of the pressure drop is across the stele and endodermis and the rest across the cortex. For the oldest portions of the root (>300 mm from the tip) almost all the pressure drop is across the endodermis and adjacent cortical cells. Hence the situation is more complex than one might conclude from experiments on young roots 90–180 mm long.

5.6 Models of Solute and Water Flux in Roots (Possible Reinterpretation of Ideas)

Space does not permit a full quantitative description of existing models of transport of water and solutes in roots, but a brief examination is useful. The differences between models involve: (1) the use of classical transport equations versus irreversible thermodynamics transport equations (Eqs. 5.2 and 5.5a vs. Eqs. 5.4 and 5.5b), (2) treating the root 'membrane' as a simple barrier versus a composite 'membrane', and (3) treating the xylem sap inside the root vessels as a single compartment or treating the root as multiple cylindrical compartments in a linear catena.

Early experimenters assumed that classical equations provided an adequate quantitative model and for many root systems this was true. However, sometimes plots of mass flux (F_m) versus ΔP were non-linear (as in Fig. 5.5A), i.e., the root conductance H_r increased with increasing flow (Lopushinsky 1964). In extreme cases plots of F_m versus $\Delta\Psi$ showed negative H_r, i.e., flow opposite to the direction of the driving force for small flows (Mees and Weatherley 1957).

In any quantitative model, it is best (most ideal) to have transport coefficients, e.g., L_r, P_s and σ, that are independent of the driving forces. The transport equations from irreversible thermodynamics promise to fulfill this ideal. Fiscus (1975) was the first to demonstrate that the application of transport equations from irreversible thermodynamics in a root model with a simple membrane and only one compartment could explain the anomalous changes in H_r. Unfortunately, Fiscus (1975) still talked about the 'apparent resistance' ($=1/H_r$) being a function of flow; this still leads to confusion today because in reality the transport coefficients for a root are constant, i.e., L_r (as well as P_s and σ) do not need to change with flow to account for a non-linear relation in plots of J_v vs. $\Delta\Psi$ or ΔP as shown by the model of Fiscus (1975).

The elucidation of the composite membrane model (e.g., Steudle 1994) has provided further insights into the complexity of solute and water flow along the root radius. The mathematical descriptions advanced by Steudle (1994) emphasize the parallel composite properties of the root–membrane complex and a single xylem sap compartment. Steudle and Peterson (1998) suggest that the parallel composite membrane model explains: (1) why root hydraulic conductivity appears to change with flow rate, (2) why there are differences between hydraulic conductivity when measured with hydrostatic pressure or osmotic pressure gradients, and (3) why roots have a low reflection coefficient, σ. In reality, however, the root–membrane is a very complex system of parallel and serial membrane components and it is unclear if a parallel model will ultimately prove to be adequate. Hence the conclusions drawn (e.g., Steudle 1994; Steudle and Peterson 1998) may be subject to alternative interpretations and will be discussed briefly below.

Root hydraulic conductivity may not increase with flow. Plots of F_m vs. $\Delta\Psi$ are frequently curvilinear and the slope of this curve, H_r, does change with F_m. However, this does not mean that root hydraulic conductivity, L_r, is changing. So the composite membrane model really does not predict changes in L_r. There are theoretical cases in which root hydraulic conductivity may change with flow *if L_r is controlled by root membrane permeability to water*. In this case the genetic expression of root aquaporins may cause a change in membrane permeability to water and hence L_r of roots (Steudle and Henzler 1995). Recently, Tsuda and Tyree (2000) have attributed the diurnal variation in L_r of sunflower roots to diurnal variation in expression of aquaporins. Recently Siefritz et al. (2002) has demonstrated that aquaporins account for about half the root conductance in tobacco.

In some root systems the value of root hydraulic conductivity is less when measured by flow induced by osmotic pressure changes (L_{ro}) versus hydrostatic pressure changes (L_{rp}). In maize roots L_{ro} is 10 times less than L_{rp} and in *Quercus robur* roots L_{ro} is 100 times less than L_{rp} (Steudle and Peterson 1998). With reference to the parallel composite membrane model, this difference between L_{ro} and L_{rp} could be explained if osmotic flow is via a transcellular pathway and pressure flow is via an apoplastic pathway and this is a plausible explanation, theoretically. An alternative possibility is that the RPP is not measuring L_{ro} correctly.

In the RPP method, L_{ro} is computed from the half-time of the initial change in pressure (P_i) following the change in external osmotic pressure, RTC_o (see T_{s1} in Fig. 5.5C). Two processes must occur immediately after a change in C_o; for example, if C_o is increased, then the solute must first diffuse to the selective (semipermeable) membrane and then water must flow out of the root to lower P_i. The time constant, T_{s1}, could be a measure of the time for the solute to diffuse to the membrane or could be a measure of the time for water to pass out of the root via the membrane, or a combination of the two. In comparable experiments using a cell pressure probe on single cells, the water permeation through the cell membrane is the limiting process. In roots, however, this may not be the case if the selective membrane is located at the endodermis (either the plasmalemma membranes of the endodermal cells or the Casparian band). Let us say that the solutes reach the endodermis by diffusion through the apoplast of the epidermis and cortex, i.e., the cell walls. The direct radial distance is 250–350 μm in maize but the actual tortuous path length is more likely to be 400–500 μm (Fig. 5.3A). The solutes have to diffuse this distance through water; the diffusion coefficients, D, for the solutes used in RPP experiments range from 1.8×10^{-9} to 1.5×10^{-9} $m^2\ s^{-1}$ for KNO_3 and NaCl, respectively, to 0.68×10^{-9} to 0.52×10^{-9} for manitol and sucrose, respectively, in pure water. In the confines of the cell walls, the values may be somewhat lower, so let us use a range of 1.5×10^{-9} to 0.4×10^{-9} $m^2\ s^{-1}$. The time, t, it takes half the molecules to diffuse a distance x is given by $t=x^2/2D$. Using all possible values of x and D above we have a predicted range of 53–313 s for the diffusion time compared with T_{s1} values of 34–690 s (Steudle et al. 1987). The diffusion times may be somewhat longer because of solute/water drag problems. Some molecules will reach the endodermis in less time than above which will initiate outwardly directed osmotic flow of water. The water flow will tend to sweep the solutes away delaying their arrival.

The reflection coefficient, σ, of the common solutes used in RPP measurements of maize range from 0.4–0.85 (Steudle and Peterson 1998). In contrast, these same solutes have reflection coefficients (σ_m) from 0.9–1.0 when measured on plasmalemma membranes of plant cells. In the parallel composite membrane model one of the parallel paths involves plasmalemma membranes and the other the Casparian band, which is presumed to have a low reflection coefficient (σ_c). According to the parallel composite membrane model, σ of the whole root would be a weighted average of the fractional cross-sectional areas and hydraulic conductivities of the two parallel paths. The point of time at which σ is evaluated is generally after 2 to 4 half-times (T_{s1}) and when solute flow approaches zero, so solute/water drag effects are also reduced. So the values of σ are probably much less prone to errors than is the evaluation of L_{ro}. However, this has never been verified by theoretical computations of the dynamics of diffusion and solute/water drag coupling inside the cortical cell walls.

In other types of experiments (Zhu et al. 1995; Schneider et al. 1997) σ has been evaluated in roots attached to transpiring plants. In this situation solute/water drag can cause substantial errors in the estimation of σ. This is because the constant inflow of water to the endodermis can raise the concentration of the solute at the endodermis (C_{oe}) to a value much more than C_o. Measurements of P_i were made with a cell pressure probe positioned in a xylem vessel of a root during steady-state transpiration and looking at the effect on P_i of increasing external solute concentration from zero to C_o. The reflection coefficient was calculated from:

$$\sigma = \frac{\Delta P_i}{\Delta \pi_o} \cong \frac{\Delta P_i}{RT\Delta C_o} \tag{5.10}$$

During this experiment, the flux of solution through the cortical apoplast to the endodermis, J^*_v, is transporting solute at a rate given by CJ^*_v at any given point where the solute concentration C is causing an accumulation of solute concentration towards the endodermis. At any given point the solute is diffusing out at a rate given by Fick's law and at steady state the solute/water drag and diffusion would balance:

$$CJ_v^* = D\frac{dC}{dx} \tag{5.11}$$

The solute/water drag will be most if all the water flux into the root passes through the cortical cell walls; in this case:

$$CJ_v^* = D\frac{dC}{dx} \tag{5.12}$$

where α is the fraction of the root surface area occupied by water in the apoplast. The wall area is about 2 % of the surface area and about half the wall is solid so $\alpha \approx 0.01$.

An approximate solution of Eq. (5.11) results by treating the cortical layer as a plane rather than a cylindrical annulus. Integration of Eq. (5.11) from $x=0$, $C=C_o$ to the endodermis where $x=\delta$ and $C=C_{oe}$ and substitution of Eq. (5.12) for J^*_v yields:

$$\frac{C_{oe}}{C_o} = \exp\left(-\frac{L_r(\sigma\Delta\pi - P_i)\delta}{\alpha D}\right) \tag{5.13}$$

Under a range of transpirational conditions (darkness and humid air to high light and dry air) the value of P_i ranged from +0.08 to −0.2 MPa in maize plants. After adding 50 mol/m^3 NaCl and observing the change in P_i, the value of σ, computed from Eq. (5.10), ranged from 0.15 to 0.9 (Schneider et al. 1997). However, Eq. (5.10) does not take into account the solute/water drag effects in Eq. (5.13). Not enough information is provided in the original publications to estimate J^*_v from Eq. (5.13), but the maize plants went from guttation flow (in darkness and humid air) to high transpiration rates (because of high light and dry air), so $(\sigma\Delta\pi - P_i)$ probably changed by 0.25 MPa, e.g., from −0.04 to −0.29 MPa, before the application of NaCl. Since transpiration rate is limited by vapor-phase resistances (through stomata), we can assume J^*_v did not change after application of NaCl. Assuming $L_r=2.4 \times 10^{-7}$ m s^{-1} MPa^{-1}, $\delta=4 \times 10^{-4}$ m, $D=1.2 \times 10^{-9}$ m^2 s^{-1} for NaCl in cell walls, and $\alpha=0.01$, Eq. (5.13) predicts $C_{oe}/C_o=1.38$ and 10.17 for $(\sigma\Delta\pi - P_i)=-0.04$ and −0.29 MPa, respectively. Space does not permit a more rigorous derivation for cylindrical geometry, but for this case $C_{oe}/C_o=1.43$ and 13.64 with the parameters above. Hence the six-fold change in σ from 0.15 to 0.9 could be accounted for by the difference between C_{oe} and C_o brought about by the solute/water drag effect even if part of the water entering the root is via the transcellular pathway, which would make J^*_v correspondingly smaller. From the above argument it would appear that σ may not be a function of water flow rate in roots.

Tyree et al. (1994) have advanced a root model, AMAIZED, consisting of n root segments with each segment being a link in a catena transport model with a simple root membrane. The AMAIZED model has been used to examine solute and water transport both for the steady state and dynamic situations where solute fluxes and water fluxes are constantly changing with time and position along the root. The model has been applied to large root systems consisting of an absorbing zone, i.e., young roots where there are radial and axial fluxes of water and solute, and in transport zones, i.e., older (perhaps woody) roots where only axial fluxes occur. A surprising prediction of AMAIZED is that roots with short absorbing zones have the same dynamics as roots with long absorbing zones if all other parameters are the same; this makes modeling of large root systems more feasible and meaningful theoretically. Tyree et al. (1994) excised large walnut root systems and attached an early version of an HPFM to the base. They measured flow versus applied pressure for stepwise changes in pressure (150 to 180 s steps). The model has been very good at predicting the dynamics of pressure-driven water flow (Fig. 5.7).

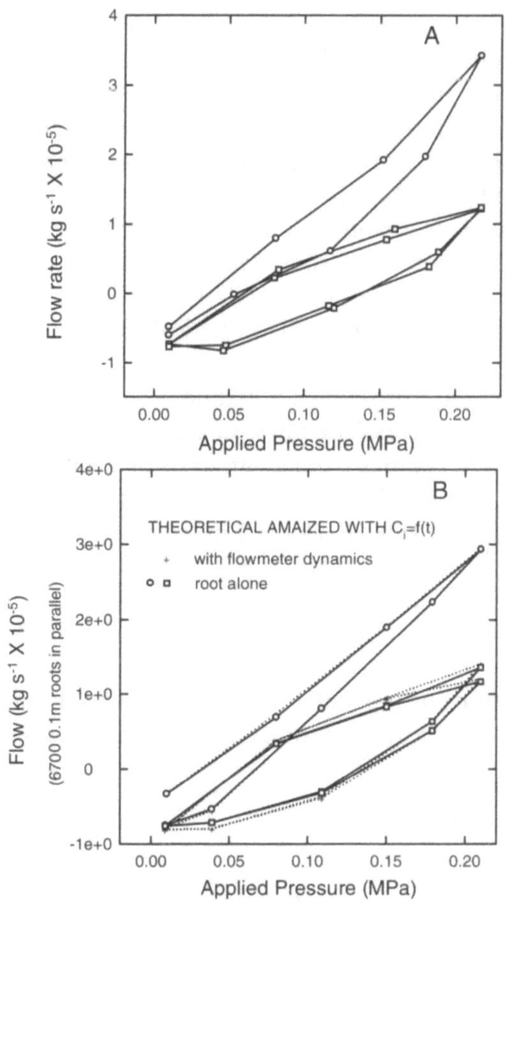

Fig. 5.7A,B. **A** Dynamic measurement of flow versus applied pressure in walnut (*Juglans regia* L. cv. Lara) roots growing in 200-l pots using a high-pressure flowmeter (HPFM). Pressure was adjusted over 150 to 180 s and flow readings were taken at the end of each period. The *arrows* indicate the direction of pressure change. *Open circles* are values measured 20 min after excising the shoots; *squares* are values measured 2 h after excising the shoot. **B** Theoretical predictions from AMAIZED, the model for dynamics of solute and water transport in roots. *Solid lines with open circles* are predictions of AMAIZED 20 min after excising the shoots; *solid lines with squares* are predictions of AMAIZED 1.5 h after excising the shoot; +'s, predictions of AMAIZED with the response time of both the root and HPFM are taken into account. Because +'s are close to open and filled circles, we can conclude that the hysteresis was due to the dynamics of the root rather than the response time of the HPFM. (Adapted from Tyree et al. 1994)

5.7 The Problem of Scaling for Root or Plant Size

The invention of the HPFM has permitted the rapid measurement of root conductance, K_r (kg s^{-1} MPa^{-1}), in both laboratory and field situations on a wide range of root systems sizes, e.g., root systems with basal diameters of 1–25 mm. The value of K_r increases with root or plant size. Root or plant size can be measured in terms of root surface area, A_r, root length, L, total root dry weight, *TRDW*, or leaf surface area, A_L. So this raises the question on how best to scale for size and what can be learned from different ways of scaling.

Since we have already established that radial root resistance is generally more than axial resistance, water uptake is likely to be limited by root surface area. Hence it is reasonable to divide K_r by A_r yielding L_r, which is a measure of root efficiency. Some roots are more efficient than others. Division of K_r by total root length (L) is not as desirable, but is justified because A_r and L are correlated approximately and L can be estimated by a low-cost, line-intersection technique rather than a high-cost, image-analysis technique.

Scaling by root mass is justified by consideration of the cost of resource allocation. Plants must invest a lot of carbon into roots to grow and to maintain them. The benefit derived from this carbon investment is enhanced scavenging for water and mineral nutrient resources. Total root dry weight (*TRDW*) is a measure of carbon investment into roots. Thus the carbon efficiency of roots might be measured in terms of $K_r/TRDW$, $A_r/TRDW$, or $L/TRDW$. Scaling by *TRDW* provides information of ecological rather than physiological importance.

Scaling of K_r by leaf surface area (A_L) provides an estimate of the 'sufficiency' of the roots to provide water to leaves. The physiological justification of scaling K_r to the leaf surface area (A_L) comes from an analysis of the Ohm's law analogue for water flow from soil to leaf (van den Honert 1948). The Ohm's law analogue describes water flow rate (F_m, kg s^{-1}) in terms of the difference in water potential between the soil (Ψ_{soil}) and the leaf (Ψ_L):

$$\Psi_{soil} - \Psi_L = (1/K_{soil} + 1/K_r + 1/K_{sh})F_m \tag{5.14}$$

where K_{soil} is the hydraulic conductance of the soil. It is usually assumed that $K_{soil} >> K_r$ and K_{sh} except in dry soils so $1/K_{soil}$ can be ignored. Leaf water potential is then approximated by:

$$\Psi_L \cong \Psi_{soil} - (1/K_r + 1/K_{sh})F_m \tag{5.15}$$

Or, if we wish to express Eq. (5.15) in terms of leaf area and average evaporative flux density (E), we have:

$$\Psi_L \cong \Psi_{soil} - (1/K_r + 1/K_{sh})A_L E \tag{5.16}$$

This equation also can be rewritten so that root and shoot conductances are scaled to leaf surface areas, i.e., to give leaf-specific shoot and root conductances, K_{sh}/A_L and K_r/A_L, respectively:

$$\Psi_L \cong \Psi_{soil} - (1/[K_r/A_L] + 1/[K_{sh}/A_L])E \qquad (5.17)$$

Meristem growth and gas exchange are maximal when water stress is small, i.e., when Ψ_L is near zero. From Eq. (5.17) it can be seen that the advantage of high K_r/A_L and K_{sh}/A_L is that Ψ_L will be closer to Ψ_{soil}. Leaf-specific stem-segment conductivities, K_L, are high in adult pioneer trees, so the water potential drop from soil to leaf is much smaller than in old-forest species (Machado and Tyree 1994). This may promote rapid extension growth of meristems in pioneers compared with old-forest species. Also, stomatal conductance (g_s) and therefore net assimilation rate are reduced when Ψ_L is too low. During the first 60 days of growth of *Quercus rubra* L. seedlings, there was a strong correlation between midday g_s and leaf-specific plant conductance, $G=K_p/A_L$, where $K_p=K_rK_{sh}/(K_r+K_{sh})$ (Ren and Sucoff 1995). This suggests that whole-seedling hydraulic conductance is limiting g_s though its effect on Ψ_L. There also is reason to believe that whole-shoot conductance limits g_s in mature trees of *Acer saccharum* Marsh (Yang and Tyree 1993). Thus, high values of K_r/A_L and K_{sh}/A_L may promote both rapid extension growth and high net assimilation rates in pioneers.

Scaling is always necessary to normalize for plant size. As seedlings grow exponentially in size we would expect an approximately proportional increase in K_r and K_{sh}. Since roots and shoot both supply water to leaves and since an increase in leaf area means in increase in rate of water loss per plant, we would expect K_r and K_{sh} to be approximately proportional to A_L.

Tyree et al. (1998) studied the growth dynamics of root and shoot hydraulic conductance in seedlings of five neotropical tree species of contrasting ecological strategy. Two species were light-demanding pioneers and three were shade-tolerant forest species. All five species were grown under the same intermediate light regime. The pioneers versus shade-tolerant species had significantly higher growth rates in terms of the rate of increase in A_L, A_r, L, $TRDW$, K_r and K_{sh}. When the scaled root conductances were compared between species, no pattern was found relating K_r/A_r or K_r/L. On the other hand, all pioneers were significantly higher in terms of $K_r/TRDW$, K_r/A_L, $A_r/TRDW$, and $L/TRDW$. The tentative conclusion to be drawn from this rather limited study is that scaling by $TRDW$ or A_L may be of more ecological significance than scaling by L and A_r. Whenever possible, it is best to use all scaling methods in ecological and physiological studies of roots, but the less used scaling methods ($TWRD$ and A_L) are clearly important and could be used on their own.

The HPFM and RPP have recently been used to study ecological aspects of root physiology, e.g., the effect of drought and mycorrhizae on L_r. Readers interested in this aspect should consult Nardini et al. (2002).

5.8 Summary and Prospects

The main resistance to water uptake in root systems appears to be the radial resistance from the fine root surface to the stele. The next biggest resistance is the axial resistance in the first few cm of root length near the apex of the roots. The axial resistance is determined by the number and diameter of vessels in any given cross section. Since the vessels are fewer in number and smallest in diameter in the first 10 cm of maize roots, the axial resistance of the first 10 cm is about 180 times more than the next 10 cm. Hence most of the root resistance is in the apical 10 cm and most of that in the root radius.

Equations describing solute and water transport in roots must account for the coupling between flows of solute and water and for the non-ideality of the osmotic forces. Hence root water and solute transport require a minimum of three parameters, L_r (root hydraulic conductance), P_s (solute permeability), and σ (reflection coefficient). Some people believe that two L_r values are needed, one (L_{rp}) that describes the conductance to pressure-driven flow and the other (L_{ro}) that describes osmotic driven flow, which is ten times or more lower. In this chapter I argue against this concept and suggest instead that measurements of L_{ro} are incorrect. The differences between L_{rp} and L_{ro} could be explained by the time it takes solutes to reach the osmotic barrier in roots, i.e., the Casparian band. Others have reported that values of σ might vary with water flux rates in roots but I argue that these conclusions might also be in error because the erroneous measurements of σ did not take into account the likely coupling of solute flux to water flux.

More work needs to be done to fully elucidate the mechanism and pathway of water and solute flux in roots and the experimental approach needs to be extended to a wider range of root types. So far, maize roots are the most fully characterized roots, but it seems unlikely to me that maize roots can be taken as a universal model of all root systems. The second exciting area concerns the role of aquapores in root hydraulic conductivity and in periodicity. Unpublished results from roots of tobacco, peach, honey locust and apple have revealed a very strong diurnal periodicity in which roots are 10 times more conductive to water at midday than at midnight (Tyree and Zimmermann 2002). Our understanding of whole-plant water relations will have to be revised after further elucidation of the periodicity of root hydraulic conductivity, because up until recently we have all assumed that roots act like constant, passive pathways for water movement.

References

Esau K (1960) Anatomy of seed plants. Wiley, New York

Fiscus El (1975) The interaction between osmotic- and pressure-induced water flow in plant roots. Plant Physiol 55:917–922

Fiscus EL (1977) Determination of hydraulic and osmotic properties of soybean root systems. Plant Physiol 59:1013–1020

Frensch J, Hsiao TC, Steudle E (1996) Water and solute transport along developing maize roots. Planta 198:348–355

Kedem O, Katchalsky A, Curran PF (1962) Permeability of composite membranes. Parts 1, 2 and 3. Trans Faraday Soc 59:1918–1953

Lopushinsky W (1964) Effects of water movement on ion movement into the xylem of tomato roots. Plant Physiol 39:494–501

Machado J-L, Tyree MT (1994) Patterns of hydraulic architecture and water relations of two tropical canopy trees with contrasting leaf phenologies: *Ochroma pyramidale* and *Pseudobombax septenatum* Tree Physiol 14:219–240

Mees GC, Weatherley PE (1957) The mechanism of water absorption by roots. I. Preliminary studies on the effects of hydrostatic pressure gradients. Proc R Soc Lond Ser B 147: 367–380

Miller DM (1985) Studies of root function in *Zea mays*. III. Xylem sap composition at maximum root pressure provides evidence of active transport into the xylem and a measurement of the reflection coefficient of the root. Plant Physiol 77:162–167

Nardini A, Salleo S, Tyree MT (2002) Ecological aspects of water permeability in roots. In: Waisel Y, Eshel A, Kafkafi U (eds) Plant roots: the hidden half. Marcel Dekker, New York, pp 683–698

Newman EI (1973) Permeability to water of five herbaceous species. New Phytol 72:547–555

Peterson CA, Murrmann M, Steudle E (1993) Location of major barriers to water and ion movement in young roots of *Zea mays* L. Planta 190:127–136

Peterson CA, Steudle E (1993) Lateral hydraulic conductivity of early metaxylem vessels in *Zea mays* L. roots. Planta 189:288–297

Ren Z, Sucoff E (1995) Water movement through *Quercus rubra* L. Leaf water potential and conductance during polycyclic growth. Plant Cell Environ 18:447–453

Schneider H, Zhu JJ, Zimmermann U (1997) Xylem and cell turgor pressure probe measurements in intact roots of glycophytes: transpiration induces a change in the radial and cellular reflection coefficients. Plant Cell Environ 20:221–229

Siefritz F, Tyree MT, Lovisolo C, Schubert A, Kaldenhoff R (2002) PIP1 plasma membrane aquaporins in tobacco: From cellular effects to function in plants. Plant Cell 14:869–876

Steudle E (1994) Water transport in roots. Plant Soil 167:79–90

Steudle E (1993) Pressure probe techniques: basic principles and application to studies of water and solute relations at the cell, tissue, and organ level. In: Smith JAC, Griffith H (eds) Water deficits: plant responses from cell to community. Bios Scientific Publishers, Oxford, pp 5–36

Steudle E, Frensch J (1996) Water transport in plants: role of the apoplast. Plant Soil 187:67–79

Steudle E, Henzler T (1995) Water channels in plants: do basic concepts of water transport change? J Exp Bot 46:1067–1076

Steudle E, Peterson CA (1998) How does water get through roots? J Exp Bot 49:775–788

Steudle E, Murrmann M, Peterson CA (1993) Transport of water and solutes across maize roots modified by puncturing the endodermis. Further evidence for the composite transport model of the root. Plant Physiol 103:335–349

Steudle E, Oren R, Schulze E-D (1987) Water transport in maize roots. Plant Physiol 84:1220–1232

Tsuda M, Tyree MT (2000) Plant hydraulic conductance measured by the high pressure flow meter in crop plants. J Exp Bot 51:823–828

Tyree MT (1999) Water relations and hydraulic architecture. In: Pugnaire FI, Valladares F (eds) Handbook of functional ecology. Marcel Dekker, New York, pp 221–268

Tyree MT, Zimmermann MH (2002) Xylem structure and the ascent of sap, 2nd edn. Springer, Berlin Heidelberg New York

Tyree MT, Patiño S, Bennink J, Alexander J (1995) Dynamic measurement of root hydraulic conductance using a high-pressure flowmeter in the laboratory and field. J Exp Bot 46:83–94

Tyree MT, Yang S, Cruiziat P, Sinclair B (1994) Novel methods of measuring hydraulic conductivity of tree root systems and interpretation using AMAIZED. Plant Physiol 104:189–199

Tyree MT, Velez V, Dalling JW (1998) Growth dynamics of root and shoot hydraulic conductance in seedlings of five neotropical tree species: scaling to show possible adaptation to differing light regimes. Oecologia 114:293–298

van den Honert TH (1948) Water transport in plants as a catenary process. Disc Farad Soc 3:146–153

Yang Y, Tyree MT (1993) Hydraulic resistance in the shoots of *Acer saccharum* and its influence on leaf water potential and transpiration. Tree Physiol 12:231–242

Zhu JJ, Zimmermann U, Thürmer F, Haase A (1995) Xylem pressure response in maize roots subjected to osmotic stress: determination of radial reflection coefficients by use of the xylem pressure probe. Plant Cell Environ 18:906–912

6 Root Growth and Function in Relation to Soil Structure, Composition, and Strength

A.G. Bengough

6.1 Introduction

Soil is an extremely complex growth medium, with an enormous biodiversity of fauna and flora, widely differing mineral and organic matter composition, and a complex multi-scaled network of three-dimensional pore space. Soil strength and physical conditions can exert a major constraint to root growth. Hard soils with massive structures restrict the accessibility of nutrients and water to plants. In many laboratory and glasshouse experiments, and in much of horticulture, plants are now cultured using agar, nutrient solution, or artificial composts, an environment very different to the natural one. We must understand the most important factors and stresses that limit root growth in soil, so that results gained in vitro can be related to agricultural and natural ecosystems. Laboratory studies should concentrate on the most important, rather than the technically most convenient, problems. These problems must also be considered in appropriate combination – for example, in many soils, water stress will be accompanied by a large increase in soil strength and hence mechanical impedance to root growth. In this chapter the impact of soil structure, composition and strength on root ecology and resource acquisition by plants is considered.

In Section 6.2, the physical structure and composition of soil is described and how it can be measured. Root growth and function are then considered separately with respect to roots growing in the bulk of the soil, and roots growing inside continuous macropores. The physical limitations to root growth are considered in Section 6.3.1, concentrating on the effects of soil strength in Section 6.3.2. Localised compression of the rhizosphere is discussed in Section 6.3.3, and the consequences for water and nutrient uptake to individual roots in Section 6.3.4. The growth of roots in macropores is considered both in terms of root elongation and distribution within these pores (Sect. 6.4.1), and in terms of the effects of root clumping on water and nutri-

Ecological Studies, Vol. 168
H. de Kroon, E.J.W. Visser (Eds.)
Root Ecology

ent uptake (Sect. 6.4.2). In Section 6.5, the likely ecological consequences of structurally degraded and hard soils are discussed, and the chapter is concluded with a forward look to areas of particular interest (Sect. 6.6).

6.2 An Introduction to Soil Structure and Some Ways to Quantify It

Soil consists of weathered particles of rock and partly decomposed organic matter, interspersed with pores containing aqueous solutions and gases. Mineral particles range in size over six orders of magnitude, from several nm for small clay particles, to several mm for sand and gravel (see Table 6.1).

The size range of soil pores is similarly large, from nm-wide interlayer spaces in clay particles to mm-sized macropores created by earthworms (Table 6.1). The pore space forms the fundamental network within which all biological processes occur in the soil, including the transport of nutrients, water and oxygen to the root surface, and the movement of root pathogens. It is difficult to quantify soil structure because of the opaque nature of most soil particles, and the large range of sizes of soil pores and particles. Techniques to quantify soil structure usually measure, some physical property of the soil (e.g. water and air flow parameters, water release characteristic, or soil strength and stability), or describe the nature and distribution of pore space within a sample. The range of techniques used is described in Burke et al. (1986) and in standard soil physics texts.

Table 6.1. Particle and pore size ranges in soil in relation to the approximate sizes of soil organisms, and the matric potentials at which soil pores drain

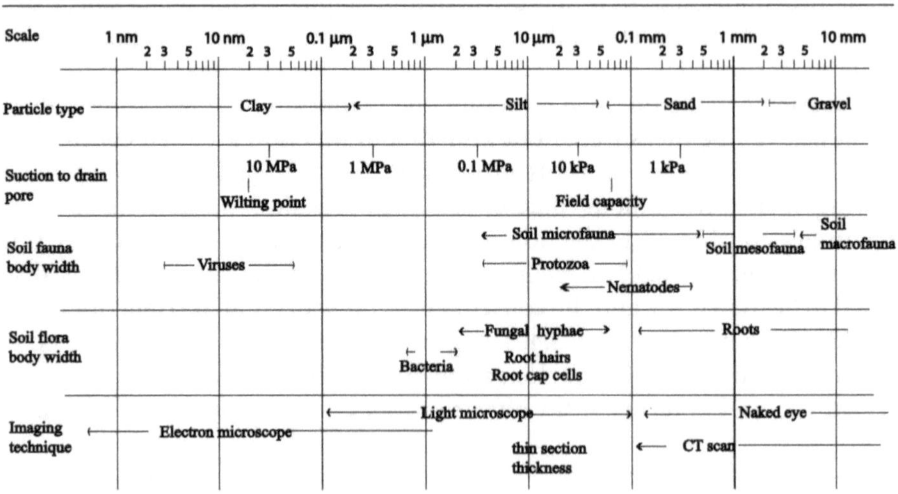

The physical property approach is appropriate if you can measure a relevant property at a scale appropriate to the application. Saturated hydraulic conductivity (K_{sat}; the flux of water per unit pressure gradient, in a saturated soil), for example, is typically measured at the cm scale or greater. Hydraulic conductivity of a tube increases with the fourth power of tube radius, and so large continuous pores, such as old root or earthworm channels, will dominate the measurement of K_{sat}. Measurement of K_{sat} at the cm scale will therefore tell a lot about the number and continuity of large pores, but relatively less about the finer pore fraction. The probability of large pore sizes and types being present increases with the size of the soil sample, and so it is more likely that a 50-cm soil core will contain continuous structural cracks and earthworm burrows than a 5-cm core. If such scaling rules are obeyed, the average K_{sat} for the 50-cm core will therefore be greater than the average K_{sat} for 5-cm cores. Quisenberry et al. (1994) studied water and chloride flow through an undisturbed cubic block of soil measuring 0.32 m. The solution was collected in a large array of identical containers beneath the block. The main flow paths were continuous macropores that were clustered together such that half of the total water was collected in fewer than 20 % of the containers, and half of the chloride collected in just 10 % of the containers.

An alternative and potentially powerful approach that can be applied at scales below 10 cm is to quantify the pore distribution directly, and use this knowledge to estimate the transport properties of the soil. To map the pore space requires either some non-invasive scanning technique or, alternatively, excavation of two-dimensional plane surfaces in the soil. Non-invasive techniques such as scanning the sample with X-rays or gamma-rays and constructing a three-dimensional image using computer tomography can be applied to soils using modified medical scanners, although the resolution is generally limited to greater than 250 μm. Dual energy scanning now makes it possible to separate the components of attenuation by water and solid matrix (Rogasik et al. 1999). Individual roots of 0.1-mm diameter have been imaged in soil using X-ray scanning, and so it is now possible to image both the root, soil matrix, and soil water in three dimensions (see review, including magnetic resonance imaging, by Asseng et al. 2000). The higher the resolution, however, the longer the scan period and the smaller the volume of sample that can be investigated – a doubling of resolution would require an eight-fold increase in scan time, and an eight-fold increase in the size of the associated data sets. If living roots are to be imaged within the soil, there is clearly an optimum balance that must be made between spatial resolution and the temporal resolution in relation to the rate of root growth.

Resin-embedding techniques can be used on soil blocks to characterise the pore fraction smaller than 100 μm (Murphy et al. 1977). Polished blocks or thin sections from these blocks are then examined to map the distribution of pore space and solid. The process of differentiating pore space can be made easier by use of a fluorescent dye in the embedding resin, and Ringrose-Voase

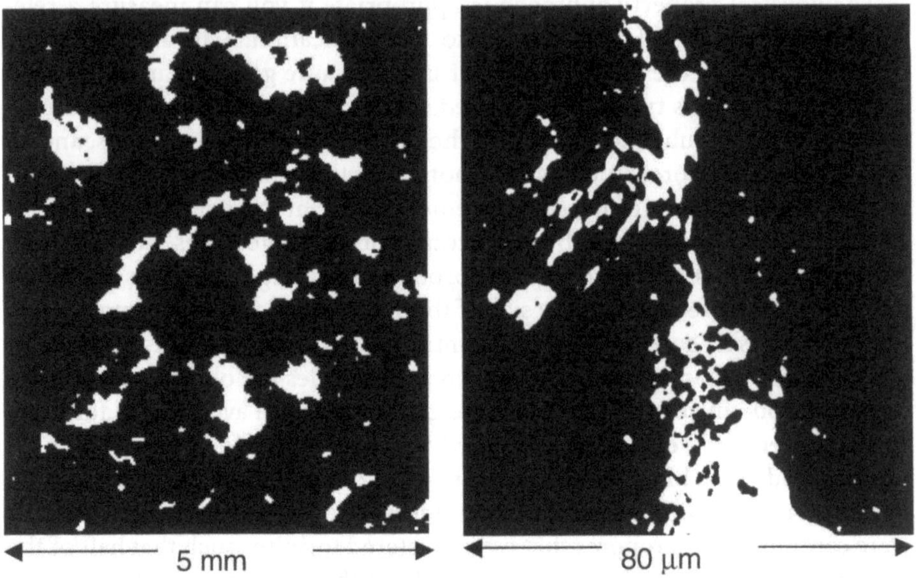

Fig. 6.1. Porosity maps taken from a single soil thin section shown at 62-fold different magnifications. The thin section was prepared from a core of fallow sandy loam soil. Pore space is *coloured white*, and soil particles are *black*. Note the complex arrangement of pore space present in the section, and the increasingly small pores that become visible at high magnification (images kindly supplied by Naoise Nunan, SCRI)

(1996) offers much useful advice on how to perform the various procedures, including the thresholding of pore space. Considerable variation is often present, even in cores packed from reconstituted sieved soils, and complex pore structures can be seen at widely differing magnifications (Fig. 6.1).

Soil pores can be classified into channels, fissures and packing pores according to their shape in cross-section (Ringrose-Voase 1996). These pore classes can be identified automatically using image analysis techniques. The elongation and irregularity of the pore is classified and plotted in two-dimensional space. The probability of a particular feature being of a given type can then be estimated.

The distribution of solid particles and pores in cross-section can also be used to calculate the fractal dimension of the soil, if the soil shows fractal scaling (Bartoli et al. 1991; Crawford et al. 1995). The quantitative description of soil structure (be it fractal, or not) and its relation to transport processes presents an exciting opportunity for theoretically exploring the nature of root–soil structure interactions: Soil structures surrounding roots can both be measured and simulated in three dimensions, allowing pore-scale simulation of flows to roots. This will allow the effects of structural heterogeneity to be investigated, and enable testing against conventional models of root uptake from continuous uniform soil.

6.3 Root Growth in Bulk Soil

The roots of any plant will generally experience a wide range of physical conditions and environments, including growth through soil pores of various sizes, aggregates and larger soil structural units under a range of moisture conditions. For simplicity we consider the factors affecting root growth and uptake separately, according to whether the roots are situated in the bulk of the soil, or in large continuous macropores. In reality, individual roots may enter and leave macropores, and so part of their length may be in the bulk soil, and part in a macropore.

6.3.1 Physical Limitations to Root Growth

Many physical, chemical, and biological factors can limit root growth in soil. Physical factors are strongly related to soil structure and composition, and include the effects of water stress (Chap. 7), poor aeration (Chap. 8), and mechanical impedance.

Water stress depends on the matric and osmotic potentials at the root surface. Water is usually thought of as being available for uptake at matric potentials greater than −1.5 MPa, the wilting point of many mesophytic plants. However, drought stress slows root growth at potentials greater than −1.5 MPa, as illustrated for maize root elongation in Fig. 6.2a. The data in this figure are from experiments performed in vermiculite or in very loosely packed soil so that the strength of the soil was independent of water content and non-limiting to root growth. A disadvantage of this approach is that the vermiculite, or soil, in the rhizosphere may be drier than in the bulk of the medium, and so the roots may experience a more negative potential than expected. Sharp et al. (1988) attempted to minimise this effect by studying elongation in young, non-transpiring seedlings, where water was required mainly for cell expansion and respiration, but the magnitude of these effects, in this and other studies, is unknown.

Poor aeration of the soil often becomes a problem when the air-filled porosity is less than 10 % of soil volume, although aeration problems can occur in wet sands with air-filled porosities of up to 20 % (e.g. Warnaas and Eavis 1972). The demand of the root for oxygen may exceed the supply rate by diffusion, creating hypoxia or anoxia in the root. This situation is common in wet compacted, or waterlogged soil where there are insufficient continuous air-filled pores to provide pathways for rapid oxygen transport. The effect of oxygen supply, as measured using a platinum electrode, on root elongation is shown in Fig. 6.2b. In many plants hypoxia causes the formation of continuous channels, or aerenchyma, in the root cortex, through either cell lysis or cell separation (see Chap. 8). These aerenchyma provide low resistance path-

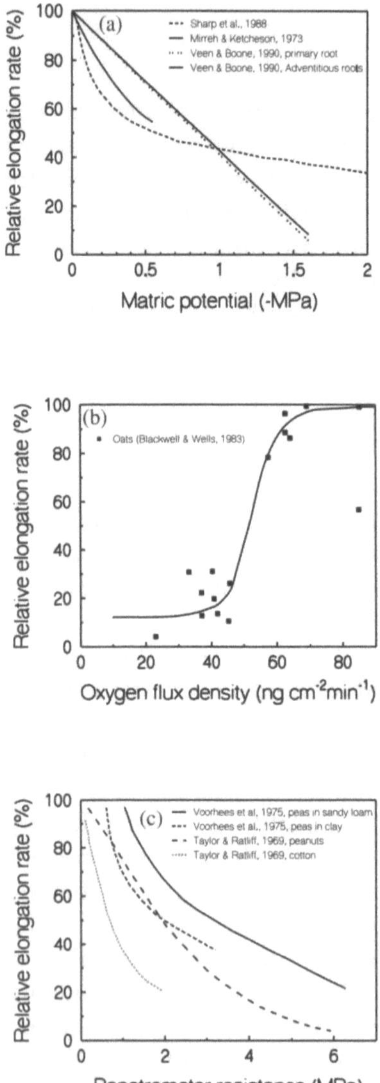

Fig. 6.2a–c. Effect of soil physical conditions on root elongation rate. Root elongation rate is shown as a fraction of the control elongation rate for **a** maize roots growing in loose soil or vermiculite at a range of matric potentials (after Mirreh and Ketcheson 1973; Sharp et al. 1988; Veen and Boone 1990); **b** oats growing in soil with a range of oxygen diffusivities (after Blackwell and Wells 1983); and **c** peas (after Voorhees et al. 1975) and peanuts and cotton (after Taylor and Ratliff 1969) in soils of different strengths

ways for oxygen transport to the meristem (see also Chaps. 8 and 13), and are particularly prevalent in species adapted to wetland habitats, such as rice – although they also occur in non-wetland cereals, including maize, wheat, and barley.

Mechanical impedance slows root elongation as the strength of the soil increases, and there are insufficient continuous channels, larger than the root diameter, for unimpeded root elongation (Fig. 6.2 c). Penetrometer resistance gives an empirical measure of the strength of the soil, but exceeds the penetration resistance experienced by roots by a factor of between two and eight times (Bengough and Mullins 1990). There are a number of reasons for this, but one of the main factors is that roots experience a much smaller component of frictional resistance to growth than do penetrometers due to the low-friction properties of root tips (Bengough and Mullins 1991): more than 80 % of the penetration resistance to a penetrometer can be frictional.

The aim of much of the work described above has been to control the root environment so that the effect of single physical limitations to root growth can be studied in isolation. Plants in the field experience a wide range of moisture conditions, from saturation to wilting point and drier. In agricultural systems the topsoil is often tilled, and subject to compaction by vehicles or animals. These influences are in addition to the many natural processes that affect the structure, such as freeze-thaw cycles, and the activity of earthworms and roots. At different times during a growing season roots will experience very different limiting factors, according to the prevailing water content and soil strength. The recently developed concept of the least limiting water range (LLWR, da Silva et al. 1994) is a valuable framework for considering which stresses are acting, and for determining the effect of management practices on the physical fertility of the soil. The LLWR is defined as the range of water content within which root growth is least limited by matric potential, mechanical impedance, and aeration. The concept requires the setting of somewhat arbitrary values of matric potential and soil strength that represent important limitations to root growth. The soil shown in Fig. 6.3a has had a

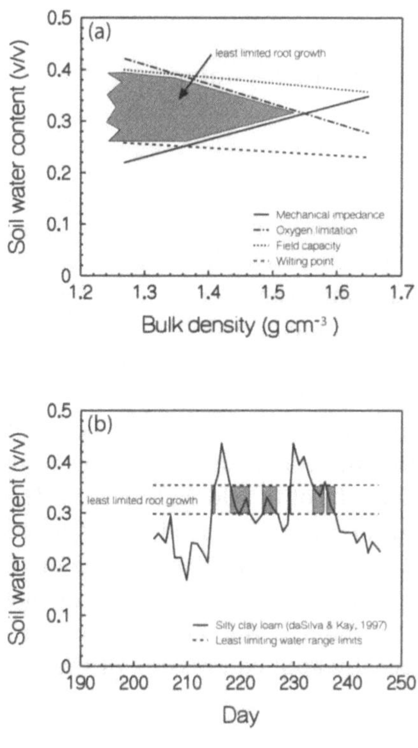

Fig. 6.3a, b. Least limiting water content range (LLWCR) **a** defined for a silt loam as a function of bulk density (after da Silva et al. 1994); the water content range for which root growth is not limited is shown by the *shaded area* for the corresponding bulk densities. **b** Variation in water content in a silty clay loam in relation to the LLWCR (after da Silva and Kay 1997); the *shaded region* indicates the days on which root growth is not limited by mechanical impedance, oxygen availability, or water availability

series of these levels defined relating to mechanical, oxygen, and drought limitations. These levels depend on both the bulk density and the water content, as bulk density influences strength, the available airfilled pore space, and the matric potential. Thus, when the soil water content lies within the shaded area, root growth is not subject to significant physical stresses. When combined with temporal variability in soil moisture, the length of the growing season that a crop spends within the LLWR can be determined (Fig. 6.3b). The number of days that the soil water content is outside the LLWR is a measure of the length of time that the plants will experience growth limitations due to adverse soil physical conditions. The frequency at which the soil water content lies outside the LLWR is affected by weather conditions and tillage treatment (da Silva and Kay 1997). The 'relative bulk density' (i.e. actual bulk density divided by the bulk density of the soil after a standard compaction treatment) was a more powerful parameter than the bulk density for separating the effects of soil management from inherent soil properties, being independent of clay and organic matter content (da Silva et al. 1997). The relative importance of the different limits to the LLWR has been evaluated for a number of Canadian topsoils – in 78–90 % of cases, soil strength was the factor limiting root growth at low water content (Topp et al. 1994, cited by da Silva et al. 1994). In the next section we consider the effects of mechanical impedance on root growth in more detail, because of its clear dependence on the mechanical fabric of the soil, and its practical importance in the field.

6.3.2 Effects of Soil Strength on Root Growth and Physiology

6.3.2.1 Growth of Root Tips in Hard Soil

Mechanically impeded roots are shorter and thicker than roots grown in loose soil (Fig. 6.4). Elongation is slowed due to a decrease in both the rate at which new cells are added onto a file (i.e. the cell flux), and a shorter final cell length. In pea roots grown in sand, for example, a penetrometer resistance of 1.5 MPa decreased the cell flux in the cortex from 5 to 2.2 cells h^{-1}, and final cell length from 0.33 to 0.15 mm (Croser et al. 1999). The elongation zone is shorter in impeded roots, because the local strain rate (i.e. the extension rate per unit length of root) is decreased towards the proximal end of the elongation zone (Croser et al. 1999). The turgor pressure in the cells of the elongation zone drives root growth. The turgor is maintained, or even increased, in impeded roots (Atwell 1990; Atwell and Newsome 1990; Clark et al. 1996; Croser et al. 2000). The slower local strain rate in mechanically impeded roots is partly due to a stiffening of the cell walls in the axial direction, at the rear of the expansion zone, and partly due to the direct externally applied pressure of the soil (Croser et al. 2000).

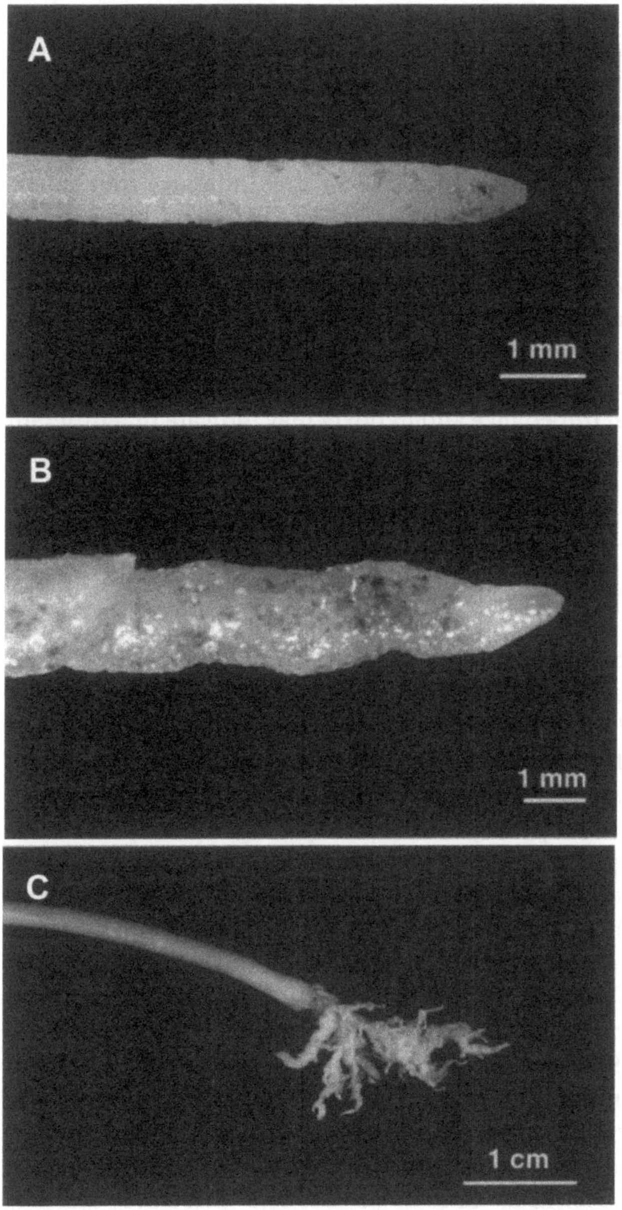

Fig. 6.4A–C. Effect of mechanical impedance on root morphology. **A** Apical region of pea root grown in sandy loam soil of bulk density 1 g cm^{-3} (penetrometer resistance 0.6 MPa). **B** Apical region of pea root grown in same soil at bulk density 1.6 g cm^{-3} (penetrometer resistance 2 MPa). Note the increased root diameter and distorted root surface; **C** Ten-day-old barley root system grown in compacted sand circulated with aerated nutrient solution. Root system is very stunted due to the large mechanical impedance (penetrometer resistance >2 MPa)

Mechanical impedance causes root diameter to increase by up to a factor of two (Materechera et al. 1991) – this is mainly due to an increase in cell diameter in the cortex, but extra cell layers are sometimes found. Cells in the outer layers of the cortex increase in diameter much more than the inner layers, probably due to the confining pressure of the cells surrounding the inner layers (Wilson et al. 1977). Impedance caused the number of cortical cell layers to increase from seven to ten in pea roots (Croser et al. 1999), whereas the number of cortical cell layers remained constant in impeded lupin roots (Atwell 1988). The increase in cell diameter may be associated with the deposition of longitudinally oriented microfibrils in the cell wall (Veen 1982) that results in the cell walls becoming stiffer axially.

Changes in the physical properties of cell walls that influence root extension in response to mechanical impedance are mediated by biochemical processes, probably involving the ubiquitous phytohormone ethylene. Ethylene synthesis increases within 1 h of applying an external mechanical confining stress to roots (He et al. 1996), and it has been shown that inhibitors of ethylene action and synthesis can counteract the effects of impedance on root elongation (Sarquis et al. 1991). He et al. (1996) suggest that the strong synergistic effect of mechanical impedance and hypoxia may indicate that different mechanisms of ethylene biosynthesis promotion operate for these two stresses.

6.3.2.2 Root Branching in Hard Soil

Root systems growing in hard soils look very stunted compared with those grown in loose soil, although there has been relatively little work to quantify the effects of impedance on root branching. This is no doubt because of the considerable time and effort required to excavate, wash, and measure intact roots grown in compacted soil, and the likelihood of damaging them.

The effect of mechanical impedance on lateral spacing in barley was studied by growing roots in growth cells of glass beads circulated with nutrient solution, to which an external pressure was applied (Goss 1977). This enabled relatively easy extraction of the root system with minimal damage. There were twice as many lateral roots per cm of seminal root axis in the impeded treatment, as compared with the unimpeded. More lateral roots per cm of main axis were found for pea and wheat seedlings grown in compacted sandy loam soil (Barley et al. 1965), and lupin roots in compacted sandy clay loam (Atwell 1988). The total number of lateral roots per main axis, however, was the same for lupins grown in compacted soil (Atwell 1988). The total number of laterals decreased by a factor of two for barley grown in the compressed glass beads (Goss 1977). Total lateral production therefore remains unchanged, or is decreased by mechanical impedance, even though main axes become more densely branched.

6.3.3 Localised Compression of Soil Around Roots

Roots influence the soil around them in several important respects. Root tips exert pressures of up to 1 MPa as they compress the soil around them (Misra et al. 1986; Clark et al. 1996), they produce border cells and mucilage, and they absorb water and nutrients from the soil. Growing roots compress the surrounding soil, unless they are growing in a continuous channel or fissure wider than the root diameter. The stress exerted by the root on the soil depends on the soil strength, which increases with increasing bulk density and decreasing water content. The stress exerted on the soil has been predicted to be greatest at the root apex, using a finite element critical-state soil mechanics model (Kirby and Bengough 2002). The stress decreases with increasing distance from the root apex, and with increasing distance from the root surface. Greatest compression of the soil therefore occurs closest to the root surface, and the variation of bulk density with distance from the root has been measured experimentally and modelled by a number of workers (discussed by Young 1998). Changes in porosity adjacent to the root surface potentially affect transport of nutrients, air and water to the root surface, and so are of particular interest.

The bulk density at the surface of maize, pea and wheat roots was between 1.7 and 1.8 g cm^{-3}, as compared with bulk densities of 1.3–1.54 g cm^{-3} in the bulk soil (Braunack and Freebairn 1988; Bruand et al. 1996). The relative compression of soil adjacent to roots was much greater than reported in the early studies of Greacen et al. (1968) who reported densities of about 1.54 g cm^{-3} adjacent to pea roots (compared with 1.50 g cm^{-3} for bulk soil). Cockroft et al. (1969) reported a 8.5 % decrease in voids ratio adjacent to roots in clay soil, though it is not possible to accurately convert this into change in bulk density because the clay was subject to swelling. One possible reason for the smaller values recorded is the relatively poorer resolution of the X-radiograph technique used, which limited spatial resolution to about 0.5 mm, and perhaps evened out the peak in density at the root surface.

The relative increase in density at the root surface, and the distance from the root surface that the soil will be compressed, depends on the root diameter and on how compressible the soil is. The volume of pore space that must be lost from the rhizosphere soil to accommodate the root is equal to the volume of the root. In models of soil compression the distance from the root that the compression extends is therefore expressed as a multiple of the root diameter (Greacen et al. 1968; Dexter 1987). The less compressible the soil, relative to its initial bulk density, the further from the root the zone of compression will extend. The variation in density around an expanding cylinder can be predicted using soil mechanics models (Greacen et al. 1968; Vesic 1972). However, these models require many input parameters of soil mechanical properties, and are difficult to apply. For this reason, Dexter (1987) proposed a simple model of soil compression around roots, relying on an approximately exponential variation in bulk density with distance from the root surface (see Fig. 6.5):

$$\eta(r) = \eta_o + (\eta_i - \eta_o)(1 - e^{-k(r-r_o)/r_o})$$

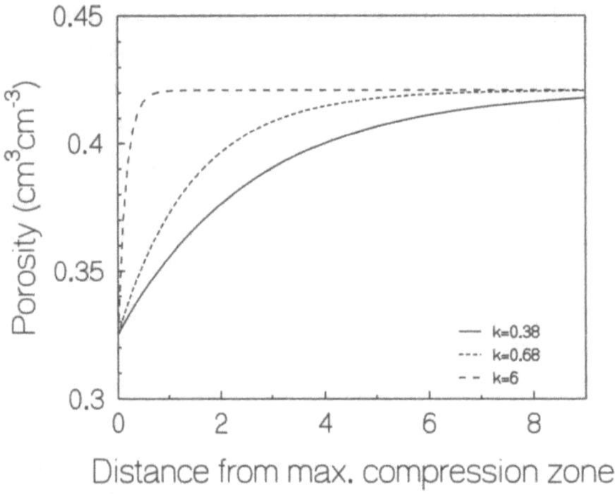

Fig. 6.5. Model of porosity as a function of distance from the zone of maximum compression showing the effect of the '*k*' value (model from Dexter 1987)

where $\eta(r)$ is the porosity at distance r from the root surface, η_o is the porosity in the zone of maximum compression at the root surface, η_i is the porosity of the bulk soil, k is a parameter describing the change in density with distance from the root surface, and r_o is the root radius. The model has been applied with regard to the fungal infection of roots, and the different distances over which pathogenic or mycorrhizal fungi can infect different hosts (Walker and Smith 1988). The model is relatively simple to use, and requires only the minimum porosity to which the soil will be compressed by the root, and the value of the constant k. The minimum porosity can be estimated by compressing the soil under a constant load (e.g. 200 kPa, Dexter 1987), whilst k was found to be between 0.34 and 0.68 from the compression around the root (Greacen et al. 1968), and around an inflating tube (Dexter and Tanner 1972). More recent studies, however, have since found much larger values of k (4.3 for maize roots growing in silty clay loam, Bruand et al. 1996; 6.5 for wheat coleoptiles in clay, Liddell 1992). The variation in k between 0.34 and 6 has a very large effect on the rate of change of porosity outside the zone of maximum compression immediately around roots (Fig. 6.5). We need to develop better ways to estimate values of k for particular soil types and conditions, and so predict rhizosphere compression more accurately. Improved prediction would enable the effects of rhizosphere compaction on fluxes of water, nutrients and gases in the rhizosphere to be modelled, and this is considered in the next section.

6.3.4 Water and Nutrient Uptake

Local compaction of soil in the rhizosphere affects transport of water and nutrients to roots, but the importance of the effect is probably smaller than that of water content and soil texture. Compaction decreases the volume of the pores in the soil of effective diameter greater than about 30 μm (Marshall et al. 1996). Compaction may decrease the continuity of the large pores, decreasing the saturated hydraulic conductivity, and changing the tortuosity of the flow paths through the water-filled pore space. Nye and Tinker (1977) reviewed the effects of changes in bulk density on diffusion in soil cores. Both increases and decreases in diffusion rates have been recorded in response to increasing bulk density, depending on soil type, water content, and diffusing ions.

The effect of bulk density has been clarified for chloride diffusion in sandy loam and silty clay soils (So and Nye 1989). The "impedance factor", f_L (a measure of the tortuosity of the path followed by the solute through the pores – the smaller f_L, the more tortuous the diffusion path), was measured for both soils at three bulk densities and at a range of water contents. At constant volumetric water content, an increase in bulk density decreased f_L in both soils: when the large pores were compressed, replacing air by solid particles, more of the water became closely associated with solid surfaces, increasing the tortuosity of the diffusion path. At constant matric potential, increasing bulk density increased f_L in both soils, due to the associated increase in volumetric water content: when the large pores were compressed, the additional water films associated with new small pores provided additional pathways for diffusion, decreasing the tortuosity. Changing bulk density from 0.8 to 1.2 g cm^{-3} in the silty clay had an effect on f_L equivalent to decreasing matric potential ten-fold. Changing bulk density had a larger effect on f_L than soil texture when the soil was wet, but a smaller effect on f_L when the soil was dry.

There is an optimum bulk density for water and nutrient uptake by crops (Boone 1988). Over-compaction increases mechanical impedance to root growth and decreases the root length available for water and nutrient uptake. Soils that are very loose have poor root-soil contact, increasing the tortuosity of the transport paths to the root surface. A several-fold increase in dry matter yield of oats, for example, occurred when a manganese-deficient soil was compacted (Passioura and Leeper 1963). The degree of soil-root contact has been measured in repacked soil using the thin-section technique (Kooistra et al. 1992). Apparent soil-root contact was very variable, but increased from 60 to 87% as soil bulk density increased from 1.1 to 1.5 g cm^{-3}. In undisturbed soils the degree of soil-root contact may be much smaller if roots are growing in continuous cracks or biopores.

6.4 Root Growth in Macropores

There have been few quantitative studies of root spatial distribution in relation to macropore distribution. Most studies consist of qualitative observations of apparent associations between roots and visible macropores, although recently thin-section techniques have been used to quantify these associations (Stewart et al. 1999). Stewart et al. (1999) found that 11–26 % of roots in established pasture were located in pore space – almost double the number expected by random chance. In Australian hard-setting soils, up to 80 % of wheat roots have been located within the 1 mm thick sheath of soil surrounding macropores (Pierret et al. 1999). The soil in this 'macropore sheath' contains a significantly different microbial population to the bulk soil, due to the combined influence of the macropore physical environment, and the close proximity of the root.

6.4.1 Root Elongation and Distribution in Macropores

Channels and fissures, including biopores formed by earthworms and decayed roots, offer pathways for unimpeded root growth. Cracks narrower than the root diameter also enable more rapid root elongation than in the bulk soil, by partially decreasing mechanical impedance (Whiteley and Dexter 1983). Channels and fissures may allow roots to bypass zones of compacted soil, and so access a larger volume of soil containing water and nutrients. Roots encounter channels and fissures partly at random, although it is possible that gradients in soil properties, such as soil strength or aeration, may bias the direction of growth towards them. Barley roots, for example, grew into artificial cylindrical channels in compacted soil with a frequency greater than that expected by random chance (Stirzaker et al. 1996).

Once a root enters a channel or fissure, it will grow along it until an opportunity arises for the tip to re-enter the bulk soil. The distance that the root will follow the channel will depend on penetration resistance of the surrounding soil, the width of the air gap, and the angle of the channel to the preferred direction of growth. The root is more likely to stay in the fissure if the walls are hard, causing the root to buckle instead of penetrate (Whiteley and Dexter 1983). The chance of a root penetrating the wall is smaller for wide vertical fissures than for narrow horizontal ones (Whiteley and Dexter 1983). Thick roots have a greater chance of crossing air gaps and penetrating macropore walls, due to their greater buckling stress than thinner roots (Whiteley et al. 1982).

When roots grow from the bulk soil into macropores, their growth rate depends on the strength of the bulk soil from which they have grown, for a period of up to 1 week (Goss and Russell 1980; Bengough and Young 1993). The size and persistence of the effect increases with the strength of the bulk soil – for example, pea roots that had grown through a layer of soil with a penetration resistance of 1.8 MPa took 3 days to recover the unimpeded elongation rate (Bengough and Young 1993). This persistence of the effect of soil strength is probably due to the stiffening of cell walls in the proximal region of the elongation zone (Croser et al. 2000). The elongation rate only recovers when new cells have been produced and expanded into the elongation region.

Channels much larger than the root diameter may provide a poor environment for root growth in wet soil (Stirzaker et al. 1996). The reasons for this are not clear, but could be associated with poor soil-root contact in the large artificial biopores, and perhaps even growth inhibitors produced by the root tip. Much better plant growth was obtained when roots were grown in a network of narrow root channels formed by decaying lucern and ryegrass roots. In zero tillage systems these networks of root and earthworm channels can counteract the effects of increasing soil strength, even though they may represent less than 1 % of the total soil volume (Ehlers et al. 1983).

Root-soil contact may be an important problem in relation to water and nutrient uptake for roots contained in biopores. Diurnal shrinkage of the root diameter by up to 40 % may occur in roots grown in biopores (Huck et al. 1970). Such shrinkage is less likely to occur when roots are in close contact with the soil (Passioura 1988). Hence, perhaps the phenomenon is mainly limited to areas where soil-root contact was initially poor.

The presence of roots in a relatively small number of continuous macropores gives rise to a clustered root distribution. Clumping of roots can also be caused, at the cm scale, by the production of many short laterals from the parent root axis (Tardieu 1988a,b; Grabarnik et al. 1998). Clumping of maize roots has been observed by plotting maps of root intersection with excavated planes, on both compacted and uncompacted clay loam soil (Tardieu 1988a). Root length density measured in 8-cm^3 volumes in the non-compacted area had a very skewed frequency distribution, with a mode of <0.1 cm cm^{-3}, a mean of 0.85 cm cm^{-3}, and a maximum of >5 cm cm^{-3} (Tardieu 1988b). The half mean distance between roots in these experiments was 2 cm, if it was assumed that the RLD was uniform. Direct calculation from root maps gave half mean distances of more than 20 cm, because of root clumping – this has important implications for the calculation of water and nutrient uptake, and this will be discussed in the next section.

6.4.2 Effect of Root Clumping on Water and Nutrient Uptake

Regularly spaced roots should, theoretically, extract water and nutrients most efficiently from a soil in which these resources are distributed uniformly. In reality, roots tend to be distributed more randomly in the cultivated topsoil, and clumped in continuous macropores in the subsoil. Passioura (1988, 1991) performed calculations of the likely effect that root clumping within different arrangements of macropores would have on water uptake. Roots that were confined to macropores were assumed to act as a single root, or plane of roots, of extent equal to the length or area of the macropores. The time taken for the root to extract 63 % of the initial water content was estimated for a dry soil in which the hydraulic conductivity was limiting uptake. If roots were solely confined to biopores, the time was between 10 and 100 days, compared with less than 3 days for evenly distributed roots in the same soil. Different shapes of soil structural unit (slabs, prisms, cubes) resulted in faster extraction of water than the biopores, although still up to an order of magnitude slower than for evenly distributed roots.

The "root position effectivity ratio" was proposed as a useful framework for comparing the effectiveness of different root distributions on uptake (Van Noordwijk et al. 1993). The starting point was a two-dimensional map of root intersections with an excavated plane in the soil. The frequency distribution of distances from randomly chosen points in the soil to the roots was determined. A set of circular annuli that conserved the frequency distribution of distances for an 'average' root was constructed (using the method of Rappoldt 1991). The cylindrical root uptake model can then be used to predict uptake from the weighted sum of each cylinder class. The equivalent length density, L^*, of regularly spaced roots that would leave the same residue in the soil after a given time is then calculated. The root position effectivity ratio, R_{per}, is then equal to L^*/L, where L is the actual root length density for the measured root distribution. The value of R_{per} ranges between unity for regularly distributed roots, tending to zero for very clumped roots. This is an attractive and potentially powerful approach, as it gives a simple ranking of how effective real root distributions are.

6.5 Ecological Consequences of Soil Structure and Strength

Soil structure and strength feature prominently in standard texts on root growth and function (e.g. Russell 1977; Wild 1988; Waisel et al. 1996), and is widely recognised to be of fundamental importance in the agronomic and farming communities. Soil structure plays a major role in determining the physical fertility of soils, and yet it receives cursory mention in important ecological texts (e.g. Grime 2001). Perhaps one reason for this is that the vast bulk of research on the consequences of soil strength and structural degradation has concentrated on agricultural crops. These species are generally grown in vast monocultures in soils fertilised with agrochemicals or animal manures.

So what are the likely ecological consequences of soils with large strength and degraded physical structures on mixed plant communities? This area is in need of further research, but some simple generalities appear possible. It is interesting to consider plant functional types as the three extremes identified by Grime – competitors (undisturbed, fertile sites), stress tolerators (infertile, undisturbed sites), and ruderals (fertile, disturbed sites). Decreased root length associated with hard soils will increase nutrient and water depletion zones around roots. This is likely to restrict nutrient and water uptake in the longer term, favouring the stress tolerators over the competitor plant strategies. In nutritionally fertile but physically degraded soils, short growth windows may present themselves where sufficient nutrients are available for the growth of ruderal types, whereas competitor types may run into water limitation if rooting depths are restricted. The systematic screening of plants under laboratory conditions (e.g. Grime et al. 1997) should allow better prediction of community performance in important classes of physically degraded soils, and is very relevant to the establishment of appropriate plant communities in the many structurally degraded urban, embankment, and industrial soils.

An example of the local separation of species in soils with very different physical conditions is the river forelands in the Netherlands (Engelaar et al. 1993). Plantago and Rumex species could be separated within a very short distance into those adapted to trampled, compacted conditions, and those able to survive on intermittently flooded soils. The formation of aerenchyma in *R. palustris* under waterlogged conditions was inhibited in compacted soils, limiting colonisation of compacted waterlogged areas by this species.

It is possible that studying natural plant communities evolved to grow in hard soils will help to identify particular traits of interest and relevance to plant breeders: it is already recognised that wild relatives of crop plants provide an important genetic resource.

6.6 Summary and Prospects

Hard soils with degraded structures present a major limitation to root growth and function. Soil strength and structure greatly influence root distribution and hence exert strong control on nutrient and water uptake by roots. The relations between soil structure and root growth and function are still poorly understood, due to the experimental difficulties in adequately studying root systems in situ. New techniques, such as computer tomography in association with X- or gamma-ray scanning, are allowing the imaging and visualisation of macropores in three dimensions (3-D) non-destructively. Image analysis and numerical simulation can be used in conjunction with the soil thin-section technique, to enable the quantitative prediction of soil hydraulic properties. The simulation of soil structures, water and nutrient transport in 3-D is computer intensive, and has only recently become possible at the pore-scale. This area should develop rapidly during the next few years, and bring new insights into how soil structure affects root uptake and function.

To exploit these advances, more quantitative data on root spatial distribution with respect to soil pore structure is required. Better understanding of root spatial distribution in the soil fabric will allow us to predict the effects of soil structure and management on plant uptake of water and nutrients. New conceptual frameworks, such as the least limiting water range (LLWR), are needed so that detailed root-scale information on the effects of soil physical stresses on root growth can be interpreted in relation to soil conditions prevailing in the field. Such frameworks will require considerable effort to evaluate, but do bridge the gap between laboratory studies and practical application in agriculture and horticulture.

Advances in understanding of the physiological responses of roots to mechanical impedance suggest there is considerable potential to manipulate the response. The identification of ethylene as a relevant phytohormone is important; however, its wide involvement in many stress responses imply that there will not be a quick-fix. Relatively little is known about the effects of soil strength on root growth in natural plant communities – up to now research has concentrated on agricultural crop plants. It may be that plant breeders can learn by studying traits in wild relatives adapted to particular harsh environments. Conceptual frameworks such as the LLWR should also allow us to evaluate the probable benefits of breeding plants with root systems better adapted to adverse soil physical conditions.

Acknowledgements. The Scottish Crop Research Institute receives grant-in-aid from the Scottish Executive Rural Affairs Department.

References

Asseng S, Aylmore L, MacFall J, Hopmans J, Gregory P (2000) Computer-assisted tomography and magnetic resonance imaging. In: Smit A, Bengough AG, Engels C, Van Noordwijk M, Pellerin S, Van de Geijn S (eds) Root methods: a handbook. Springer, Berlin Heidelberg New York, pp 343–364

Atwell BJ (1988) Physiological responses of lupin roots to soil compaction. Plant Soil 111:277–281

Atwell BJ (1990) The effect of soil compaction on wheat during early tillering. II. Concentrations of cell constituents. New Phytol 115:17–41

Atwell BJ, Newsome JC (1990) Turgor pressure in mechanically impeded lupin roots. Aust J Plant Phys 17:49–56

Barley KP, Farrell, DA, Greacen, E L (1965) The influence of soil strength on the penetration of a loam by plant roots. Aust J Soil Res 3:69–79

Bartoli F, Philippy R, Doirisse M, Niquet S, Dubuit M (1991) Structure and self-similarity in silty and sandy soils – the fractal approach. J Soil Sci 42:167–185

Bengough AG, Mullins CE (1990) Mechanical impedance to root growth – a review of experimental techniques and root growth responses. J Soil Sci 41:341–358

Bengough AG, Mullins CE (1991) Penetrometer resistance, root penetration resistance and root elongation rate in 2 sandy loam soils. Plant Soil 131:59–66

Bengough AG, Young IM (1993) Root elongation of seedling peas through layered soil of different penetration resistances. Plant Soil 149:129–139

Blackwell MV, Wells EA (1983) Limiting oxygen flux densities for oat root extension. Plant Soil 73:129–139

Boone FR (1988) Weather and other environmental factors influencing crop responses to tillage and traffic. Soil Til Res 11: 283–324
Braunack MV, Freebairn DM (1988) The effect of bulk density on root growth. Int Soil Til Res Org 11th Int Conf Edinburgh 1:25–30
Bruand A, Cousin I, Nicoullaud B, Duval O, Begon JC (1996) Backscattered electron scanning images of soil porosity for analyzing soil compaction around roots. Soil Sci Soc Am J 60:895–901
Burke W, Gabriels D, Bouma J (1986) Soil structure assessment, Balkema, Rotterdam
Clark LJ, Whalley WR, Dexter AR, Barraclough PB, Leigh RA (1996) Complete mechanical impedance increases the turgor of cells in the apex of pea roots. Plant Cell Environ 19:1099–1102
Cockroft B, Barley KP, Greacen EL (1969) The penetration of clays by fine probes and root tips. Aust J Soil Res 7:333–348
Crawford JW, Matsui N, Young IM (1995) The relation between the moisture-release curve and the structure of soil. Eur J Soil Sci 46:369–375
Croser C, Bengough AG, Pritchard J (1999) The effect of mechanical impedance on root growth in pea (*Pisum sativum*). I. Rates of cell flux, mitosis, and strain during recovery. Physiol Plant 107:277–286
Croser C, Bengough AG, Pritchard J (2000) The effect of mechanical impedance on root growth in pea (*Pisum sativum*). 2. Cell expansion and wall rheology during recovery. Physiol Plant 109:150–159
da Silva AP, Kay BD (1997) Effect of soil water content variation on the least limiting water range. Soil Sci Soc Am J 61:884–888
da Silva AP, Kay BD, Perfect E (1994) Characterization of the least limiting water range of soils. Soil Sci Soc Am J 58:1775–1781
da Silva AP, Kay BD, Perfect E (1997) Management versus inherent soil properties effects on bulk density and relative compaction. Soil Til Res 44:81–93
Dexter AR (1987) Compression of soil around roots. Plant Soil 97:401–406
Dexter AR, Tanner DW (1972) Soil deformations induced by a moving cutting blade, and expanding tube and a penetrating sphere. J Agric Eng Res 17:371-375
Ehlers W, Kopke U, Hesse F, Bohm W (1983) Penetration resistance and root-growth of oats in tilled and untilled loess soil. Soil Til Res 3:261–275
Engelaar WMHG, Vanbruggen MW, Vandenhoek WPM, Huyser MAH, Blom CWPM (1993) Root porosities and radial oxygen losses of rumex and plantago species as influenced by soil pore diameter and soil aeration. New Phytol 125:565–574
Goss MJ (1977) Effects of mechanical impedance on root growth in barley (*Hordeum vulgare* L.). I. Effects on the elongation and branching of seminal root axes. J Exp Bot 28:96–111
Goss MJ, Russell RS (1980) Effects of mechanical impedance on root growth in barley (*Hordeum vulgare* L.). III. Observation on the mechanism of response. J Exp Bot 31:577–588
Grabarnik P, Pages L, Bengough AG (1998) Geometrical properties of simulated maize root systems: consequences for length density and intersection density. Plant Soil 200:157–167
Greacen EL, Farrell DA, Cockroft B (1968) Soil resistance to metal probes and plant roots. Trans 9th Congr ISSS 1:769–779
Grime JP (2001) Plant strategies, vegetation processes, and ecosystem properties. Wiley, Chichester
Grime JP, Thompson K, Hunt R et al (1997) Integrated screening validates primary axes of specialisation in plants. Oikos 79: 259–281
He CJ, Finlayson SA, Drew MC, Jordan WR, Morgan PW (1996) Ethylene biosynthesis during aerenchyma formation in roots of maize subjected to mechanical impedance and hypoxia. Plant Physiol 112:1679–1685

Kirby JM, Bengough AG (2002) Influence of soil strength on root growth – experiments and analysis using a critical state model. Eur J Soil Sci 53:1–9

Huck MG, Klepper B, Taylor HM (1970) Diurnal variations in root diameter. Plant Physiol 45:529–530

Kooistra MJ, Schoonderbeek D, Boone FR, Veen BW, Vannoordwijk M (1992) Root-soil contact of maize, as measured by a thin-section technique. 2. Effects of soil compaction. Plant Soil 139:119–129

Liddell CM (1992) Experimental approach to measure the compression of soil around emerging wheat coleoptiles. Soil Biol Biochem 24:471–477

Marshall TJ, Holmes JW, Rose CW (1996) Soil physics, 3rd edn. Cambridge University Press, Cambridge

Materechera SA, Dexter AR, Alston AM (1991) Penetration of very strong soils by seedling roots of different plant species. Plant Soil 135:31–41

Mirreh HF, Ketcheson JW (1973) Influence of soil water matric potential and resistance to penetration on corn root elongation. Can J Soil Sci 53:383–388

Misra RK, Dexter AR, Alston AM (1986) Maximum axial and radial growth pressures of plant roots. Plant Soil 95:315–318

Murphy CP, Bullock P, Biswell KJ (1977) The measurement and characterisation of voids in soil thin sections by image analysis. 1. Principles and techniques. J Soil Sci 28:498–508

Nye PH, Tinker PB (1977) Solute movement in the soil-root system, 1st edn. Blackwell, Oxford

Passioura JB (1988) Water transport in and into roots. Ann Rev Plant Physiol Plant Mol Biol 39:245–265

Passioura JB (1991) Soil structure and plant-growth. Aust J Soil Res 29:717–728

Passioura, JB, GW Leeper (1963) Soil compaction and manganese deficiency. Nature 200:29–30

Pierret A, Moran CJ, Pankhurst CE (1999) Differentiation of soil properties related to the spatial association of wheat roots and soil macropores. Plant Soil 211:51–58

Quisenberry VL, Phillips RE, Zeleznik JM (1994) Spatial distribution of water and chloride macropore flow in well-structured soil. Soil Sci Soc Am J 58:1294–1300

Rappoldt C (1991) The application of diffusion models to aggregated soil. Soil Sci 150:645–661

Ringrose-Voase AJ (1996) Measurement of soil macropore geometry by image analysis of sections through impregnated soil. Plant Soil 183:27–47

Rogasik H, Crawford JW, Wendroth O, Young IM, Joschko M, Ritz K (1999) Discrimination of soil phases by dual energy x-ray tomography. Soil Sci Soc Am J 63:741–751

Russell RS (1977) Plant root systems. Their function and interaction with the soil. McGraw Hill, Maidenhead, 298 pp

Sarquis JI, Jordan WR, Morgan PW (1991) Ethylene evolution from maize (*Zea mays* L.) seedling roots and shoots in response to mechanical impedance. Plant Physiol 96:1171–1177

Sharp RE, Silk WK, Hsiao TC (1988) Growth of the primary maize root at low water potentials. I. Spatial distribution of expansive growth. Plant Physiol 87:50–57

So HB, Nye PH (1989) The effect of bulk-density, water-content and soil type on the diffusion of chloride in soil. J Soil Sci 40:743–749

Stirzaker RJ, Passioura JB, Wilms Y (1996) Soil structure and plant growth: impact of bulk density and biopores. Plant Soil 185:151–162

Stewart JB, Moran CJ, Wood JT (1999) Macropore sheath: quantification of plant root and soil macropore association. Plant Soil 211, 59–67

Tardieu F (1988a) Analysis of the spatial variability of maize root density. I. Effect of wheel compaction on the spatial arrangement of roots. Plant Soil 107:259–266

Tardieu F (1988b) Analysis of the spatial variability of maize root density. II. Distances between roots. Plant Soil 107:267–272

Taylor HM, Ratliff LH (1969) Root elongation rates of cotton and peanuts as a function of soil strength and water content. Soil Sci 108:113–119

Topp, GC, Galganov YT, Wires KC, Culley JLB (1994) Nonlimiting water range (NLWR): an approach for assessing soil structure. Soil quality evaluation program Technical Report 2, Centre for Land and Biological Resources Research, Agriculture and Agrifood Canada, Ottawa

Van Noordwijk M, Brouwer G, Harmanny K (1993) Concepts and methods for studying interactions of roots and soil structure. Geoderma 56:351–375

Veen BW (1982) The influence of mechanical impedance on the growth of maize roots. Plant Soil 66:101–109

Veen BW, Boone FR (1990) The influence of mechanical resistance and soil-water on the growth of seminal roots of maize. Soil Til Res 16:219–226

Vesic AS (1972) Expansion of cavities in infinite soil mass. J Soil Mech Found Div 98:265–290

Voorhees WB, Farrell DA, Larson WE (1975) Soil strength and aeration effects of root elongation. Soil Sci Soc Am Proc 39:948–953

Waisel Y, Eshel A, Kafkafi U (1996) Plant roots: the hidden half. Marcel Dekker, New York, 1002 pp

Walker NA, Smith SE (1988) Effect of soil compression on estimates of rhizosphere width – comparing Ferriss equation with Gilligan. Phytopathology 78:253–255

Warnaas BC, Eavis BW (1972) Soil physical conditions affecting seedling root growth. II. Mechanical impedance, aeration and moisture availability as influenced by grain-size distribution in silica sands. Plant Soil 36:623–634

Whiteley GM, Dexter AR (1983) Behavior of roots in cracks between soil peds. Plant Soil 74:153–162

Whiteley GM, Hewitt JS, Dexter AR (1982) The buckling of plant roots. Physiol Plant 54:333–342

Wild A (1988) Russell's soil conditions and plant growth. Longman Scientific & Technical, Harlow, 991 pp

Wilson AJ, Robards AW, Goss MJ (1977) Effects of mechanical impedance on root growth in barley, *Hordeum vulgare* L. II. Effects on cell development in seminal roots. J Exp Bot 28:1216–1227

Young IM (1998) Biophysical interactions at the root-soil interface: a review. J Agric Sci 130:1–7

7 Adaptation of Roots to Drought

W.J. Davies and M.A. Bacon

7.1 Introduction

In this chapter we examine the morphological, physiological and biochemical adaptations of roots to drought and discuss how roots perceive soil drying and communicate such information to the shoots.

7.1.1 Soil Drying – a Composite Stress

Soil drying places a number of different constraints on the growth and functioning of roots and most of these are relatively ill defined. This is because of the highly heterogeneous nature of the rooting environment, the delicate nature of the relationship between roots and soil structure and the difficulty of investigating root growth and functioning without disrupting this relationship. When soil water availability decreases, roots may respond to water status changes in soil water status, soil mechanical strength, changed water flux into the root, localised drying, modification of nutrient mobility from soil to root, or a combination of these effects.

7.1.1.1 Changes in Soil Water Status

The ease with which soil gives up water to the root is measured in terms of water potential; this is given units of pressure, to describe the hydrostatic 'pull' of the soil on water. As the water content of a soil declines, its water potential will decrease (i.e. become more negative), as water molecules become less freely available and more tightly bound to the soil particles. The growth and functioning of the root will respond to the soil water potential in the rhizosphere immediately adjacent to the root, as opposed to the actual water content of the bulk soil on a volume-for-volume basis. The water potential of a particular soil is dependent on its water content and its specific physical properties.

Ecological Studies, Vol. 168
H. de Kroon, E.J.W. Visser (Eds.)
Root Ecology

These will also influence the heterogeneity and distribution of water through the soil and, because water moves very slowly though drying soil, there may be very substantial gradients in soil water potential over a few millimetres adjacent to a root.

A primary response to soil drying is reduced water content within the root. In the absence of any physiological adaptation, reduced water content will result in a reduced ability to sustain an adequate turgor pressure within root cells and maintain root growth. This hydrostatic pressure is required to drive the expansion of root cells, by deforming the cell wall surrounding the cell. Although the intrinsic role of turgor in regulating growth is currently debated (see e.g. Kramer 1988), there is no doubt that reduced water availability will reduce turgor and therefore growth rate, if no physiological or biochemical adaptations to the growth process occurs. As will be discussed later, the active regulation of cell wall properties and cellular turgor are key adaptations of roots growing under reduced water availability.

7.1.1.2 The Pathway of Water Movement

It is now well understood that root resistances to water uptake will vary as the pathway for water movement into roots can vary between species and as the fluxes of water vary (Steudle and Petersen 1998; Chap. 5). Most roots develop an endodermis with water-proofing polymers in the cell walls. This forces water and ions moving radially in roots to cross a membrane somewhere in the radial pathway, providing substantial resistance to uptake. However, at higher fluxes, more water may move through alternative pathways. One feature of root structure that has received attention recently is the existence in some roots of an apoplastic bypass for water flux (see Hartung et al. 2002). The existence of a radial pathway for water and ions into roots where a membrane does not influence flux will mean that different driving forces for water uptake will dominate. This will also impact on the effects of transpiration flux on the delivery of hormones and other xylem-borne solutes to the shoot (Freundl et al. 1998). In a similar manner to the endodermis, some roots develop a Casperian band in the hypodermis and this will also affect the pathway and magnitude of radial ion and water flux. Although water movement into and through the root is predominantly regulated by physical forces, we are now beginning to understand that metabolically regulated water channels in roots (aquaporins) can fine tune water fluxes, perhaps in response to transpiration demand (Tyerman et al. 1999). All of these variables will influence root water status in drying soil and will therefore modify growth and other aspects of root functioning.

7.1.1.3 Other Variables

As soil dries, uptake of mineral nutrients will also be modified (e.g. Shaner and Boyer 1976) and soil strength may also be increased. Changes in plant growth and functioning as soil dries may be at least in part attributable to changes in the physical properties of the soil or to limited nutrient uptake, but, again, the extent of these effects is largely undefined. Lips (1997) has shown that when nitrate supply to plants is limiting due to soil drought then biochemical nitrate reduction can be shifted from the shoot to the root with the result that xylem composition and pH can vary dramatically. These variables can be important root signals regulating shoot growth and functioning (see below).

It is clear therefore that the study of root adaptation to soil drying may involve responses to a combination of different factors, the prevalence of which will depend on the specific properties of the soil in which a plant grows. We must therefore be cautious in extrapolating our findings of apparent adaptation, when we look at such factors in isolation.

In this chapter, we will first consider the gross morphological adaptations which occur in root systems in response to soil drying and identify common adaptations in a variety of natural and managed systems. This will then lead onto a discussion of the complex physiological and biochemical adaptation of roots, uncovered by an expanding body of detailed research, before considering how plant growth regulators may regulate such adaptation. The chapter concludes with the consideration of roots as the primary sensors of soil water availability and their pivotal role in the generation of root-borne signals that enable the plant to perceive and respond to the continuous variation in water availability. In this way we hope to both demonstrate the sophistication of the mechanisms that have evolved to allow sustained access to water when its availability declines and illustrate the role of roots in the control of plant water balance.

7.2 Growth of Roots in Drying Soil

7.2.1 Morphological Adaptations to Drying Soil

Root growth is most commonly reduced as soil dries but often not as much as shoot growth with the result that the root-to-shoot ratio of plants commonly increases at low soil water potentials (Fig. 7.1). There are a very few data that suggest that root growth can actually be increased by soil drying. Those that do (e.g. Sharp and Davies 1979) attribute such effects to a stress of particular magnitude which results in increased availability of assimilates to roots, as

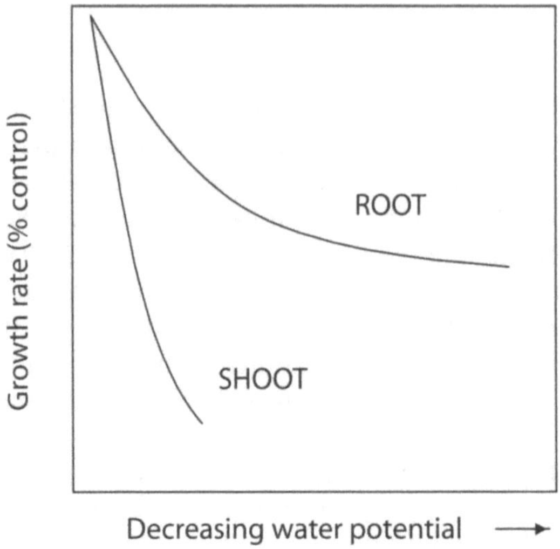

Fig. 7.1. Generalised response of root and shoot growth rates in response to an increasing water deficit around the roots

shoot growth is limited by water deficit in the absence of any effect on carbon gain.

One of the common responses to soil drying is that roots show enhanced geo-tropism (e.g. Sharp and Davies 1985). An increased rooting depth can significantly increase water uptake by root systems even when a relatively few roots are involved. The adaptive significance of responses of this kind is only clear if plants are competing in natural communities for different soil water resources. There is nothing to be gained by plants in a monoculture investing increased carbohydrate into deeper rooting when all the plants in the stand are competing for the same reserves of soil water.

As soil water supply is restricted, we also commonly observe changes in the diameter of roots. At low water potential, in substrates with a low mechanical impedance (i.e. roots can penetrate the substrate easily), roots have been observed to thin, an adaptation presumably to commit limited carbohydrate supply to extension growth and allow plants to explore new and deeper water reserves (Sharp et al. 1988). However, in most soils with high mechanical impedance, roots have been shown to swell as soil dries, particularly behind the root apex (Spollen et al. 2000). The prevalence of this phenomenon could allow roots to continue to penetrate the soil as its mechanical impedance increases on drying. Some species of plants are apparently well adapted to compacted soils. Different species will clearly have differing abilities to penetrate soil as mechanical impedance increases. A sustained ability to penetrate soils of increasing impedance may be related to a capacity of these plants to generate high turgors in root tips (see Richards and Greacen 1986; Atwell and Newsome 1990). Roots of many plants in compacted soils are restricted to cracks in the soil structure (see also Chap. 6). As a result, roots will often be clumped in these fissures causing substantial localised drying, even when the water content of the bulk soil is still quite substantial. This can have substantial implications for root signalling (see below). Tardieu et al. (1992) have argued that clumped roots will generate root signals at comparatively high soil water contents, with the signals reflecting the reduced access that the plants have to available soil water.

7.2.2 Physiological Adaptation of Roots to Soil Drying

A true understanding of how root growth is modified and regulated during periods of drought can only be achieved by focusing attention within the root elongation zone. Figure 7.2A illustrates the typical effect of a soil drying episode or low water potential treatment on the spatial distribution of growth within the apical root elongation zone of a generalised primary root. As root

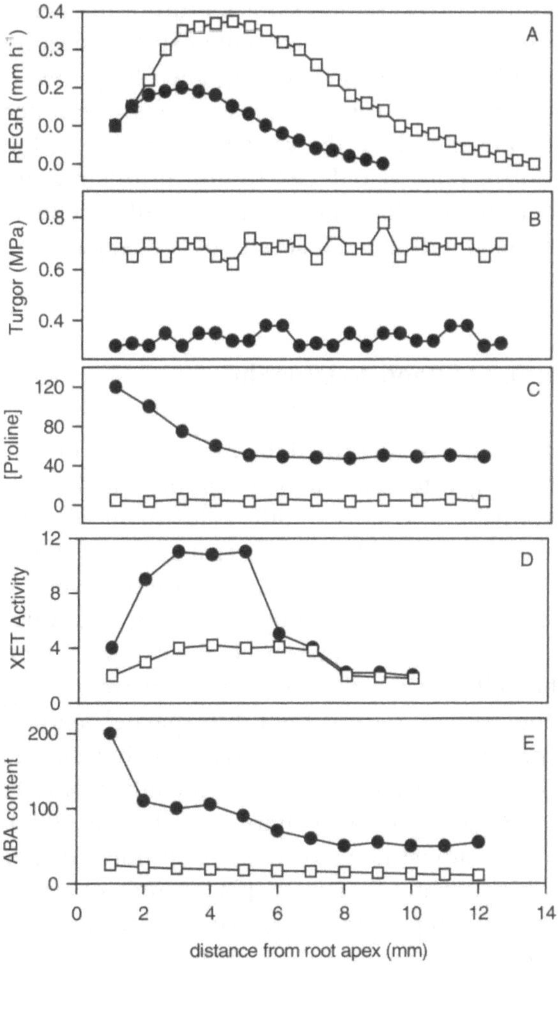

Fig. 7.2A–E. Typically observed differences in the distribution of **A** growth rates (after Sharp et al. 1988); **B** cellular turgor (after Spollen and Sharp 1991); **C** proline content; **D** xyloglucan endotransglycosylase activity (after Wu et al. 1994); and **E** ABA content (after Saab et al. 1990), within the elongation zone of a primary root growing in high (□) or low water potential (●) media

water potential declines, maximal growth rates move apically (i.e. towards the root tip) and local growth rates throughout the majority of the zone are reduced (e.g. Sharp et al. 1988). One exception to this is the very apical region of the elongation zone. In this region, local growth rates are maintained at levels recorded under well-watered or high water potential conditions. The sustained rates of local expansion close to the root apex translate into a sustained, but reduced, rate of root elongation.

Over the last 10–15 years, Sharp and coworkers have presented a large body of work that addresses the regulation of growth of the primary root of the maize seedling during drought. Early in this work it was recognised that any causal growth analysis required investigation of the region of the root that was actively expanding (see Spollen et al. 1993). At a whole-organ level, growth rate slows when the root is exposed to a low water potential. Many plants show this slowing in growth rate when water supply is limited and it is often explained by a presumed reduction in root turgor in the elongating zone. There are in fact very few direct measurements of root turgor in drying soil. Spollen and Sharp (1991) have demonstrated that, in their system, despite the existence of active solute regulation (see e.g. Voetberg and Sharp 1991), turgor throughout the growing zone of the root declines as substrate water potential declines (Fig. 7.2B). However, cells in the very apical region of the roots continue to expand at control rates at reduced turgor. In this region, therefore, the fall in turgor did not restrict local growth rate. A fall in cellular turgor cannot therefore explain the variation in local rates of root expansion, at least in the root apex (Fig. 7.2A,B).

7.2.3 The Biochemical Adaptation of Roots to Drought

Early follow-up work from that which first identified the maintenance of apical growth at low water potential suggested that increases in proline deposition within this region may contribute to an osmotic adjustment of cells exhibiting growth maintenance (Sharp et al. 1990; Voetberg and Sharp 1991). Accumulation of proline (Fig. 7.2C), an imino acid, will decrease the osmotic potential and consequent water potential of a growing cell (i.e. its accumulation will increase the ability of that cell to pull in water via the osmotic effect of proline dissolved within the cell). This active 'osmotic adjustment' can maintain a flux of water into the cell and permit the full or partial maintenance of cellular turgor – the primary driver of expansive growth. Osmotic adjustment of roots is reported in a variety of species, including woody species (Osonubi and Davies 1978) and grasses (Voliare and Thomas 1995). Although the potential for osmotic adjustment to maintain cellular turgor and growth is compelling, its significance has been placed into question (Munns 1988). If sugars and other compatible solutes such as proline are being used to sustain favourable cellular turgors, their diversion from other parts of the plant's metabolism, or their increased synthesis, may compromise the overall carbon/protein allocation to other parts of the plant's metabolism, including cell wall synthesis. Nevertheless, solute regulation in roots does sustain some cell growth at low water potential (Sharp et al. 1990; Voetberg and Sharp 1991). In contrast, however, solute regulation in leaves often seems to be a consequence of a limitation of cell growth and the most effective shoot solute regulators are often those species which show the greatest growth limitations to soil drying (Kuang et al. 1990).

The observation that growth rates within roots are maintained at reduced turgor leads us to suggest that the structure of the cell wall may change in some way to allow it to yield or deform more easily at lower turgor. This has led researchers to investigate the effects of reduced water availability on the activity of several proposed enzymatic regulators of cell wall extensibility, particularly wall-associated xyloglucan endotransglycosylase (XET), expansins (cell wall looseners) and peroxidases (cell wall stiffeners). There are now many informative, correlations of such in vitro activities with distinct growth profiles within the root elongation zone but correlations do not demonstrate causal relationships and the use of molecular tools will be required to establish cause and effect.

Xyloglucan endotransglycosylase (XET) was one of the first enzymes to be implicated in the regulation of root cell expansion at low water potentials and is found in a broad range of plant species from simple liverworts (e.g. *Marchantia polymorpha*) to broad-leaved deciduous trees (e.g. *Acer pseudoplatanus*; see Fry et al. 1992). Interestingly, some of the highest activities of this enzyme are found in the gramineae (including *Zea mays*, *Holcus lanatus* and *Bromus erectus*), which actually possess relatively small amounts of the enzyme's cell wall substrate. Regulated cleavage and reforming of biochemical bonds within the cell wall by this enzyme provides a potential mechanism by which controlled expansion can take place. Indeed, Wu et al. (1994) have reported enhanced extractable XET in the region of the root that exhibited growth maintenance at low water potential and reduced turgor (Fig. 7.2D).

More recently, a similar body of evidence points towards another group of cell wall enzymes (called expansins) as catalysers of cellular expansion. These enzymes, with acidic pH optima, appear to disrupt hydrogen bonding between cell wall polymers which maintain the integrity of the cell wall structure and catalyse the irreversible extension of plant cell walls. Wu et al. (1996) have demonstrated an increase in cell wall yielding in the very apex of the maize root at low water potential which correlated well with an increase in expansin protein levels and wall susceptibility to expansins.

7.2.4 Regulation of the Morphological, Physiological and Biochemical Responses of Roots to Soil Drying

7.2.4.1 A Role for Abscisic Acid?

The plant growth regulator abscisic acid (ABA) is increasingly implicated in coordinating the responses of roots and shoots to a change in soil water availability. A large body of evidence demonstrates that this plant growth regulator accumulates in the roots in response to drying soil and travels from the roots in the xylem vessels, to the shoots, acting as a root-to-shoot signal of variation in soil water availability. The potential for ABA to act as a root signal has been investigated in a range of plant species, including woody angiosperms and gymnosperms (e.g. Jackson et al. 1995; Auge et al. 2000; Comstock 2002), grasses (e.g. Puliga et al. 1996), and broad-leafed herbaceous plants (e.g. Neales et al. 1991). The full nature of the signalling mechanism and the effects of ABA on shoot growth are still under study. However, ABA clearly plays a central role in mediating a restriction in transpirational water loss via its potent effect on the guard cells that surround the stomatal pores on the leaf surface (see Hetherington 1998).

As the soil dries, the rate of shoot growth is commonly reduced more than the rate of root growth, leading to the frequently-reported change in root-to-shoot ratio (Fig. 7.1). This often-observed phenomenon is also thought to be coordinated by plant growth regulators such as ABA. Saab et al. (1990) showed that ABA accumulation within the root elongation zone of primary maize roots (Fig. 7.2E) was required for the maintenance of apical root growth. The action of ABA can be modified chemically using fluridone (an inhibitor of carotenoid and ABA biosynthesis) or by using mutant plants which do not possess key enzymes in the ABA biosynthetic pathway. Work from Sharp's laboratory has shown that inhibition of ABA synthesis in the whole plant, via fluridone treatment or use of the maize mutant *vp5*, prevents both accumulation of ABA and the maintenance of primary maize root growth at low water potential (Sharp et al. 2002). It is difficult to argue unequivocally that this is a direct effect of ABA on root cell expansion, as ABA synthesis throughout the whole plant was restricted, but these data provide powerful evidence for the involvement of ABA in growth regulation at low tissue water potentials. Evidence within the literature suggests that ABA can restrict cellular expansion in many situations, particularly in shoots (see Sharp et al. 2002). Its role in maintaining expansion, at least in the root apex, would therefore suggest that ABA may have very contrasting modes of action in different organs or parts of organs. It is also clear from Sharp's work that the nature and magnitude of the ABA effect on growth is critically dependent on the water status of the plant tissue with, for example, the growth-maintaining effect of ABA on roots at low water potential, not apparent at high water potentials.

It is difficult to measure anything but the bulk amount of ABA in the root but a more useful measurement to establish a causal link to growth regulation would be the concentration or amount at the active site for growth regulation. Bacon et al. (1998) established that the concentration of ABA at its site of action in relation to growth of leaves could be tightly regulated by apoplastic pH (i.e. the pH of the fluid bathing the exterior surface of growing cells). So much so that no **accumulation** of ABA is necessary to explain the regulation of growth by ABA as water availability is reduced. A simple increase in the pH of the apoplast is sufficient to increase the residency time of ABA within the apoplast before it is taken up into the symplast. In roots, if we presume the active site for growth regulation to be the apoplast surrounding the root cells, we should consider the concentration of ABA at such active sites as a more appropriate variable to quantify. Such quantification might be achieved using techniques pioneered by Outlaw's laboratory (see e.g. Lu et al. 1997). This concentration is not necessarily reflected by the total amount of ABA accumulating in such tissue. Although, in Saab's work, ABA appeared to accumulate to the greatest extent in the region where growth was maintained; apoplastic pH has been observed to be significantly lower in this region of the root than in the rest of the elongation zone (Winch and Pritchard 1999). We may therefore speculate that, although the total amount of ABA in this region is high, the actual concentration at the active site for growth regulation may indeed be no different, or even lower than that elsewhere in the elongation zone.

The increasing availability of genetic mutants in several species has led to their frequent use as a tool for the investigation of cellular mechanisms of growth regulation. ABA biosynthesis inhibitors and mutants have been used to inhibit ABA accumulation through the entire root elongation zone. In these investigations, however, prevention of ABA accumulation could not be achieved in a localised manner. The possibility therefore that a restriction in ABA accumulation elsewhere in the elongation zone, or indeed elsewhere in the plant, may explain the dependence of apical growth maintenance on ABA accumulation cannot be excluded. The possible indirect effects may involve ABA-regulated gene expression, which control intrinsic 'house-keeping' functions within the plant or disruption of whole-plant water flux in an ABA-deficient plant. Transformation with root-specific anti-sense biosynthesis genes (i.e. genes artificially inserted into the plant genome in the reverse direction, to prevent expression of the target gene) may help to elucidate whether the effect of ABA on root growth maintenance is direct or indirect.

ABA would certainly appear to play a key role in coordinating the whole-plant response to drought. In some way, its accumulation in roots sustains a reduced level of root expansion to exploit deeper, unexplored reserves of water. Its transport from the roots to the shoots will minimise water loss via its effects on the guard cells surrounding the stomatal pores and may also restrict the rate of new leaf area produced.

7.2.4.2 A Role for Ethylene?

ABA biosynthesis mutants exhibit excessive ethylene evolution (Sharp et al. 2000). For this and other reasons, ABA is therefore thought to suppress the synthesis of ethylene particularly at low water potentials (e.g. Sharp et al. 2000). Ethylene (C_2H_2) is a gaseous plant growth regulator implicated in a wide range of physiological processes. Low concentrations of ethylene stimulate root elongation and a thinning of the root, whereas runaway ethylene synthesis, in the absence of ABA, inhibits root elongation and causes swelling just behind the root apex (Spollen et al. 2000). An accumulation of ABA within the root has been shown to suppress enhanced ethylene synthesis and be, at least in part, responsible for the maintenance of apical root elongation (Spollen et al. 2000). While data are scarce, it is tempting to suggest that, at low soil mechanical impedances (i.e. when no increase in soil strength occurs as the substrate dries), ABA may suppress ethylene synthesis and cause the root to thin. At high mechanical impedance, the effect of ethylene must predominate to generate appropriate root morphology to penetrate soil with an increasing mechanical impedance. Such a suggestion may explain the differences in observed changes in root morphology discussed earlier. Indeed, the suppression of ethylene perception within germinating tomato seedlings with silver thiosulphate (STS) treatment prevents root penetration into soil (Zacarias and Reid 1992) and ethylene-insensitive tomato mutants, such as NR, behave in a similar manner and are unable to penetrate the growth media (Clark et al. 1999). Such observations suggest that ethylene may have an important role to play in providing the root with the ability to effectively penetrate into substrates. In the context of drought, we would speculate that ethylene may be particularly important in mediating changes in root morphology, if soil mechanical impedance increases with decreasing soil water content.

One note of caution must be added to these observations. As with investigations of the role and action of ABA in root growth adaptation at low water potential, most have been carried out with germinating seedlings. Although the fundamental processes of cellular elongation remain the same, the adaptive significance displayed and the mode of action of these plant growth regulators may differ in mature plants.

7.3 Perception and Signalling of Soil Drying by Roots

7.3.1 Roots as Sensors of Soil Water Status

As well as roots functioning as the major site for water and nutrient uptake, the root system of a plant also has the unique function of perceiving soil water

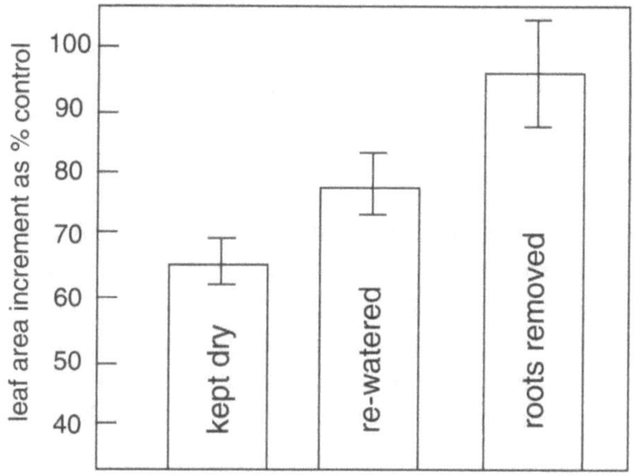

Fig. 7.3. The effect of a split root experiment on the expansion rate of apple tree leaves. The graph shows the effect of sustaining a soil drying treatment to one half of the root system, re-watering the roots or removing the roots in contact with drying soil, on leaf expansion rate relative to a well-watered control. (After Gowing et al. 1990)

status and generating the signals that communicate this information to the shoots. The ability of the roots to perceive soil drying and produce a 'measure' of this change, which is communicated to the shoots, can be demonstrated in 'split root' experiments. In an experiment with apple trees, Gowing et al. (1990) demonstrated that by splitting the root system of apple saplings and allowing the soil around one half of the system to dry, the rate of leaf expansion measured on an area basis and the rate of leaf production was restricted. Re-watering of the roots in dry soil, or most importantly their removal, restored leaf expansion and production rates (Fig. 7.3). As half of the root system was allowed to dry the soil, the other half of the system was watered sufficiently to ensure that the water status of the plant was maintained at a level equivalent to that measured in well-watered plants. A change in water availability could not therefore explain the observed responses. This experiment suggested that there was some form of signal carried to the shoot by the xylem, produced by the roots in contact with the drying soil. The signal was described as positive root signal (i.e. something moved from the root to the shoots), because removal of roots in contact with drying soil, removed the inhibition of leaf expansion.

7.3.1.1 Abscisic Acid as a Root Signal

A large amount of work suggests that abscisic acid acts not only as a regulator of root growth, but also as the key root-signal in the process which communicates information on variation of soil water availability to the shoots. Large increases in ABA concentration in the xylem of plants with roots in drying soil have been reported in a variety of species (see Hartung and Davies 1991). Conjugated forms of the hormone may also move through the xylem and act as a root signal (Hartung et al. 2002).

Although the ABA-dependent mechanism by which stomatal water loss is restricted under drought is now well understood, debate continues over whether or not ABA induces the observed reductions in growth in plants in drying soil. Munns and Cramer (1996) doubted that enough ABA could enter the xylem from the roots in order to explain the observed reduction in leaf expansion. Grafting experiments such as those reported by Holbrook et al. (2002) have raised further doubts. These experiments suggest that ABA-deficient roots grafted on ABA-sufficient shoots exert very little effect on growth and functioning of shoots, although contrasting results have been provided by Borel et al. (2001). The key observation in this regard, however, appears to be that of Bacon et al. (1998) who demonstrated unequivocally that even those concentrations of ABA typically measured in well-watered plants could exert a potent effect on growth if the apoplastic pH surrounding growing cells was increased. Significant increases in xylem sap pH as the soil dries (Fig. 7.4) have been recorded in xylem sap expressed from stems cut close to the roots (Bacon et al. 1998). We may therefore speculate that the increase in xylem sap pH may also be a root-borne signal but that the effect of pH is exerted through modified ABA distribution. Wilkinson (1999) has identified such changes in xylem sap pH in response to soil water availability in a diverse range of crop and native species, including *Phaseoulus vulgaris* (French bean), *Anastatica hierchuntica* (rose of Jericho), *Helianthus annuus* (sunflower) and *Commelina communis* (dayflower).

It remains to be confirmed why xylem sap pH increases as soil dries. Reduced root ATPase activity (the enzyme activity responsible for maintaining

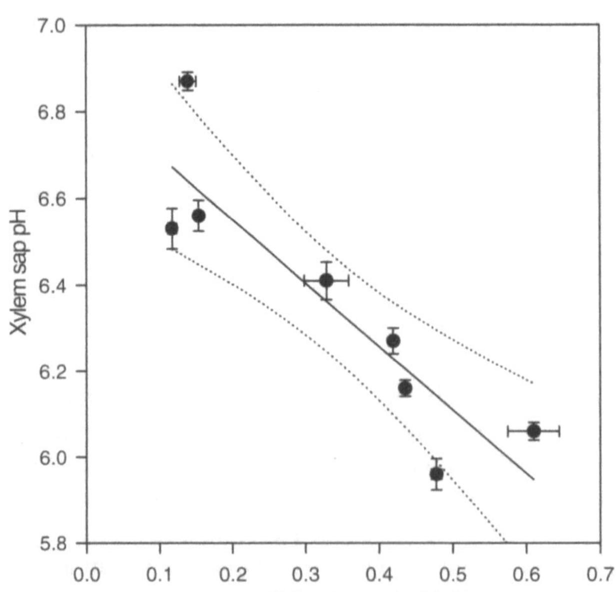

Fig. 7.4. The correlation between xylem sap pH and soil water content. Sap was expressed from the base of barley plants, close to the root/shoot junction. (After Bacon et al. 1998)

the acidic pH of the apoplast) as the soil dries is commonly observed and could presumably have a profound effect on the pH of the xylem sap originating from the roots. A second possibility involves changes in nitrate reductase activity. This enzyme is responsible for converting nitrate (NO_3–) to nitrite (NO_2–) in the root and shows marked changes in activity in stressed plants. Wilkinson and Davies (2002) have discussed how these and other stress-induced changes in the ionic composition of xylem sap expressed from the roots could explain the measured increases in xylem sap pH as the roots of a plant dry the soil in which they are growing. Such changes in ionic sap composition may also influence the pattern of ionic exchange with the walls of the root xylem and hence moderate xylem sap pH and will affect exchange of other charged molecules (e.g. ABA) with the xylem parenchyma (e.g. Hartung et al. 2002).

Increases in xylem sap pH from the root can be detected within hours, while significant rises in root-borne ABA may occur only after several days in contact with drying soil. In this way we may speculate that the rapid rise in pH quickly directs the low levels of ABA which are always present within the plant to sites of action – instigating stomatal closure, shoot growth inhibition and maintenance of root elongation (Fig. 7.5). As the drought develops, the synthesis and transport of ABA from the roots to the shoots provides a more pronounced signal. This *de novo* synthesis of ABA may also have a greater role in protecting the plant from dehydration as opposed to regulating the uptake and loss of available water.

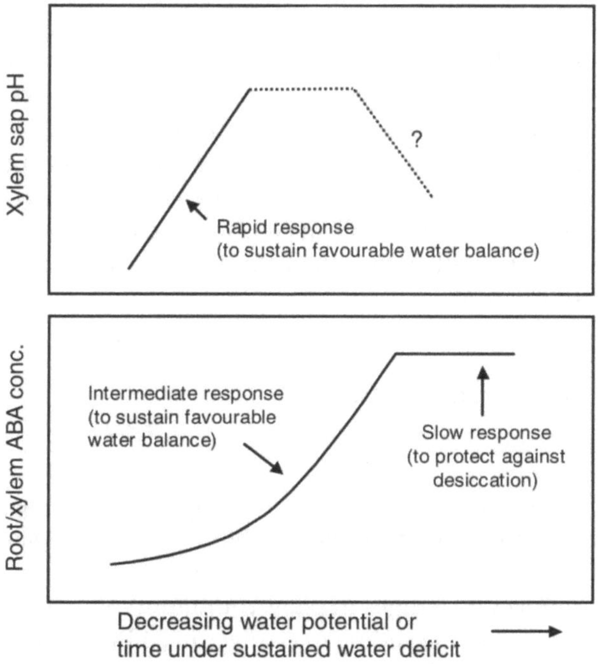

Fig. 7.5. Hypothesised interaction of a xylem sap pH and ABA signal emanating from the roots to sustain a favourable water balance and protect against desiccation

7.3.1.2 Ethylene as a Root Signal of the Effects of Soil Drying and Soil Compaction

Evidence in the literature suggests that ethylene has little or no effect on stomatal behaviour, although Tissera and Ayres (1986) have shown that the application of ethephon to roots to enhance ethylene levels within a plant has a substantial effect on the stomatal conductance of plants infected with powdery mildew. Ethylene can act as both a promoter and an inhibitor of growth and, as we have seen above, can play an important role in adapting the morphology of roots developing under water deficit.

Hussain et al. (1999) have identified a role of ethylene in the regulation of growth of tomato plants during compaction stress using wild-type genotypes and transgenics with a reduced capacity to produce ethylene. Shoot growth of both genotypes was comparable in uncompacted soil and in uniformly compacted soil. Plants with roots split between two containers (one containing compacted and the other uncompacted soil) showed different responses however. The wild-type genotype showed reduced growth while $AC01_{AS}$ (the low ethylene producing transgenic) showed growth rates comparable to control plants in uncompacted soil. Reduced growth rates in wild types were accompanied by enhanced accumulation of ethylene, suggesting that the ability of the transgenic to sustain growth was related to its lower ethylene production. Interestingly, excising wild-type roots in compacted soil restored shoot growth rate, as did treatment of half-compacted plants with silver ions, a treatment shown to block ethylene action. Taken together, these results suggest that ethylene has a key role as a root signal when roots of tomato plants encounter compacted soil. It is unlikely however that ethylene is the actual signal present within the xylem. A precursor such as 1-aminocyclopropane-1-carboxylic acid (ACC) is a likely candidate for the role of root-borne signal under such circumstances. There are clear interactions between soil drying and soil compaction. As many soil types dry, their bulk density increases. Indeed, the primary adaptation of roots to soil drying may actually be in response to this increase in soil strength rather than the absolute decline in soil water availability (see also Chap. 6).

7.3.1.3 Adaptive Advantages of Chemically Based Signalling of Soil Drying

Davies and Gowing (1999) have argued that mechanisms of chemical signalling of the kind described above can account for the highly developed responses of some species to very mild soil drying. These responses can apparently feed through to important changes in plant community structure and function, with individual species responding to changes in soil water status as small as a few tens of kPa. While little directly comparative work has been undertaken, the existence of such chemically based signalling systems in different species would appear to correlate well with the ability to sustain favourable water balances as soil dries. Puliga et al. (1996) demonstrated significant differences in the accumulation of ABA within the leaf elongation zones of several Mediterranean grass species including *Festuca arundinacea*, *Eragrostis curvula* and *Sporobolus stafianus* and could relate such variation to differences in the susceptibility of the species to drought. In particular, *Festuca arundinacea* exhibited a highly sensitive leaf growth rate to soil drying and significant increases in ABA concentrations in the absence of any change in leaf water status, both changes that might contribute to a capacity to survive severe soil drying.

The extent to which chemically based signalling mechanisms are developed in different species may determine which species close their stomata and restrict leaf development before any measurable change in leaf water status, e.g. *Malus* and *Zea mays* (Tardieu et al. 1996). Such a 'water-saving' or 'isohydric' strategy is likely to prevent any significant and potentially damaging changes in leaf water status. Conversely, many species appear to adopt a so-called anisohydric strategy, responding to a change in soil water status via a change in leaf hydraulic status (e.g. sunflower). While it is tempting to suggest that species prevailing in environments subject to frequent periods of soil drying (e.g. Mediterranean grasslands) would possess highly developed chemically based signalling mechanisms, compared to those prevalent in water-sufficient environments, little comparative evidence exists to date. Our own recent observations in commercial tomato cultivars would lead us to suggest that sustained yielding in crops experiencing soil drying is associated with increased xylem-borne ABA concentrations, highly sensitive stomata and little change in leaf water potential (Davies et al. 2000).

There would also appear to be some important differences in the root signalling mechanisms that have evolved in herbaceous and woody species. There are now several reports in the literature which apparently suggest that chemical signalling systems are not as well developed in woody plants and that much regulation of growth and functioning may be hydraulic rather than chemical (e.g. Mencuccini et al. 2000; Comstock 2002). Recent work by Wilkinson and Davies (2002) suggests that in some woody species, xylem sap pH may not increase in response to soil drying and some species may actually show some acidification of sap at low soil water potentials. The basis of these changes has not been explained but such significant variation in the pH response to soil drying may be expected to have very significant effects on plant community structure and functioning.

The significance of root-borne signalling of soil water status will also be clearly dependent on the overall growth form of a plant species. Small plant species with shoots only a few centimetres aboveground will be well coupled with the root environment. Any signals passing from the root to the shoot will take only a few seconds or minutes. In the case of grass species, the leaves of which may be several centimetres tall, the distance from the root to the leaf growth zone may only be a few centimetres. Again, we would speculate that the transport of chemical signals from the roots to the shoots would be dynamic control syystem. How effective are root-borne signals in communicating information from the roots to shoots that may be tens, if not hundreds, of metres away, as is the case in many trees? In tall trees, xylem sap modified in the root may take several days to reach the leaves and, clearly, the potential for modification of a root-borne signal transported over several metres is substantial. In such species, signalling from the root and responses to the arial environment will integrate to produce a desired whole-plant response to stress.

7.3.2 Signals from the Soil

Little attention has been given to the possibility that chemical signals may be generated in drying soil, taken up by the roots and then moved to the shoots in the transpiration stream. Hartung et al. (1996) established that the concentration of ABA dissolved in the soil solution ranged from 0.6–2.8 nM, a concentration range predicated by computer simulation models to prevent ABA release from root hairs to the soil (Slovik et al. 1995). Such concentrations of ABA in the soil would appear to the important for long distance chemical signalling in a whole range of species, including desert plants such as *Anastatica hierochuntica*, several grasses (e.g. *Secale cereale*), large dicots (e.g. *Helianthus*), and other herbaceous species (e.g. *Arctotheca calendula*).

Changes in the water status of the roots and/or the accumulation of abscisic acid in roots will modify the activity of ion channels in the plasma membranes of the roots. For example, Roberts and Snowman (2000) have shown that ABA accumulation and water stress reduce potassium transport to the xylem and that potassium ions accumulate in the roots as a result.

The advantages of potassium accumulation for osmoregulation in roots should be obvious but the importance of reduced potassium transport in response to soil drying has not been fully considered. Extracellular potassium is likely to influence both stomatal behaviour (Amodeo et al. 1996) and growth (Van Volkenburgh 1999) and it therefore seems likely that modified transport of this ion can also act as a 'measure' of soil drying. Uptake and transport of several other ions can also be modified by soil drying and some of these changes can be quite sensitive to small changes in soil water availability. For example, small moisture-deficit-related changes in the transport of nitrate ions can act as a sensitive regulator of leaf growth of sunflower (Palmer et al. 1996).

Very interestingly, Holbrook and co-workers have shown that the concentration of potassium ions moving through the xylem can influence the hydraulic conductivity of the transport pathway, perhaps by affecting the nature of the pit membranes (Zwieniecki et al. 2001). This means that a root-sourced chemical signal can influence the properties of the water transport pathways through the root and therefore influence the hydraulic signalling between the roots and shoots.

Freundl et al. (1998) have recently highlighted species differences in the proportion of water moving into the root system along an apoplastic pathway, as discussed above. A water transport pathway that is dominated by the so-called apoplastic bypass has been identified in *Zea mays* while a greater proportion of the transpiration stream moves through the symplast in *Helianthus*. Apoplastic water flow raises the possibility that soil-sourced chemicals may gain free access to the transpiration stream if the Casparian band is equally permeable to water and solutes. Hartung et al. (1996) have reported concentrations in the soil of ABA and its conjugates of up to 10 nM. While this soil-sourced ABA may enter the plant and be incorporated into the signalling pathway, it is probably more important in its contribution to minimising the concentration gradient and potential dissociation of ABA across the root surface into the soil (Slovik et al. 1995). It seems likely therefore that soil-sourced chemicals may have a role to play in signalling, particularly in plants with an apoplastic bypass.

7.4 Summary and Prospects

We know relatively little about the regulation of growth and functioning of roots in drying soil. This is partly because the changes that take place in the soil as it dries can only be quantified at a relatively gross scale. Steep gradients in water and nutrient status are difficult to define with precision and cracks in the soil and other heterogeneities contribute to uncertainties over soil strength. Spatial variation in root growth and its controlling variables has been defined but only in growing media with low mechanical impedance. There has been good progress made in understanding the chemical regulation of plant growth and functioning in response to soil drying and the increasing availability of new genetic tools will help in our search for understanding in this area. Nevertheless, there is still a real need for cellular and molecular physiology to proceed within the context of an investigation of whole-plant functioning. Drought stress responses are a whole-plant phenomenon and we cannot expect to achieve real progress in this area unless we accept this fact. This is no more evident than in the recent report by Kavi Kishor et al. (1995) that overexpression of a proline-synthesising gene conferred a greater ability to osmotically adjust at low water potential. Indeed, such transformed plants did perform better at low substrate water potential; however, close examination of the plants' water relations revealed that this was not a function of solute regulation. The increasing ease of plant transformation offers new and powerful ways to elucidate and manipulate plant adaptation to stress. However, we must not lose sight of the need for integrated analysis of plant adaptation to drought at the molecular, cellular, tissue, whole-plant and community levels.

References

Auge RM, Green CD, Stodola AJW, Saxton AM, Olinick JB, Evans RM (2000) Correlations of stomatal conductance with hydraulic and chemical factors in several deciduous tree species in natural habitats. New Phytol 145:183–500

Amodeo G, Talbot LD, Zeiger E (1996) Use of potassium and sucrose by onion guard cells during a daily cycle of osmoregulation. Plant Cell Physiol 37:575–579

Atwell BJ, Newsome JC (1990) Turgor pressure in mechanically impeded lupin roots. Aust J Plant Physiol 17:49–56

Bacon MA, Wilkinson S, Davies WJ (1998) pH-regulation cell expansion in ABA-dependent in droughted plants. Plant Physiol 118:1507–1515

Borel C, Frey A, Marion-Poll A, Tardieu F Simmoneau T (2001) Does engineering abscisic acid biosynthesis in *Nicotiana plumbaginifolia* modify stomatal response to drought? Plant Cell Environ 24:477–49

Clark DG, Gubrium EK, Barett JE, Nell TA, Klee HJ (1999) Root formation in ethylene-insensitive plants. Plant Physiol 121:53–59

Comstock J (2002) Hydraulic and chemical signalling in the control of stomatal conductance and transpiration. J Exp Bot(in press)

Davies WJ, Gowing DJG (1999) Plant responses to small perturbations in soil water status. In: Press MC, Scholes J, Barker M (eds) Plant physiological ecology. Blackwell, Oxford, pp 67–89

Davies WJ, Bacon MA, Thompson DS, Soeih W, Gonnzalez-Rodrigues L (2000) Regulation of leaf and fruit growth in plants growing in drying soil: exploitation of a plants chemical signalling system and hydraulic architecture to increase the efficiency of water use in agriculture. J Exp Bot 51:1617–1626

Freundl E, Steudle E, Hartung W (1998) Water uptake by roots of maize and sunflower affects the radial transport of abscisic acid and its concentration in the xylem. Planta 207:8–19

Fry SC, Smith RC, Renwick KF, Martin DJ, Hodge SK (1992) Xyloglucan endotransglycosylase a new cell-wall loosening enzyme activity from plants. Biochem J 282:821–828

Gowing DJ, Davies WJ, Jones HG (1990) A positive root-sourced signal as an indicator of soil drying in apple. *Malus domestica* Borkh. J Exp Bot 41:1535–1540

Hartung W, Davies WJ (1991) Drought-induced changes in physiology and ABA. In: Davies WJ, Jones HG (eds) Abscisic acid. Bios Scientific Publishers, Oxford

Hartung W, Sauter A, Turner N, Fillery I, Heilmeier H (1996) Abscisic acid in soils: what is its function and which factors and mechanisms influence its concentration? Plant Soil 184:105–110

Hartung W, Sauter A, Hose E (2002) Abscisic acid in the xylem: where does it come from and where does it go? J Exp Bot 27:27–32

Hetherington A (1998) Plant physiology: spreading the drought warning. Curr Biol 8:R911–R913

Holbrook NM, Shashidar VR, James RA, Munns R (2002) Stomatal control in tomato with ABA-deficient roots: response of grafted plants to soil drying. J Exp Bot 53 (in press)

Hussain A, Black CR, Taylor LB, Roberts JA (1999) Soil compaction: a role for ethylene in regulating leaf expansion and shoot growth in tomato? Plant Physiol 121:1227–1237
Jackson GE, Irvine J, Grace J, Khalil IA (1995) Abscisic acid concentrations and fluxes in droughted conifer saplings. Plant Cell Environ 18:13–22
Kavi Kishor PBK, Hong ZL, Miao GH, Hu CAA, Verma DPS (1995) Overexpression of delta-pyrroline-5-carboxylate syntheses increases proline production and confers osmotolerance in transgenic plants. Plant Physiol 108:1387–1394
Kramer P (1988) Changing concepts regarding plant water relations. Plant Cell Environ 11:565–568
Kuang JB, Turner NC, Henson IE (1990) Influence of xylem water potential on leaf elongation and osmotic adjustment of wheat and lupin. J Exp Bot 41:217–221
Lips HS (1997) The role of inorganic nitrogen ions in plant adaptation processes. Russ J Plant Physiol 44:421–431
Lu P, Outlaw WH, Smith BG, Freed GA (1997) A new mechanism for the regulation of stomatal aperture in intact leaves – accumulation of mesophyll-derived sucrose in the guard-cell wall of *Vicia faba*. Plant Physiol 114:109–118
Mencuccinni M, Mambelli S, Comstock J (2000) Stomatal responsiveness to leaf water status in common bean (*Phaseolus vulgaris* L.) is a function of time of day. Plant Cell Environ 23:1109–1118
Munns R (1988) Why measure osmotic adjustment? Aust J Plant Physiol 15:717–726
Munns R, Cramer G (1996) Is co-ordination of leaf and root growth mediated by abscisic acid? Plant Soil 185:33–49
Neales TF, Mcleod AL (1991) Do leaves contribute to the abscisic acid present in the xylem sap of droughted sunflowers. Plant Cell Environ 14:979–986
Osonubi O, Davies WJ (1978) Solute accumulation in leaves and roots of woody plants subjected to water stress. Oecologia 32:323–332
Palmer SJ, Berridge DM, McDonald AJS, Davies WJ (1996) Control of leaf expansion in sunflower (*Helianthus annuus* L) by nitrogen nutrition. J Exp Bot 47:359–368
Puliga S, Vazzana C, Davies WJ (1996) Control of crops leaf growth by chemical and hydraulic influences. J Exp Bot 47:529–537
Richards BG, Greacen EL (1986) Mechanical stresses on an expanding cylindrical root analog in antigranulocytes media. Aust J Soil Res 24:393–404
Roberts SK, Snowman BN (2000) The effect of ABA on channel mediated K^+ transport across higher plant roots. J Exp Bot 51:1585–1594
Saab IN, Sharp RE, Pritchard J, Voetberg GS (1990) Increased endogenous abscisic acid maintains primary root-growth and inhibits shoot growth of maize seedlings at low water potentials. Plant Physiol 93:1329–1336
Shaner DL, Boyer JS (1976) Nitrate reductase activity in maize leaves. II. Regulation by nitrate flux at low water potential. Plant Physiol 58: 505–509
Sharp RE, Davies WJ (1979) Solute regulation and growth by roots and shoots of water-stressed maize plants. Planta 147:43–49
Sharp RE, Davies WJ (1985) Root growth and water uptake by maize plants in drying soil. J Exp Bot 36:1441–1456
Sharp RE, Silk WK, Hsiao TC (1988) Growth of the maize primary root at low water potentials. 1. Spatial distribution of expansive growth. Plant Physiol 87:50–57
Sharp RE, Hsiao TC, Silk WC (1990) Growth of the maize primary root at low water potentials. 2. Role of growth and deposition of hexose and potassium in osmotic adjustment. Plant Physiol 93:1337–1346
Sharp RE, LeNoble ME, Else MA, Thorne ET, Gherardi F (2000) Endogenous ABA maintains shoot growth in tomato independently of effects on plant water balance: evidence for an interaction with ethylene. J Exp Bot 51:1575–1584
Sharp RE, LeNoble ME (2002) ABA, ethylene and the control of shoot and root growth under water stress. J Exp Bot 53:33–37

Slovik S, Daeter W, Hartung W (1995) Compartmental redistribution and long-distance transport of abscisic acid (ABA) in plants as influenced by environmental changes in the rhizosphere – a biomathematical model. J Exp Bot 46:881–894

Spollen WG, Sharp RE (1991) Spatial distribution of turgor and root growth at low water potentials. Plant Physiol 96:438–443

Spollen WG, Sharp RE, Saab IN, Wu Y (1993) Regulation of cell expansion in roots and shoots at low water potentials. In: Smith JAC, Griffiths H (eds) Water deficits. Bios Scientific Publishers, Oxford, pp 37–52

Spollen WG, LeNoble ME, Samuels TD, Bernstein N, Sharp RE (2000) Abscisic acid accumulation maintains primary root elongation at low water potential by restricting ethylene production. Plant Physiol 122:967–976

Steudle E, Petersen CA (1998) How does water get through roots? J Exp Bot 49:775–788

Tardieu F, Zhang J, Katerji N, Bathenod O, Palmer S, Davies WJ (1992) Xylem ABA controls the stomatal conductance of field grown maize subjected to soil compaction or soil drying. Plant Cell Environ 15:193–197

Tardieu F, Lafarge T, Simonneau T (1996) Stomatal control by fed or endogenous ABA in sunflower: interpretation of correlations between leaf water potential and stomatal conductance in anisohydric species. Plant Cell Environ 19:75–84

Tissera P, Ayres PG (1986) Endogenous ethylene affects the behaviour of stomata in epidermis isolated from rust infected fava bean. New Phytol 104:3–6

Tyerman SD, Bohnert HJ, Maurel C, Steudle E, Smith JAC (1999) Plant aquaporins: their molecular biology, biophysics and significance for plant water relations. J Exp Bot 50:1055–1071

Voetberg GS, Sharp RE (1991) Growth of the maize primary root at low water potentials. 3. Role of increased proline deposition in osmotic adjustment. Plant Physiol 96:1125–1130

Volaire F, Thomas H (1995) Effects of drought on water relations mineral uptake water-soluble carbohydrate and survival of two contrasting populations of cocksfoot (*Dacytylis glomerata* L). Ann Bot 75:513–524

VanVolkenburgh E (1999) Leaf expansion – an integrating plant behaviour. Plant Cell Environ 22:1463–1473

Wilkinson S (1999) pH as a stress signal. Plant Growth Regulation 29:87–99

Wilkinson S, Davies WJ (2002) ABA-based chemical signalling: the co-ordination of responses to stress in plants. Plant Cell Environ 25:195–210

Winch S, Pritchard J (1999) Acid-induced wall loosening is confined to the accelerating region of the root growing zone. J Exp Bot 50:1481–1487

Wu YJ, Spollen WG, Sharp RE, Hetherington PR, Fry SC (1994) Root-growth maintenance at low water potentials – increased activity of xyloglucan endotransglycosylase and its possible regulation by ABA. Plant Physiol 106:607–615

Wu YJ, Sharp RE, Durachko DM, Cosgrove DJ (1996) Growth maintenance of the maize primary root at low water potentials involves increases in cell-wall extension properties, expansin activity and cell wall susceptibility to expansion. Plant Physiol 111:765–772

Zacarias L, Reid MS (1992) Inhibition of ethylene action prevents root penetration through compressed media in tomato (*Lycopersicon esscülentum*) seedlings. Physiol Plant 86:301–307

Zwieniecki MA, Melcher PJ, Hobrook NM (2001) Hydrogel control of xylem hydraulic resistance in plants. Science 291:1059–1062

8 Physiology, Biochemistry and Molecular Biology of Plant Root Systems Subjected to Flooding of the Soil

M.B. Jackson and B. Ricard

8.1 Introduction

Despite the fact that water uptake is a prime function of roots, excess water in soil seriously damages root growth and function and can prove fatal. Certain physical properties of water that prevent adequate gas exchange at the root surface are primarily responsible. Not only is the direct entry of oxygen by gas diffusion largely prevented, but, in addition, the normal disposal of physiologically active gases and volatiles is equally hampered. Oxygen shortage, resulting from low diffusion coefficients in flooded soils, is exacerbated by competition for oxygen from aerobic microorganisms that inhabit soil and rhizosphere. Oxygen depletion also favors growth of certain bacteria that can diminish nitrate availability to roots and, over longer time periods, produce highly soluble and toxic Mn^{2+} and Fe^{2+} and eventually even convert SO_4^{2+} to the respiratory poison H_2S (see Chap. 13). In the longer term, increased incidence of soil-borne fungal and bacterial diseases can be additional problems in saturated soils. This chapter reviews morphological, physiological, metabolic and molecular responses to flooding. It largely ignores the impact on germination and of total plant submergence. The focus is on effects of oxygen shortage and mechanisms of avoiding internal anoxia. Since oxygen-deficient roots influence shoot adaptation by long-distance signalling (Jackson 2002), this aspect is also considered briefly.

8.2 Inhibition of Root Growth by Partial Oxygen Shortage

The major effect of partial oxygen shortage due to soil flooding is the stunting of growth in root length and mass (Fig. 8.1A,B). The highest external concentration of oxygen below which extension growth and oxygen consumption

Ecological Studies, Vol. 168
H. de Kroon, E.J.W. Visser (Eds.)
Root Ecology

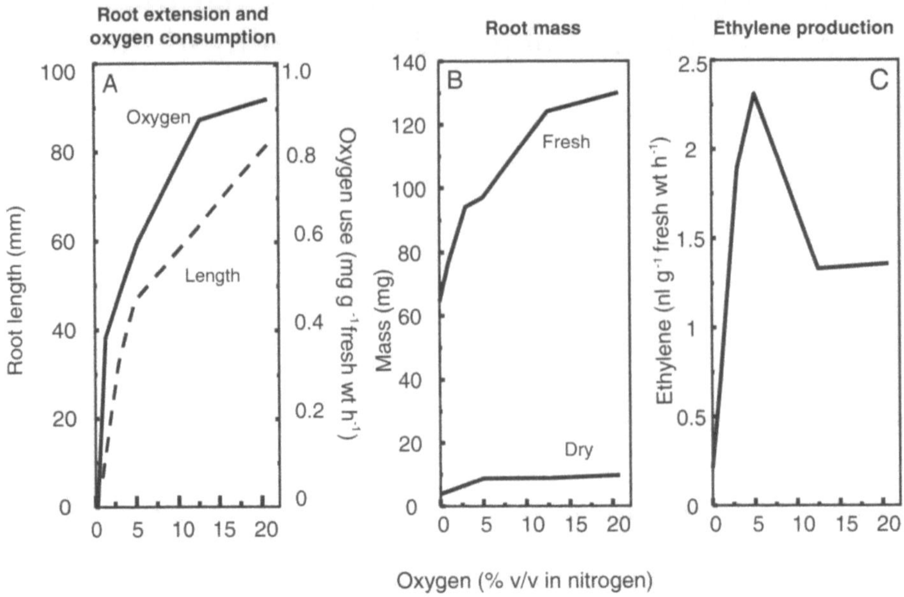

Fig. 8.1. Effect of a range of oxygen concentrations applied for 3 days to root systems of 4-day-old barley seedlings on A root extension and oxygen consumption, B final fresh and dry mass, and C ethylene production by the whole root system (from Jackson et al. 1984, in part)

suffer compared to roots supplied with air has been termed the critical oxygen pressure for extension (COPe) or respiration (COPr) by Berry and Norris (1949). COPe can be greater than COPr, presumably because of lower affinities for oxygen of growth-related enzymes compared to respiratory cytochrome oxidase. For example, many dioxygenase and *p*-450 monooxygenase enzymes use molecular oxygen to synthesise physiologically important substances such as gibberellin hormones, ethylene (ACC oxidase), flavonoids and fatty acids (lipoxygenase). The large difference between the high COPr compared to the concentration of oxygen calculated to inhibit cytochrome oxidase itself can be accounted for by oxygen consumption *en route* as the gas diffuses deeper into root tissue and by physical resistance to this inward movement. Several predictions made on the basis of this notion have been verified experimentally: external COP is higher under warmer conditions that promote faster respiration and a higher COPr is measured in apical rather than older regions of a root axis. Therefore, root tips become anoxic before older regions and inner cells of root axes become anoxic before outer cells, thereby creating an anoxic core.

Inhibition of root growth by oxygen shortage may have an ethylene component because small but not extinguished oxygen concentrations stimulate ethylene production by root apices of some species, notably barley and *Zea mays* (Fig. 8.1C; Brailsford et al. 1993). A covering of water can be expected to entrap the ethylene and thus enhance its impact on extension (Konings and Jackson 1979).

8.3 Possible Causes of Severe Growth Inhibition and Cell Death in the Absence of Oxygen

Short exposure (a few minutes or hours) to the complete absence of oxygen is sufficient to kill tips of non-acclimated roots. However, surprisingly small amounts of external oxygen (e.g. 0.006–0.01 mol m^{-3} in solution) may suffice to prevent death and allow recovery on return to aerated conditions. Cause of death is generally attributed to (1) insufficient ATP to satisfy demand, and (2) injury from products generated by anaerobic metabolism. These possibilities are examined below.

8.3.1 ATP Supply and Demand

The chemical energy generated by anaerobic roots comes mainly from glycolysis. The final yield of two ATPs from each glucose molecule entering the pathway is only about 5 % of that generated by the aerobic oxidation of glucose via the Krebs cycle and oxidative phosphorylation by the electron transport chain (Fig. 8.2). The fuelling of glycolysis requires a continuous supply of glucose and of NAD^+. Lack of ATP and the sucrose and glucose needed to support ATP synthesis (see below) can be causes of injury and death. Root tips can survive anoxia temporarily for lengths of time that vary with species and degree of conditioning (e.g. prior exposure to partial oxygen shortage for several hours – see Sect. 8.4). Duration of survival may depend on how long supply and demand can be kept in balance while maintaining cell integrity. A balance might sometimes be achieved by adoption of reversible quiescence. This may explain tolerance of willow root tips (*Salix viminalis*) to flooded soil (Jackson and Attwood 1996). The opposite strategy would be to raise the rate of anaerobic respiration sufficiently to pay the energy debt.

Experimental evidence for energy starvation as a cause of root tip death is considerable. Marked decreases in ATP after anoxia imposition have been linked with root death. Treatments that improve survival (e.g. hypoxic pretreatment with 3 % oxygen for several hours) raise ATP levels and promote glycolysis and fermentation. Similarly, mutations that inhibit glycolysis by strongly suppressing alcohol dehydrogenase (ADH) activity depress fermentation and shorten root survival time. However, ATP supply may be less critical than is often thought. Xia et al. (1995) suggest that only a small amount of ATP is required to prolong survival of hypoxically acclimated roots. Thus, a key effect of hypoxic training could also be a suppression of ATP-consuming processes (a form of quiecence).

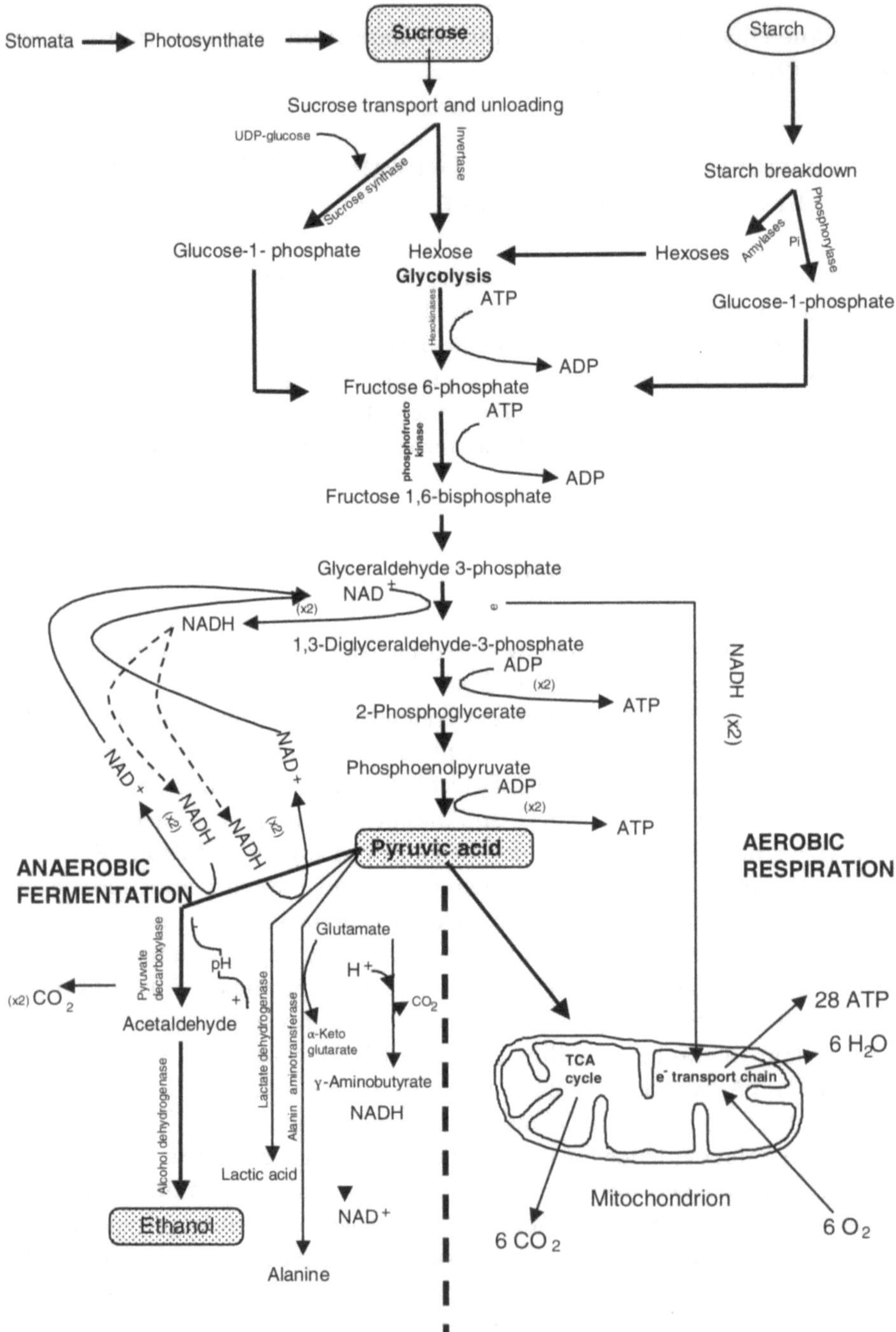

Fig. 8.2. Summary diagram depicting various pathways from sucrose that generate energy or dispose of protons generated by metabolic oxidation in the presence or absence of oxygen

ATP generation by glycolysis depends on a supply of glucose and its precursors. Reports of long-term anaerobic survival at warm temperatures are almost always for roots given external hexose experimentally. These effects are coupled with some evidence of very slow metabolic activity setting-in after a period of enhanced fermentation. This early window of faster fermentation presumably sets up the longer-term adjustment to an absence of oxygen. During the highly active early phase, seedling roots given hexose ferment more vigorously in association with preservation of mitochondrial structure (Vartapetian et al. 1977). In the absence of such feeding, endogenous sugar decreases sharply during anoxia, possibly as a consequence of depressed sucrose unloading at root tips (Saglio 1985).

Poor ability to mobilise starch reserves may cause hexose starvation under anoxia because of slow hydrolyzing activity (Perata et al. 1997). Even when anoxic roots are given extra hexose they often die in a few days, indicating a cause of death other than simply substrate-starved depression of glycolysis. Clearly, roots deprived of oxygen can die for reasons not connected to hexose starvation.

8.3.2 Self-Injury from Products of Anaerobic Metabolism

The view that ethanol generated by fermentation explains death from anoxia is no longer widely accepted (discussed in Vartapetian and Jackson 1997). An alternative culprit is poisoning by an excess of protons in the cytoplasm (Roberts et al. 1982), leading to a potentially fatal drop of 0.5–0.6 pH units (Saint-Ges et al. 1991). Treatments raising cytoplasmic acidity (e.g. CO_2 enrichment) enhance root damage from anoxia (Roberts et al. 1984a,b), while treatments that limit acidosis improve survival (Roberts et al. 1985; Xia and Roberts 1996; Chang et al. 2000). Early death from cytoplasmic acidosis may be prevented by the roots themselves switching from lactate to alcoholic fermentation (Davies et al. 1974; Roberts et al. 1984a). Experimental support for the importance of this switch has been obtained by using a permeant weak base to offset acidification and dampen ethanol biosynthesis (Fox et al. 1995). Similarly, in mutants that are unable to run ethanolic fermentation, lactate continues to be produced resulting in earlier death in association with cytoplasmic acidification (Roberts et al. 1984b). However, the true picture concerning the role of proton poisoning is undoubtedly more complex. Ratcliffe (1995) discusses the various pathways and processes that may serve to regulate cytoplasmic pH. These include: lactate excretion (Xia and Saglio 1992), excretion of protons (Saint-Ges et al. 1991) and maintenance of proton-consuming pathways such as those mediated by nitrate reductase (an enzyme that is promoted by oxygen depletion and pH decrease). Low cytoplasmic pH is thus one likely cause of anoxic damage. However, the source of the damaging protons (e.g. the vacuole) and the mechanisms that regulate their concentration remain unclear.

8.4 Hypoxic Acclimation to Anoxia

Short-term tolerance of roots can be improved by a hypoxic pre-treatment (acclimation). As little as 6 h of partial oxygen shortage lengthens survival of anoxia by 8 to 72 h (Saglio et al. 1988). Whole rice seedlings can likewise be trained (Ellis and Setter 1999). Pre-treatment with abscisic acid (ABA) also prolongs anaerobiosis tolerance (Hwang and VanToai 1991). A great deal is known about the biochemistry and molecular biology that underlies the effect. This will now be summarised.

8.4.1 Oxygen Sensing and Signal Transduction

Improving tolerance by hypoxic pre-treatment implies that oxygen levels can be sensed and that the activation of a signal transduction pathway leads to the induction of genes and, ultimately, synthesis of the proteins that underpin the tolerance. These proteins may include the so-called anaerobic proteins (ANPs). Some microorganisms and mammals sense changes in oxygen tension with haem proteins (reviewed by Franklin Bunn and Poyton 1996). Plants also have non-symbiotic haemoglobins that were proposed by Appleby et al. (1988) to be the elusive oxygen sensors. This now seems unlikely because haemoglobin binds too strongly to oxygen for the extent of binding to act as a sensitive measure of oxygen tension (Arrendondo-Peter et al. 1997; Duff et al. 1997). Further, expression studies have revealed that induction of haemoglobin mRNA by low oxygen is too slow to play a part in regulating expression of genes coding for enzymes such as ADH and LDH that play a part in hypoxic acclimation (Taylor et al. 1994).

One known component of the oxygen-sensing signal transduction pathway is Ca^{2+} (Subbaiah et al. 1994a); oxygen depletion causing a sharp increase in cytosolic Ca^{2+} that precedes anaerobic gene expression (Subbaiah et al. 1994b) and originates in mitochondria (Subbaiah et al. 1998). Studies of mutants defective in regulating the expression of the *ADH* gene (Conley et al. 1999) may help identify other signalling components.

8.4.2 Regulation of Gene Expression

Regulation of anaerobic gene expression involves proteinaceous *trans*-acting factors (transcription factors) that bind to sites present in the promoter regions (*cis*-elements). These may constitute response elements required for gene activation or repression. Walker et al. (1987) were the first to describe a possible anaerobic response element (ARE) for *ADH1* (in maize). A similar ARE has been found in many anaerobically regulated genes. Analysis of three

ADH promoter sequences confirmed the functional role of the ARE (Olive et al. 1991; Dolferus et al. 1994). In addition to the ARE, other *cis* elements are involved in regulating anaerobic gene expression. In vivo DMS footprinting is a technique that helps visualise the binding of putative protein transcription factors to regulatory DNA sequences. It has been used to locate at least four sites in maize and arabidopsis *ADH* promoters (including the ARE) that bind transcription factors. Two sites are moderately conserved half G-boxes which are common to the promoters of many stress-induced genes and also those regulated by UV/visible light.

Several transcription factors have been identified which bind to the ARE and the G-boxes. Myb factors known to bind to ARE-like elements are involved in the induction of the arabidopsis *ADH1* gene (Hoeren et al. 1998) and anoxia affects expression of several myb-related genes (Magaraggia et al. 1997). A G-box factor (GBF) has been found that binds to the maize *ADH1* promoter in vitro (deVetten and Ferl 1995). The complex it forms with the G-box is associated with a 14-3-3 protein that is itself induced by hypoxia (deVetten et al. 1992). 14-3-3 proteins are brain regulatory proteins involved in signalling pathways and may be ubiquitous in eukaryotes. In mammals, they bring kinase-related signals to the GBF. The finding that a plant GBF can interact with a 14-3-3 protein that phosphorylates and binds Ca^{2+} (Lu et al. 1992) reinforces the proposed link between anoxic gene expression and Ca^{2+} signalling. Overall, these results indicate a complex interplay of several *cis* elements interacting with a galaxy of proteins whose activities may be further regulated by phosphorylation and changes in calcium levels.

8.4.3 Selective Gene Expression and Enzyme Synthesis

Careful control of oxygen supply has shown that maximal induction of ANP gene expression is accomplished in cells that are only partially oxygen deficient rather than fully anoxic. Increased synthesis of the ANPs occurs only during the hypoxic period preceding anoxia (Saglio et al. 1999). These proteins are crucial for acclimation (Chang et al. 2000). Because protein synthesis is extremely slow during anoxia, the relative activities of most ANPs are actually lower than in well-aerated tips (Bouny and Saglio 1996). Therefore, ANPs should probably be termed hypoxically induced proteins (HIPs). HIPs comprise enzymes involved in energy metabolism, pH regulation, aerenchyma formation, protective functions and several other activities. The synthesis of these enzymes and the arrest of synthesis of other proteins is known collectively as the anaerobic response. It is regulated at transcriptional and post-transcriptional levels (Rowland and Strommer 1986; Fennoy and Bailey-Serres 1995). Post-transcriptional regulations include changes in transcript stability and efficiency of ribosome loading (Fennoy and Bailey-Serres 1995). Regulation of RNase activity could play a role in conserving non-translating ribosomes and poorly translated mRNAs during low-oxygen stress (Fennoy et al. 1997). A variety of mechanisms for the selective translation of anaerobic rather than aerobic mRNA have been described. These include interactions within the sequence of ADH mRNA (Bailey-Serres and Dawe 1996) and the phosphorylation of proteins involved in the translation machinery (Webster et al. 1991; Manjunath et al. 1999).

Recently, high-throughput analysis of gene transcript and protein profiles has been applied to hypoxically stressed rice and maize. Serial analysis of gene expression (SAGE; e.g. Matsumura et al. 1999) allowed the study of almost 6000 rice genes while two-dimensional isoelectric focusing (IEF) SDS-PAGE linked to matrix-assisted laser desorption/ionization delayed-extraction reflectron time-of-flight (MALDI-DE-TOF) mass spectrometry permitted over 200 proteins to be analysed in maize root tips (Chang et al. 2000). In contrast to the SAGE analysis, several of the 42 proteins up-regulated by hypoxia in maize root tips were the familiar metabolic enzymes such as ADH, PDC, and aconitase. The presence of many others suggests a complex network of interacting proteins that eventually may explain the acclimation process.

8.4.4 Metabolic Basis of Improved Tolerance to Anoxia

The major consequence of hypoxic acclimation is the improvement of cellular energy status. Hypoxic acclimation is accompanied by enhanced synthesis of a large number of glycolytic and fermentative enzymes; however, the necessity for such synthesis remains unclear. Certainly these enzymes are required for anaerobic energy production. Null mutants of ADH from three monocotyledons all have a reduced tolerance to anaerobic stress. However, only a very low level of ADH activity is sufficient for flooding tolerance (Johnson et al. 1994). Similarly, in sucrose synthase double mutants of maize, the low enzyme activity observed after acclimation was still sufficient to improve survival following hypoxic pre-treatment (Ricard et al. 1998). Most of the hypoxically induced enzymes appear to be present in well-oxygenated tissues in amounts more than sufficient to account for the ATP actually produced by anaerobic metabolism. Indeed, attempts to improve anoxic tolerance through the overexpression of individual ANPs have, generally, been disappointing. For instance, the moderate increase in ethanolic flux resulting from the overexpression of bacterial *PDC* in tobacco did not increase anoxia tolerance but, instead, resulted in premature cell death (Tadege et al. 1998), although some success was reported in rice over-expressing a rice gene coding for PDC (Quimio et al. 2000).

8.4.4.1 Sugar Transport and Degradation

Fermentable carbohydrate from seed reserves or leaves must move to sink tissues via the phloem. Oxygen shortage could affect energy metabolism by interfering with transport. All membrane-based sugar transporters so far isolated function as proton symporters and are thus energy-dependent. However, high sucrose concentrations within phloem cells would, theoretically, be sufficient, in the absence of chemical energy, to move sucrose along the concentration gradient from cell to cell by diffusion through plasmodesmata. In roots, the available evidence supports a mostly symplastic route for sugar entry into cells (for a review, see Kühn et al. 1999). Yet, although long-distance transport of sugar is not affected by anoxia, maize root tips are still killed. Hypoxic pre-treatment permits survival (Saglio 1985), possibly by protecting plasmodesmata from the collapse that anoxia normally causes (Saglio et al. 1999). Enzymes such as sucrose synthase that utilise sucrose as initial substrate must also be important in hypoxic acclimation.

8.4.4.2 Glycolytic and Fermentative Enzymes

The only enzymes of glycolysis and fermentation with activities close to those of in vivo fermentation rates are hexokinase (HXK) and PDC. An increase in HXK activity upon hypoxic acclimation of maize root apices has been proposed to explain the corresponding increase in glycolytic flux of excised maize root tips fed with glucose (Bouny and Saglio 1996). A role for HXK in regulating glycolysis is supported by its induction in anoxia-tolerant seedlings of *Echinochloa phyllopogon*, but not in anoxia-sensitive *E. crus-pavonis* (Fox et al. 1996).

Transcription of *PDC* is induced by low-oxygen tensions and enzyme activity rises with the associated lowering of cytosolic pH in anoxic tissues (Andrews et al. 1994; Rivoal et al. 1997). Ethanol production can be manipulated experimentally by modifying cytoplasmic pH in ways consistent with the low-pH-mediated activation of PDC (Fox et al. 1995) and can even be induced in the presence of oxygen through constitutively high expression of bacterial PDC in tobacco pollen (Bucher et al. 1995). Overexpression of PDC increased ethanol production under anoxia in tobacco roots but resulted in more damage to roots (Tadege et al. 1998). These results led to the proposal that (1) ethanolic flux is regulated by the concentration of pyruvate rather than by pH, and (2) that survival is not enhanced by abnormally fast ethanolic flux.

8.4.5 Cytoplasmic Acidosis

A second category of hypoxically induced enzymes that might be involved in acclimation includes those influencing cytoplasmic pH. Such enzymes include those responsible for alanine synthesis (Good and Crosby 1989). Other proton-consuming pathways activated by hypoxia include α-aminobutyric acid (GABA) synthesis (Ford et al. 1996). Nitrate may prolong the survival of anoxia in maize roots (Roberts et al. 1985) possibly by suppressing cytoplasmic acidification (Fan et al. 1997). However, questions have been raised about the quantitative significance of these proton-consuming pathways.

The involvement of lactate production in cytosolic acidification has been debated for many years. The role of lactic acid in initiating acidification is increasingly questioned. Its accumulation is now thought to cause runaway acidification leading to cell death (reviewed by Ratcliffe 1995, 1999). Outward lactate transport may be important in certain plant tissues (Rivoal and Hanson 1993) by helping to maintain a less acidic cytoplasm.

8.4.6 Other Routes to Tolerance

Other hypoxically induced proteins that have a protective function include scavengers of active oxygen species generated by re-oxygenation after anaerobic stress (Monk et al. 1987; Biemelt et al. 1998). Anaerobically inducible superoxide dismutase (SOD) is thought to reduce damage to membranes from such free radicals by converting superoxide radicals to hydrogen peroxide (Monk et al. 1987). A second free radical scavenging enzyme, ascorbate peroxidase, has been reported to be inducible by hypoxia but other such enzymes (monodehydroascorbate reductase, dehydroascorbate reductase and glutathione reductase) were only slightly influenced by hypoxia and strongly decreased in activity by anoxia (Biemelt et al. 1998).

Of great potential significance is the temporary induction of a large family of genes after only 1–2 h without oxygen (Huq and Hodges 1999). Although with no significant homology to any known genes or proteins, these early-induced genes may turn out to have regulatory functions. Thus, a large number of enzymes and proteins, only some of which are directly involved in metabolic adjustments, are induced by hypoxic treatment. They underline the complex nature of the molecular and biochemical responses by roots to partial oxygen shortage. They demonstrate an increasing understanding of the molecular and enzymic changes associated with hypoxic acclimation. Exactly which of these changes are fundamental to acclimation and the oxygen-sensing mechanisms that precede them remain unknown.

8.5 Aerenchyma and Avoidance of Anoxia

Tolerance of anoxia by root tips is normally restricted to only a few days at the most. To survive for longer, oxygen must enter the root interior or some form of protective quiecence set in. In wetland plants, oxygen entry is aided by the presence of interconnected intercellular gas-filled spaces that can extend from the shoot into the root tip (Ashford and Allaway 1995). These spaces are normally created by cell separations (Jackson and Armstrong 1999) that create a pathway of small resistance to oxygen movement by diffusion along concentration gradients (Armstrong 1979). Transport over longer distances can be achieved by pressurised mass-flow (Armstrong et al. 1996). This depends on the existence of (1) perforated partitions that are more porous to inward gas diffusion than outward mass flow, and (2) an outlet of low resistance (Armstrong and Armstrong 1994). The maximum root length that diffusion alone can support is about 30 cm (Armstrong et al. 1991). Thomson et al. (1990) showed that, without aerenchyma, roots longer than 100 mm are fatally damaged by an oxygen-free external medium.

Gas space beyond a small background continuum is usually formed from cell separations resulting from different rates of cell expansion, division and wall adhesion (schizogeny) and/or by the selective death of cells (lysigeny, Jackson and Armstrong 1999; Seago et al. 2000). In rice and maize, aerenchyma is created by spatially regulated cell death, which can be regulated by oxygen supply, hormones, nitrate shortage (Konings and Verschuren 1980) or mechanical impedance (reviewed by Jackson and Armstrong 1999). The main external signal promoting aerenchyma formation in maize roots is an oxygen supply restricted to approximately 3–12.5% (Jackson et al. 1985). How partial oxygen shortage is sensed and transduced are not well-understood (see above). The overall sequence of events can be divided into three stages: (1) partial oxygen shortage promoting ethylene biosynthesis (Drew et al. 1979; Brailsford et al. 1993); (2) trapping of ethylene within the roots by the water covering; and (3) induction of programmed cell death in target cells of the cortex by ethylene to create longitudinally interconnected gas spaces that replace the original files of cells (Gunawardena et al. 2001a). The sensing of ethylene is relayed by a signal transduction cascade that probably involves Ca^{2+}, protein kinases (He et al. 1996) and cytoplasmic acidification (Kawai et al. 1998). Cell wall changes are also surprisingly rapid (<0.5 days; Webb and Jackson 1986; Gunawardena et al. 2001b) and closely accompany cytoplasmic changes that are strongly reminiscent of animal cell apoptosis. The cell wall changes involve de-esterification of pectins detected using monoclonal antibodies (Gunawardena et al. 2001b). These are followed by debris clean-up mediated by hydrolytic enzymes such as cellulase (Grineva and Bragina 1993; He et al. 1994), xylanases (Grineva et al. 2000) and possibly xyloglucan *endo*-transglycosylase (XET; Saab and Sachs 1996).

8.6 Stem Hypertrophy, Adventitious Rooting and Related Phenomena

Other developmental effects promoting internal aeration of roots include accelerated shoot extension (reviewed in Voesenek and Blom 1999) and/or increased uprightness of submerged leaves should flooding deepen to include parts of shoot (Grimoldi et al. 1999). These responses to submergence maintain or regain access to aerial or dissolved oxygen, or to light for the generation of photosynthetic oxygen. This oxygen may then diffuse readily to the root elongation zone within aerenchyma (Waters et al. 1989). Swelling of submerged portions of the lower shoot (Nunez-Elisea et al. 1999) and the appearance of hypertrophic lenticels (Kawase 1974; Nunez-Elisea et al. 1999) may also facilitate oxygen entry into the aerenchyma of nearby adventitious roots. The swellings are probably a consequence of cell expansion promoted by endogenous ethylene trapped in the submerged tissue by water (Kawase 1974; Wample and Reid 1979).

A more widespread response is the emergence, from the shoot base, of aerenchymatous adventitious roots to replace longer and deeper roots damaged or killed by anoxia (Voesenek et al. 1992; Vignolio et al. 1999). In cereals such as maize, ethylene promotes the outgrowth of root primordia from the stem base (Jackson et al 1981). Some species can re-orientate root extension to a more horizontal (*Rumex palustris*, Laan et al. 1989) or even upright direction (salix and eucalyptus, Pereira and Kozlowski 1977; *Ludwegia peploides*, Ellmore 1981) and thus towards zones where oxygen may be more readily available. The mechanisms controlling root re-orientation are unknown but presumably act by modifying the normal gravity response. This may involve relocation of stretch channels to end walls (Pickard and Ding 1992). Ethylene can sometimes influence the direction of root elongation (Goodlass and Smith 1979). Whatever the mechanism, developing a near-surface root system is an important anoxia avoidance strategy (Justin and Armstrong 1987).

8.7 Signalling by Oxygen-Deficient Roots

Evidence of root-to-shoot signalling can be seen in land plants such as peas (*Pisum sativum*), tomato (*Lycopersicon esculentum*) and castor bean (*Ricinus communis*) during the first hours of soil flooding (Jackson 2002). Responses to these signals can reduce the damage to the shoot resulting from loss of normal supplies of basic resources from the roots, especially water. Epinastic leaf curvature (a downward re-orientation of whole leaves and leaflets involving growth promotion on the upper, i.e. adaxial surface) and stomatal closure are two well-known responses to signals from flooded roots. Both reactions commence within a few hours of the start of flooding. They moderate evapotranspiration and thus suppress a loss of leaf hydration that can result from decreased root hydraulic conductance in flooded plants. This lowering of root conductance is a response to oxygen shortage and/or increased CO_2. It may be the outcome of effects on the resistance of water channels (aquaporins) or of a decrease in their population density, possibly as a consequence of depressed gene transcription (Birner and Steudle 1993; Clarkson et al. 2000).

Epinastic curvature remains the clearest example of a response to root-to-shoot signalling via the transpiration stream (Fig. 8.3). Severely hypoxic or anaerobic roots of tomato generate a positive message (i.e. increased signal delivery) that stimulates epinastic leaf curvature within a few hours in association with faster ethylene biosynthesis in the responding petioles and elsewhere in the shoot (Jackson 1997). Bradford and Yang (1980) showed that the

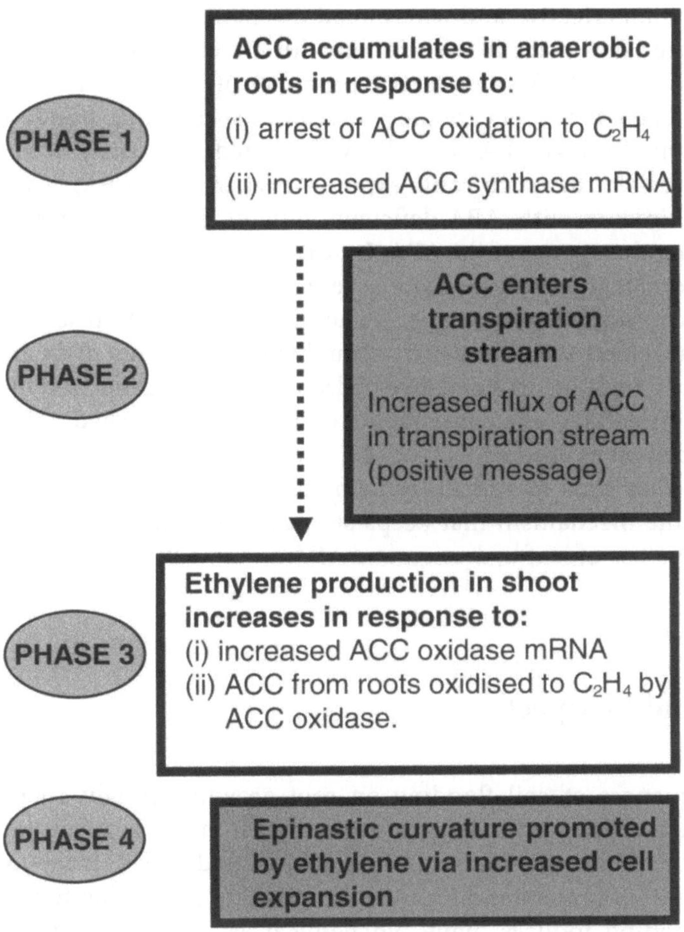

Fig. 8.3. Steps in root-to-shoot signalling that induce leaf epinastic curvature in flooded tomato plants

signal is 1-aminocyclopropane-1-carboxylic acid (ACC), the precursor of ethylene. The amounts are demonstrably sufficient to support the faster rates of shoot ethylene production (Else and Jackson 1998) that, in turn, cause epinastic growth. The release of larger amounts of ACC in xylem sap by anaerobic roots has two probable causes. The first is a blocking of ACC oxidation to ethylene in root cells. This leads to the accumulation of ACC that is drawn into the shoot as a transpiration stream solute. The second likely cause of increased ACC in oxygen-deficient root is an up-regulation, within 1 h, of one member (*LE-ACS7*) of the six-gene family of ACC synthases present in tomato (Shiu et al. 1998). An unidentified root signal also enhances activity in the petioles of ACC oxidase, the enzyme converting ACC to ethylene (English et al. 1995). An increase in the expression of an ACC oxidase gene in the shoot underlies this increased enzyme activity (English, Roberts, Lycett and Jackson, unpubl.).

Stomata of many species close in response to waterlogging (e.g. tomato: Jackson et al. 1978; soybean: Sojka 1985; woody species: Pereira and Kozlowski 1977). It is an effective and widespread mechanism for curtailing transpiration. Closure takes place in association with increases, in leaves, of the hormone abscisic acid (ABA), an effective promoter of stomatal closure. However, it is improbable that roots are the source of this ABA since flooded roots export substantially less ABA than do well-aerated ones (Else et al. 1996). It remains possible that such roots export a precursor of ABA; however, grafting experiments with ABA-deficient mutants suggest that the shoots themselves are the source of the ABA. In *Ricinus communis*, early loss of root hydraulic conductance is especially severe. The resulting decreased hydration of leaves is sufficient to trigger accumulation of ABA in leaves (Else et al. 2001), an effect well known in droughted plants. The effect in flooded ricinus plants amounts to a negative hydraulic message generated by oxygen-deficient roots created as a consequence of abnormally low hydraulic conductances induced by oxygen shortage. The signal that induces stomatal closure in other species such as tomato remains unclear (Else et al. 1996). Also unresolved is the mechanism that keeps stomata closed long after the effects of flooding on loss of leaf hydration and ABA accumulation have subsided.

8.8 Summary and Prospects

The drastic consequences of soil flooding on root growth strongly affect species distribution and the productivity of farm crops in many parts of the world. Identifying the damaging factors in flooded soil and how damage to roots ensues is central for understanding and manipulating adaptation to the stress. Although flooding induces many biochemical as well as biological changes in soil, oxygen shortage is of overriding importance. The metabolic basis of injury from anoxia is generally attributed to the imbalance between ATP supply and demand (resulting from energy and sugar starvation) or self-poisoning by products of anaerobic metabolism (leading to low cytoplasmic pH). Short-term, true anoxia tolerance is based on the capability to sense declining oxygen levels, thereby activating a signal transduction pathway which leads to the repression of genes involved in dispensable reactions and, conversely, activation of genes which improve energy levels or limit cytoplasmic acidification. This results in a prolongation of survival without oxygen. However, changes in energy metabolism alone cannot explain long-term plant tolerance to flooding and submergence. Developmental changes that allow plants to avoid anoxia are especially important in the medium and long term. These include aerenchyma formation, expansion of lenticel, stem hypertrophy, replacement root production, re-orientation of root extension and re-oxidation of the rhizosphere. Root adaptation to flooding also involves the generation of signals that, within only a few hours, render the shoot system more resilient to deprivations arising from damage to basic root functions on which aboveground parts depend.

The study of root systems in flooded soils is moving away from classic plant physiology and biochemistry and increasingly harnesses genomics, proteomics and metabolomics. These global approaches are helping in the search for the cellular bases of oxygen sensing, causes of injury in oxygen-starved cells and mechanisms of acclimation to oxygen shortage. However, promising answers to these key questions will then require testing as the findings of basic cell biologists are re-worked in ecophysiological and agrophysiological contexts. Consequently, a resurgence of physiological and biochemical enquiry can be expected in the medium term.

References

Andrews DL, MacAlpine DM, Cobb BG, Johnson JR, Drew MR (1994) Differential induction of mRNAs for the glycolytic and ethanolic fermentative pathways by hypoxia and anoxia in maize seedlings. Plant Physiol 106:1575–1582

Appleby CA, Bogusz D, Dennis ES, Trinick MJ, Peacock WJ (1988) A role for haemoglobin in all plant roots? Plant Cell Environ 11:359–367

Armstrong W (1979) Aeration in higher plants. In: Woolhouse HW (ed) Advances in botanical research, vol 7. Academic Press, London, pp 225–332

Armstrong J, Armstrong W (1994) A physical model involving nucleopore membranes to investigate the mechanism of humidity-induced convection in *Phragmites australis*. Proc R Soc Edinb 102B:529–540

Armstrong W, Beckett PM, Justin SHFW, Lythe S (1991) Modelling, and other aspects of root aeration by diffusion. In: Jackson MB, Lambers H, Davies DD (eds) Plant life under oxygen deprivation. Ecology, physiology and biochemistry. SPB Academic, The Hague, pp 267–282

Armstrong J, Armstrong W, Beckett PM, Halder JE, Lythe S, Holt R, Sinclair A (1996) Pathways of aeration and the mechanisms and beneficial effects of humidity- and Venturi-induced convections in *Phragmites australis*. Aquat Bot 54:177–198

Arredondo-Peter R, Hargrove MS, Sarth G, Moran JF, Lohrman J, Olson JS, Klucas RV (1997) Rice hemoglobins: gene cloning, analysis, and O_2-binding kinetics of a recombinant protein synthesized in *Escherichia coli*. Plant Physiol 115:159–1266

Ashford AE, Allaway WG (1995) There is a continuum of gas space in young plants of *Avicennia marina*. Hydrobiologia 295:1–3

Bailey-Serres J, Dawe K (1996) Both 5′ and 3′ sequences of maize *adh1* mRNA are required for enhanced translation under low-oxygen conditions. Plant Physiol 112:685–695

Berry LJ, Norris WE Jr (1949) Studies of onion root respiration. I. Velocity of oxygen consumption in different segments of root at different temperatures as a function of partial pressure of oxygen. Biochim Biophy Acta 3:593–606

Biemelt S, Keetman U, Albrecht G (1998) Re-aeration following hypoxia or anoxia leads to activation of the antioxidative defense system in roots of wheat seedlings. Plant Physiol 116:651–658

Birner TP, Steudle E (1993) Effects of anaerobic conditions on water and solute relations, and on active-transport in roots of maize (*Zea-mays* L.) Planta 190:474–483

Bouny M, Saglio P (1996) Glycolytic flux and hexokinase activities in anoxic maize root tips acclimated by hypoxic pretreatment. Plant Physiol 111:187–194

Bradford KJ, Yang SF (1980) Xylem transport of 1-aminocyclopropane-1-carboxylic acid, an ethylene precursor, in waterlogged plants. Plant Physiol 65:322–326

Brailsford RW, Voesenek LACJ, Blom CWPM, Smith AR, Hall MA, Jackson MB (1993) Enhanced ethylene production by primary roots of *Zea mays* L. in response to subambient partial pressures of oxygen. Plant Cell Environ 16:1071–1080

Bucher M, Brander KA, Sbicego S, Mandel T; Kuhlemeier C (1995) Aerobic fermentation in tobacco pollen. Plant Mol Biol 28:739–750

Chang WPP, Huang L, Shen M, Webster C, Burlingame AL, Roberts JKM (2000) Patterns of protein synthesis and tolerance of anoxia in root tips of maize seedlings acclimated to a low oxygen environment, and identification of proteins by mass spectrometry. Plant Physiol 122:295–317

Clarkson DTC, Carvajal M, Henzler T, Waterhouse RN, Smyth AJ, Cooke DT, Steudle E (2000) Root hydraulic conductance: diurnal aquaporin expression and the effects of nutrient stress. J Exp Bot 51:61–70

Conley TR, Peng H-P, Shih M-C (1999) Mutations affecting induction of glycolytic and fermentative genes during germination and environmental stresses in *Arabidopsis*. Plant Physiol 119:599–607

Davies DD, Grego S, Kenworthy P (1974) The control of the production of lactate and ethanol by higher plants. Planta 118:297–310

deVetten NC, Ferl RJ (1995) Characterization of a maize G-box binding factor that is induced by hypoxia. Plant J 7:589–601

deVetten NC, Lu G, Ferl RJ (1992) A maize protein associated with the G-box binding complex has homology to brain regulatory proteins. Plant Cell 4:1295–1307

Dolferus R, Jacobs M, Peacock WJ, Dennis ES (1994) Differential interactions of promoter elements in stress responses of *Arabidopsis Adh* gene. Plant Physiol 105:1075–1087

Drew MC, Jackson MB, Giffard S (1979) Ethylene promoted adventitious rooting and development of cortical air spaces (aerenchyma) may be adaptive responses to flooding in *Zea mays* L. Planta 147:83–88

Duff SMG, Wittenberg JB, Hill RD (1997) Expression, purification, and properties of recombinant barley (*Hordeum* sp.) hemoglobin: optical spectra and reactions with gaseous ligands. J Biol Chem 272:16746–16752

Ellis MH, Setter TL (1999) Hypoxia induces anoxia tolerance in completely submerged rice seedlings. J Plant Physiol 154:219–230

Ellmore GS (1981) Root dimorphism in *Ludwigia peploides* (Onagraceae): structure and gas content of mature roots. Am J Bot 68:557–568

Else MA, Jackson MB (1998) Transport of 1-aminocyclopropane-1-carboxylic acid (ACC) in the transpiration stream of tomato (*Lycopersicon esculentum*) in relation to foliar ethylene production and petiole epinasty. Aust J Plant Physiol 25:453–458

Else MA, Tiekstra AE, Croker SJ, Davies WJ, Jackson MB (1996) Stomatal closure in flooded tomato plants involves abscisic acid and a chemically unidentified anti-transpirant in xylem sap. Plant Physiol 112:239–247

Else MA, Coupland D, Dutton L, Jackson MB (2001) Decreased root hydraulic conductivity reduces leaf water potential, initiates stomatal closure and slows leaf expansion in flooded plants of castor oil (*Ricinus communis*) despite diminished delivery of ABA from the roots in xylem sap. Physiol Plant 111:46–54

English PJ, Lycett, GW, Roberts JA, Jackson MB (1995) Increased 1-aminocyclopropane-1-carboxylic acid oxidase activity in shoots of flooded tomato plants raises ethylene production to physiologically active levels. Plant Physiol 109:1435–1440

Fan TWM, Higashi RM, Frenkiel TA, Lane AN (1997) Anaerobic nitrate and ammonium metabolism in flood-tolerant rice coleoptiles. J Exp Bot 48:1655–1666

Fennoy SL, Bailey-Serres J (1995) Posttranscriptional regulation of gene expression in oxygen-deprived roots of maize. Plant J 7:287–295

Fennoy SL, Jayachandran S, Bailey-Serres J (1997) RNase activities are reduced concomitantly with conservation of total cellular RNA and ribosomes in O_2-deprived seedling roots of maize. Plant Physiol 115:1109–1117

Ford YY, Ratcliffe RG, Robins RJ (1996) Phytohormone induced GABA production in transformed root cultures of *Datura stramonium*: an in vivo ^{15}N NMR study. J Exp Bot 47:811–818

Fox GG, McCallan NR, Ratcliffe RG (1995) Manipulating cytoplasmic pH under anoxia: a critical test of the role of pH in the switch from aerobic to anaerobic metabolism. Planta 195:324–330

Fox TC, Green BJ, Drew MC, Kennedy RA, Rumpho ME (1996). Anaerobic expression of hexokinase in shoots of *Echinochloa phyllopogan* and *Echinochloa crus-pavonia*. Plant Physiol 111S:254 (Abstr.)

Franklin Bunn H, Poyton RO (1996) Oxygen sensing and molecular adaptation to hypoxia. Physiol Rev 76:839–885

Good AG, Crosby WL (1989) Anaerobic induction of alanine aminotransferase in barley root tissue. Plant Physiol 90:1305–1309

Goodlass G, Smith KA (1979) Effects of ethylene on root extension and nodulation of pea (*Pisum sativum* L.) and white clover (*Trifolium repens* L.). Plant Soil 51:387–395

Grimoldi AA, Insausti P, Roitman GG, Soriano A (1999) Responses to flooding intensity in *Leontodon taraxacoides*. New Phytol 141:119–128

Grineva GM, Bragina TV (1993) Formation of adaptations to flooding in corn – structural and functional parameters. Russ J Plant Physiol 40:583–587

Grineva GM, Bragina TV, Platonov AV (2000) Ethylene-induced activation of hydrolytic enzymes in adventitious maize roots under conditions of advancing flooding. Dok Bot Sci 373–375:23–25 (Russian version – Dok Akad Nauk. 374:393–396)

Gunawardena HLAN, Pearce DME, Jackson MB, Hawes CR Evans DE (2001a) Characterization of programmed cell death during aerenchyma formation induced by ethylene or hypoxia in roots of maize (*Zea mays* L.). Planta 212:205–214

Gunawardena HLAN, Pearce DMS, Jackson MB, Hawes CR, Evans DE (2001b) Rapid changes in cell wall pectic polysaccharides are closely associated with early stages of aerenchyma formation, a spatially localised form of programmed cell death in roots of maize (*Zea mays* L.) promoted by ethylene. Plant Cell Environ 24:1369–1375

He C-J, Drew MC, Morgan PW (1994) Induction of enzymes associated with lysigenous aerenchyma formation in roots of *Zea mays* L. during hypoxia or nitrogen starvation. Plant Physiol 105:861–865

He C-J, Finlayson SA, Drew MC, Jordan WR, Morgan PW (1996) Ethylene biosynthesis during aerenchyma formation in roots of maize subjected to mechanical impedance and hypoxia. Plant Physiol 112:1679–1685

Hoeren FU, Dolferus R, Wu Y, Peacock WJ, Dennis ES (1998) Evidence for a role of atMYB2 in the induction of the *Arabidopsis* alcohol dehydrogenase gene (*ADH1*) by low oxygen. Genetics 149:479–490

Huq E, Hodges TK (1999) An anaerobically inducible early (*aie*) gene family from rice. Plant Mol Biol 40:91–601

Hwang SY, VanToai TT (1991) Abscisic acid induces anaerobiosis tolerance in corn. Plant Physiol 97:593–597

Jackson MB (1997) Hormones from roots as signals for the shoots of stressed plants. Trends Plant Sci 2:22–28

Jackson MB (2002) Long-distance signalling from roots to shoots assessed: the flooding story. J Exp Bot 53:175–181

Jackson MB, Attwood PA (1996) Roots of willow (*Salix viminalis* L.) show marked tolerance to oxygen shortage in flooded soils and in solution culture. Plant Soil 187:37–45

Jackson MB, Armstrong W (1999) Formation of aerenchyma and the processes of plant ventilation in relation to soil flooding and submergence. Plant Biol 1:274–287

Jackson MB, Gales K, Campbell DJ (1978) Effect of waterlogged soil conditions on the production of ethylene and on water relationships of tomato plants. J Exp Bot 29:183–193

Jackson MB, Drew MC, Giffard SC (1981) Effects of applying ethylene to the root system of *Zea mays* L. on growth and nutrient concentration in relation to flooding. Physiol Plant 52:23–28

Jackson MB, Dobson CM, Herman B, Merryweather A (1984) Modification of 3,5-diiodo-4-hydroxybenzoic acid (DIHB) activity and stimulation of ethylene production by small concentrations of oxygen in the environment. Plant Growth Regul 2:251–262

Jackson MB, Fenning TM, Drew MC, Saker LR (1985) Stimulation of ethylene production and gas space (aerenchyma) formation in adventitious roots of *Zea mays* by small partial pressures of oxygen. Planta 165:486–492

Johnson JR, Cobb G, Drew MC (1994) Hypoxic induction of anoxia tolerance in roots of *Adh1* null *Zea mays* L. Plant Physiol 105:61–67

Justin SHFW, Armstrong W (1987) Anatomical characteristics of roots and plant response to soil flooding. New Phytol 105:465–495

Kawai M, Samarajeewa PK, Barrero RA, Nishiguchi M, Uchimiya H (1998) Cellular dissection of the degradation pattern of cortical cell death during aerenchyma formation of rice roots. Planta 204:277–287

Kawase M (1974) Role of ethylene in induction of flooding damage in sunflower. Physiol Plant 31:29–38

Konings H, Jackson MB (1979) A relationship between rates of ethylene production by roots and the promoting or inhibiting effects of exogenous ethylene and water on root elongation. Z Pflanzenphysiol 92:385–397

Konings H, Verschuren G (1980) Formation of aerenchyma in roots of *Zea mays* in aerated solutions, and its relation to nutrient supply. Physiol Plant 49:265–270

Kühn C, Barker L, Bürkle L, Frommer W-B (1999) Update on sucrose transport in higher plants. J Exp Bot 50:935–953

Laan P, Berrevoets MJ, Lythe S, Armstrong W, Blom CWPM (1989) Root morphology and aerenchyma formation as indicators for the flood-tolerance of *Rumex crispus*. J Ecol 77:693–703

Lu G, DeLisle AJ, deVetten NC, Ferl RJ (1992) Brain proteins in plants: an *Arabidopsis* homolog to neurotransmitter pathway activators is part of a DNA binding complex. Proc Natl Acad Sci USA 9:11390–11494

Magaraggia F, Solinas G, Valle G, Giovinazzo G, Coraggio I (1997) Maturation and translation mechanisms involved in the expression of a myb gene of rice. Plant Mol Biol 35:1003–1008

Manjunath S, Williams AJ, Bailey-Serres J (1999) Oxygen deprivation stimulates Ca^{2+}-mediated phosphorylation of mRNA cap-binding protein elF4E in maize roots. Plant J 19:21–30

Matsumura H, Nirasawa S, Terauchi R (1999) Transcript profiling in rice (*Oryza sativa* L.) seedlings using serial analysis of gene expression (SAGE). Plant J 20:719–726

Monk LS, Fagerstedt KV, Crawford RMM (1987) Superoxide dismutase as an anaerobic polypeptide. A key factor in recovery from oxygen deprivation in *Iris pseudacorus*? Plant Physiol 85:1016–1020

Nunez-Elisea R, Schaffer B, Fisher JB, Colls AM, Crane JH (1999) Influence of flooding on net CO_2 assimilation, growth and stem anatomy of *Annona* species. Ann Bot 84:771–780

Olive MR, Peacock WJ, Dennis ES (1991) The anaerobic responsive element contains two GC-rich sequences essential for binding nuclear protein and hypoxic activation of the maize *Adh* promoter. Nucleic Acids Res 19:7053–7060

Perata P, Gugliemninetti L, Alpi A (1997) Mobilization of endosperm reserves in cereal seeds under anoxia. Ann Bot 79:49–56

Pereira JS, Kozlowski TT (1977) Variations among woody angiosperms in response to flooding. Physiol Plant 41:184–192

Pickard BG, Ding JP (1992) Gravity sensing in higher plants. Adv Compart Environ Physiol 10:81–110

Quimio CA, Torrizo LB, Setter TL, Ellis M, Grover A, Abrigo EM, Oliva NP, Ella ES, Carpena AL, Ito O, Peacock WJ, Dennis E, Datta SK (2000) Enhancement of submergence tolerance in transgenic rice overproducing pyruvate decarboxylase. J Plant Physiol 156:516–521

Ratcliffe RG (1995) Metabolic aspects of the anoxic response in plant tissue. In: Smirnoff N (ed) Environment and plant metabolism: flexibility and acclimation. Bios Scientific Publishers, Oxford, pp 111–127

Ratcliffe RG (1999) Intracellular pH regulation in plants under anoxia. In: Egginton S, Taylor EW, Raven JA (eds) Regulation of tissue pH in plants and animals, a reappraisal of current techniques. Cambridge University Press, Cambridge, pp 193–213

Ricard B, VanToai T, Chourey P, Saglio P (1998) Evidence for the critical role of sucrose synthase for anoxic tolerance of maize roots using a double mutant. Plant Physiol 116:1323–1331

Rivoal J, Hanson AD (1993) Evidence for a large and sustained glycolytic flux to lactate in anoxic roots of some members of the halophytic genus *Limonium*. Plant Physiol 101:553–560

Rivoal J, Thind S, Pradet A, Ricard B (1997) Differential induction of pyruvate decarboxylase subunits and transcripts in anoxic rice seedlings. Plant Physiol 114:1021–1029

Roberts, JKM, Wemmer D, Ray PM, Jardetsky O (1982) Regulation of cytoplasmic and vacuolar pH in maize root tips under different experimental conditions. Plant Physiol 69:1344–1347

Roberts JKM, Callis J, Wemmer D, Walbot V, Jardetzky O (1984a) Mechanism of cytoplasmic pH regulation in hypoxic maize root tips and its role in survival under hypoxia. Proc Nat Acad Sci USA 81:3379–3383

Roberts JKM, Callis J, Jardetzky O, Walbot V, Freeling M (1984b) Cytoplasmic acidosis as a determinant of flooding intolerance in plants. Proc Natl Acad Sci USA 81:6029–6033

Roberts JKM, Andrade FH, Anderson IC (1985) Further evidence that cytoplasmic acidosis is a determinant of flooding intolerance in plants. Plant Physiol 77:492–494

Rowland LJ, Strommer JN (1986) Anaerobic treatment of maize roots affects transcription of *Adh1* and transcript stability. Mol Cell Biol 6:3368–3372

Saab IN, Sachs MM (1996) A flooding-induced xyloglucan *endo*-transglycosylase homolog in maize is responsive to ethylene and associated with aerenchyma. Plant Physiol 112:385–391

Saglio PH (1985) Effect of path or sink anoxia on sugar transport in roots of maize seedlings. Plant Physiol 77:285–290

Saglio PH, Drew MC, Pradet, A (1988) Metabolic acclimation to anoxia induced by low (2–4 kPa) partial pressure oxygen pretreatment (hypoxia) in root tips of *Zea mays*. Plant Physiol 86:61–66

Saglio P, Germain V, Ricard B (1999) The response of plants to oxygen deprivation: role of enzyme induction in the improvement of tolerance to anoxia. In: Lerner HR (ed) Plant responses to environmental stresses. Marcel Dekker, New York, pp 373–393

Saint-Ges V, Roby C, Bligny R, Pradet A, Douce R (1991) Kinetic studies of the variations of cytoplasmic pH, nucleotide triphosphates (^{31}P NMR) and lactate during normoxic and anoxic transitions in maize root-tips. Eur J Biochem 200:477–482

Seago JL Jr, Peterson CA, Kinsley LJ, Broderick J (2000) Development and structure of the root cortex in *Caltha palustris* L. and *Nymphaea odorata* Ait. Ann Bot 86:631–640

Shiu OY, Oetiker JH, Yip WK, Yang SF (1998) The promoter of *LE-ACS7*, an early flooding-induced 1-aminocyclopropane-1-carboxylate synthase gene of the tomato, is tagged by a *Sol3* transposon. Proc Natl Acad Sci USA 95:10334–10339

Sojka RE (1985) Soil-oxygen effects on two determinate soybean isolines. Soil Sci 140: 333–343

Subbaiah CC, Zhang J, Sachs MM (1994a) Involvement of intracellular calcium in anaerobic gene expression and survival of maize seedlings. Plant Physiol 105:369–376

Subbaiah CC, Bush DS, Sachs MM (1994b) Elevation of cytosolic calcium precedes anoxic gene expression in maize suspension cultured cells. Plant Cell 6:1747–1762

Subbaiah CC, Bush DS, Sachs MM (1998) Mitochondrial contribution to the anoxic Ca^{2+} signal in maize suspension-cultured cells. Plant Physiol 118:759–771

Tadege M, Brandle R, Kuhlemeier C (1998) Anoxia tolerance in tobacco roots: effect of overexpression of pyruvate decarboxylase. Plant J 14:327–335

Taylor ER, Nie XZ, MacGregor AW, Hill RD (1994) A cereal hemoglobin gene is expressed in seed and root tissues under anaerobic conditions. Plant Mol Biol 24:853–862

Thomson CJ, Armstrong W, Waters I, Greenway H (1990) Aerenchyma formation and associated oxygen movement in seminal roots of wheat. Plant Cell Environ 13:395–403

Vartapetian BB, Jackson, MB (1997) Plant adaptations to anaerobic stress. Ann Bot 79:(Suppl A) 3–20

Vartapetian BB, Andreeva IN, Kozlova GI, Agapova LP (1977) Mitochondrial ultrastructure in roots of mesophyte and hydrophyte at anoxia and after glucose feeding. Protoplasma 91:243–256

Vignolio OR, Fernandez ON, Macerira NO (1999) Flooding tolerance in five populations of *Lotus glaber* Mill. (syn. *Lotus tenuis* Waldst. et. Kit.). Aust J Agric Sci 50:555–559

Voesenek LACJ, Blom CWPM (1999) Stimulated shoot elongation: a mechanism of semiaquatic plants to avoid submergence stress. In: Lerner HR (ed) Plant responses to environmental stresses: from phytohormones to genome reorganization. Marcel Dekker, New York, pp 431–448

Voesenek LACJ, Van der Sman AJM, Harren FJM, Blom CWPM (1992) An amalgamation between hormone physiology and plant ecology: a review of flooding resistance and ethylene. J Plant Growth Regul 11:171–188

Walker J, Howard E, Dennis E, Peacock W (1987) DNA sequences required for anaerobic expression of the maize *alcohol dehydrogenase I* gene. Proc Natl Acad Sci USA 84:6624–6628

Wample RL, Reid DM (1979) The role of endogenous auxins and ethylene in the formation of adventitious roots and hypocotyls in flooded sunflower plants (*Helianthus annuus* L.). Physiol Plant 45:219–226

Waters I, Armstrong W, Thompson CJ, Setter TL, Adkins S, Gibbs J, Greenway H (1989) Diurnal changes in radial oxygen loss and ethanol metabolism in roots of submerged and non-submerged rice seedlings. New Phytol 113:439–451

Webb J, Jackson MB (1986) A transmission and cryo-scanning electron microscope study of the formation of aerenchyma (cortical gas-filled space) in adventitious roots of rice (*Oryza sativa*). J Exp Bot 37:832–841

Webster C, Gaut RL, Browning KS, Ravel JM, Roberts KJM (1991) Hypoxia enhances phosphorylation of eukaryotic initiation factor 4A in maize root tips. J Biol Chem 226:23341–23346

Xia J-H, Roberts JKM (1996) Regulation of H^+ extrusion and cytoplasmic pH in maize root tips acclimated to a low-oxygen environment. Plant Physiol 111:227–233

Xia J-H, Saglio PH (1992) Lactic acid efflux as a mechanism of hypoxic acclimation of maize root tips to anoxia. Plant Physiol 100:40–46

Xia J-H, Saglio P, Roberts JKM (1995) Nucleotide levels do not critically determine survival of maize root tips acclimated to a low-oxygen environment. Plant Physiol 108:589–595

9 Root Competition: Towards a Mechanistic Understanding

H. de Kroon, L. Mommer and A. Nishiwaki

9.1 Introduction

Current evidence suggests that competition belowground is a ubiquitous phenomenon in many natural and semi-natural types of vegetation. In reviewing 23 studies that separated root and shoot competition under controlled conditions, Wilson (1988) found that root competition was usually the more intense form of interaction. Even in relatively productive soils where plants produce large canopies and competition for light may be expected to be an important factor, root interactions may still play a significant role. Several studies (Belcher et al. 1995; Gerry and Wilson 1995; Cahill 1999) have confirmed the significance of belowground interactions in competition under field conditions.

A number of reviews over the past 15 years have paid attention to competition underground (Casper and Jackson 1997; Schwinning and Weiner 1998) or aspects of nutrient acquisition in competitive settings (Caldwell and Richards 1986; Grime et al. 1991; Caldwell 1988; Chap. 2). In this review we will integrate the perspectives in these papers in an attempt to synthesise the current mechanistic understanding of belowground competition. We will begin by reviewing empirical studies in which belowground competitive ability has been correlated with specific plant traits. The remainder of the review is aimed at providing a mechanistic basis for these findings and clarifying some of the conflicting results that have appeared. To reach this goal, first, the mechanisms of root-to-root interactions will be examined, including both indirect effects through nutrient or water depletion and direct allelochemical interference. Secondly, we hypothesise on the root distribution patterns that are likely to follow from these interactions and the empirical evidence for different rooting patterns in the field. Thirdly, we analyse the nature of the competitive interactions between plant individuals that result from the responses

Ecological Studies, Vol. 168
H. de Kroon, E.J.W. Visser (Eds.)
Root Ecology

(i.e. symmetric versus asymmetric competition) as well as the competitive dynamics in the shorter and longer term.

9.2 What Traits Confer Belowground Competitive Ability?

The general understanding in most papers on belowground competition is that successful competitors have a root system that is able to rapidly proliferate in resource-rich volumes of soil, depleting the resources before competing plants do. This notion arguably arose from Grime's concept of competitors in his triangular model of primary plant strategies (Grime 1979). In this view, successful competitors have high rates of resource capture above- and belowground with active foraging mechanisms that locate resource-acquiring structures (leaves, root tips) in resource-rich areas (high-light patches, nutrient hotspots) within the plant's vicinity (Campbell et al. 1991; Grime et al. 1991). Strong competitors typically have high growth rates and thus a high demand for mineral nutrients and water. As a result, zones of resource depletion rapidly develop in soil patches where roots cluster, and redistribution of roots and rapid proliferation in undepleted parts of the soil volume are required to sustain this high demand.

What evidence exists to support the proposal that rapid and extensive root proliferation confers belowground competitive ability? Perhaps the most detailed test can be found in the work of Caldwell and co-workers on belowground interactions in the cold-desert plant community of the Great Basin in the USA. Here, individual plants of the dominant shrub *Artemisia tridentata* grow interspersed with bunchgrasses, of which *Pseudoroegneria spicata* is one of the abundant native species. An increasingly common invader is *Agropyron desertorum*, a bunchgrass similar to *Pseudoroegneria* in many characteristics but a superior competitor in this system. Competition is very likely to be primarily belowground in the Great Basin area, in which water becomes an increasingly scarce resource towards the summer in the calcareous soils that are poor in nutrients in general and phosphate in particular. Earlier studies showed that *Agropyron* exhibited root growth earlier in the season than *Pseudoroegneria* resulting in more rapid water extraction (Eissenstat and Caldwell 1988a,b). The greater root growth of *Agropyron* at the low temperatures of the early cold-desert growing season, and its greater rooting densities, achieved by virtue of having thinner roots rather than having higher root biomass, were two characteristics that seemed particularly important for its competitive success.

The results on the acquisition of phosphate by these cold-desert species were less consistent with their competitive hierarchy. One study on P acquisition from enriched soil microsites did demonstrate that *Agropyron* responded to P enrichment by a combination of root morphological and physiological changes leading to earlier and greater acquisition than in *Pseudoroegneria* and *Artemisia* (Black et al. 1994). Exactly what root characteristics were responsible for this result remained unclear (Jackson et al. 1990; Caldwell et al. 1991b). *Artemisia* shrubs were 4–10 times more effective in P acquisition per unit root length than the grasses and here also the underlying causes remain unclear (Van Auken et al. 1992). Summarising his extensive studies, Caldwell (1994) concluded that there is no simple relationship between root investment in resource-rich soil patches and acquisition of soil resources. Larigauderie and Richards (1994) also found similar root length growth rates and other morphological responses among seven cold-desert grass cultivars and concluded that aspects of root proliferation are not directly related to productivity and competition.

Other studies have also been unable to relate belowground competitive ability to nutrient procurement from enriched soil patches (Kosola and Gross 1999) or evenly spread fertiliser (Theodose et al. 1996) in an old-field and alpine plant community, respectively. The competitive invasive grass *Bromus madritensis* in the Mojave Desert did acquire more phosphate from root interspaces shared with native shrubs (Yoder and Nowak 2000), but the rooting characteristics by which this was achieved remained unclear. Recently, Leuschner et al. (2001) showed that the higher competitive ability of beech compared to oak correlated with the higher fine root biomass of the former species in the organic nutrient-rich soil profile of the forest floor, but did not examine nutrient acquisition. Other studies have quantified differences in rooting characteristics, including root proliferation, between species but it is not clear how they are correlated with differences in competitive ability (Mou et al. 1997; Huante et al. 1998).

We conclude that the evidence for the contention that high root proliferation leads to high nutrient acquisition and results in a high belowground competitive ability is far from conclusive.

9.3 Mechanisms of Root–Root Interactions

9.3.1 Indirect Effects Through Resource Depletion

Extensive work by Nye, Tinker, Barber and others in the 1970s and 1980s revealed how roots deplete their immediate soil environment through uptake of water and nutrients (summarised in Caldwell and Richards 1986; Caldwell 1988). The effects of nutrient uptake are dependent on diffusion in the soil. Soil diffusion rates may differ several orders of magnitude between different nutrients, with nitrate as a highly mobile nutrient and phosphate as a more immobile nutrient, and ammonium and potassium with more intermediate diffusion rates. Consequently, depletion zones for nitrate are wide and may readily overlap between neighbouring roots, but differences in nutrient concentration are small. Conversely, overlap between depletion zones for phosphate will occur only at very high root densities, but are characterised by steep concentration gradients. In any case, a more even distribution of the roots in the soil will minimise overlap in depletion zones and hence competition between roots. From the overwhelming influence of soil diffusion rates one may also infer that the timing and location of root growth in vacant soil patches are more important than the physiological potentials of the roots (i.e. nutrient uptake capacity) for the capture of available nutrients.

Local enrichment of the soil alters root behaviour, as summarised by Hutchings and John (Chap. 2). Both the number of lateral root primordia as well as the length increment of the lateral roots may be substantially enhanced. Local nitrate concentration has been shown to act as a stimulus, at least for lateral root growth (Zhang and Forde 1998), and the same is probably true for other anions (Robinson 1994). The converse situation is that branching and growth of roots are retarded when they enter sectors of the soil that are depleted as a result of uptake by neighbouring roots. Evidence to date suggests that roots exert these effects on their neighbours as a result of the low nutrient concentrations to which they deplete the soil in their vicinity, not because of the rapid early uptake that they may realise in vacant soil patches.

The combined effects of root responses in depleted and enriched sectors of the soil volume are that the root system will preferentially develop in the enriched soil parts (Fransen et al. 1999b). This is not only the result of a localised response of individual roots, but may also be affected by a coordinated activity of the whole plant (Gersani and Sachs 1992; Fitter 1994; Hutchings and de Kroon 1994; Robinson 1994). It is probably the nutrient status of the whole plant that mediates the local response, with more intense local root development in enriched patches if the plant is more starved and if the patches are smaller (Zhang and Forde 1998). For example, Gersani and Sachs (1992) showed with pea seedlings that more lateral roots were initiated in high nutrient solution when another part of the root system was either removed or deprived of nutrients. Also, mature plants of the perennial grass *Holcus lanatus* developed a higher relative root biomass in rich soil when growing in a split treatment with poor soil, compared with plants under corresponding uniformly rich conditions (Fransen and de Kroon 2001; Fig. 9.1). However, plants grown in a heterogeneous treatment with a similar patch contrast, but overall poorer soil, did not exhibit this enhanced root development. In this experiment, a lower relative root biomass of *H. lanatus* developed in the poor side of the split treatment, compared with a corresponding uniformly poor soil, suggesting that the plant may curtail root growth in nutrient-poor areas if other roots have located more favourable areas. In other experiments root growth in poor soils was rather independent of the behaviour of other parts of the root system (Robinson 1994; Fransen et al. 1998) and

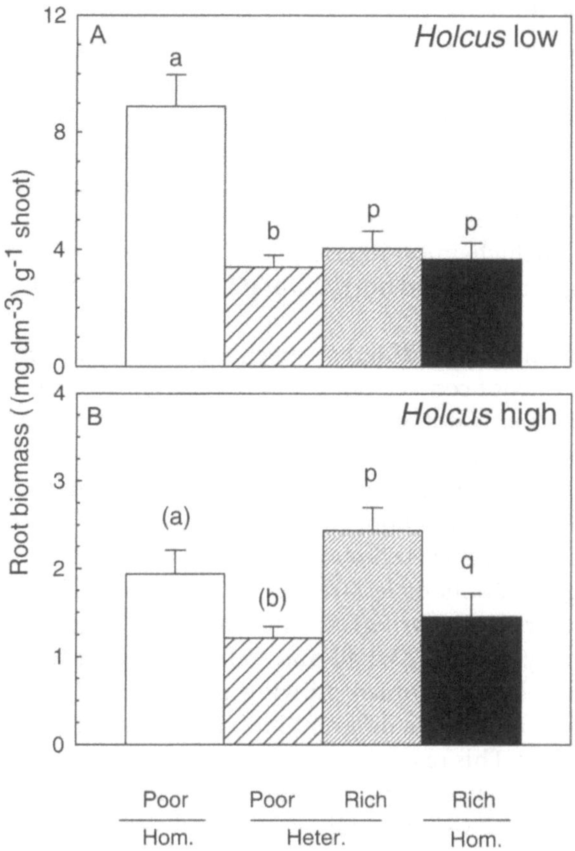

Fig. 9.1. Mean (±SE) root biomass per gram shoot biomass in soil cores taken in *Holcus lanatus* plots at the end of a second growing season, at A low and B high nutrient availability. For A initial N-mineral values were 8.8 and 4.6 mg kg^{-1} for the rich and poor soil, respectively; for B these values were 16.3 and 8.8 mg kg^{-1}. For all comparisons, *bars with the same letter* within each species × nutrient level combination are not significantly different at $P>0.05$; (*a*) and (*b*) denote significant differences at $P=0.051$. B shows selective root placement of *Holcus lanatus* in the nutrient-rich part of the heterogeneous treatment. From A it can be concluded that this response does not occur when overall nutrient availability is lower. (Adapted from Fransen and de Kroon 2001)

the evidence that patch exploitation occurs at the expense of root growth elsewhere is still inconclusive.

9.3.2 Direct Chemical Interactions

Schenk et al. (1999) give an excellent summary of the direct interactions between roots and distinguish between two different classes of interaction. The first comprises allopathic effects involving the exudation of toxic substances and the non-specific inhibitory effects on neighbouring root development. The second involves non-toxic chemical signals that may specifically affect the roots of neighbours.

Some of the most intriguing examples of both types of interaction have been reported by Mahall and Callaway (1991, 1992, 1996) with the desert shrubs *Larrea tridentata* and *Ambrosia dumosa*. *Larrea* influences neighbouring roots by allelochemical substances.

Applying a root chamber set-up, as depicted in Fig. 9.2A, Mahall and Callaway (1992) showed that root elongation is inhibited within the rhizosphere up to a centimetre's distance of a *Larrea* root. Mixing activated carbon with the substrate alleviated the inhibition, showing that a readily diffusible organic substance was responsible for the inhibition, and not the depletion of water or nutrients around the target roots (Mahall and Callaway 1992). The signal is non-specific, affecting sister roots of the same plant, roots of other *Larrea* plants and roots of other species (*Ambrosia*) in a similar manner. Exudation of allelochemicals by roots is a widespread phenomenon (Chap. 10) and many more examples of allelochemical interferences have now been documented, especially in the agricultural and horticultural literature (Schenk et al. 1999).

Ambrosia roots use a fundamentally different mechanism to detect and avoid other roots. Only root-to-root contact, not root proximity, invokes root growth inhibition. This inhibition is not altered by adding activated carbon, indicating that a contact-mediated chemical signal is involved rather than a diffusible organic chemical. Further, the inhibition is specific and the patterns are complex (Mahall and Callaway 1996). Roots of *Ambrosia* plants are inhibited upon contact with a root of another *Ambrosia* individual if both individuals are from the same region. Roots of less-related plants from different regions, or roots of different species (*Larrea*), are not inhibited, suggesting that the signal is at least partly genetic. Mahall and Callaway also reported that sister roots of the same plant continued growing upon contact, while growth was reduced if roots belonged to a different individual with the same genotype as the target individual. This result must be interpreted with caution because, in the set-up of Fig. 9.2A, two roots from the same individual plant cannot be forced to encounter each other while growing in perpendicular directions, and therefore the contact between these roots must have been less intense than in the case of two roots from physiologically separate individuals.

Few other examples of the operation of signals exist and they may operate differently. Using similar experimental root boxes as those used by Mahall and Callaway (1991; Fig. 9.2A), Nishiwaki (unpubl.) studied root-to-root recognition in the grass *Miscanthus sinensis*. Here the recognition mechanism was entirely genetic. Root growth was hardly affected when roots contacted roots from plants of a different genotype, but was significantly inhibited when roots contacted a root of a separate, cloned individual with the same genotype (Fig. 9.2B). Somewhat comparable results have been obtained by Falik et al. (unpubl.) with pea seedlings. In slanted chambers, roots were forced to grow adjacent to roots of the same individual on one side (self roots) and roots of another individual on the other side (non-self roots). Lateral root number and root length growth were significantly higher towards the non-self than the self roots. The difference was smaller when the self roots belonged to a genetically identical but physically separated individual, suggesting that the self/non-self recognition mechanisms involved a physiological mechanism in addition to a genetic effect.

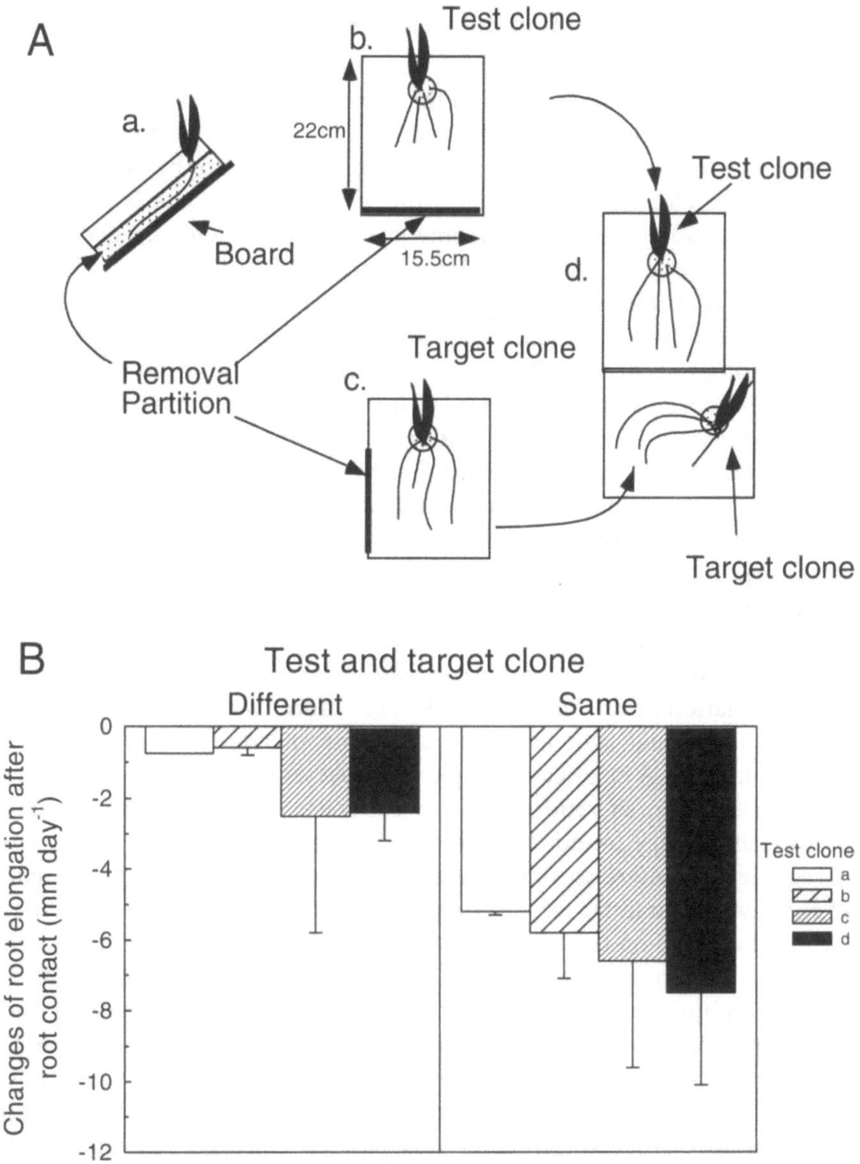

Fig. 9.2A, B. **A** Root chamber design and arrangement to study root interaction mechanisms (adapted from Mahall and Callaway 1991), consisting of flat rectangular chambers, placed at a 45° angle so that roots grow down along a Plexiglas viewing window. *a* Lateral view. The board is removable so that the roots can be observed. *b, c* Face view of test and target chambers. By removing the partitions the chambers are connected in such a way that the roots of the test clone grow into the rhizosphere of the target clones. *d* Face view of test and target chambers when connected and partitions were removed. **B** Changes in root elongation of test clones of *Miscanthus sinensis* after contact with roots of different clones, or with roots of the same clone (Nishiwaki, unpubl.). Four different clones were tested. The stronger growth reduction with roots of the same clone was statistically significant ($F_{1,13}$=5.50, P=0.036)

9.4 Root Distributions as a Consequence of Root–Root Interactions

Based on the interactions described in the previous section, what root distributions are to be expected? If roots reduce the growth and branching of neighbouring roots, either by water and nutrient depletion or by exudation of (non-specific) allelochemicals, segregation of individual roots is to be expected, irrespective of the identity (same or different individuals, genotypes or species) of these roots. Due to methodological difficulties, few people have looked at root distribution at such a small scale. In experiments with cold-desert species, Caldwell et al. (1991a, 1996) examined fine-scale root distributions by freezing soil cores in situ and identifying grass and shrub roots at cut surfaces of the frozen soil cores under UV light. Shrub and grass roots indeed tended to be separated at the millimetre to centimetre scale. The specific mechanism remained unclear. Nutrient depletion around roots was ruled out because there was no effect of experimental nutrient addition on the degree of root segregation. Companion studies provided indications that roots of the two grass species (*Pseudoroegneria spicata* and *Agropyron desertorum*) interacted by specific recognition mechanisms (Krannitz and Caldwell 1995; Huber-Sannwald et al. 1996) but no such evidence was found for interactions between the grasses and the shrub *Artemisia tridentata*.

Root segregation at the whole-plant level is more commonly investigated than at the individual root level, especially for species from arid regions (Schenk et al. 1999). A classical example of a species forming root territories is *Larrea tridentata*, as reported in an excavation study by Brisson and Reynolds (1994). They found that the root systems of mature *Larrea* shrubs had irregular circumferences and were often displaced relative to the stem. These root systems had far less overlap than circular root systems of similar area would have had on the basis of random root developments (see Fig. 2.4, Chap. 2). This pattern may be explained by the non-specific allelopathic influences between individuals that prevent roots of one plant from growing into the soil volume occupied by the roots of a neighbour. In addition, very little of the plot surface area remained unoccupied, suggesting that plants had redirected their foraging activities into non-occupied soil patches. The data do not allow us to distinguish whether this pattern emerged by whole-plant coordinated compensatory growth of roots into bare soil patches or unaltered root growth rates at sides without competitors, but modelling results are in support of the former explanation (Brisson and Reynolds 1997).

Similar results as for *Larrea* have been obtained for the root system distributions in 3-year-old monocultures of two tree species (Mou et al. 1995) but here the underlying mechanism remains unclear. These patterns of root avoidance may also be expected simply on the basis of soil resource depletion (see Chap. 2). Based on an experiment in which split-rooted pea plants were given a choice between a variable number of competitor pea plants or a pot without competitors, Gersani et al. (1998) suggested that the root densities in different soil patches increased until they reached a similar uptake rate per unit root biomass. Such a response would lead to a maximisation of resource uptake rate at the level of the whole plant and conform to Fretwell's (1972) "ideal free distribution". This general theory on distribution patterns predicts that individuals (i.e. individual roots in this case) distribute themselves unequally in different habitats (i.e. soil patches) in such a way that each habitat will contain a fraction of the population (i.e. total root mass) proportional to the fraction of resources that the habitat supplies. As a result, the marginal resource uptake rate by each individual in each of the habitats becomes equal, irrespective of the initial richness of the habitats. Model results support these empirical results and the relevance of this theory for root behaviour (Gleeson and Fry 1997).

What root distributions might be expected in cases in which specific root recognition mechanisms are involved, as in the cases of *Ambrosia* or *Miscanthus* described above? Because root growth of different *Ambrosia* individuals is inhibited upon contact, root segregation between individuals is also to be expected here. Consistent with this expectation, mature *Ambrosia* shrubs have a regular distribution in the field (Schenk and Mahall 2002). As mature shrubs split to become physiologically separated, the root recognition mechanism of *Ambrosia* will create separate root territories for these different fragments, thus avoiding scramble competition between these genetically identical clones (Schenk et al. 1999).

A different pattern may be predicted if intermingling of roots of genetically different individuals occurs while overlap between self roots is avoided. Gersani et al. (2001) recently reported on competition experiments with soybean that are consistent with these predictions. They compared the growth and biomass allocation of two plants of the same species in two situations. First, the plants shared the soil volume of two pots (both were "fence-sitters" as above) and, second, each plant had its own pot of soil (both were "owners"). Note that in both cases the plants together had access to the same quantity of soil resources. In the case of competing (fence-sitting) plants, the root biomass of both plants was higher and the fruit yield was lower compared with the plants that were grown in their own pot (owners; Fig. 9.3). Apparently, the roots were able to distinguish roots of the same plant from those of a neighbour. Continued proliferation of roots in the presence of a neighbour resulted in higher root investments with decreasing returns in terms of nutrient uptake, which explains the lower yield of the plants in competition. Gersani et al. (2001) related these responses to the economic principle of "the tragedy of the commons" in which the activities of competing individuals increase their own marginal benefits (nutrient uptake in this case) but decrease the value of the common good (nutrient availability), at a disadvantage to all. The competitive struggle may be indeed inevitable because a plant that restrains itself and does not proliferate its roots in the zone of influence of its neighbour will acquire fewer nutrients than a neighbour that does proliferate roots.

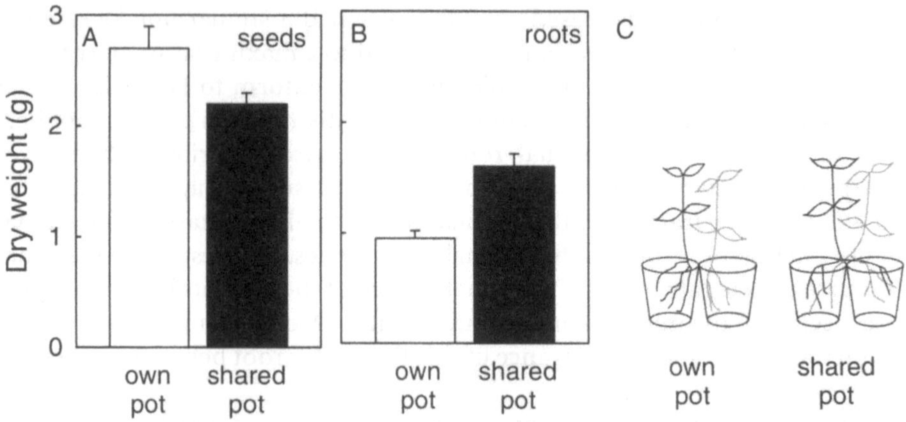

Fig. 9.3A–C. Mean (±SE) dry weight of seeds **(A)** and roots **(B)** of *Glycine max* that were growing as owner of a pot or sharing a pot with another individual with in total a similar amount of nutrients. From the graphs it is clear that root competition between individual soybean plants results in a tragedy of the commons: lower seed weight is produced from the same amount of nutrients, because of increased root growth. (Redrawn from Gersani et al. 2001)

Note that plant foraging responses reported in Hutchings and John (Chap. 2) and above are entirely compatible with any of these root distribution patterns. Much depends on the volume of unoccupied soil available to the plants and the soil patchiness. As plants grow, resources in rich soil volumes and vacant soil will be occupied first. When rooting densities increase and the root systems start to interfere, the plants will start to develop root territories or their root systems start to intermingle, depending on the specific root-to-root interaction mechanisms that come into play.

In conclusion, the available evidence indicates that there are two general root distribution patterns when root systems interact. Either root avoidance between individuals leading to the formation of plant root territories or, conversely, intense root intermingling. Both may involve non-specific interference mechanisms such as resource depletion or allelochemical interactions, as well as specific signals by which roots may recognise other roots of the same physiological individual or the same genotype upon contact. Our understanding of the interplay between these different mechanisms, how they result in specific rooting patterns, and under what conditions they operate is still in its infancy. Schenk et al. (1999) suggested that root territories are more commonly found with species under resource-poor and dry conditions. Consistent with this generalisation, Kleijn and de Kroon (unpubl.) found no indication that roots of the productive grass *Elymus repens* changed their growth direction in response to the presence of *Holcus lanatus* plants in parts of the tray in which they were growing.

9.5 Belowground Competition as a Consequence of Root Distribution Patterns

9.5.1 Symmetric Competition for Space

If root segregation occurs and root territories develop, both small and large plants have access to their own soil volume from which they can extract soil-based resources. Competition is essentially for space with resources secured from within the space occupied. Larger plants cannot pre-empt the contested resources at the expense of smaller plants. This leads to symmetric competition, in which plants acquire resources in proportion to their size (Weiner 1990; Schwinning and Weiner 1998). Size symmetry of competition has profound consequences for population dynamics and persistence of individuals. These include: size differences between individuals are relatively constant at high densities, density-dependent mortality is low and because resources are relatively equally shared between individuals, plants within the population also perform relatively equally (Weiner 1990).

Brisson and Reynolds (1997) produced a revealing spatial model of belowground competition, inspired by their study of *Larrea* territories, which exposes the implications of root territorial interactions. In their model, "compensatory" plants were able to enhance their growth into neighbour-free areas in proportion to the extent to which their root growth was arrested by the presence of neighbouring roots. "Noncompensatory" model plants did not expand anymore as they contacted roots of neighbouring plants (Fig. 9.4A). Mortality of individuals was dependent on the degree of plant interference, with a lower survivorship for plants that were less able to compensate. The model results were in support of the notion of size-symmetric competition because the size inequalities, the relative size-differences within the population, remained small as competition proceeded (Fig. 9.4B). This was true for both modelled plant types, although resources were more evenly shared between individuals in monocultures of compensatory plants, and mortality rates were much lower than in monocultures of noncompensatory plants. In mixed populations, the compensatory plants were competitively superior to the noncompensatory plants that were readily outcompeted (Fig. 9.4C). Here some degree of asymmetry was introduced among the individuals because the compensatory plants had much better access to the bare soil patches than the noncompensatory plants. These results show how symmetric competition for space may result in efficient space filling (Fig. 9.4D) and a relatively equal share of resources between individuals, as seen with *Larrea* (Brisson and Reynolds 1994). They also show that differences in foraging characteristics between individuals (compensatory growth) may have profound and immediate consequences for their persistence, though this outcome was much influenced by the specific mortality function. If individuals beyond a certain

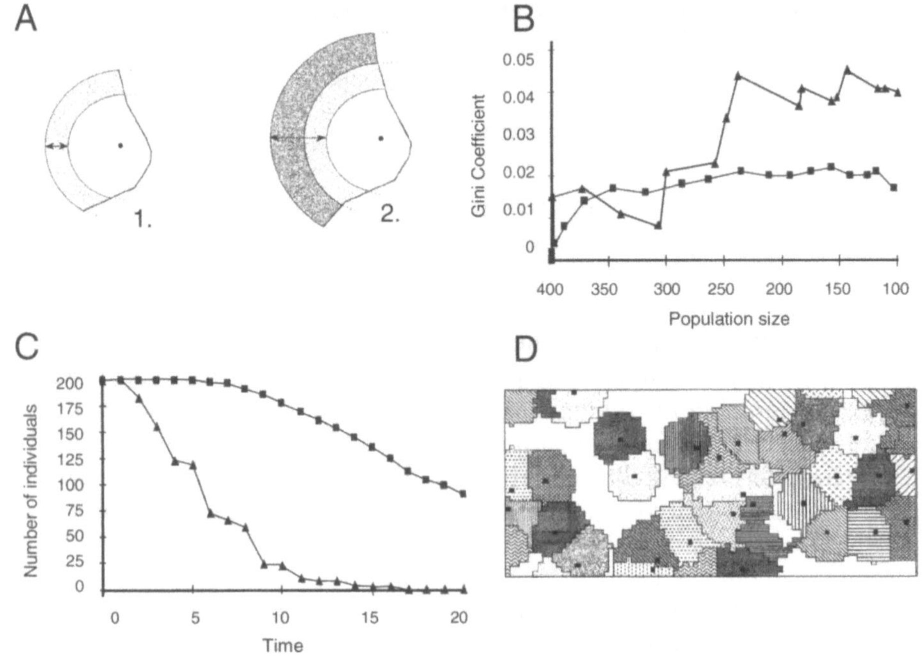

Fig. 9.4A–D. Root territoriality and compensatory root growth and its consequences in the model study of Brisson and Reynolds (1997). **A** Increases in size of the root system for a plant without compensatory growth in the presence of neighbours (*1*), and in a plant with compensatory growth in the presence of neighbours (*2*). The *white part* represents the size of the root system at t=x. The *arrow* represents the extent of radial growth in one time step. The *lighter part of the circle* represents the radial increase under noncompensatory conditions, the *darker part* the extra space filling because of compensatory growth. **B** Changes in size inequalities as a function of population size, for simulated monocultures with compensatory (*squares*) or noncompensatory plants (*triangles*). **C** Changes in population size for a mixed population starting with an equal number of plants with compensatory (*squares*) and noncompensatory growth (*triangles*). The latter plants are readily outcompeted. **D** The spatial arrangement of individual rooting systems for a small population of plants capable of compensatory growth after some simulated time. At t=0 the population was randomly distributed (*dots*)

size can survive with the resources supplied by the soil area that they have occupied, the mortality differences between the compensatory and noncompensatory plants would soon vanish, although the former would on average still have larger territories than the latter.

9.5.2 Symmetric or Asymmetric Competition for Nutrients

When root systems intermingle and belowground space is essentially shared between competing individuals, size-symmetric competition for nutrients may also be the rule (Gerry and Wilson 1995; Casper and Jackson 1997; Weiner et al. 1997; Berntson and Wayne 2000; Cahill and Casper 2000; Fransen et al. 2001). The reason is not only that the larger plants acquire nutrients that are inaccessible to the smaller plants, but also that the smaller plants gain resources in proportion to their size that the larger plants cannot take up. This situation contrasts with size-asymmetric competition in which larger plants receive a disproportionately greater share of the resources than smaller plants. Asymmetric competition is generally observed in competition for light as a unidirectional resource in which the larger plants shade the smaller plants.

It has been suggested that this picture of symmetric competition belowground may change in heterogeneous habitats (Casper and Jackson 1997; Schwinning and Weiner 1998). Larger plants may reach nutrient-rich patches and deplete them of nutrients before smaller plants can gain access, providing them with a disproportionate advantage in terms of nutrient acquisition and resulting in asymmetric competition. However, if the smaller plants were better foragers than the larger plants, the smaller plants would obtain a disproportionate share of the resources, a situation referred to as "negative asymmetry" (Weiner et al. 1997). To date experimental evidence for asymmetric competition belowground is still scarce (see below).

9.5.3 The Dynamics of Competition

The distribution of roots in the soil is a dynamic process. Soil properties may vary enormously from patch to patch and this variation has a temporal component as well, with rich patches becoming depleted and new rich patches appearing. In addition, in temperate ecosystems there is a seasonal pattern in which most nutrients are released in early spring and become more depleted in the summer (Olff et al. 1994). Moreover, roots have limited lifespans, which implies that small volumes of soil are vacated by dying roots and become ready to be recolonised. Root turnover may differ greatly among plant species (Eissenstat and Yanai 1997; Van der Krift and Berendse 2002; Chap. 3) and hence this will affect the spatio-temporal dynamics of soil volume occupancy.

In order to understand the nature of competition belowground and the root characteristics that enhance competitive ability given these extremely dynamic patterns, it is useful to separate two different situations (Grubb 1994). In the first situation, a large part of the soil volume is still devoid of roots and competing plants expand their root systems into the bare soil. This is the usual initial situation in many competition experiments, but it may also occur temporarily in nature, for example, after disturbances. Disturbed soils are usually rich in nutrients (see Fransen and de Kroon 2001) and plants compete for the transient soil resource. In the second situation, roots have more or less filled belowground space and the nutrients are depleted. Root turnover and ongoing mineralisation of organic material then provide the nutrients for which plants compete.

In the first situation, and especially when enriched soil patches are present, rapid root proliferation and a concomitant rapid pre-emption of the transient nutrient resource confer a competitive advantage, as has been experimentally demonstrated with grass seedlings (Hodge et al. 1999; Robinson et al. 1999). This advantage solved an outstanding paradox (Robinson 1996) in that many plant species exhibit much greater root proliferation than required to capture the nutrients in enriched patches, especially if these nutrients are mobile (see Chap. 2). These results also accord with the simulations of Brisson and Reynolds (1997) where the better foraging compensatory plants competitively excluded the noncompensatory plants, as explained above. Further evidence has recently been obtained by Bliss et al. (2002) who showed in two-species mixtures that the better root foragers grew larger in heterogeneous soils than in homogeneous soils, while the reverse was true for the weaker foraging species. Likewise, it has been suggested that nutrient heterogeneity may enhance the asymmetry of competition as large plants have a higher chance of encountering and exploiting enriched soil patches (Casper and Jackson 1997; Schwinning and Weiner 1998), but experimental evidence is lacking thus far (see Casper and Cahill 1996, 1998; but see below).

In the long term, when the soil volume has largely been filled and nutrient-rich patches have been depleted, a different situation may emerge. Few studies have looked at the processes that then occur. We suggest that the nature of competitive dynamics will then depend on the nutrient renewal rates of the soil and the turnover of the roots of the competing species. Only when enriched soil patches reappear frequently and root turnover is high will root proliferating ability be competitively superior in the long term. In such cases, small soil patches are vacated repeatedly through root senescence and nutrients become available irregularly in time and space. Rapid root proliferation would enable the re-occupation of vacant soil and the capture of nutrients before competitors. This scenario may be realistic in some ecosystems. Herbaceous species with high growth rates are characteristic of nutrient-rich habitats, where soils have high nutrient renewal rates. These species tend to have higher root proliferation rates (Fransen et al. 1998) and shorter root life spans (higher root turnover; Van der Krift and Berendse 2002) than slower-growing species from more nutrient-poor habitats. Only in the presumably highly dynamic situation of such nutrient-rich soils are the high long-term costs of root proliferation and short root lifespans (Fransen and de Kroon 2001) likely to be balanced by resource acquisition.

As in the transient situation described above, asymmetric competition may be expected in such dynamic soils. Asymmetric competition predicts that the larger plants readily gain advantage and outcompete the smaller plants, but belowground these transitions may be slow. Nishiwaki (unpubl.) studied the genotypic diversity in *Miscanthus sinensis* grasslands of different ages. Even in a 1000-year-old grassland, different clones were growing interspersed in a fine pattern in a plot of $2 \times 2\ m^2$. Experimental work had shown, however, that the roots of these clones readily intermingled and thus competitive replacement of the weaker by the stronger genotypes might have been expected especially over the very long term. The slow replacement may have been due to a number of reasons. First, competitive abilities among the clones may be relatively equal (see Fig. 9.2B). Second, if initial plant densities are low so that roots of an individual occupy a given soil volume before they start to intermingle, most root interactions are between the roots of the same individual rather than between the roots of different individuals. Such clumping may slow down competitive replacement considerably (cf. de Kroon et al. 1992; Pacala 1997; Stoll and Prati 2001). Finally, root turnover and nutrient renewal rates may be low in these *Miscanthus* grasslands. This would also slow down the dynamics belowground.

The final suggestion already indicates that the competitive dynamics will be much slower in soils that have low nutrient renewal rates and with plants that have long root life spans. Depletion of the soil to very low nutrient concentrations may be more important than opportunistic proliferation into vacant and enriched soil patches. This would prevent the proliferation of competitor roots into the occupied territory and safeguard the soil resources that slowly renew within this zone of influence.

The results of our recent competition experiment with two grass species (Fransen et al. 2001) illustrate some of the shifts of competitive relationships that may occur from the initial phase of interaction, with very bare soil, to the longer-term in which most of the soil volume is occupied. *Festuca rubra* and *Anthoxanthum odoratum* plants were grown in monocultures and mixtures, both in homogeneous and heterogeneous soils. Earlier experiments (Fransen et al. 1998, 1999a) had identified *Festuca* as the species with the larger and denser root system, but *Anthoxanthum* as better able to acquire patchily distributed resources, though not through root proliferation but through higher nutrient uptake rates. In the first year of the experiment, *Festuca* gained advantage over *Anthoxanthum* both in the homogeneous and the heterogeneous soils (Fig. 9.5), consistent with the suggestion above that plants with larger root systems are competitively superior when much vacant undepleted soil is present. Larger shoot size inequalities in the heterogeneous treatments also suggested that the degree of size asymmetry was higher under patchy soil conditions.

These competitive relationships changed in the second year of the experiment when most of the soil volume was occupied. While *Festuca* remained the superior competitor in the homogeneous soil, it lost its initial advantage to *Anthoxanthum* in the heterogeneous soil (Fig. 9.5). The advantage of *Festuca* was not sustained, perhaps because in the long term the benefits of high root densities in enriched patches faded and may even have resulted in a growth disadvantage due to patch depletion and nutrient losses associated with the

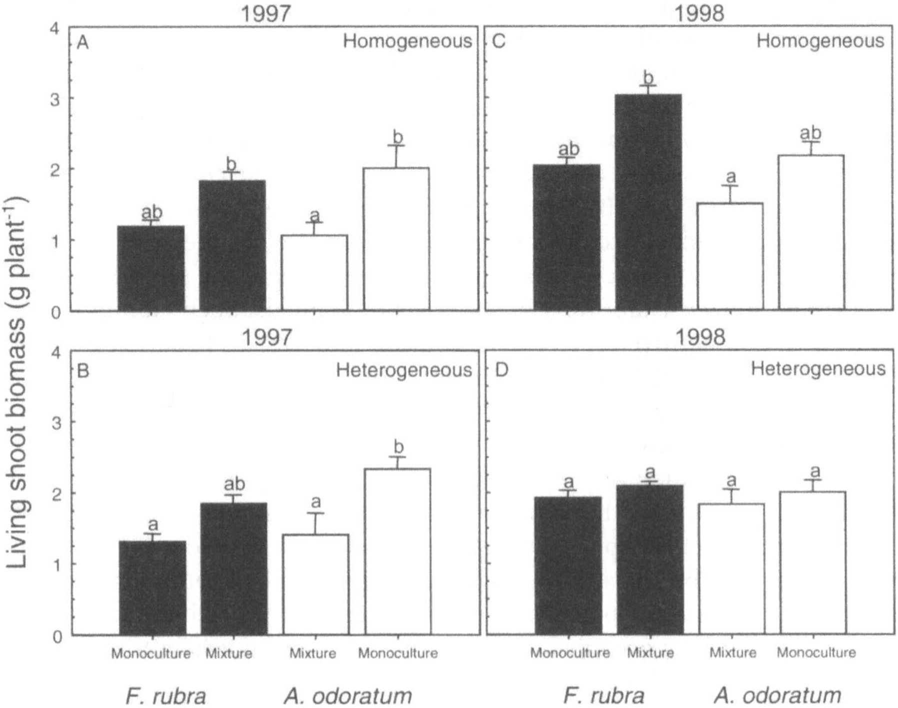

Fig. 9.5A–D. Mean (±SE) living shoot biomass per plant of *Festuca rubra* (*black bars*) and *Anthoxanthum odoratum* (*white bars*) in homogeneous and heterogeneous (*checkerboard*) soils after a first (**A, B**) and second (**C, D**) growing season. *Festuca* appeared to be the better competitor in the first growing season, but it lost its advantage to *Anthoxanthum* in the heterogeneous soil by the end of the second growing season. (Adapted from Fransen et al. 2001; see text for further details)

turnover of roots (Fransen and de Kroon 2001). In contrast to the hypothesis of asymmetric competition, shoot size inequalities did not increase further in the second year of the experiment, although mean shoot sizes were significantly larger in the second year. Apparently, once patches have been occupied, size differences are consolidated or even reduced, and competition between plants of different sizes becomes more symmetrical. The properties of *Anthoxanthum*, namely higher root physiological plasticity that permitted capture of the low nutrient supplies from the slowly mineralising patches of organic material, resulted in a competitive advantage belowground on the longer term.

9.6 Summary and Prospects

The general contention with which this review started was that many papers on belowground competition consider successful competitors as species that have root systems that are able to rapidly proliferate in resource-rich areas of soil, depleting the resources before competitors. Our analysis of root-to-root interactions, the root distributions that follow from these interactions and their dynamics in the shorter and longer run showed that rapid root proliferation will confer competitive advantage only in a subset of the conditions that may be encountered in nature. This may explain the generally weak correlation between root proliferation ability and belowground competitive ability that emerges from a number of empirical studies. In many situations, plants that are able to deplete the soil to low nutrient levels, preventing the roots of neighbours from entering their root territories, may be much more effective competitors in the long run. In addition, chemical interactions, either through allelopathic substances or highly specific self/non-self recognition mechanisms, may play an important role, but have not received the recognition that they deserve in the competition literature. More experimental studies are needed to look into the processes and responses in the long run and to reveal what characteristics under what conditions confer competitive ability underground.

Acknowledgements. We are grateful to Omer Falik, Mike Hutchings, David Kleijn and Jochen Schenk for valuable comments and suggestions.

References

Belcher JW, Keddy PA, Twolan-Strutt L (1995) Root and shoot competition intensity along a soil depth gradient. J Ecol 83:673–682
Berntson GM, Wayne PM (2000) Characterizing the size dependence of resource acquisition within crowded plant populations. Ecology 81:1072–1085
Black RA, Richards JH, Manwaring JH (1994) Nutrient uptake from enriched soil microsites by three great basin perennials. Ecology 75:110–122
Bliss KM, Jones RH, Mitchell RJ, Mou PP (2002) Are competitive interactions influenced by spatial nutrient heterogeneity and root foraging behavior? New Phytol 154:409–417
Brisson J, Reynolds JF (1994) The effect of neighbors on root distribution in a creosotebush (*Larrea tridentata*) population. Ecology 75:1693–1702
Brisson J, Reynolds JF (1997) Effects of compensatory growth on population processes: a simulation study. Ecology 78:2378–2384
Cahill JF (1999) Fertilization effects on interactions between above- and belowground competition in an old field. Ecology 80:466–480
Cahill JF, Casper BB (2000) Investigating the relationship between neighbor root biomass and belowground competition: field evidence for symmetric competition belowground. Oikos 90:311–320

Caldwell MM (1988) Plant root systems and competition. In: Greuter W, Zimmer B (eds) Proceedings of the XIV International Botanical Congress. Koeltz, Königstein, pp 385–404

Caldwell MM (1994) Exploiting nutrients in fertile soil microsites. In: Caldwell MM, Pearcy RW (eds) Exploitation of environmental heterogeneity by plants. Ecophysiological processes above- and belowground. Academic Press, San Diego, pp 325–347

Caldwell MM, Richards JH (1986) Competing root systems: morphology and models of absorption. In: Givnish TJ (ed) On the economy of plant form and function. Cambridge University Press, Cambridge, pp 251–273

Caldwell MM, Manwaring JH, Durham SL (1991a) The microscale distribution of neighbouring plant roots in fertile soil microsites. Funct Ecol 5:765–772

Caldwell MM, Manwaring JH, Jackson RB (1991b) Exploitation of phosphate from fertile soil microsites by three Great Basin perennials when in competition. Funct Ecol 5:757–764

Caldwell MM, Manwaring JH, Durham SL (1996) Species interactions at the level of fine roots in the field: influence of soil nutrient heterogeneity and plant size. Oecologia 106:440–447

Campbell BD, Grime JP, Mackey JML, Jalili A (1991) The quest for a mechanistic understanding of resource competition in plant communities: the role of experiments. Funct Ecol 5:241–253

Casper BB, Cahill JF (1996) Limited effects of soil nutrient heterogeneity on populations of *Abutilon theophrasti* (Malvaceae). Am J Bot 83:333–341

Casper BB, Jackson RB (1997) Plant competition underground. Annu Rev Ecol Syst 28:545–570

Casper BB, Cahill JF (1998) Population-level responses to nutrient heterogeneity and density by *Abutilon theophrasti* (Malvaceae): an experimental neighborhood approach. Am J Bot 85:1680–1687

de Kroon H, Hara T, Kwant R (1992) Size hierarchies of shoots and clones in clonal herb monocultures: do clonal and non-clonal plants compete differently? Oikos 63:410–419

Eissenstat DM, Caldwell MM (1988a) Competitive ability is linked to rates of water extraction. A field study of two aridland tussock grasses. Oecologia 75:1–7

Eissenstat DM, Caldwell MM (1988b) Seasonal timing of root growth in favorable microsites. Ecology 69:870–873

Eissenstat DM, Yanai RD (1997) The ecology of root lifespan. Adv Ecol Res 27:1–60

Fitter AH (1994) Architecture and biomass allocation as components of the plastic response of root systems to soil heterogeneity. In: Caldwell MM, Pearcy RW (eds) Exploitation of environmental heterogeneity by plants. Ecophysiological processes above- and belowground. Academic Press, San Diego, pp 305–323

Fransen B, de Kroon H (2001) Long-term disadvantages of selective root placement: root proliferation and shoot biomass of two perennial grass species in a 2-year experiment. J Ecol 89:711–722

Fransen B, de Kroon H, Berendse F (1998) Root morphological plasticity and nutrient acquisition of perennial grass species from habitats of different nutrient availability. Oecologia 115:351–358

Fransen B, Blijjenberg J, de Kroon H (1999a) Root morphological and physiological plasticity of perennial grass species and the exploitation of spatial and temporal heterogeneous nutrient patches. Plant Soil 211:179–189

Fransen B, de Kroon H, de Kovel C, van den Bosch F (1999b) Disentangling the effects of selective root placement and inherent growth rate on plant biomass accumulation in heterogeneous environments: a modelling study. Ann Bot 84:305–311

Fransen B, de Kroon H, Berendse F (2001) Soil nutrient heterogeneity alters competition between two perennial grass species. Ecology 82:2534–2546

Fretwell SD (1972) Populations in a seasonal environment. Princeton University Press, Princeton

Gerry AK, Wilson SD (1995) The influence of initial size on the competitive responses of six plant species. Ecology 76:272–279

Gersani M, Sachs T (1992) Development correlations between roots in heterogeneous environments. Plant Cell Environ 15:463–469

Gersani M, Abramsky Z, Falik O (1998) Density-dependent habitat selection in plants. Evol Ecol 12:223–234

Gersani M, Brown JS, O'Brien EE, Maina GM, Abramsky Z (2001) Tragedy of the commons as a result of root competition. J Ecol 89:660–669

Gleeson SK, Fry JE (1997) Root proliferation and marginal patch value. Oikos 79:387–393

Grime,JP (1979) Plant strategies and vegetation processes. John Wiley, Chichester

Grime JP, Campbell BD, Mackey JML, Crick JC (1991) Root plasticity, nitrogen capture and competitive ability. In: Atkinson D (ed) Plant root growth. An ecological perspective. Blackwell, London, pp 381–397

Grubb PJ (1994) Root competition in soils of different fertility: a paradox resolved? Phytocoenologia 24:495–505

Hodge A, Robinson D, Griffiths BS, Fitter AH (1999) Why plants bother: root proliferation results in increased nitrogen capture from an organic patch when two grasses compete. Plant Cell Environ 22:811–820

Huante P, Rincon E, Chapin FS (1998) Foraging for nutrients, responses to changes in light, and competition in tropical deciduous tree seedlings. Oecologia 117:209–216

Huber-Sannwald E, Pyke DA, Caldwell MM (1996) Morphological plasticity following species-specific recognition and competition in two perennial grasses. Am J Bot 83:919–931

Hutchings MJ, de Kroon H (1994) Foraging in plants: the role of morphological plasticity in resource acquisition. Adv Ecol Res 25:159–238

Jackson RB, Manwaring JH, Caldwell MM (1990) Rapid physiological adjustment of roots to localized soil enrichment. Nature 344:58–60

Kosola KR, Gross KL (1999) Resource competition and suppression of plants colonizing early successional old fields. Oecologia 188:69–75

Krannitz PG, Caldwell MM (1995) Root growth responses of three Great Basin perennials to intra- and interspecific contact with other roots. Flora 190:161–167

Larigauderie A, Richards JH (1994) Root proliferation characteristics of seven perennial arid-land grasses in nutrient-enriched microsites. Oecologia 99:102–111

Leuschner C, Hertel D, Coners H, Buttner V (2001) Root competition between beech and oak: a hypothesis. Oecologia 126:276–284

Mahall BE, Callaway RM (1991) Root communication among desert shrubs. Proc Natl Acad Sci USA 88:874–876

Mahall BE, Callaway RM (1992) Root communication mechanisms and intracommunity distributions of two Mojave Desert shrubs. Ecology 73:2145–2151

Mahall BE, Callaway RM (1996) Effects of regional origin and genotype on intraspecific root communication in the desert shrub *Ambrosia dumosa* (Asteraceae). Am J Bot 83:93–98

Mou P, Jones RH, Mitchell RJ, Zutter B (1995) Spatial distribution of roots in sweetgum and loblolly pine monocultures and relations with above-ground biomass and soil nutrients. Funct Ecol 9:689–699

Mou P, Mitchell RJ, Jones RH (1997) Root distribution of two tree species under a heterogeneous nutrient environment. J Appl Ecol 34:645–656

Olff H, Berendse F, de Visser W (1994) Changes in nitrogen mineralization, tissue nutrient concentrations and biomass compartmentation after cessation of fertilizer application to mown grassland. J Ecol 82:611–620

Pacala SW (1997) Dynamics of plant communities. In: Crawley MJ (ed) Plant ecology. Blackwell, Oxford, pp 532–555
Robinson D (1994) The responses of plants to non-uniform supplies of nutrients. New Phytol 127:635–674
Robinson D (1996) Resource capture by localized root proliferation: Why do plants bother? Ann Bot 77:179–185
Robinson D, Hodge A, Griffiths BS, Fitter AH (1999) Plant root proliferation in nitrogen-rich patches confers competitive advantage. Proc R Soc Lond B 266:431–435
Schenk HJ, Mahall BE (2002) Positive and negative plant interactions contribute to a north-south-patterned association between two desert shrub species. Oecologia 132:402–410
Schenk HJ, Callaway RM, Mahall BE (1999) Spatial root segregation: are plants territorial? Adv Ecol Res 28:145–180
Schwinning S, Weiner J (1998) Mechanisms determining the degree of size-asymmetry in plant competition. Oecologia 113:447–455
Stoll P, Prati D (2001) Intraspecific aggregation alters competitive interactions in experimental plant communities. Ecology 82:319–327
Theodose TA, Jaeger CH, Bowman WD, Schardt JC (1996) Uptake and allocation of ^{15}N in alpine plants: implications for the importance of competitive ability in predicting community structure in a stressful environment. Oikos 75:59–66
Van Auken OW, Manwaring JH, Caldwell MM (1992) Effectiveness of phosphate acquisition by juvenile cold-desert perennials from different patterns of fertile-soil microsites. Oecologia 91:1–6
Van der Krift TAJ, Berendse F (2002) Root life spans of four grass species from habitats differing in nutrient availability. Funct Ecol 16:198–203
Weiner J (1990) Asymmetric competition in plant populations. Trend Ecol Evol 5:360–364
Weiner J, Wright DB, Castro S (1997) Symmetry of below-ground competition between *Kochia scoparia* individuals. Oikos 79:85–91
Wilson JB (1988) Shoot competition and root competition. J Appl Ecol 25:279–296
Yoder CK, Nowak RS (2000) Phosphorus acquisition by *Bromus madritensis* ssp. rubens from soil interspaces shared with Mojave Desert shrubs. Funct Ecol 14:685–692
Zhang HM, Forde BG (1998) An *Arabidopsis* MADS box gene that controls nutrient-induced changes in root architecture. Science 279:407–409

10 Root Exudates: an Overview

INDERJIT and L.A. WESTON

10.1 Introduction

Roots of many weed and crop species contribute biologically active chemicals into the environment known as root exudates. Root exudates are known to influence growth and establishment of crop and weed species, and these are released from living root systems. Many perennial woody and herbaceous plants have deep and extensive root/rhizome subterranean systems, which can produce prolific amounts of root exudates over long periods of time. Root exudates contribute many types of organic compounds to the rhizosphere. In addition to simple and complex sugars and growth regulators, root exudates contain different classes of primary and secondary compounds including amino acids, organic acids, phenolic acids, flavonoids, enzymes, fatty acids, nucleotides, tannins, steroids, terpenoids, alkaloids, polyacetylenes, and vitamins (Table 10.1; Rovira 1969; Schönwitz and Ziegler 1982; Rice 1984; Uren 2000). Uren (2000) suggested that the amount of root exudates produced varies with the plant species, cultivar, the age of the plant, and substrate and stress factors.

In particular, exudates influence growth and establishment of other plant species (allelopathy), microflora and microfauna, availability of inorganic ions, and may have other less well-documented ecophysiological functions. In studies performed with cereals, a portion of plant photosynthates (between 5–21 %) is typically released via root exudation (Haller and Stolp 1985; Flores et al. 1996). Depending on the size of a plant's root system and its response to biotic and abiotic factors, sizeable quantities of root exudates may be released over time into the soil rhizosphere. Root exudates can directly influence the soil rhizosphere and its composition through (1) their effects on soil-borne pests and pathogens, and other microflora and microfauna, (2) their influence on soil nutrients and resulting microbial ecology, and (3) their allelopathic activities on other plant species. Root exudation can often be modified by abiotic and biotic factors, as well as physical, chemical and biological soil factors.

Ecological Studies, Vol. 168
H. de Kroon, E.J.W. Visser (Eds.)
Root Ecology

Table 10.1. Root exudates and their functions

Chemical	Function	Reference
Citrate, malate (organic acids)	P mobilization from sparingly soluble Fe, Al and Ca phosphate	Gardener et al. (1983)
L-Tryptophan (amino acid)	Inhibits root growth of different plant species	Kat-Noguchi et al. (1994)
Aspartic acid (amino acid)	Stimulates the growth of *Phytopthora megasperma* f. sp. *medicaginis*	El-Hamalawi and Erwin (1986)
Proline, aspartic acid and valine (amino acids)	Inhibitory to nematodes	Tanda et al. (1989)
Glucose and fructose (sugars)	Replace phosphate from apatite rocks and di- and tri-calcium phosphate	Moghimi et al. (1978)
Quercetin, luteolin (flavonoids)	Simulate growth rate of *Rhizobium meliloti*, a chemoattractant; establishment of vesicular-arbuscular mycorrhizal symbioses	Hartwig et al. (1991); Poulin et al. (1993)
p-Benzoquinone (quinone)	Herbicides	Netzley et al. (1988)
Sorgoleone (quinone)	Reduces growth of weed seedlings	Einhellig and Souza (1992)
DIBOA (2,4-dihydroxy-1. 4-benzoxazin-3-one) (hydroxamic acid)	Inhibits root growth of wild oat	Perez and Ormeno-Nunez (1991)
DIMBOA (2,4-dihydroxy-7-methoxy-1, 4-benzoxazin-3-one) (hydroxamic acid)	Wheat allelochemical	Wu et al. (2000)
Quercitrin (flavonoid)	Inhibits seedling growth of mustard	Inderjit and Dakshini (1994)
Extracellular root proteins	Adaptation to nutrient-limitation (positive effects)	Flores et al. (1996)
Uronic acid, deoxy sugars (rhamnose and fructose)	Provide favorable environment for the growth of beneficial microorganisms such as the free-living N_2-fixing bacterium *Azospirillium lipoferum*	Umali-Garcia et al. (1980)

Roots are also the main source of organic material within soil and affect its abiotic and biotic properties. The article will focus on the following key points with respect to root exudates:

1. Composition and diversity of root exudates.
2. Role of root exudates in potential allelopathic effects.
3. Other effects of root exudates, including their impact on the microbial community and the availability of soil nutrients.

10.2 Examples of Root Exudation

Allelopathy can be defined as the direct or indirect harmful effect of one plant on another through the production of chemical compounds that are released into the environment, including root exudates. Allelopathic interference by root exudates has been well documented (Rovira 1969; Smith 1976; Rice 1984). As an example, Mahall and Callaway (1992) investigated the rooting interference mechanisms of two codominant shrubs, *Larrea tridentata* and *Ambrosia dumosa*, upon each other in the Mojave Deserts of California. They used activated carbon to demonstrate root-mediated allelopathy of *Larrea*. It was found that activated carbon significantly reduced the inhibition of growth of neighboring roots by *Larrea* roots. These authors concluded that root interaction between *Larrea* and *Ambrosia* cannot be solely explained by resource competition, and suggested the involvement of inhibitory and readily diffusible chemicals exuded by *Larrea* roots. Another classical example of root-mediated allelopathic interactions is that of black walnut (*Juglans nigra*). Black walnut releases juglone, a toxic hydroquinone compound, from its living root system, bark and nut hulls over time. Walnut has been reported to severely inhibit growth and establishment of certain solanaceous and ericaeous species as well as many ornamentals and forbs (Jose and Gillespie 1998). Juglone is known to be a potent inhibitor of respiration in sensitive species (Dana and Lemer 1990). It may persist in the soil for several months following removal of a mature tree from the landscape. Certain species and cultivars of ornamentals are much more sensitive to the presence of juglone than others, indicating differential sensitivity or selectivity of the phytotoxic constituents. Fisher (1978) also showed that edaphic conditions and soil moisture play an important role in the interactions between walnut roots and other species.

Recent studies with winter wheat (*Triticum aestivum*) have shown that different wheat accessions contain multiple phenolic compounds, which are released by the living roots as well as decomposing residues of roots and shoots. Ongoing studies by Wu et al. (2000) have reported the ability of wheat roots to exude significant quantities of bioactive phenolics that subsequently suppress the root growth of annual ryegrass (*Lolium rigidum*) seedlings. Phe-

Table 10.2. Quantification of phenolic acids and DIMBOA (2,4-dihydroxy-7-methoxy-1,4-benzoxazin-3-one) in root exudates collected from aqueous agar medium previously supporting growth of wheat seedlings for 17 days (from Wu et al. 2000). Means not followed by the same letters are significantly different at the 5 % level within a row

Allelochemical*	Tasman	AUS#18060	L 1512-2721	HY-65
p-Hydroxybenzoic acid	8.39[a]	7.34[a]	3.68[d]	4.12[b]
Vanillic acid	7.82[a]	9.88[b]	6.23[c]	3.47[d]
cis-*p*-Coumaric acid	0.23[a]	1.67[b]	1.40[b]	0.14[a]
Syringic acid	19.32[a]	13.41[b]	23.20[a]	8.03[c]
cis-Ferulic acid	1.51[a]	2.13[a]	5.13[b]	0.57[a]
trans-*p*-Coumaric acid	6.54[a]	6.60[a]	7.71[b]	3.53[c]
DIMBOA	34.22[a]	nd	79.08[b]	nd
trans-Ferulic acid	10.19[a]	12.82[ab]	14.79[b]	3.17[c]

* Identified as trimethylsilyl (TMS) derivative and reported in μg/l of water agar.
nd, Not detected.

nolic acids and hydroxamic acids such as 2,4-dihydroxy-7-methoxy-1,4-benzoxazin-3-one (DIMBOA) were quantified from the root exudates released into aqueous agar medium previously supporting growth of wheat seedlings for 17 days (Table 10.2). This study clearly demonstrates the exudation of phenolic acids and DIMBOA by 17-day-old wheat seedlings; however, further studies are needed to demonstrate the release of phenolic acids and DIMBOA from wheat roots directly into a soil setting.

To summarize the above discussion, it can be concluded that root exudates play an important role in growth and yield of crop species. Certain chemical components of root exudates have the potential to act as natural herbicides. Clearly, there is a need for more research in this emerging area, especially regarding the development of ecologically relevant bioassays and extraction methods for small quantities of exudates and the understanding of the biochemical pathways involved in the production of bioactive root exudates.

10.3 Methods of Measuring Root Exudation

Measuring root exudation is critical to fully understand the role of root exudates in allelopathy and root ecology research. It is important to distinguish between living root exudates and root extracts. Chemicals released from living roots in a laboratory or field situation are defined as root exudates. On the other hand, root extracts represent a mixture of chemicals extracted from dried roots or residues with aqueous or organic solvents. In this article, we will focus on root exudates; further discussion of root extracts is beyond the scope of this chapter.

In 1982, Tang and Young developed a technique to collect allelochemicals from the undisturbed root system of bigalta limpograss (*Hemarthria altissima*). Pots were filled with crushed basaltic rocks and a 2:1 (v/v) sand/rock mixture. Containers wrapped with aluminum foil were heat-sterilized for 2 days at 100 °C. After planting stolon cuttings, each pot was irrigated with a 10 % Hoagland solution (100 ml/day). Pots without limpograss cuttings were utilized as the control. After 40 days, the limpograss exhibited well-developed root systems. After washing XAD-4 resin with organic solvents, XAD-4 packing was placed into the column as a slurry (see Tang and Young 1982) and the column was attached so that the circulating Hoagland's solution containing leachate from roots was collected onto the column at a rate of 1 l/h. The column was detached after 3 days and washed with water followed by methanol. Different fractions (acidic, basic and neutral) were collected and bioassayed for activity upon lettuce (*Lactuca sativa*) germination. Radicle length was measured after 48 h of incubation. This trapping technique for root exudate collection, however, had certain problems. For example, the medium used for growth was inert and required heat sterilization. Soil chemistry and microbial ecology impacts were not addressed due to the complexity of experimental design required to eliminate confounding effects.

In 1984, Young developed a simple donor-receptor pot system to study the effects of root exudates upon higher plant growth. His system maximized the production of root exudates under controlled conditions and allowed for the collection and bioassay of these exudates upon receptor species. Root exudates of asparagus (*Asparagus officinalis*) were collected using the model suggested by Tang and Young (1982). A vermiculite-nutrient culture was used to investigate whether asparagus root exudates had autotoxic effects. In one experiment, he studied the effects of root exudates upon three cultivars of asparagus. He found significant reductions in number and height of stalks of asparagus seedlings when treated with exudates produced by these cultivars. Young (1984) also studied effects of asparagus root exudates on growth and nutrient uptake of asparagus in vermiculite cultures. Due to difficulties in collection and extraction of root exudates in soil settings, research conducted has generally not been able to identify or correlate the presence of bioactive constituents in soil with root exudation and subsequent allelopathy. Therefore, the design of pot systems, hydroponic designs or other root production systems for collection and analysis of bioactive components of root exudates is necessary at this point to gain insight into this challenging area.

Major progress was made in the isolation and identification of five related hydroquinone compounds produced by living sorghum (*Sorghum bicolor*) roots due to the development of a unique capillary mat system for production of large quantities of living seedling roots and their respective exudates (Czarnota et al. 2001; Fig. 10.1). The capillary mat system consists of placement of 100 g of sorghum seed between two double layers of cheesecloth (60 cm long), placed on an aluminum screen. The fine mesh screen is then placed over

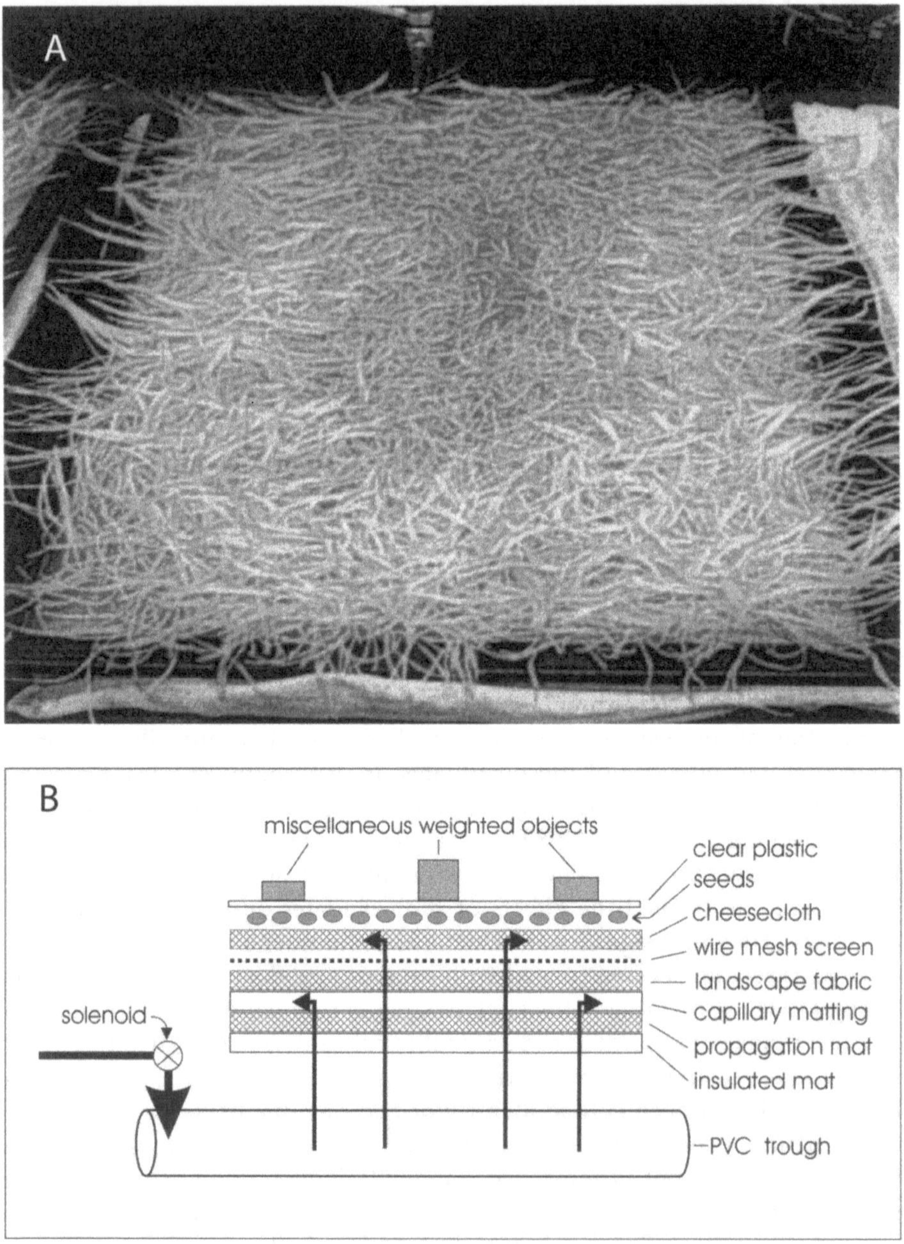

Fig. 10.1. **A** Capillary mat system used to generate seedling roots for collection of bioactive root exudates. Note presence of large quantities of healthy seedling roots on mat surface. **B** Diagram of assembly of capillary mat system used for root exudate collection from seedling monocots. See text for further explanation. (Czarnota 2001)

a layer of ridge insulation, capillary mat and weed mat (or landscape barrier). The landscape fabric is a permeable barrier that prevents evaporation of the moisture from the matting but allows gas penetration to still occur. Other absorptive fabrics may also be used. The capillary mat, weed mat and cheesecloth are placed into a recirculating water trough. A solenoid regulates the water level consistently in the trough. The water moves to the matting by capillary flow and just results in a moist mat with no recirculation. The entire system is covered with plastic and topped with small plastic weights to hold the system in place and keep the screen in contact with the capillary mat.

After 96 h, sorghum roots are cut off and extracted with a solvent, methylene chloride. No sorgoleone occurs in the trough. The capillary mat system enables the production of large quantities of exudates-producing roots within 10–14 days, without contamination by fungal or bacterial pathogens. The capillary mat system is used to extract large amounts of sorgoleone, the major chemical constituent in sorghum root exudates. This capillary mat system has also enabled us to easily and cost effectively produce exudates in 10–14 days from a variety of monocots.

When one considers the activity of root exudates on a variety of organisms, difficulties are often encountered in performing bioassays due to the complexity in working in a soil setting, collecting sufficient quantities of exudate for further study, and designing appropriate bioassays for assessment of activity on diverse rhizosphere organisms. In addition, relatively few studies have been performed to accurately characterize the chemical constituents of exudates produced by the plant species of interest. Future research efforts should focus on the root and root hairs as interesting systems for the production of bioactive secondary products (Czarnota 2001).

10.4 Fate and Movement of Exudates in Soil

Allelopathic compounds, produced as root exudates, generally enter the rhizosphere soon after their release (Inderjit 2001). These compounds may be released in large quantities, but are subject to physical (sorption), chemical (metal oxidation) and biological (microbial degradation) changes within the rhizosphere (Cheng 1995; Dalton 1999; Huang et al. 1999; Inderjit and Dakshini 1999). The process of chemical oxidation or breakdown of root exudates may result in loss of their phytotoxicity and may occur before these allelochemicals reach the roots of target plant species (Lehmann et al. 1987; Cheng 1995). In addition to chemical breakdown, microbial breakdown can limit the soil persistence of many secondary products. If chemically intact, allelochemicals may be transported in the soil as a solute dissolved in the aqueous soil solution which can move in soil suspension either by mass flow or by diffusion along a concentration gradient. Cheng (1995), however, ruled out the possibility of mass flow of allelochemicals in rhizosphere settings because roots typically absorb free water that moves toward the roots from the bulk soil reservoir, and which does not generally contain allelochemicals. After entering the soil, root allelochemicals may be present in free, irreversibly or reversibly bound forms. All these forms can play important roles in allelopathy (Rice 1984).

Sorghum bicolor produces a mixture of long-chain hydroquinones from its living root system. Sorgoleone itself is the primary constituent of the exudate, which is released as golden yellow droplets from living root hairs of *Sorghum* spp. (Fig. 10.2; Netzley et al. 1988; Czarnota et al. 2001). Sorgoleone has been investigated extensively for use as a biocontrol agent by Chang et al. (1986) and Netzley et al. (1988), which is explained below. While discussing the significance of sorgoleone in allelopathic interactions, Inderjit and Weston (2000) have pointed out the need to focus future research to determine the significance of root exudates in allelopathy in a natural soil setting. Further studies were designed to evaluate the soil activity and persistence of sorgoleone in sterile and non-sterile soil conditions (Czarnota et al. 2001). Sorgoleone does persist at detectable levels for up to 8 weeks following soil incorporation, in both sterile and non-sterile soils, but the majority undergoes rapid degradation during the first week following soil incorporation. After

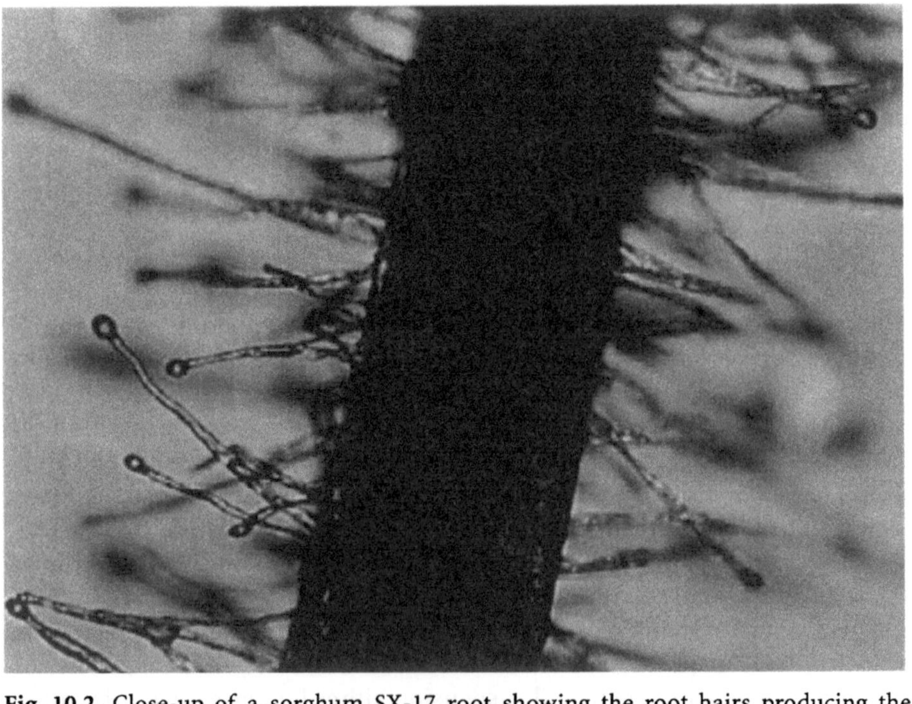

Fig. 10.2. Close-up of a sorghum SX-17 root showing the root hairs producing the golden brown exudate containing sorgoleone; 100 ×. (Czarnota et al. 2001)

incorporation and incubation, inhibition of selected weed species by root uptake was noted at concentrations of 10 ppm or greater, depending on the sensitivity of the species evaluated. Juglone, another hydroquinone, has also been shown to persist in a natural soil setting for up to 6 months following removal of black walnut trees and roots from a landscape site (Dana and Lemer 1990; Jose and Gillespie 1998). These studies show that root exudates may persist at levels rendering potent inhibition of sensitive neighboring species, even in the presence of an active soil microbial community.

10.5 Case Study: Root Exudation by Sorghum

Sorghum and other related sorghum species have long been known to suppress weeds when used as cover crops in nursery and agronomic settings (Weston et al. 1989; Weston 1996). In addition, johnsongrass (*Sorghum halepense*) has been reported to show strong allelopathic activity in the field, where it exists often in monoculture as a rhizome-producing perennial grass (Nicollier et al. 1983). Sorgoleone, the major constituent of root exudates of *Sorghum bicolor*, has been shown to possess strong allelopathic activity (Chang et al. 1986; Netzley et al. 1988; Einhellig et al. 1993; Nimbal et al. 1996a,b; Gonzalez et al. 1997). Rasmussen et al. (1992) reported that sorgoleone disrupts mitochondrial functions in soybean (*Glycine max*) and corn (*Zea mays*). Czarnota et al. (2001) also found that sorgoleone and other structurally related quinones were as potent as cyanide as inhibitors of mitochondrial respiration. Sorgoleone can also be considered as an effective photosynthetic herbicide, as it exhibits a second mode of action because of its potential to inhibit electron transfer between Q_A and Q_B within photosystem II (Gonzalez et al. 1997) by effectively outcompeting for the plastoquinone binding site in photosystem II (Czarnota et al. 2001; Table 10.3). Symptomology observed in sensitive species includes chlorosis and necrosis, similar to that observed with sensitive indicators in killed sorghum residues (Geneve and Weston 1988). Czarnota et al. (2001) performed extensive studies on localization of production of sorghum root exudates within the root hair cell itself. Sorgoleone and other related constituents are produced exclusively in living sorghum root hairs which are individual cells containing large proportions of endoplasmic reticulum and golgi bodies. It is highly likely that sorgoleone is synthesized in association with the endoplasmic reticulum and is then transported across the cell where it accumulates in large quantities between the plasmalemma and cell wall of individual root hairs (Fig. 10.3). Within 3 h of seed germination, newly formed living root hairs of *S. bicolor* were observed to exude golden yellow droplets of sorgoleone. Per gram root tissue, johnsongrass is found to release maximum amounts of root exudates when compared with different sorghum accessions (Table 10.4). Interestingly, john-

Table 10.3. Comparison of inhibitory activity upon electron transport in respiration and photosynthesis with sorgoleone, produced from root exudates of *S.bicolor* × *S. sudanese*, and known inhibitors of respiration and photosynthesis. (Czarnota 2001)

	Inhibitor concentration required for 50 % inhibition of O_2 evolution (μM)	
	Photosynthesis[a]	Respiration[b]
Sorgoleone	0.10	<8.0
Atrazine	0.20	–
Cyanide	–	8.0

[a] Isolated spinach thyllakoids were used for photosynthetic studies.
[b] Isolated potato mitochondrial membranes were used for respiration studies.

Table 10.4. Weight of dried root exudate produced by different sorghum accessions or species. (Czarnota 2001)

Sorghum accession	Exudate produced/root fresh weight (mg/g)[a]
SX-15	1.3
SX-17	1.6
8446	1.6
855-F	1.8
Shattercane	0.5
Johnsongrass	14.8
Della	2.3

[a] Exudate produced refers to the average of six replicates collected.

songrass exudes proportionately more sorgoleone (>10-fold greater concentrations) than *Sorghum bicolor* or sorghum sudangrass (*S. bicolor* × *Sorghum sudanese*) accessions. Root exudation may be associated with the ability of the perennial weed johnsongrass to form dense monocultures over time. More recently, Yang et al. (2001) isolated several genes associated with the production of sorgoleone in *S. bicolor* root hairs. One key gene involved in sorgoleone biosynthesis is expressed only in root hair tissues of *Sorghum* spp., and not in sorghum shoots or tissues of unrelated species. As further research on the expression of genes involved in root exudation progresses, it is likely that we will discover that numerous structural and regulatory genes are involved in this complex process. The isolation and identification of key genes and gene products in sorghum and other bioactive exudate-producing species will further our understanding of the regulation of the root exudation process as a whole.

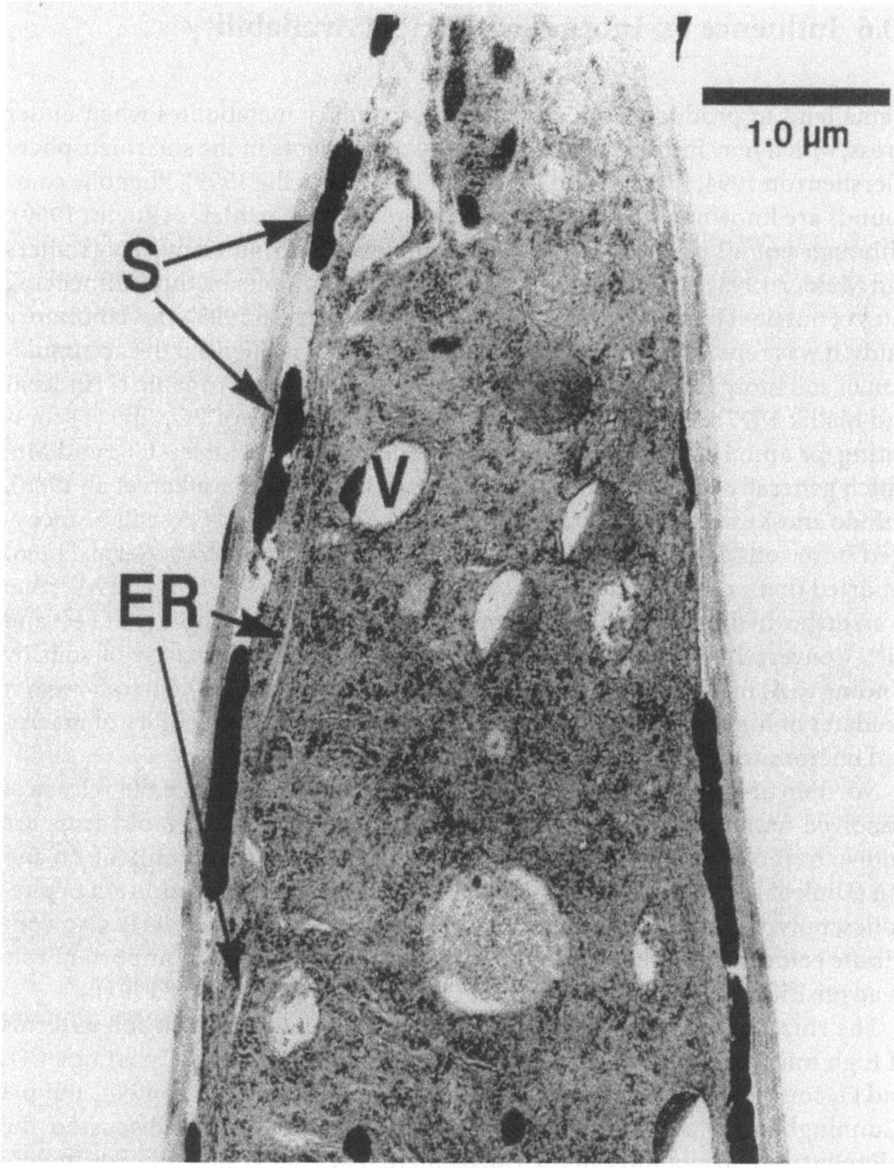

Fig. 10.3. Transmission electron micrograph of a sorghum SX-17 root showing exudate (in *black*) deposited between the plasma membrane and cell wall. *S* Root exudates; *ER* endoplasmic reticulum; *V* vacuole. (Czarnota et al. 2001)

10.6 Influence on Inorganic Nutrient Availability

Plants tend to produce large amounts of secondary metabolites when under stress, which may influence the availability of nutrients in the soil rhizosphere (Gershenzon 1994; Inderjit and Del Moral 1997; Einhellig 1999). Phenolic compounds are known to be prevalent in many plant root exudates (Rovira 1969). Although not all phenolic acids form complexes with nutrient ions (Kuiters and Mulder 1993), many are known to form these complexes, thus influencing rates of nutrient turnover in soil (Schlesinger 1991; Appel 1993). In a laboratory study, it was reported that, in general, phenolic acids influenced the accumulation of soil inorganic ions such as Al, Fe, Mn, and PO_4^-, and organic N (Inderjit and Mallik 1997). Phenolics may increase the availability of PO_4^- by: (1) competing for anion absorption sites, and (2) binding with soluble Al, Fe and Mn, which generally bind to phosphate (Tan and Binger 1986; Kafkafi et al. 1988). Shindo and Kuwatsuka (1977) reported that higher Al^{3+} and Fe^{3+} can be recovered from soils as amounts of protocatechuic acid increase. Vance et al. (1986) reported that protocatechuic acid can form complexes with Fe^{3+} and Al^{3+}, due to an ortho-hydroxy functional group, resulting in higher solubility of Fe^{3+} and Al^{3+}. Conversely, phenolics are also reported to reduce Al toxicity of soils by binding to Al in less soluble forms (Hue et al. 1986; Appel 1993). Therefore, root exudates of higher plants may have strong effects on the availability of macro- and micronutrients in rhizosphere settings.

Northup et al. (1995) hypothesize that polyphenols influence the release of dissolved organic nitrogen and mineral nitrogen in soils. Organic acids are known to mobilize P, complex with Fe, and influence the solubility of Zn and Mn (Dinkelaker et al. 1989; Marschner and Romheld 1996). In addition to phenolics, polyphenolics and other organic compounds, root exudates also contribute ectoenzymes such as acid phosphatases which play an important role in acquisition of organically bound P (Tarafdar and Marschner 1994).

The rhizosphere, or root–soil interface, is different from bulk soil in terms of high microbial activity, presence of root exudates, nutrient dynamics, CO_2 and O_2 concentration, osmotic and redox potential, moisture content, and pH (Cunningham et al. 1996). Marschner and Romheld (1996) discussed the micronutrient availability in the rhizosphere. They suggested that low molecular weight root exudates as well as CO_2 produced by root respiration and eventual creation of H^+/HCO_3- are the main plant factors responsible for changes in rhizosphere pH. The availability of micronutrients is clearly influenced by soil pH, which can be affected, as already mentioned, by root exudates. For example, availability of Zn, Mn, and to some extent Fe, increases with decreases in pH while the availability of Mo decreases with decreasing soil pH.

In response to P-, Fe- and N-deficient conditions, white lupin (*Lupinus albus*) develops clusters of rootlets called proteoid roots that contribute to

Table 10.5. Soil pH, citrate concentration and DTPA extractable iron, manganese, and zinc in bulk soil and proteoid root rhizosphere soil of 13-week-old white lupin (*Lupinus albus* L.) grown in phosphorus-deficient calcareous soil. (Marschner and Romheld 1996)

		DTPA extractable [μmol (kg soil)$^{-1}$]			
Soil	pH (H_2O) [μmol (g soil)$^{-1}$]	Citrate	Fe	Mn	Zn
Bulk	7.5	nd	34	44	2.8
Proteoid root	4.8	47.7	251	244	16.8

nd, Not detected.

lower soil pH and mobilize soluble phosphate and influence the solubility of Fe and Mn (Dinkelaker et al. 1995; Marschner and Romheld 1996). Marschner and Romheld (1996) reported that soil around proteoid roots had significantly higher levels of citrate compared with bulk soil (Table 10.5). The increase in P availability in the lupin rhizosphere is likely due to the release of citrate by proteoid roots (Gardner et al. 1983). Dinkelaker et al. (1989) reported that proteoid roots of white lupin excrete citric acid rather than citrate. The proteoid root zone soil contained significantly greater amounts of P, Mn and Fe (Dinkelaker et al. 1989; Johnson et al. 1994; Marschner and Romheld 1996).

Under phosphate-deficient conditions, it is believed that roots exude greater amounts of organic acids in general (Hoffland et al. 1989; Johnson et al. 1994; Flores et al. 1996). Besides influencing nutrient acquisition by plants and microorganisms, organic acids from root exudates may also influence metal detoxification, microbial growth and podzolization. Several phenolic acids are known to react with metal oxides (e.g. Fe and Mn) and soil clays to form humic acid like substances, and thus their role in humic acid formation has been suggested (Lehmann et al. 1987).

Genotypic differences in nutrient acquisition could be related to root size, morphology and physiological status of roots (Marschner 1998; Chap. 2). Root exudate production may, in fact, be regulated by nutrient status in the soil rhizosphere and induced in response to deficiencies of a particular nutrient. Tawaraya et al. (1998) reported that hydrophobic compounds present in root exudates of onion (*Allium cepa*) grown under P-deficient conditions increase appressorium formation and thus root colonization by the arbuscular mycorrhiza fungus, *Gigaspora margarita*, leading to subsequently enhanced P availability. Root exudates from corn can also solubilize insoluble sources of iron (Elgala and Amberger 1988).

The gaseous environment in the rhizosphere may also influence root exudation and subsequent soil chemistry. Under elevated atmospheric CO_2 conditions, root exudates are proposed to actively contribute to soil organic carbon content and thus lead to greater long-term soil carbon storage. Uselman et al. (2000) reported that warm conditions are likely to enhance root exudation of soil organic carbon over time, which may result in greater long-term soil carbon storage. As our ability to study root exudates and identify soil microbial populations in rhizosphere settings is enhanced, our knowledge of how exudates impact soil chemistry and nutrient availability will also increase dramatically.

10.7 Influence on Soil Organisms

Flavonoids in root exudates of many legume species have been shown to act as powerful chemical signals in a number of plant–microbe interactions, such as the initiation or stimulation of germination of pathogenic fungal species like *Fusarium soloni* and *Verticillium dahlaie* (Mol 1995; Mol and Van-Riessen 1995; Ruan et al. 1995). Chaves et al. (1996) investigated exudates of five soybean (*Glycine max*) cultivars and found that they stimulated apothecia formation in *Whetzelinia.*

Allelochemicals in root/rhizome exudates have been shown to possess antimicrobial and antinematicidal activities. Kobayashi et al. (1996) reported that the antimicrobial activity of wheat root exudates is due to the presence of phenolic acids, sugars, amino acids, and possibly other compounds. Nishimura et al. (2000) identified a new sesquiterpenoid, 8α-angeloyloxycichoralexin, from the rhizomes of chicory (*Chichorium intybus*), with potent antifungal activities against *Pericularia oryzae*, *P. sasakii* and *Alternaria kikuchiana*. They also observed nematicidal activities with the purified compound and the rhizomes of chicory itself. The pharmaceutical industry is currently studying the bacteriocidal and antifungal properties of many secondary products produced from higher plants. As we discover new chemical constituents contained in bioactive root exudates, greater interest in their pharmacological properties is certain.

10.8 Other Roles of Root Exudates

In addition to their role in allelopathy and inorganic ion availability, root exudates play important roles in stimulating nodule formation in N-fixing legumes (Long 1989; Schultze et al. 1994). Flavonoids appear to be most important in the legume-*Rhizobium* symbiosis (Flores et al. 1996; Chap. 12). Roots of alfalfa (*Medicago sativa*) exudate flavonoids, chalcones and isoflavonoids when inoculated with *Rhizobium meliloti* (Maxwell et al. 1989;

Dakora et al. 1993a). Several of these compounds were shown to regulate NodD1 and NodD2 proteins in *R. melioti* and thus impact the process of nodule formation by enhancing Nod genes. Root exudates may also directly influence the growth of the rhizobium bacteria. Hartwig et al. (1991) identified several flavonoids from seeds and roots of alfalfa, and reported that flavonoids with hydroxyl substituents on the 5 and 7 ring positions enhanced the growth rate of *R. meliloti* strains. However, 5-deoxyflavonoid, isolated from alfalfa root exudates, did not enhance the growth of *R. meliloti*, indicating flavonoid specificity for each *Rhizobium* species.

The root exudates of inoculated legumes differ in chemical composition from those of non-inoculated plants. Compared with uninoculated plants, root exudates of alfalfa inoculated with symbiotic *R. meliloti* contained isoflavonoids medicarpin and formononetin-7-*O*-(6″-*O*-malonylglycoside), which were absent in control exudates (Dakora et al. 1993a). These authors (Dakora et al. 1993b) also showed that root exudates of common bean (*Phaseolus vulgaris*) inoculated with symbiotic *Rhizobium leguminosarum* bv *phaseoli* contained high levels of daidzein and coumestral, compared with the non-inoculated control.

Woody legumes also produce various flavonoids within their root exudates. Scheidemann and Wetzel (1997) identified five different flavonoids from root exudates of black locust (*Robinia pseudoacacia*). It was shown that these flavonoids have the capacity to induce β-galactosidase activity in certain *Rhizobium* species and illustrates their importance in nodulation of black locust, a legume tree.

Rhizobacteria also appear to chemically transform naturally occurring flavonoids in root exudates (Rao and Cooper 1995; Steele et al. 1999). Sterile root exudates of *Lotus pedunculatus* were reported to contain a variety of related flavones (Steele et al. 1999). Changes in the flavonoid content of these exudates were observed when roots were inoculated with *Mesorhizobium loti*. It was concluded that rhizobacteria (*M. loti* and *Rhizobium leguminosarum*) can make specific alterations to the exudates from host roots, thereby altering their activity and potential to serve as chemosignals or stimulants.

Root exudates have also been investigated for their role in biological control of weeds (Rice 1995). For example, root exudates of tomato (*Lycopersicon esculentum*), flax (*Linum usitatissimum*) and sorghum (*Sorghum vulgare*) stimulated the germination of broomrape (*Orobanche rumosa*), a serious parasitic weed. Different species of another parasitic weed, *Striga* (*S. hermonthica*, *S. gesneroides* and *S. asiatica*), require specific chemicals produced by host roots for initiation of seed germination and haustoria formation. Through haustoria, *Striga* attaches to host roots and causes severe damage to the host plant (Butler 1995). A sesquiterpene derivative, strigol, was identified as one potent *Striga* germination stimulant, isolated from cotton (*Gossypium hirsutum*) root exudates (Cook et al. 1972). The isolation and identification of these bioactive germination stimulants may lead to the development of natural and synthetic products for use in biocontrol of weeds and is currently an area of active research in weed biology.

10.9 Summary and Prospects

Root exudates are important sources of organic compounds and secondary products, some of which exhibit strong biological activity. Root exudates are commonly produced by numerous plant species with great variation observed in chemical constituents within exudates. In addition to allelopathic effects upon neighboring plants, root exudates may also affect the status of the soil microbial community, the availability of nutrients and the behavior of parasitic plants. At low soil concentrations, some of these effects may be positive, through their action as chemostimulants of soil microflora and fauna. The net effect of root exudates on plant–plant interactions is less clear, but several examples of potent bioherbicidal activity of exudates in surrounding plants have been well documented. Root exudates can influence the growth of the plant itself, neighboring plant species, and the rhizosphere community. An ecosystem-level approach, integrating numerous populations of organisms, is needed to truly understand the major ecophysiological function of root exudates. In addition, a number of environmental stresses influence root exudation and the ultimate function of exudates in the soil rhizosphere.

Recent advances in our ability to identify and detect the chemical constituents within root exudates will enable us to more thoroughly evaluate the presence of these compounds in the soil rhizosphere and determine the important roles they play in plant–plant or plant–microbial interactions. Greater attention must now be given to the fate of root exudates and their chemical constituents in the rhizosphere. We are entering an era when sensitive isolation and detection capabilities will allow us to closely monitor persistence of secondary products. Further studies evaluating the biological activity of the chemical constituents present in root exudates, their production under various environmental conditions, as well as their persistence in the rhizosphere are required to fully understand the diverse roles these compounds play in rhizosphere ecology.

Acknowledgements. We sincerely thank Dr. Mark Czarnota for use of his photo figures in this manuscript and Dr. Paul Weston for his assistance in creation of figures utilized. We also appreciate the editorial comments of Professors Hans de Kroon and Eric Visser.

References

Appel HM (1993) Phenolics in ecological interactions: the importance of oxidation. J Chem Ecol 19:1521–1552

Butler LG (1995) Chemical communication between the parasitic weed *Striga* and its crop host: a new dimension in allelochemistry. In: Inderjit, Dakshini KMM, Einhellig FA (eds) Allelopathy: organisms, processes and applications. American Chemical Society, Washington, DC, pp 158–168

Chang M, Netzley DH, Butler LG, Lynn DG (1986) Chemical regulation of distance: characterization of the first natural host germination stimulant for *Striga asiatica*. J Am Chem Soc 108:7558–7560

Chaves MS, Martinelli JA, Loch LC (1996) Effect of root exudates of soybeans on carpogenic germination of *Whetzelinia sclerotiorum* sclerotia. Summa Phytopathol 22:256–258

Cheng HH (1995) Characterization of the mechanisms of allelopathy: modeling and experimental approaches. In: Inderjit, Dakshini KMM, Einhellig FA (eds) Allelopathy: organisms, processes, and applications. American Chemical Society, Washington, DC, pp 132–141

Cook CE, Whichard LD, Wall ME, Egley GH, Coggon P, Luhan PA, McPhail AT (1972) Germination stimulants. I. The structure of strigol. A potent seed germination stimulant for witchweed (*Striga lutea* Lour.). J Am Chem Soc 94:6198–6199

Cunningham SD, Anderson TA, Schwab AP, Hsu FC (1996) Phytoremediation of soils contaminated with organic pollutants. Adv Agro 56:55–114

Czarnota MA (2001) Sorghum (*Sorghum* spp.) root exudates: production, localization, chemical composition and mode of action. PhD Thesis, Cornell University, Ithaca, NY

Czarnota MA, Paul RN, Dayan F, Nimbal CI, Weston LA (2001) Mode of action, localization of production, chemical nature and activity of sorgoleone: a potent PS II inhibitor produced in Sorghum spp. root exudates. Weed Technol 15(4):813–825

Dakora FD, Joseph CM, Phillips DA (1993a) Alfalfa (*Medicago sativa* L.) root exudates contain isoflavonoids in the presence of *Rhizobium meliloti*. Plant Physiol 101:819–824

Dakora FD, Joseph CM, Phillips DA (1993b) Common bean root exudates contain elevated levels of daidzein and coumestrol in response to *Rhizobium* inoculation. Mol Plant Microbe Interact 6:665–668

Dalton B R (1999) The occurrence and behavior of plant phenolic acids in soil environment and their potential involvement in allelochemical interference interactions: methodological limitations in establishing conclusive proof of allelopathy. In: Inderjit, Dakshini KMM, Foy CL (eds) Principles and practices in plant ecology: allelochemical interactions. CRC Press, Boca Raton, pp 57–74

Dana MN, Lemer BR (1990) Black walnut toxicity. Purdue University Extension Bulletin. HO-193. West Laffayette, IN

Dinkelaker B, Romheld V, Marschner H (1989) Citric acid secretion and precipitation of calcium citrate in the rhizosphere of white lupin (*Lupinus albus* L.). Plant Cell Environ 12:285–292

Dinkelaker B, Hengeler C, Marschner H (1995) Distribution and function of proteoid roots and other root clusters. Bot Acta 108:183–200

Einhellig FA (1999) An integrated view of allelochemicals amid multiple stresses. In: Inderjit, Dakshini KMM, Foy CL (eds) Principles and practices in plant ecology: allelochemical interactions. CRC Press, Boca Raton, pp 479–494

Einhellig FA, Souza IF (1992) Phytotoxicity of sorgoleone found in grain sorghum root exudates. J Chem Ecol 18:1–11

Einhellig FA, Rasmussen JA, Hejl AH, Souza IF (1993) Effects of root exudate sorgoleone on photosynthesis. J Chem Ecol 19:369–375

Elgala AM, Amberger A (1988) Root exudate and the ability of corn to utilize insoluble sources of iron. J Plant Nut 11:677–680

El-Hamalawi ZA, Erwin DC (1986) Components in alfalfa root extract and root exudates that increase oospore germination of *Phytophthora megasperma* f. sp. *medicaginis*. Phytopathology 76:508–513

Fisher RF (1978) Juglone inhibits pine growth under certain moisture regimes. Soil Sci Soc Am J 42:801–803

Flores HE, Weber C, Puffett J (1996) Underground plant metabolism: the biosynthetic potential of roots. In: Waisel Y, Eshel A, Kafkafi U (eds) Plant roots: the hidden half, 2nd edn. Marcel Dekker, New York, pp 931–956

Gardner WK, Barber DA, Parbery DG (1983) The acquisition of phosphorus by *Lupinus albus* L.. 3. The probable mechanism by which phosphorus movement in the soil/root interface is enhanced. Plant Soil 170:107–124

Geneve RL, Weston LA (1988) Growth reduction of eastern redbud *(Cercis canadensis* L.) seedlings caused by interaction with a sorghum-sudangrass hybrid (sudex). J Environ Hortic 6:24–26

Gershenzon J (1994) Metabolic costs of terpenoids accumulation in higher plants. J Chem Ecol 20:1281–1328

Gonzalez VM, Kazimir J, Nimbal C, Weston LA, Cheniae GM (1997) Inhibition of a photosystem II electron transfer reaction by the natural product sorgoleone. J Agric Food Chem 45:1415–1421

Haller T, Stolp H (1985) Quantitative estimation of root exudation of maize plants. Plant Soil 86:207–216

Hartwig UA, Joseph CM, Phillips DA (1991) Flavonoids released naturally from alfalfa seeds enhance growth rate of *Rhizobium meliloti*. Plant Physiol 95:797–803

Hoffland E, Findenegg GR, Nelemans JA (1989) Solubilization of rock phosphate by rape. Plant Soil 113:161–165

Huang PM, Wang MC, Wang MK (1999) Catalytic transformation of phenolic compounds in the soil. In: Inderjit, Dakshini KMM, Foy CL (eds) Principles and practices in plant ecology: allelochemical interactions. CRC Press, Boca Raton, pp 287–306

Hue NV, Craddock GR, Adams F (1986) Effect of organic acids on aluminum toxicity in subsoils. Soil Sci Soc Am J 50:28–34

Inderjit (2001) Soils: environmental effect on allelochemical activity. Agron J 93:79–84

Inderjit, Dakshini KMM (1994) Allelopathic potential of phenolics from the roots of *Pluchea lanceolata*. Physiol Plant 92:571–576

Inderjit, Dakshini KMM (1999) Bioassays for allelopathy: interactions of soil organic and inorganic constituents. In: Inderjit, Dakshini KMM, Foy CL (eds) Principles and practices in plant ecology: allelochemical interactions. CRC Press, Boca Raton, pp 35–44

Inderjit, Del Moral R (1997) Is separating resource competition and allelopathy from allelopathy realistic? Bot Rev 63:221–230

Inderjit, Mallik AU (1997) Effect of phenolic compounds on selected soil properties. For Ecol Manag 92:11–18

Inderjit, Weston LA (2000) Are laboratory bioassays for allelopathy suitable for prediction of field responses? J Chem Ecol 26:2111–2118

Johnson JF, Allan DL, Vance CP (1994) Phosphorus stress-induced proteoid roots show altered metabolism in *Lupinus albus*. Plant Physiol 104:657–665

Jose S, Gillespie AR (1998) Allelopathy in black walnut (*Juglans nigra* L.) alley cropping. I. Spatio-temporal variation in soil juglone in a black walnut-corn (*Zea mays* L.) alley cropping system in the midwestern USA. Plant Soil 203:191–197

Kafkafi U, Bar-Yosef B, Rosenberg R, Sposito G (1988) Phosphorus adsorption by kaolinite and montmorillonite: II. Organic anion competition. Soil Sci Soc Am J 52:1585–1589
Kato-Noguchi H, Mizutani J, Hasegawa K (1994) Allelopathy of oats. II. Allelochemical effect of L-tryptophan and its concentration in oat root exudates. J Chem Ecol 20:315–319
Kobayashi A, Kim MJ, Kawazu K (1996) Uptake and exudation of phenolic compounds by wheat and antimicrobial components of the root exudate. Z Naturforsch 51:527–533
Kuiters AT, Mulder W (1993) Water-soluble organic matter in forest soils: I. Complexing properties and implications for soil equilibria. Plant Soil 152:215–224
Lehmann RG, Cheng HH, Harsh JB (1987) Oxidation of phenolic acids by iron and manganese oxides. Soil Sci Soc Am J 51:352–356
Long SR (1989) Rhizobium-legume nodulation: life together in the underground. Cell 56:203–214
Mahall BE, Callaway RM (1992) Root communication mechanisms and intracommunity distributions of two Mojave Desert shrubs. Ecology 73:2145–2151
Marschner H (1998) Role of root growth, arbuscular mycorrhiza, and root exudates for the efficiency in nutrient acquisition. Field Crops Res 56:203–207
Marschner H, Romheld V (1996) Root-induced changes in the availability of micronutrients in the rhizosphere. In: Waisel Y, Eshel A, Kafkafi U (eds) Plant roots: the hidden half, 2nd edn. Marcel Dekker, New York. pp 557–579
Maxwell CA, Hartwig UA, Joseph CM, Phillips DA (1989) A chalcone and two related flavonoids released from alfalfa roots induce nod genes of *Rhizobium meliloti*. Plant Physiol 91:842–847
Moghimi A, Tate ME, Oades JM (1978) Characterization of rhizosphere products especially 2-ketogluconic acid. Soil Biol Biochem 10:283–287
Mol L (1995) Effect of plant roots on the germination of microsclerotia of *Verticillium dahliae*: II. Quantitative analysis of the luring effects of crops. Eur J Plant Pathol 101:679–685
Mol L, Van-Riessen HW (1995) Effect of plant roots on the germination of microsclerotia of *Verticillium dahliae*: I. Use of root observation boxes to assess differences among crops. Eur J Plant Pathol 101:673–678
Netzley DH, Riopel JL, Ejeta G, Butler LG (1988) Germination stimulants of witchweed (*Striga asiatica*) from hydrophobic root exudate of sorghum (*Sorghum bicolor*). Weed Sci 36:441–446
Nicollier JF, Pope DF, Thompson AC (1983) Biological activity of durrin and other compounds from Johnsongrass (*Sorghum halepense*). J Agric Food Chem 31:744–748
Nimbal CI, Pedersen JF, Yerkes CN, Weston LA, Weller SC (1996a) Phytotoxicity and distribution of sorgoleone in grain sorghum germplasm. J Agric Food Chem 44:1343–1347
Nimbal CI, Yerkes CN, Weston LA, Weller SC (1996b) Herbicidal activity and site of action of the natural product sorgoleone. Pestic Biochem Physiol 54:73–83
Nishimura H, Kondo Y, Nagasaka T, Satoh A (2000) Allelochemicals in chicory and utilization in processed foods. J Chem Ecol 26:2233–2241
Northup RR, Yu Z, Dahlgren RA, Vogt KA (1995) Polyphenol control of nitrogen release from pine litter. Nature 377:227–229
Perez FJ, Ormeno-Nunez J (1991) Root exudates of wild oats: allelopathic effect on spring wheat. Phytochemistry 30:2199–2202
Poulin MJ, Bel-Rhlid R, Piche Y, Chenevery R (1993) Flavonoid released by carrot (*Daucus carota*) seedlings stimulate hyphal development of vesicular-arbuscular mycorrhizal fungi in the presence of optimal CO_2 enrichment. J Chem Ecol 19:2317–2327

Rao JR, Cooper JE (1995) Soybean nodulating bacteria modify *nod* gene inducers daidzein and genistein to yield aromatic products that can influence gene-inducing activity. Mol Plant-Microbe Interact 8:855–862

Rasmussen JA, Heil AM, Einhellig FA, Thomas TA (1992) Sorgoleone from root exudate inhibits mitochondrial functions. J Chem Ecol 18:197–207

Rice EL (1984) Allelopathy. Academic Press, Orlando, FL

Rice EL (1995) Biological control of weeds and plant diseases: advances in applied allelopathy. University of Oklahoma Press, Norman, OK

Rovira AD (1969) Plant root exudates. Bot Rev 35:35–59

Ruan Y, Kotraiah V, Straney DC (1995) Flavonoids stimulate spore germination in *Fusarium solani* pathogenic on legumes in a manner sensitive to inhibitors of cAMP-dependent protein kinase. Mol Plant-Microbe Interact 8:929–938

Scheidemann P, Wetzel A (1997) Identification and characterization of flavonoids in the root exudate of *Robinia pseudoacaia*. Trees 11:316–321

Schlesinger WH (1991) Biogeochemistry: an analysis of global change. Academic Press, San Diego

Schönwitz R, Ziegler H (1982) Exudation of water-soluble vitamins and of some carbohydrates by intact roots of maize seedlings (*Zea mays*) into a mineral nutrient solubilization. Z. Pflanzenphysiol 107:7–14

Schultze M, Kondorosi E, Ratet P, Buire M, Kondorosi A (1994) Cell and molecular biology of *Rhizobium*-plant interactions. Int Rev Cytol 156:1–75

Shindo H, Kuwatsuka S (1977) Behavior of phenolic substances in decaying process of plants. III. Degradation pathway of phenolic acids. Soil Sci Plant Nutr 21:227–238

Smith WH (1976) Character and significance of forest tree root exudates. Ecology 57:324–331

Steele H, Werner D, Cooper JE (1999) Flavonoids in seed and root exudates of *Lotus pedunculatus* and their biotransformation in *Mesorhizobium loti*. Physiol Plant 107:251–258

Tan K, Binger A (1986) Effect of humic acid on aluminum toxicity in corn plants. Soil Sci 141:20–25

Tanda AG, Atwal AS, Bajaj YPS (1989) In vitro inhibition of root-knot nematode *Meloidogyne incognita* by sesame root exudates and its amino acids. Nematologica 35:115–124

Tang CS, Young CC (1982) Collection and identification of allelopathic compounds from the undisturbed root system of bigalta limpograss (*Hemarthria altissima*). Plant Physiol 69:155–160

Tarafdar JC, Marschner H (1994) Phosphatase activity in the rhizosphere of VA mycorrhizal wheat supplied with inorganic and organic phosphorus. Soil Biol Biochem 26:387–395

Tawaraya K, Hashimoto K, Wagatsuma T (1998) Effect of root exudate fractions from P-deficient and P-sufficient onion plants on root colonisation by the arbuscular mycorrhizaal fungus *Gigaspora margarita*. Mycorrhiza 8:67–70

Vance GF, Mokma DL, Boyd SA (1986) Phenolic compounds in soils of hydrosequences and developmental sequences of spodosols. Soil Sci Soc Am J 50:992–996

Umali-Garcia M, Hubbell DH, Gaskins MH, Dazzo FB (1980) Association of *Azospirillum* with grass roots. Appl Environ Microbiol 39:219–226

Uren NC (2000) Types, amounts, and possible functions of compounds released into the rhizosphere by soil-grown plants. In: Pinto R, Varanini Z, Nannipieri P (eds) The rhizosphere: biochemistry and organic substances at the soil-plant interface. Marcel Dekker, New York, pp 19–40

Uselman SM, Qualls RG, Thomas RB (2000) Effect of increased atmospheric CO_2, temperature and soil N availability on root exudation of dissolved organic carbon by a N-fixing tree (*Robinia pseudoacacia* L.). Plant Soil 222:191–202

Weston LA (1996) Utilization of allelopathy for weed management in agroecosystems. Agron J 88:860–866

Weston LA, Harmon R, Mueller S (1989) Allelopathic potential of sorghum–sudangrass hybrid (sudex). J Chem Ecol 15:1855–1865

Wu H, Haig T, Pratley J, Lemerle D, An M (2000) Distribution and exudation of allelochemicals in wheat *Triticum aestivum*. J Chem Ecol 26:2141–2154

Yang X, Scheffler B, Weston LA (2001) Analysis of gene expression related to sorgoleone production using m-RNA differential display. Proc WSSA 41:37

Young CC (1984) Autointoxication in root exudates of *Asparagus officinalis* L. Plant Soil 82:247–253

11 Mycorrhizas

F.A. Smith, S.E. Smith and S. Timonen

11.1 Introduction

Mycorrhizas ('fungus-roots': also called mycorrhizae) of one sort or another are a normal part of root function and ecology for the vast majority of terrestrial plants. Because the formation of mycorrhizas can affect the rates of growth and eventually reproduction of plants they are a potential selection factor that can influence the composition of plant communities and hence, presumably, the evolutionary success of individual plant species. Fossil mycorrhizas occur in Devonian rocks 400 million years old, so present-day mycorrhizal symbioses are the result of a very long period of coevolution between plants and soil fungi.

About 90% of seed plants belong to families in which most species form mycorrhizas (Trappe 1987; Molina et al. 1992). Roots of most ferns also form mycorrhizas and analogous symbioses occur in the gametophytes of many bryophytes and pteridophytes (for references, see Smith and Smith 1997). The few angiosperm families that are often regarded as constitutively non-mycorrhizal or mainly non-mycorrhizal include Amaranthaceae, Chenopodiaceae, Brassicaceae (Cruciferae), Cyperaceae and Juncaceae, and several others (see Brundrett 1991). The inability of *Arabidopsis* (Brassicaceae) to form mycorrhizas deprives researchers of much invaluable molecular biology! There are, however, many exceptions within these 'non-mycorrhizal' families (Newman and Reddell 1987; Trappe 1987; Arriola et al. 1997), just as there are 'mycorrhizal' families with non-mycorrhizal taxa, such as the non-mycorrhizal genus *Lupinus* in the highly mycorrhizal Leguminosae. The relative success of mycorrhizal versus non-mycorrhizal plants in both natural and managed environments is thought to be fundamentally important in plant community ecology (Read 1993; Janos 1995).

Many plant species cannot complete their life cycle in their normal habitats without forming mycorrhizas; these are conventionally called 'obligately mycorrhizal' plants. Other plant species are 'facultatively mycorrhizal'; they are capable of forming mycorrhizas but can complete their life cycles without

Ecological Studies, Vol. 168
H. de Kroon, E.J.W. Visser (Eds.)
Root Ecology

forming them (Janos 1980, 1995). Most of the obligately mycorrhizal plant species can complete their life cycles when grown in artificially fertilized soils but others (especially some tropical species) cannot. Roots of the latter seem to have transferred to their fungal partners the capacity to acquire adequate supplies of soil nutrients (Janos 1995). Molina et al. (1992) suggested that about 70 % of angiosperms are obligately mycorrhizal, 12 % facultatively mycorrhizal, and 18 % constitutively non-mycorrhizal. The success of mycorrhizal plants in a community thus depends on the ecological setting, including soil nutrient status, types of neighboring plants, presence or absence of fungal inoculum, etc.

Mycorrhizas are usually described as mutualistic symbioses because most classes involve supply of nutrients to the plant via the fungus in exchange for photosynthate. However, such exchange of resources does not seem to be coupled either directly or indirectly. It is not surprising, therefore, that mycorrhizal symbioses do not always give large positive growth responses and that sometimes the fungus seems functionally parasitic (see Smith 1980; Koide 1991a; Johnson et al. 1997). Where myco-heterotrophic (achlorophyllous) plants obtain their supply of organic carbon from mycorrhizal fungi (see Leake 1994) the existence of mutualism becomes very tenuous – it is not clear why the fungus colonizes the heterotrophic plant at all. Given the evidence for transfer of organic C via mycorrhizal fungi from a plant to another autotrophic plant (of the same or a different species), the cost-benefit analysis becomes even more complex, as does the physiology underlying the traffic of solutes.

Different types of mycorrhizas have many developmental stages in common. These include: (1) growth of fungal hyphae from germinating spores or other sources of inoculum such as previously colonized plant roots, (2) recognition of a plant root as a potential partner by the fungus, and vice versa, (3) evasion or inhibition of attack/defence mechanisms of the partners (as compared to pathogenic associations), (4) extensive colonization of the root, (5) development of fungal–plant interfaces that are stable over periods of days, weeks, or longer, and (6) development of external fungal hyphae that spread extensively into the soil environment. Functional compatibility of the symbioses involves prolonged reciprocal transfer of resources between the partners and responses in terms of plant nutrition, growth and reproduction. The end of a functional mycorrhizal symbiosis involves senescence and disorganization of fungal structures within the roots, formation of fungal spores, and senescence of roots. These processes are described in detail in the literature (e.g. Allen 1992; Smith and Read 1997). Changes in expression of plant or fungal genes that are associated with mycorrhizal development and function are an active area of research beyond the scope of this chapter (Franken and Requena 2001; Martin 2001).

Time-scales for these stages of development differ. Recognition can occur within hours and colonization within days. Although some fungal structures turn over quite rapidly (scale of days or weeks), individual mycorrhizas can live for months or years, depending on the class of mycorrhiza and the plant involved. Over a scale of years there can be effects on plant community structure and (by extrapolation over many millennia) effects on plant evolution.

Many of the themes covered elsewhere in this volume in relation to non-mycorrhizal roots are relevant to – or can be influenced by – formation of mycorrhizas. This chapter is mainly aimed at plant ecologists who are unfamiliar with mycorrhizas. We cover four topics: (1) classification and distribution of mycorrhizas, (2) basic structure and major functional features, (3) the 'mycorrhizosphere', including associations with soil bacteria, and (4) ecological functioning of mycorrhizas. Background material is well covered in the literature (e.g. Allen 1992; Smith and Read 1997;Varma and Hock 1999; van der Heijden and Sanders 2002). Important reviews concerning the ecological significance of mycorrhizas are by St. John and Coleman (1983), Brundrett (1991), Read (1993) and Janos (1995).

11.2 Classification and Root Structures

The major classes of mycorrhizas are defined by the symbiotic structures and taxa of plants and fungi (see Table 11.1). The plants are easily identified but this may not be the case for the fungi. Traditional methods of fungal identification via fruit bodies (where formed) and spore types are now being rapidly overtaken by DNA fingerprinting that is throwing much light on the fungal side of the symbioses (e.g. Bruns and Gardes 1993; Helgason et al. 1999; Sharples et al. 2000).

11.2.1 Arbuscular Mycorrhizas

Arbuscular mycorrhizas (AM), originally known as vesicular-arbuscular mycorrhizas (VAM), are by far the most common mycorrhizal class, occurring in about 80 % of terrestrial plants. The fungi have been placed in the Zygomycota, order Glomales, but it has recently been proposed that they belong in a new phylum, the Glomeromycota (Schüßler et al. 2001). They have never been cultured successfully in the absence of a plant partner. The fungi show quite broad specificity in that one species or isolate can form AM symbioses with many plant species, and a plant that forms AM can do so, even simultaneously, with different AM fungal species. However, different AM fungi do not necessarily colonize the roots of a host to the same extent when present individually in soil, and do not necessarily give similar responses in terms of growth or nutrient concentration.

There are two major morphological types of AM, *Arum* and *Paris* types, named after the plants in which these were first described (Gallaud 1905; for

Table 11.1. Characteristics of the major classes of mycorrhizas (based on Table 1 of Smith and Read 1997). Fungal taxa are abbreviated from Zygomycota, Ascomycota and Basidiomycota; plant taxa from Angiospermae, Gymnospermae, Pteridophyta and Bryophyta. Presence or absence of structures shown as + or –; (+) or (–) are rare conditions

	Arbuscular (AM)	Ecto (ECM)	Ectendo	Ericoid (ERM)	Arbutoid	Monotropoid	Orchid
Fungal taxa	Zygo	Basidio Asco (Zygo)[e]	Basidio Asco	Asco	Basidio	Basidio	Basidio
Plant taxa	Angio Gymno Pterido Bryo	Angio Gymno	Angio Gymno	Angio: Ericales Bryo	Angio: Ericales	Angio: Ericales	Angio: Orchidaceae
Achlorophylly	– or (+)[a]	–	–	–	–	+	+[f]
Intracellular interfaces							
Hyphal coils or pegs	+ or –[b]	–	+	+	+	+	+
Arbuscules	+ or (–)[c]		–	–	–	–	–
Extracellular interfaces							
Intercellular hyphae	+ or –[d]	+	+	–	+	+	–
Fungal sheath	–	+	+ or –	–	+ or –	+	–

[a] AM occurrence relatively rare in achlorophyllous angiosperms.
[b] Extensive cortical hyphal coils in *Paris*-type AM.
[c] Reports of absence of arbuscules in some *Paris*-type AM.
[d] Hyphae in intercellular spaces in *Arum*-type AM but not in *Paris*-types.
[e] In some Australian plants.
[f] All orchids are achlorophyllous as seedlings but most are green as adults.

detailed discussion, see Smith and Smith 1997). They are illustrated in Fig. 11.1. In *Arum* types, intraradical fungal hyphae grow through cortical intercellular spaces and then penetrate cortical cell walls to form the multi-branched arbuscules that appear especially well designed for nutrient transfer. The fungus still remains separated from the host by an apoplast, so to talk of 'intracellular' growth can be misleading. However, we shall use the word to contrast with fungal growth between cells: see Table 11.1. In *Paris*-type AM, intraradical hyphae are not intercellular but occur as intracellular coils or branches, some of which develop arbuscules. Some plants form structures with features of both types (Smith and Smith 1997). Storage vesicles, containing lipid and protein, can be intercellular or intracellular and occur irrespective of AM morphology but they are not formed by all AM fungi, which is why 'AM' is now widely favored over 'VAM'. *Arum*-type colonization is very rare in lower plants and gymnosperms and occurs in less than half of the 100 or so AM angiosperm families that have been examined at this level (Smith and Smith 1997). Both types and intermediate structures occur in herbaceous angiosperms, shrubs and trees, but the last mainly form *Paris*-types. The dis-

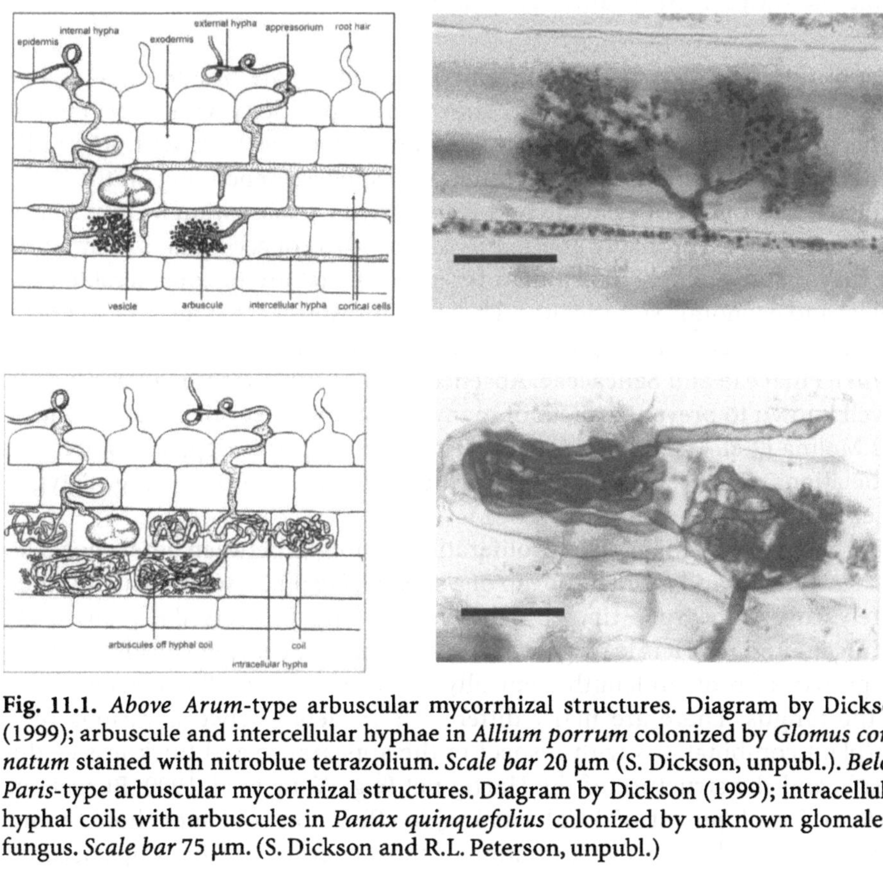

Fig. 11.1. *Above Arum*-type arbuscular mycorrhizal structures. Diagram by Dickson (1999); arbuscule and intercellular hyphae in *Allium porrum* colonized by *Glomus coronatum* stained with nitroblue tetrazolium. *Scale bar* 20 μm (S. Dickson, unpubl.). *Below Paris*-type arbuscular mycorrhizal structures. Diagram by Dickson (1999); intracellular hyphal coils with arbuscules in *Panax quinquefolius* colonized by unknown glomalean fungus. *Scale bar* 75 μm. (S. Dickson and R.L. Peterson, unpubl.)

tribution of the two types and intermediates among plant taxa, and evidence that individual AM fungi form *Arum*-type AM in some hosts and *Paris*-type AM in others, has suggested that the plant primarily determines AM morphology. The simplest explanation is that *Arum*-type AM are formed in roots that have extensive intercellular channels, whereas *Paris*-type AM occur in roots with only small or discontinuous spaces (Brundrett and Kendrick 1990a,b; Smith and Smith 1997). However, tomato *(Lycopersicon esculentum)* can form either type, depending on the AM fungus (Cavagnaro et al. 2001). Further experimental studies are needed to help sort out the amount of control on AM structure that is exerted by plant and fungus and possible functional differences between the types.

Formation of AM can result in quite subtle effects on root morphology and architecture of their hosts, including changes in branching patterns and lengths of secondary and tertiary roots (e.g. Hetrick 1991; Hooker et al. 1992; Chap. 2). This can have significant implications for the ability of AM plants to acquire nutrients independently of direct acquisition by the AM fungi.

11.2.2 Ectomycorrhizas and Ectendomycorrhizas

Most of the fungi that form ectomycorrhizas (ECM) are basidiomycetes, but some are ascomycetes. They mostly show broad specificity with respect to the plants but some are genus-specific (see Molina et al. 1992). As with AM, individual ECM associations can give different responses depending on the fungi involved, and one root system can be colonized by many fungi. Only a very small proportion (probably about 3 %: Meyer 1973) of seed plants form ECM but they include widely distributed forest trees that are of considerable ecological and economic significance. The main forest trees that form ECM are in the Betulaceae, Dipterocarpaceae, Fagaceae, Leguminosae, Myrtaceae (*Eucalyptus*), Pinaceae and Salicaceae. Absence of ECM fungi from plantation soils is well known to prevent growth of many forest trees (especially exotic ones), and Molina et al. (1992) suggested that most ECM plants depend on obligate associations. Some shrubs and a few herbs (e.g. in the Cistaceae, Polygonaceae, and Rosaceae) also form ECM.

Formation of ECM involves proliferation of fungus between outer root cells (the Hartig net) and rapid build-up of a sheath (Fig. 11.2), which usually completely covers the root tip. The sheath not only protects the plant from pathogens and restricts leakage of substrates into the soil, but can also restrict the growth of roots in length. Typically, there is no intracellular penetration by the fungus. There are many differences in detail between various fungus/plant combinations with respect to the appearance and thickness of the sheath, and the structure of the Hartig net (e.g. Wong et al. 1990; Burgess et al.1994). Development of ECM has major effects on root morphology and architecture, with extensive formation of short lateral roots. Root tips become

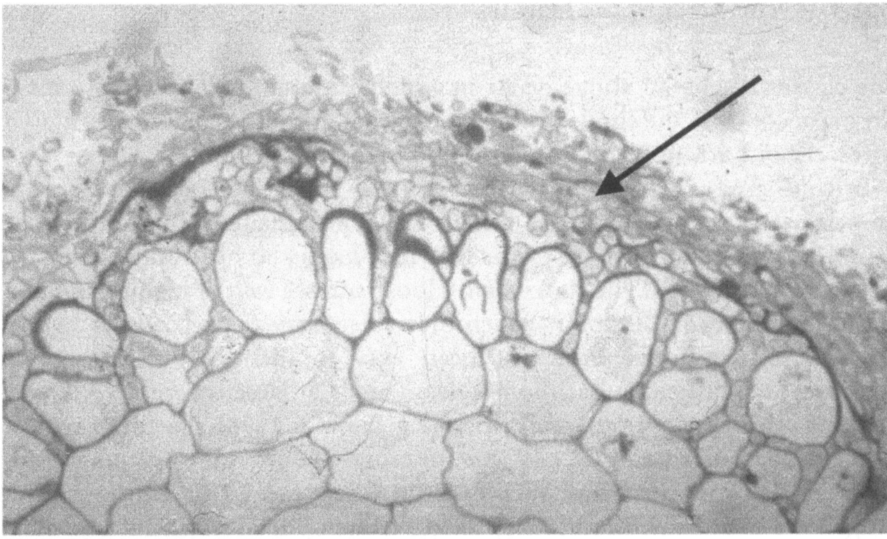

Fig. 11.2. Ectomycorrhiza. Partial cross section of root of *Eucalyptus pilularis* colonized by *Pisolithus tinctorius*. *Arrow* shows external fungal sheath and Hartig net of intercellular hyphae penetrating between cortical cells. (A.E. Ashford and R.L. Peterson, unpubl.)

swollen and growth of root hairs is inhibited, resulting in a coralloid or 'Christmas-tree'-like structure. The frequency of these structures compared with non-colonized roots depends on the plant species and ECM fungus, and the environmental conditions. The plants can have spurts of root growth during which root tips burst through established sheaths (see Smith and Read 1997 for details and references).

Ectendomycorrhizas (Table 11.1) have characteristics of ECM (though in some cases lacking a sheath) and also consistently high intracellular penetration by fungal hyphae to form branched structures. They occur mainly in conifers such as *Pinus* and the fungi involved, usually ascomycetes, can form typical ECM on other plants, e.g. *Picea* and *Betula* (Scales and Peterson 1991a,b). This is another demonstration of control by the plant over the intraradical structures formed by individual fungi. In addition, there may be environmental influence, because ectendomycorrhizas can form on otherwise ECM plants growing in disturbed soils, e.g. in forest nurseries (Laiho 1965; Mikola 1965).

11.2.3 Mycorrhizas of the Ericales

The classes discussed above occur in widely different taxonomic groups of plants. In contrast, the single order Ericales contains plants that form different types: ericoid, arbutoid and monotropoid mycorrhizas (Table 11.1).

Ericoid mycorrhizas (ERM) occur in the Ericaceae, Epacridaceae and Empetraceae, all of which are important in heathland communities. The extremely fine roots of these plants lack root hairs and have only one or two layers of cortical cells. The fungi form a loose surface weft of hyphae and penetrate the epidermal cells, in which they form very dense coils of hyphae (Fig. 11.3). The fungi have not been easy to identify but are mainly ascomycetes, with some basidiomycetes. This is an area where DNA fingerprinting is proving particularly helpful (e.g. Liu et al. 1998). The fungi are ecologically specialized and do not seem to cross-colonize other plants to form other kinds of mycorrhizas, such as ECM (see Read 2000). However, ERM fungi have been identified in bryophyte gametophytes (Duckett and Read 1995).

Three genera of the Ericaceae (*Arbutus*, *Arctostaphylos* and *Pyrola*) form arbutoid mycorrhizas. These rather resemble ectendomycorrhizas in that there is a sheath (sometimes sparse or disorganized, as in *Pyrola*), a Hartig net and intracellular coils of hyphae that are restricted to the epidermis. Fungi (basidiomycetes) that form arbutoid mycorrhizas can form ECM on other

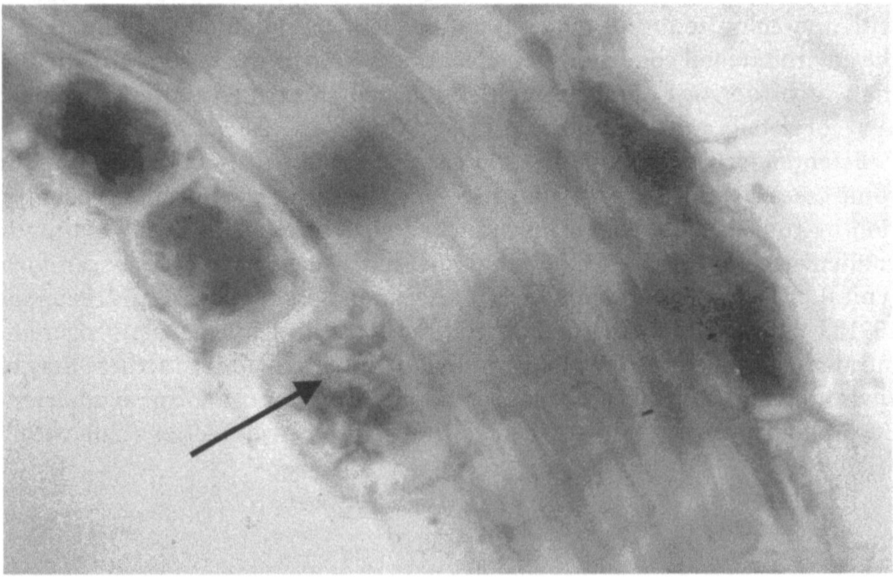

Fig. 11.3. Ericoid mycorrhiza. Root of *Epacris* sp. *Arrow* shows cortical cell layer colonized by unknown ericoid mycorrhizal fungus. (V. Gianinazzi-Pearson and S.E. Smith, unpubl.)

hosts. There have also been reports of formation of ECM and AM on plants of the Ericaceae (Molina et al. 1992), further illustrating mycorrhizal diversity in this family. Ten genera in the Monotropaceae, a family of heterotrophic plants in the Ericales, form mycorrhizas with a sheath and Hartig net from which fungal pegs grow into the epidermal cells and eventually burst open to produce a sac of fungal material of unknown function. The fungi are again basidiomycetes and some can form ECM with other plants. Molina et al. (1992) regard ectendo-, arbutoid and monotropoid mycorrhizas as host-mediated variants of ECM but, as in Table 11.1, they are usually placed in separate classes.

11.2.4 Orchid Mycorrhizas

There are thousands of orchid species distributed widely throughout many plant ecosystems. All orchids undergo a prolonged seedling stage that is heterotrophic irrespective of whether or not the adult becomes autotrophic. Having been placed for many years in the form genus *Rhizoctonia*, many of the fungi (basidiomycetes) are now being placed in a range of genera. Formation of a stable symbiosis as the minute orchid seeds germinate involves formation of dense intracellular coils of hyphae (Fig. 11.4). The degree of specificity between fungi and plants is not resolved, in part because of the taxonomic difficulties with both symbionts. However, at least some of the fungi can form ECM with other plants: an extreme example of plant control of mycorrhizal structures (Warcup 1985a; Sen et al. 1999; McKendrick et al. 2000).

11.2.5 Surprises in Store?

There is considerable blurring around the edges of the classes of mycorrhizas summarized in Table 11.1. Otherwise typical ECM roots sometimes have very limited penetration of cells by the fungi. Their sheaths may be thin and patchy, sometimes due to the plant/fungus combination but also possibly influenced by ecological factors (Kope and Warcup 1986). Hartig nets can be rudimentary or even absent, as in *Pisonia grandis* (Ashford and Allaway 1982). A few plant species can form ECM and AM, depending on growth stage or soil type or both: important examples are *Alnus*, *Eucalyptus*, *Populus*, *Salix* and possibly even *Pinus*. Unusual ECM are formed in roots of *Eucalyptus*, *Pinus*, *Quercus* and *Melaleuca* by a culturable zygomycetous fungus, '*Glomus tubiforme*' (Warcup 1985b), now *Densospora tubaeformis* (McGee 1996). Other ('endogonaceous') fungi of the Zygomycota form ECM with a wide range of Australian plants (Warcup 1980, 1990).

There will no doubt be more cases where unusual structures (e.g. Allen et al. 1999) or unexpected mycorrhizal fungi turn up in roots, and the fungi will

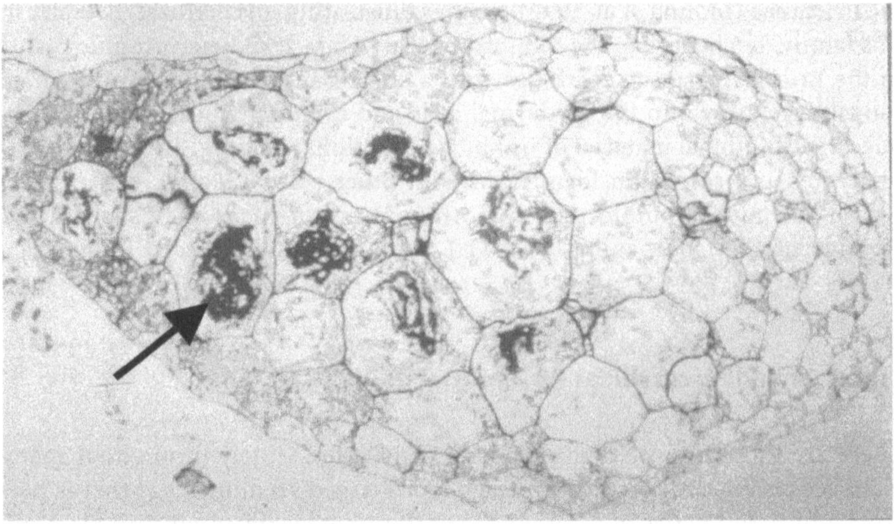

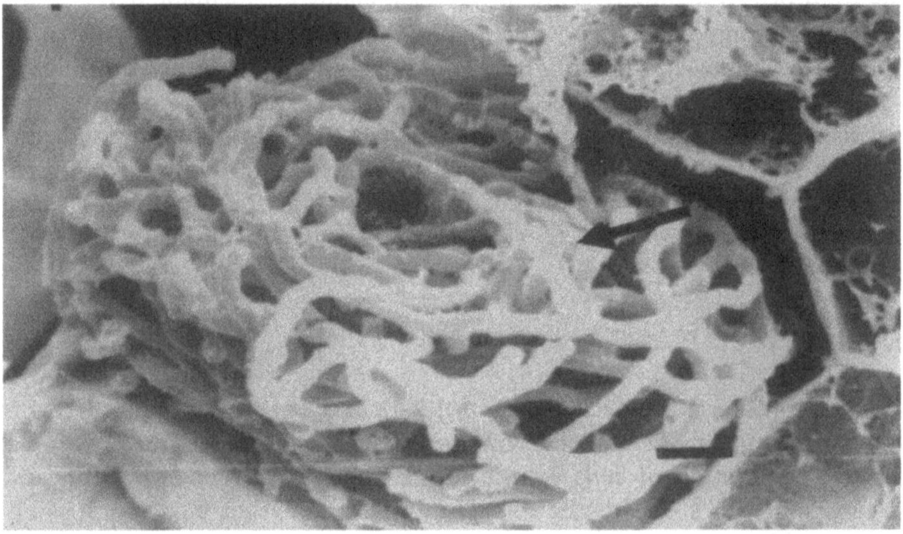

Fig. 11.4. Orchid mycorrhiza. *Above* Protocorm of *Goodyera repens*. *Arrow* shows intracellular hyphal coils of *Ceratobasidium cereale* (from Peterson and Currah 1990). *Below* Scanning electron micrograph of intracellular hyphal coils (*arrow*) of *Rhizoctonia* sp. in *Orchis morio*. *Scale bar* 0.5 μm. Micrograph by H. Beyrle and R.L. Peterson; from Smith and Smith (1996b)

eventually be identified by DNA fingerprinting. For example, the 'dark septate' fungi that form mycorrhiza-like structures in roots of alpine plants and many others worldwide (Read 1993) are now grouped in the Ascomycota (Jumpponen and Trappe 1998; Schadt et al. 2001).

11.2.6 Fungus–Plant Interfaces and Interactions

All mycorrhizal symbioses depend on the establishment of intracellular or intercellular interfaces between the symbionts, as summarized in Table 11.1 and shown in Figs. 11.1–11.4. These interfaces allow exchange of resources (see Sect. 11.5) and possibly signalling molecules (Barker and Tagu 2000) between the partners. Irrespective of differences in structural detail, all interfaces involve apoplastic compartments which contain cell wall material modified from that otherwise produced by the fungi and roots (Dexheimer and Pargney 1991; Martin and Hilbert 1991). To what extent the interfaces, whether intercellular or intracellular, are functionally isolated from the cortical apoplast as a whole is not clear. Certainly, intracellular interfaces such as arbuscules or coils occupy apoplasts of very small volume compared with fungal material in intercellular spaces. There are reports of 'AM' (glomalean mycorrhizas), especially *Paris*-types, that apparently lack arbuscules (see Smith and Smith 1997; Imhof 1999). Absence of arbuscules would not hinder function if the other interfaces allow efficient resource transfer, as in the other mycorrhizal classes that never form arbuscules.

Lastly, it may still be necessary to modify the concept of a mycorrhiza, given the existence of fungus–root associations that can result in improved plant growth in ways that may not involve nutrient transfer via the fungus. For example, the soil fungus *Piriformospora indica*, now known to be a basidiomycete quite closely related to *Rhizoctonia*, increases growth of a range of crop plants. It can apparently promote growth in the absence of colonization, as shown with filtrate from fungal cultures. This improved growth presumably involves plant hormonal responses that increase nutrient uptake by the roots (Varma et al. 1999).

11.3 Mycorrhizal Plant Communities and Their Distribution

Individual plant communities and even major terrestrial biomes are dominated by different classes of mycorrhizas. This has led to the concept of mycorrhizal guilds of plants that utilize common soil resources. Read (1984, 1991, 1993) has summarized recognizable patterns in distribution of the classes according to plant biomes in different latitudinal and altitudinal ranges. He emphasizes the importance of soil type, predominant form of soil N, the potential for limitation by soil P or N, and the nature and quantity of external fungal biomass (Table 11.2). Table 11.2 is essentially a transect down an idealized mountain, and also shows the parallels with an idealized transect along decreasing latitude, ignoring deserts and tropical forests. Boundaries between plant ecosystems are of course not as sharp as suggested by Table 11.2. Myc-

Table 11.2. Postulated relationships between latitude or altitude, climate, soil and mycorrhizal type, together with development of external hyphae; excluding tropical forests. Adapted from Read (1984); see also Fig. 15.13 of Smith and Read (1997)

	Decrease in altitude or latitude→			
Altitudinal range	Nival	Alpine heath	Forests[b] Coniferous→Deciduous	Grassland
Latitudinal range	Polar	Arctic tundra (heath understorey)	Forests Boreal→Deciduous (herb understorey)	Grassland
Main soil type	Lithosol	Peat→Podsol	Brown forest soil	Grassland soil, often sandy and semi-arid
Humus depth and type	Zero	Thick: Peat→Mor ('raw')	Thinner: Mor→Mull ('raw') (mixed/decaying)	Thin/absent
Main form of N	–[a]	Organic N	NH_4→NH_4/NO_3	NO_3
Major growth-limiting nutrient	–[a]	N	N or P	P
Main mycorrhizal type	None	ERM	ECM→ECM + AM	AM
Biomass of external hyphae	None	Low biomass (fine hyphae near roots)	High biomass (ECM hyphae often organized into strands)	High biomass near roots (individual hyphae)

[a] Plant growth limited by factors other than soil nutrients.
[b] Excludes tropical forests.

orrhizal colonization is generally believed to play a relatively unimportant part in plant communities at high altitudes or latitudes, including communities that contain herbaceous plants related to those which at lower altitudes or latitudes are normally mycorrhizal (especially AM). The reason may not lie solely in lack of fungal inoculum, but also in the physical characteristics and nutrient status of the soils (Read 1993). Just beyond the nival zone some alpine sedge and grass communities include root associations with the dark septate fungi. In order of descending altitude or latitude come heathland communities dominated by ERM plants, the boreal and temperate forests dominated by ECM plants, then grasslands and savannas dominated by AM plants. Not shown in Table 11.2 are subtropical and tropical forests. These are also mostly dominated by AM plants (Read 1993; Janos 1995).

In reality, however, most plant communities contain more than one class of mycorrhiza. Arctic and alpine tundra contain some AM and ECM as well as ERM plants and (especially in alpine sites) plants colonized by the dark septate fungi (Gardes and Dahlberg 1996). Mixed tree–shrub–herb communities, such as boreal and temperate woodlands and many Mediterranean-type ecosystems worldwide, also contain a mixture of AM, ECM and other classes (Read 1991; see also Brundrett 1991). Ericoid mycorrhizas occur not only in soils high in organic matter, but also in sandy soils that are low in nutrients (Allen et al. 1995). Examples are in the Mediterranean-type ecosystems in southern Africa (Ericaceae) and Australia (Epacridaceae). Many Ericaceae of the European and southern Californian Mediterranean shrublands form arbutoid mycorrhizas. Tropical forests dominated by the ECM dipterocarps can have co-dominant AM trees and AM understoreys. Their soils are typically P-limited, but in some cases N-limitation may also be important. The various types of mycorrhizal communities reflect many issues of physiology and ecology of mycorrhizal roots including spatial distribution and longevity, specificity of associations, etc.

In apparent contrast to the many mycorrhiza-rich plant ecosystems, extreme habitats that include salt-marshes or other wetlands are conventionally thought to be dominated by non-mycorrhizal plants. This is a dangerous generalization because AM fungi occur in at least some of these habitats (Khan 1974, 1993; Clayton and Bagyaraj 1984; Juniper and Abbott 1993). Much more research is required in this area, especially with respect to attempts to restore plant growth in salinized soils with unknown levels of mycorrhizal propagules.

11.4 The Mycorrhizosphere

11.4.1 External Hyphae

The success of mycorrhizal roots as organs for acquisition of soil nutrients depends greatly on fungal hyphae that extend from the root well beyond the normal rhizosphere to form large 'mycorrhizospheres' that include the distant region exploited by the hyphae (Fig. 11.5). AM fungal hyphae can extend more than 10 cm away from roots. Some ECM fungi easily span meters and their

Fig. 11.5. Seedling of *Pinus* colonized by *Suillus bovinus* and grown on natural forest soil in an observation chamber (approx 20 cm across). *Upper arrowhead* shows mycelial strands; *lower arrowhead* shows the advancing hyphal front (D.J. Read, unpubl.)

hyphal length is of the order of 10^3–10^4 cm/cm of root (Finlay and Söderström 1992; Read 1992). In contrast, external hyphae of ERM fungi extend less than 1 cm into soil from the fine roots of their hosts. The extensive foraging capacity (especially of AM and ECM fungi) potentially gives great advantage to a plant compared with non-mycorrhizal ones (Olsson et al. 2002). External hyphae penetrate into small crevices in the soil that are inaccessible to roots. They can also produce a wide array of degrading and chelating enzymes that help in nutrient extraction and uptake and are not produced by plants (see Sect. 11.5.1.1).

The architecture of external mycelium varies according to the fungus and environmental conditions. Hyphae of ECM and ectendomycorrhizas commonly form thicker cords and strands with good conducting capacity through already foraged areas, with fine hyphae growing into new areas. The ability of hyphae to fan out from the root surface aids in nutrient acquisition even where there are localized sites of colonization on root surfaces or intensively colonized bunches of root tips as in ECM and ectendomycorrhizas. Localized patches of hyphae develop in areas where soil nutrients are also localized (Read 1993). Absorption of nutrients from relatively large soil volumes helps prevent the development of nutrient deficiencies around roots. Individual mycorrhizal fungi produce different patterns of development of external hyphae, resulting in different abilities to absorb nutrients and translocate them to the host. Consequently, individual mycorrhizal fungi in association with a single host can exploit different soil niches (e.g. Jakobsen et al. 1992; Boddington and Dodd 1999; Smith et al. 2000, all using AM fungi; Erland and Finlay 1992; Timonen et al. 1997, with ECM fungi). This may allow more efficient acquisition of nutrients when roots are colonized by mixed populations of mycorrhizal fungi (Koide 2000).

Another role of external hyphae is of course reproduction and propagation of the fungus, including spore production on the hyphae or in fruit bodies in mycorrhizal classes where the latter occur. There is also direct hyphal colonization of adjacent seedlings of the same or different species that lie in the same mycorrhizal class or in different classes that involve common fungi. The relative importance of external hyphae and spores in colonizing plants will depend on the ecological setting, and particularly on presence or absence of extensive perennial vegetation. When seedlings connect into an existing hyphal network there will be faster colonization and faster nutrient uptake than can occur via spore germination or growth from hyphal fragments in response to appropriate environmental triggers, such as soil wetting. Hyphae that grow out from root fragments from short-lived plants form an additional source of inoculum (Tommerup and Abbott 1981).

11.4.2 The Soil Environment

The physicochemical properties of mycorrhizospheres differ markedly from those of adjacent bulk soil and can have profound indirect effects on the growth of the mycorrhizal plant and any neighbors (whether mycorrhizal or not) that share the mycorrhizosphere. The degree of modification of the properties will depend on the fungal biomass and the type of mycorrhiza. For example, the sheath of fungal material that surrounds ECM roots will prevent or greatly decrease exudation of organic material that occurs from roots generally and provides substrates for growth of many types of soil microorganisms in the immediate rhizosphere.

Changes in physicochemical properties produced by external hyphae include soil stabilization and formation of soil aggregates far beyond what can be achieved by roots alone. Mycorrhizas (especially ECM and ERM) also increase weathering of soil minerals and breakdown of soil organic matter (Sect. 11.5.1.1). Mycorrhizal fungi have been shown to have important roles in these regards in most of the ecosystems shown in Table 11.2 and others, including the tropics, sand-dunes and deserts, as well as in sites of soil rehabilitation (see Brundrett 1991; Leake and Read 1997; Smith and Read 1997 for references). These changes in soil properties produce many benefits in fertility, erosion control and water content. In addition, there will be pH changes caused by release of organic acids and uptake of soil nutrients within the mycorrhizosphere, with consequent effects on availability of other nutrients.

As in (non-mycorrhizal) rhizospheres the populations of soil bacteria, actinomycetes and fungi in mycorrhizospheres are usually very different from those in the bulk soil, and there are many possible interactions with mycorrhizal fungi. As the latter are directly in receipt of plant photosynthate they do not compete with saprophytic fungi and soil bacteria for carbon from organic material. As well as releasing some of the nutrients from organic material, mycorrhizal fungi are likely to compete effectively for soluble nutrients (N, P, etc.) released by saprophytic fungi and soil bacteria.

11.4.3 Bacteria Associated with Mycorrhizal Fungi

Mycorrhizospheres provide many unique niches for soil bacteria that directly or indirectly affect root ecology. Most of the research on rhizosphere bacteria has been done in agricultural ecosystems where mycorrhizal fungi are generally not the dominating microorganisms, due to disturbance and to pesticide and fertilizer treatments. This work has provided tools to isolate, identify and manipulate some of the soil bacteria, but has also skewed the direction of research to focus on a few easily culturable bacterial types that are common in agricultural systems. Molecular techniques are now providing information about the other, over 95%, of species that are presently non-culturable. Because of the great diversity and problems in identification, much of the research has been aimed at understanding functional properties of microbial populations rather than their identity. There are large populations of mycorrhiza-associated bacteria (MABs) that can increase mycorrhizal benefits by facilitating nutrient availability and uptake, producing plant hormones, helping mycorrhizal fungi to protect plant roots from pathogens and increasing mycorrhizal colonization (see Azcón-Aguilar and Barea 1992). Garbaye (1994) proposed seven different alternative mechanisms by which bacteria can facilitate formation of mycorrhizas: (1) production of plant cell wall softening enzymes; (2) enhancement of root-fungus recognition; (3) enhancement of fungal growth by excretion of organic carbon sources (e.g. malic and citric acids) utilized by the fungi; (4) elimination of fungal toxins in soil; (5) production of chelates that mediate nutrient uptake; (6) stimulation of germination of fungal propagules; and (7) competition with bacteria that inhibit fungal colonization. Considering the plethora of bacterial species and mixed populations, all the different alternatives may be utilized in one combination or another even in a single mycorrhizosphere.

The MABs that have been most vigorously studied are those that associate with AM fungi. According to the results of Andrade et al. (1997), MABs have a distinct distribution pattern in the mycorrhizosphere, with Pseudomonads thriving close to the roots and *Bacillus* spp. and Arthrobacters preferentially located further away in the hyphosphere. AM-associating bacteria have been shown to stimulate spore germination, fungal growth and mycorrhizal development as well as to solubilize rock P and produce plant growth regulators (see Azcón-Aguilar and Barea 1992; Garbaye 1994). Many dual inoculation trials of AM fungi and bacteria, particularly of P-solubilizing and N_2-fixing bacteria, have been carried out in pursuit of effective applications for agricultural uses, and indeed most of the published studies have shown beneficial effects on plant growth (Garbaye 1994). Bacteria-like organisms occur within hyphae and spores of several types of AM fungi and some of them phylogenetically belong to the *Burkholderia* group (Bianciotto et al. 1996; Perotto and Bonfante 1997). These bacteria are obligately endosymbiotic, but whether they are beneficial to the fungus or merely parasites remains to be investigated.

Bacteria associated with ECM fungi have been mainly studied in tree nurseries but a few studies have been carried out in forests or with forest soils. Distribution patterns of these bacteria follow the same trends as with AM roots, with Pseudomonads thriving close to the roots and *Bacillus* spp. in the hyphosphere (Timonen et al. 1998). Copious numbers of diverse bacteria are associated with some ectomycorrhizas whereas others seem devoid of bacteria (Foster and Marks 1966; Nurmiaho-Lassila et al. 1997). Numbers of culturable bacteria increase towards the root surface, although not as dramatically as in non-mycorrhizal plants, thus supporting the theory put forward by Söderström (1992) that external mycelium has a special role in distributing plant-derived carbon beyond the inner rhizosphere. This is further supported by the preference of culturable MABs for simple carbohydrates, often sugars that are present in the ECM fungi. In contrast, the bacteria in non-colonized soils are more tuned to consuming organic acids and proteins, at least in organic soils (Timonen et al. 1998). As with AM, ECM-associated bacteria can stimulate growth and colonization efficiency of some ECM fungi while suppressing growth of others (Bowen and Theodorou 1979; Duponnois et al. 1993). Some bacteria also display plant and fungal growth-promoting properties, such as breaking down fungitoxic polyphenols and producing siderophores that help in nutrient uptake and P solubilization from minerals (Leyval and Berthelin 1986; Duponnois and Garbaye 1990; Varese et al. 1996). Bacterial microfilms on ECM mycelial patches in soil high in organic material may in fact be responsible for as much decomposition as the fungus itself (Sarand et al. 1998). Nitrogen-fixing bacteria have been isolated from mycorrhizas and N_2 fixation observed in external mycelial mats (e.g. Griffiths et al. 1990; Li et al. 1992). ECM-associated bacteria may display further plant growth promotion via production of plant hormones (Strzelczyk and Pokjska-Burdziej 1984; Duponnois 1992). Some nursery applications already successfully utilize the dual plant-growth-promoting effect of ECM fungi and their associated bacteria (see Garbaye 1994).

The bacterial consorts of other mycorrhizal classes have so far received very little attention but there is no doubt that bacteria have a role. For example, bacteria have an effect on formation of orchid mycorrhizas that may be due to auxin production (Wilkinson et al.1994).

11.5 Functional Bases of Mycorrhizal Symbioses

11.5.1 Transfer of Nutrients and Carbon

11.5.1.1 Individual Plants

As already indicated, the functional basis of individual mycorrhizal symbioses lies mainly in transfer of resources from fungus to plant, and vice versa, with considerable variation in detail between the different classes. There is a huge and growing literature about nutrients that are absorbed from soil by mycorrhizal fungi and on the resulting plant growth responses. However, it is unwise to jump to the conclusion (Read 2000) that acquisition of nutrients (in general) by most plants growing in natural ecosystems occurs via their mycorrhizal fungi. Most of the work has been done with AM and ECM, and emphasis has been on P and N as major growth-limiting nutrients in many plant ecosystems in which these classes occur. Compartmented systems that separate roots from hyphae have been used to show uptake and transfer of P and N, supplied in inorganic or organic forms, and the convenient tracers ^{32}P and ^{15}N have been extensively used (Smith and Read 1997; Mäder et al. 2001). However, movement of tracer cannot give an indication of the amount of a solute transferred unless the specific activity (labelled solute/total solute) is known. (It is possible in theory for the net flux of a solute across a mycorrhizal interface to be in a direction contrary to the tracer flux.) There is much less information about other potentially scarce nutrients, although a limited amount of work has been done with Zn and Cu, especially with AM plants (see Smith and Read 1997 for many references).

Using extracellular enzymes such as phosphatases and proteases, ECM and ERM fungi can extract nutrients from organic substrates (Dighton 1991; Chalot and Brun 1998; Perez-Moreno and Read 2000). Evidence that AM fungi can acquire P or N from external organic sources is limited and controversial (see Leake and Read 1997), but at least one AM fungus can hydrolyze organic P and transport the resulting inorganic P to a host (Koide and Kabir 2000). The concept of tight cycling of nutrients between litter layers and plants via mycorrhizal fungi (see Dighton 1991) is ecologically extremely important. Breakdown of such cycling following clearance of natural vegetation can have major consequences for subsequent plant productivity. Unfortunately, little is known about the physiology of nutrient uptake by ectendomycorrhizal, arbutoid and monotropoid mycorrhizal plants.

Transfer of organic C from autotrophic hosts to AM, ECM and ERM fungi and the subsequent formation of fungal carbohydrates and lipids have been demonstrated with ^{14}C- and ^{13}C-labeling and, more recently, with ^{13}C nuclear magnetic resonance spectroscopy (Pfeffer et al. 2001; Bago et al. 2002). Conversely, transfer of organic C from sources within soil via fungal hyphae to heterotrophic orchids has been demonstrated using various external sources, including ^{14}C-labelled substrates and compartmented systems (McKendrick et al. 2000). Like the mycorrhizas of heterotrophic orchids, monotropoid mycorrhizas must also have distinct physiology, with transfer of organic C into the heterotrophic hosts. Smith and Read (1997) extensively discuss all the aspects of plant nutrition outlined above.

11.5.1.2 Linked Plants

Some myco-heterotrophic plants (e.g. monotropoid plants, orchids and others scattered through the angiosperms) depend on transfer of organic C via external hyphae from a 'donor plant', as opposed to acquisition of organic C from soil (Leake 1994). In addition, studies with radioactive C show that hyphal links allow transfer of organic C between autotrophic AM plants (e.g. Francis and Read 1984; Grime et al. 1987; Watkins et al. 1996) and ECM plants (e.g. Finlay and Read 1986a; Finlay 1989; Simard et al. 1997a,b). The need for analyses of tracer fluxes is very relevant in this context. For example, transfer of ^{14}C between plants through hyphae may result from movement of amino acids or amines in the fungus with little net transfer. The detailed work by Simard et al. (1997a,b) gives compelling evidence that movement of organic C between autotrophic ECM plants may be physiologically significant, but even this is regarded by Robinson and Fitter (1999) as unproven. Some studies have shown that most of the tracer C remains in the roots of the 'receiver' rather than being used for shoot growth, raising the question of whether the organic C leaves the fungus. However, even if it does not, it may still provide a cost saving to the apparent 'receiver', again provided that there is net movement. Robinson and Fitter (1999) concluded that carbon transfer between autotrophic plants is probably irrelevant to the plants but important to the fungi: 'the fungi move carbon according to their own carbon demands, not those of their autotrophic hosts'.

There is also tracer evidence for hyphal transfer of nutrients such as P or N between autotrophic plants (e.g. Whittingham and Read 1982; Finlay and Read 1986b; Hamel et al. 1991; Graves et al. 1997). Again, there are problems in interpretation (for P, see Newman 1988) and the quantitative significance compared with direct uptake from soil is not resolved. The quotation by Robinson and Fitter (see above) applies equally well to possible fungal demands for P and N. Despite the problems of interpretation, resource sharing via hyphal links within mycorrhizal 'guilds' of plants requires much more study. It has implications for competitive interactions of plants and allows the possibility of one autotrophic plant parasitizing another (Smith and Smith 1996a; Robinson and Fitter 1999).

11.5.2 Non-nutritional Factors

It is unwise to ignore aspects of mycorrhizas that relate to the ability of plants to tolerate stresses other than nutrient deficiencies. One is the potential for fungal hyphae to transfer water, a factor of obvious importance in semi-arid or arid ecosystems. On the whole, the evidence for transfer of water in significant quantities is more compelling for ECM plants than for AM plants, because the cross-sectional areas of the external hyphae of ECM fungi are much greater than for AM fungi (see Smith and Read 1997 for references). There may certainly be indirect effects that result in mycorrhizal plants having better water relations. These are discussed in detail by Augé (2001) and include both plant physiology and plant–soil interactions. For example, growth of external fungal hyphae may improve contact between the soil and roots. A second, and very different, benefit from at least some mycorrhizal symbioses (AM and ECM) is the ability to increase resistance or tolerance of the plants to root pathogens that include bacteria, fungi and nematodes (Duchesne 1994; Linderman 1994; Graham 2001). There is increasing interest in the alleviation of metal toxicity in plants by mycorrhizal colonization. Several ECM associations have been shown to decrease uptake and transfer of toxic metals including Zn and Ni. There is evidence that exposure to metals in soils can lead to selection for resistance in both AM and ECM fungi (Haselwandter et al. 1994; Pfleger et al. 1994; Leyval et al. 1997; Colpaert et al. 2001). Interactions between uptake of toxic metals and essential nutrients, and influence of soil factors including pH and Ca, are complex and beyond the scope of this chapter.

11.6 Diversity in Plant Growth Responses

We shall use the term 'mycorrhizal responsiveness' (MR) as a descriptor of *positive* responses by the plant. This is often defined in terms of increased weight when compared with non-mycorrhizal controls, but can be also be defined in terms of other benefits such as increased nutrient content or reproductive capacity. 'Mycorrhizal dependency' is an alternative term. We shall not concern ourselves with subtle distinctions in definitions as used by others (e.g. Reeves 1988; Janos 1995). Obligately mycorrhizal plant species by definition have capacity for high MR, while facultatively mycorrhizal plants species have low MR, sometimes none. The MR is influenced by the genome of both the plant and the mycorrhizal fungus, and will depend on environmental conditions and time of harvest. Influence of genomes has been shown by differences in MR in different plant species (or varieties) colonized by individual mycorrhizal fungi, and in individual plant species colonized by different fungi. Such studies with AM symbioses include one with *Hieracium pilosella*, *Bromus erecta* and *Festuca ovina* colonized by four AM fungal species or a mixture of these (van der Heijden et al. 1998a), and one with eight wild accessions and two cultivars of *Lycopersicon esculentum* colonized by a single AM fungus (Bryla and Koide 1990). Table 11.3 lists important factors that can increase MR in autotrophic plants, including properties of the fungus, the interfaces(s) between the partners (already discussed), and the root or root system. Structural and architectural factors that relate to root and fungus architecture and growth are distinguished from physiological processes that directly relate to resource exchange. However, there is clearly interdependence because a low flux of solute (defined as rate of transport per unit area) can result in transport of large quantities if the area for transport is large, and vice versa. In simple terms, a plant that only achieves low influx of the growth-limiting nutrient from soil in the absence of mycorrhizal colonization should show high MR, provided that other extraneous factors do not prevent this. Extraneous environmental and ecological factors that can influence MR are also shown in Table 11.3 and are discussed later. Some of the factors in Table 11.3 relate to the balance between benefits and costs. Benefits of

Table 11.3. Factors that may increase mycorrhizal responsiveness of autotrophic plants. Factors in italics are 'physiological', relating to resource acquisition and transfer

Fungus	Interface(s)	Root	Growth environment
External hyphae: Fast colonization High growth rate High extension into soil *High nutrient influx capacity* *High nutrient translocation*	Fast development Large area of contact High longevity *High organic C flux from roots* *High nutrient flux to roots*	Short length (low root/shoot ratio) Little branching Large diameter Few or short root hairs Selectively flexible root/shoot ratio[a] Inability to modify rhizosphere[b]	*Low soil nutrient availability* *High light intensity* Other 'suitable' soil conditions[c] Interplant connections[d] Interactions with rhizosphere bacteria Prevention of infection by root pathogens
Internal hyphae: High growth rate *Fast nutrient delivery to interface(s)*		*Low nutrient influx capacity* *Fast organic C delivery to interface(s)*	Low plant density[e]

[a] Inflexible in response to low soil nutrient levels, not to mycorrhizal colonization.
[b] Rhizosphere modification to increase nutrient availability.
[c] 'Suitable': good soil aeration, low salinity, etc.
[d] Increased benefits to receiver plants.
[e] Low plant density in soils with low nutrient availability (see text).

resource transfer to the plant in relation to the costs of maintaining the mycorrhizal symbioses will play a large part in the effects on plant growth and eventually success at the community level (Fitter 1991; Koide 1991a).

11.6.1 Carbon Costs of Mycorrhizal Symbioses

It is not easy to obtain reliable quantitative estimates of mycorrhizal fungal biomass and respiration. Nevertheless, the proportion of photosynthate used in fungal growth and respiration is usually estimated to lie in the range 10–20 % of the total for AM plants, with values ranging up to about 40 % for some ECM plants (see St. John and Coleman 1983; Brundrett 1991; Smith and Read 1997; Chap. 4). These costs to the plants are widely accepted as a major cause of poor or zero growth responses and also 'negative MR', i.e., growth depressions (both temporary and permanent) in mycorrhizal plants compared with non-mycorrhizal controls. In experimental studies this situation occurs especially where soil nutrient levels are so high that uptake by non-colonized roots is adequate to meet the plant's demand. Under such situations some plants reduce costs by decreasing colonization but others do not. This response can differ between mycorrhizal fungi in association with a single plant species (e.g. Thomson et al. 1986, comparing AM fungi). Growth depressions caused by drain of organic C to the fungus can be exacerbated by low light intensity, as in some glasshouses and growth-cabinets. Decreases in root/shoot ratios associated with low light intensities and changes in mycorrhizal colonization under low light further complicate cost-benefit analysis under such conditions (Johnson et al. 1997).

Smith and Smith (1996a) raised the possibility that some mycorrhizal fungi might act as 'cheaters' (Soberon and Martinez del Rio 1985) by draining the plant of resources while providing minimal nutritional benefits. Johnson et al. (1997) adopted a more wide-ranging approach in surveying the functioning of mycorrhizal associations (mainly of autotrophic plants) along a mutualism-parasitism continuum. The conclusion was that 'parasitic' mycorrhizal associations might be induced environmentally, developmentally at different stages in the plant's life cycle, or possibly genotypically in the sense of constitutive and prolonged cheating by the fungus irrespective of growth conditions. There is no evidence for the occurrence of individual mycorrhizal fungi that constitutively cheat *all* their potential host plants.

11.6.2 Growth Rates, Nutrient Demand and Mycorrhizal Responsiveness

Because rapid plant growth requires a larger supply of soil nutrients per unit time than slow growth, it is sometimes argued (e.g. Koide 1991a) that MR should be greater in fast growers. Koide et al. (1988) showed that this argument held for cultivated oat (*Avena sativa*) versus slow-growing wild oat (*A. fatua*), both with AM, but many cultivated cereals and grasses show small MR. In contrast, it can be argued that inherent slow growers should show large MR because slow growth of (non-mycorrhizal) roots will result in a greater probability of localized nutrient depletion. Wild herbaceous plants that grow naturally in soils low in nutrients often have lower relative growth rates, even when growing in high-nutrient conditions, than do plants that occur naturally in nutrient-rich soils or many cultivated plants (Koide 1991a; see also Chapin 1980; Tilman 1988). However, most wild plants (apart from some ruderals) that grow in nutrient-poor soils are mycorrhizal and many can show high MR, at least in the laboratory. The difficulties in sorting out the conflicting rationales about growth rates and MR arise from uncertainty about the physiological basis of inherent fast growth versus slow growth even in the absence of mycorrhizal complications (Lambers and Poorter 1992).

Koide (1991a) tackled conceptual problems about MR by extending the approach taken by Nye and Tinker (1977) to plant demand for nutrients. He focused on P uptake by AM plants, and suggested that where P is the growth-limiting nutrient, the maximum extent to which mycorrhizal colonization can improve plant performance is a function of the P deficit of the plant, i.e., the difference between P demand (for maximal performance) and P supply at any given time. Koide considered performance mainly as absolute growth rate (rate of increase in weight: d*W*/d*t*), but went on to discuss P needs for maximum reproduction and fitness more generally. He emphasized that there is temporal variation in P required for maximal performance, given ontogenetic changes. Further, plants may at times take up P at rates greater than optimal (an apparent example of 'luxury' uptake), in order to store it for future needs, e.g., seed production. Koide et al. (1999, 2000) have suggested the use of a 'phosphorus efficiency index' [PEI: (d*P*/d*t*)/*P*] as a measure of the efficiency with which a plant uses its P to acquire more P and have shown that AM colonization of lettuce greatly increases its PEI. Smith (2000) pursued this approach by describing relative growth rate (RGR) in terms of plant root parameters that are associated with nutrient uptake, as follows:

$$\underset{(\mathrm{RGR})}{\left[(\mathrm{d}W/\mathrm{d}t)/W\right]} = \underset{(\mathrm{NI})}{\left[(\mathrm{d}U/\mathrm{d}t)/R_A\right]} \times \underset{(\mathrm{NUE})}{(\mathrm{d}W/\mathrm{d}U)} \times \underset{(\mathrm{SRA})}{(R_A/R_W)} \times \underset{(\mathrm{RWR})}{(R_W/W)} \qquad (11.1)$$

where U is moles of a nutrient (e.g. P), R_A is root surface area and R_W is root weight. The terms in Eq. (11.1) are then respectively: RGR, net nutrient influx (NI), nutrient use efficiency (NUE), specific root area (SRA) and root weight ratio (RWR). Analogous terms can be derived on the basis of root length rather than surface area, because uptake of immobile nutrients, including P, is often considered in terms of 'inflow' (rate per unit length) rather than influx (rate per unit surface area). For a given nutrient, NUE will be maximal where the nutrient level in soil is growth-limiting. Where nutrients are being taken up in luxury amounts, NUE may be very low or zero and, if nutrients accumulate to toxic concentrations, NUE may even become negative.

Strategies that non-mycorrhizal plants have evolved for dealing with low nutrient conditions can be considered in terms of Eq. (11.1). Increasing NI under low-nutrient conditions can be achieved by changes in membrane transport parameters, i.e. increasing the substrate-affinity (decreasing Km), or increasing activity (Vmax) of the membrane transporter(s) for the nutrient. This can involve increased activity of transporters via feedback regulation of influx, or increased expression of transport-related genes, or both (Schachtman et al. 1998). In fact, these physiological strategies alone will not be very effective where nutrients (e.g. P) are quite immobile in soil and there is significant depletion adjacent to the transport sites (Silberbush and Barber 1984; Clarkson 1985). An effective physiological strategy of some constitutively non-mycorrhizal plants growing in soils low in P is to release organic acid to solubilize P. This helps to maintain the substrate concentration at the transport sites close to that in the bulk soil (see Marschner 1995), but has a cost in loss of photosynthate by the plant.

High values for SRA and RWR are characteristic features of plants, whether mycorrhizal or not, in low-nutrient soils. High SRA typically involves long, thin roots with many root hairs. High RWR is expressed more conventionally as high root/shoot ratio. Some plants show rapid changes in root architecture when plants are transferred from high-nutrient growth media to low-nutrient media or when roots arrive in a nutrient-rich patch of soil (Lambers and Poorter 1992; Robinson 1994). However, the plant will be at risk if high root biomass is achieved at the expense of photosynthetic (shoot) biomass. Root hairs could be included in Eq. (11.1) as a component of root surface area, but there are further relevant architectural properties of roots. For example, branching patterns will not be revealed by measurements of SRA or specific root length (see Hetrick 1991).

Mycorrhizal colonization can improve RGR under low-nutrient conditions by a combination of architectural and physiological strategies that involve both symbionts. A large component of NI in mycorrhizal plants will be provided by the fungal hyphae that are an extremely large but (almost literally) invisible component of the SRA. (As defined, the latter does not include the fungal surface area.) Relatively low nutrient depletion in the large mycorrhizosphere means that the concentration available to the membrane transporter(s) remains close to that in the bulk soil. Not surprisingly, therefore, there is potential for high MR. As can be implied from Table 11.3, plants that form AM but show little if any (positive) MR include those with high SRA. Such plants, especially herbs and shrubs, usually show flexibility by increasing SRA in response to low soil nutrient levels, irrespective of the presence of mycorrhizas. Good examples are many grasses and cereals (e.g. Koide 1991a). Flexibility in root development relates to other factors, including mycorrhiza formation itself. There are many examples where mycorrhizal plants produce relatively smaller root systems (low RWR) than the non-mycorrhizal equivalents. Organic C that is 'saved' is in effect diverted to the fungus and which, with its enormous surface area in soil, can access nutrients more efficiently than the equivalent plant root biomass.

Analysis of growth and nutrient uptake in terms of Eq. (11.1) has not been undertaken as far as we are aware, although some studies with AM have come close. For example, Manjunath and Habte (1991) showed that 65 % of the variability in MR of selected *Leucaena* and *Sesbania* species was explained by root weight alone. Other important root characteristics included root density in soil (cm cm^{-3}) and the incidence and length of root hairs.

11.7 Plant–Fungal Interactions at the Community Level

11.7.1 Plant Density, Competition and Succession

Mycorrhizal symbioses can strongly influence plant community structure in many ways (see Brundrett 1991; Janos 1995; Sanders et al. 1999). Interactions between plants depend greatly on MR of individual plant species, as influenced by the properties of the mycorrhizal fungi that are present, and environmental conditions such as those summarized in Table 11.3. High plant densities (plant number per unit soil volume) in low-nutrient soils can greatly decrease MR of AM plants below values found at low plant densities (Koide 1991b; Hartnett et al. 1993; West 1996; Facelli et al. 1999). As shown in Table 11.4, this effect arises because the AM plants have higher density-dependent (intraspecific) competition than the non-mycorrhizal plants for P, the growth-limiting nutrient. Higher levels of soil P can reverse the effects of density, i.e., MR is low or absent when plants are at low density, but higher at high density (Hartnett et al. 1993). This will occur where higher soil P satisfies demand by mycorrhizal plants at low and high density, but satisfies demand by non-mycorrhizal plants only at low density. Where plants are facultatively mycorrhizal, effects of density on MR will be less pronounced or absent (Hartnett et al. 1993). Differences in growth rates complicate the interactions between growth rates and MR, as shown by Hartnett et al. (1993) and Hetrick et al. (1989, 1994) with three AM prairie grasses (Table 11.5), with soils low in

Table 11.4. Effects of plant density on mycorrhizal responsiveness and intraspecific competition in *Trifolium subterraneum* colonized by *Gigasporita margarita* (data from Facelli et al. 1999). Plants harvested after 8 weeks; AM colonization independent of density. Standard errors of means mostly less than 10%. Mycorrhizal responsiveness calculated as (*M*–*NM*/*M*), from mean dry weights of individual mycorrhizal (*M*) and equivalent non-mycorrhizal (*NM*) plants. Relative competition intensity calculated as (*S*–*C*)/*S*, from mean dry weights of plants grown singly (*S*) and of individual plants in pots containing 3, 6 or 18 plants (*C*). Asterisks show means not significantly different from zero; n/a: not applicable

Treatment (plants/pot)	Plant wt (g/plant)	Root/shoot ratio	Mycorrhizal responsiveness	Competition intensity
1 NM	0.28	0.72	n/a	n/a
3 NM	0.25	0.89	n/a	0.1*
6 NM	0.26	0.67	n/a	0.1*
18 NM	0.15	0.84	n/a	0.5
1 M	2.31	0.68	7.3	n/a
3 M	0.84	0.70	2.4	0.6
6 M	0.44	0.68	0.7	0.8
18 M	0.20	0.89	0.3*	0.9

P. When the grasses were mycorrhizal, their intraspecific competition reflected the growth rates at low density rather than their MR *per se*. Strong intraspecific competition can help explain apparent lack of benefits of AM colonization for dense plant associations in the field (see Fitter 1991).

Effects of AM colonization on interspecific competition have received a great deal of attention (e.g. Miller and Allen 1992; Hartnett et al. 1993; Hartnett and Wilson 1999; Marler et al. 1999; Hartnett and Wilson 2002). Formation of AM can increase the ability of plant species with high MR to outcompete facultatively mycorrhizal or constitutively non-mycorrhizal plants (Grime et al. 1987; van der Heijden et al. 1998b; Hartnett and Wilson 1999). The outcomes again depend on inherent growth rates of competing species at low densities (Table 11.5). Grime et al. (1987) and van der Heijden et al. (1998b) found that plant (floristic) diversity was decreased when AM were suppressed, but Hartnett and Wilson (1999) and O'Connor et al. (2002) found the reverse. As suggested by Bergelson and Crawley (1988), the different effects of AM on diversity will depend greatly on whether the dominant species are facultatively or obligately mycorrhizal. In the former case suppression of AM will favor the dominant plant species at the expense of subordinate obligately mycorrhizal species, while, in the latter case, suppression of AM will favor growth of subordinate species that are facultatively mycorrhizal. (For further discussion that specifically relates to grasslands, see Hartnett and Wilson 2002.)

Relatively little is known about the influence of mycorrhizal colonization on competition between long-lived woody plant species, whether AM or ECM, or mixtures of ECM and AM plants. Pedersen et al. (1999) showed that ECM

Table 11.5. Summary of relationships between growth, mycorrhizal responsiveness and competition, in terms of dry weight per plant, in three prairie grass species (*Andropogon gerardii, Koeleria pyramidata* and *Elymus canadensis*), all grown in soil with low phosphate. Growth rates and mycorrhizal responsiveness are for single plants per pot. *M* Mycorrhizal treatments; *NM* non-mycorrhizal treatments. Data collated from Hartnett et al. (1993) and Hetrick et al. (1989, 1994)

Plant species	Growth rate (single plant)	Mycorrhizal responsiveness (single plant)	Intraspecific competition	Interspecific competition
A. gerardii (*A.g*)	NM: Very low M: High	High	NM: Nil M: Strong	NM on *E.c*: Low M on *E.c*: Moderate NM on *K.p*: Nil M on *K.p*: Strong
E. canadensis (*E.c*)	NM: Moderate M: Moderate	Nil	NM: Moderate M: Moderate	NM on *A.g*: Nil M on *A.g*: Moderate
K. pyramidata (*K.p*)	NM: Low M: Low	Nil	NM: V. weak M: V. weak	NM on *A.g*: Nil M on *A.g*: Nil

colonization of slash pine increased its P uptake when competing with non-mycorrhizal grass. Competition was dependent on external P, with the grass performing better at low P. Perry et al. (1989) found that ECM mediation of competition between coniferous tree species was complex and concluded that generalizations should be made cautiously.

Possible sharing of resources, both organic carbon and mineral nutrients, via hyphal links between plants is repeatedly proposed as a factor involved in interactions between mycorrhizal plants at the community level (e.g., Grime et al. 1987; Hartnett and Wilson 1999). Such sharing – if quantitatively significant (see Sect. 11.5.1.2) – would greatly complicate understanding of causes of competition between mycorrhizal plants (Miller and Allen 1992; Hartnett and Wilson 2002). It is obvious that there will be differences between the linked plants in the demand they make on the hyphal network for inorganic nutrients.

Roles of mycorrhizas in plant succession have received increasing attention in recent years, particularly with respect to recovery of plant ecosystems from disturbance or establishment on bare substrates. Under such conditions, and in a wide range of plant ecosystems (especially forests), non-mycorrhizal or facultatively mycorrhizal plants often become replaced by obligately mycorrhizal plants (Janos 1980; 1995; Brundrett 1991; Read 1993; Marler et al. 1999). Such succession is in accord with Odum's (1969) general strategies of ecosystem development and reflects decrease in mineral nutrient content of soil solutions, due to increasing proportions of minerals held in biota or to leaching. In the absence of fungal inoculum there is of course no opportunity for establishment of obligately mycorrhizal plants and competition between constitutively non-mycorrhizal plants and facultatively mycorrhizal plants is important.

In low-nutrient soils, establishment of obligately mycorrhizal plants ensures that high levels of soil propagules are maintained, with consequent advantages for future generations of mycorrhizal plants, for example, in the event of a localized or temporary gap in the plant community following disturbance (Janos 1980, 1995). However, disturbance that increases soil nutrient content (e.g. regular burning) can favor re-emergence of communities dominated by facultative mycorrhizal plants, as in some tropical grasslands (Janos 1980). More generally, build-up of organic matter in plant ecosystems can favor ECM or ERM symbioses and aid in tight nutrient cycling (Read 1993). If plants within one mycorrhizal class (e.g. ECM) occupy much of the space available for roots, representatives of other classes will be at a disadvantage and the propagules of their fungal associates will decline in numbers. In contrast, in ecosystems with a well-established mixture of mycorrhizal classes, there may be different soil niches that allow compartmentation in space and time in acquisition of soil resources via the various mycorrhizal fungi. For example, in mixed ECM/AM associations, the former exploit the surface organic layers while the latter proliferate in the mineral soil (Read 1993).

11.7.2 The Mycorrhizal Fungal Community

Van der Heijden et al. (1998a,b) measured effects on plant growth of different AM fungal species obtained from the natural grassland in which their experimental plants coexisted. Van der Heijden et al. (1998b) concluded that plant responses (and diversity) increased as numbers of AM fungi in mixtures was increased. Wardle (1999) has suggested that this conclusion is marred by flaws in experimental design, but this does not affect the important demonstration that different AM fungi obtained from the same plant community produce different MR of individual plant species. This evidence for the importance of AM fungal biodiversity greatly strengthens much previous evidence that AM fungi derived from different sources have very different effects on nutrition acquisition and MR in individual plants (e.g., Johnson et al. 1997; Jakobsen et al. 2002). In other words, lack of specificity in terms of AM colonization does not mean that outcomes in terms of function are equal (see also Table 11.3). Specificity between some ECM fungi and their hosts is well established, and specificity is probably more widespread among mycorrhizal fungi in general than usually appreciated (Molina et al. 1992).

The population density and composition of the mycorrhizal fungal community is greatly affected by the plant community (Janos 1980; Brundrett 1991; Molina et al. 1992; Eom et al. 2000; Bever 2002), but little is known about the basis of competitive interactions among mycorrhizal fungi. Widden (1997) has suggested that competitive strategies of fungi (in general) mirror those of plants, i.e. there are categories such as ruderal, highly combative, and stress-tolerant fungi. This aspect of ecological theory may well apply to mycorrhizal fungi.

11.8 Summary and Prospects

Wide diversity of microbial populations in soils gives the immobile and relatively long-living plants considerable flexibility to utilize the complex resources of soils and to adjust to changing environmental conditions. Mycorrhizas are key components of the plant–root–soil continuum in most plant ecosystems because they can affect nutrition and growth of plants directly and indirectly, and can improve soil structure and stability. The roles of mycorrhizas in many natural plant ecosystems are becoming more widely recognized by plant ecologists and better understood. The importance of mycorrhizas is also quite well appreciated in commercial forest ecosystems worldwide. In contrast, much work with agro-ecosystems ignores mycorrhizas. Sometimes this seems reasonable, as where levels of applied fertilizer are high and crop plants inherently show small MR, as with many temperate cereals. However, there is now emphasis on decreasing fertilizer input in intensive agriculture in Europe and elsewhere, and there are many developing countries where high fertilizer input is currently impracticable for economic or logistic reasons. New strategies for improving plant productivity, both by conventional plant breeding and molecular biology, are starting to take into account potential improvements in terms of operation of key nutrient transporters and root architecture – the two major factors included in our 'belowground growth analysis' (Eq. 11.1). Given limitations of these factors as they relate to immobile nutrients (e.g. P) for the vast majority of plants, it seems essential that mycorrhizas are not overlooked in the new programs (Zhu et al. 2001). Nor should mycorrhizal fungi, or other beneficial mycorrhizosphere organisms, be forgotten where the aim of genetic modification of plants is to prevent colonization by root pathogens, otherwise the beneficial microbial babies may be thrown out with the pathogenic bathwater.

Future developments in mycorrhizal research will increasingly depend on interdisciplinary approaches. A good example is a recent study of mycorrhizal versus saprophytic status of soil fungi that combined isotopic analysis ($\delta^{13}C$ and $\delta^{15}N$) with molecular phylogeny (Hobbie et al. 2001). DNA fingerprinting will make identification and exploitation of 'difficult' mycorrhizal fungi from diverse habitats also become much easier. Techniques of molecular biology are being rapidly incorporated into nearly all approaches to mycorrhizal research, and we anticipate exciting advances in identifying key processes in recognition or non-recognition between potential symbionts and in subsequent physiological function and interactions. 'Molecular mycorrhizal ecology' is already becoming well established (e.g., Read 2000; Dahlberg 2001; Franken and Requena 2001).

As we have indicated, the most challenging area to investigate experimentally is that of the influence of mycorrhizal symbioses in competition and other interactions between plants in different ecosystems. There is a need for more experimental studies with mycorrhizal plant populations and communities, e.g. in micro- or mesocosms, that can take into account the roles of mycorrhizas in satisfying demands for nutrients by competing plants. It is also necessary to understand how plant biodiversity in mixed plant communities is influenced by whether the plants are facultatively or obligately mycorrhizal, and by plant growth rates and nutrient demand under different environmental conditions. (The last includes increasing atmospheric CO_2: see Fitter et al. 2000; Treseder and Allen 2000.) The many interactions between mycorrhizal fungi and soil organisms that are involved in maintenance of biodiversity belowground must also be better understood.

In conclusion, we believe that the starting point of a study of root ecology in the 'real world' underground should be the question: are the roots of the plant(s) mycorrhizal? If they are – and there is a probability of about 90 % that they will be in most plant ecosystems – our next question is: can the mycorrhizas realistically be ignored? We think that it is very unwise to do so.

Acknowledgements. We are very grateful to colleagues who have over the years broadened our minds about the ecological implications of mycorrhizas. In particular, we thank Alastair Fitter, Jim Graham, Iver Jakobsen, Dave Janos, Nancy Johnson, Roger Koide, Hans Lambers, Ed Newman, David Read, Robin Sen, Bengt Söderström and – especially – Evelina Facelli and Patrick O'Connor. We apologize for the omission of much relevant recent material due to space constraints. Experimental work by FAS and SES is funded by the Australian Research Council and the Cooperative Research Centres for Soil and Land Management and Molecular Plant Breeding. We all thank the Academy of Finland for financial support that made possible S.T.'s visit to Adelaide, and Lauri Kaila for helpful comments.

References

Allen MF (ed) (1992) Mycorrhizal functioning: an integrative plant-fungal process. Chapman and Hall, London

Allen EB, Allen MF, Helm DJ, Trappe JM, Molina R, Rincon E (1995) Patterns and regulation of mycorrhizal plant and fungal diversity. Plant Soil 170:47–62

Allen MF, Egerton-Warberton LM, Allen EB, Karen O (1999) Mycorrhizae in *Adenostoma fasculatum* Hook.& Arn.: a combination of unusual ecto- and endo-forms. Mycorrhiza 8:225–228

Andrade G, Mihara KL, Linderman RG, Bethlenfalvay GJ (1997) Bacteria from the rhizosphere and hyphosphere soils of different arbuscular mycorrhizal fungi. Plant Soil 192:71–79

Arriola L, Niemira B, Safir GR (1997) Border cells and arbuscular mycorrhizae in four Amaranthaceae species. Phytopathology 87:1240–1242

Ashford AE, Allaway WG (1982) A sheathing mycorrhiza on *Pisonia grandis* R. Br. (Nyctaginaceae) with development of transfer cells rather than a Hartig net. New Phytol 90:511–519

Augé RM (2001) Water relations, drought and vesicular-arbuscular mycorrhizal symbiosis. Mycorrhiza 11:3–42

Azcón-Aguilar C, Barea JM (1992) Interactions between mycorrhizal fungi and other rhizosphere microorganisms. In: Allen MF (ed) Mycorrhizal functioning: an integrative plant-fungal process. Chapman and Hall, London, pp 163–198

Bago B, Pfeffer PE, Zipfel W, Lammers P, Shachar-Hill Y (2002) Tracking metabolism and imaging transport in arbuscular mycorrhizal fungi. Plant Soil 244:189–197

Barker SJ, Tagu D (2000) The roles of auxins and cytokinins in mycorrhizal symbioses. J Plant Growth Regul 19:144–154

Bergelson JM, Crawley MJ (1988) Mycorrhizal infection and plant species diversity. Nature 334:202

Bever JD (2002) Host-specificity of AM fungal population growth rates can generate feedback on plant growth. Plant Soil 244:281–290

Bianciotto V, Bandi C, Minerdi D, Sironi M, Tichy HV, Bonfante P (1996) An obligately endosymbiotic mycorrhizal fungus itself harbors obligately intracellular bacteria. Appl Environ Microbiol 62:3005–3010

Boddington CL, Dodd JC (1999) Evidence that differences in phosphate metabolism in mycorrhizas formed by species of *Glomus* and *Gigaspora* might be related to their life-cycle strategies. New Phytol 142:531–538

Bowen GD, Theodorou C (1979) Interactions between bacteria and ectomycorrhizal fungi. Soil Biol Biochem 11:119–126

Brundrett M (1991) Mycorrhizas in natural ecosystems. Adv Ecol Res 21:171–313
Brundrett M, Kendrick B (1990a) The roots and mycorrhizas of herbaceous woodland plants: I. Quantitative aspects of morphology. New Phytol 114:457–468
Brundrett M, Kendrick B (1990b) The roots and mycorrhizas of herbaceous woodland plants: II. Structural aspects of morphology. New Phytol 114:469–480
Bruns TD, Gardes M (1993) Molecular tools for the identification of ectomycorrhizal fungi-taxon-specific oligonucleotide probes for suilloid fungi. Mol Ecol 2:233–242
Bryla DR, Koide RT (1990) Role of mycorrhizal infection in the growth and reproduction of wild vs. cultivated plants. II. Eight wild accessions and two cultivars of *Lycopersicon esculentum* Mill. Oecologia 84:82–92
Burgess T, Dell B, Malajczuk N (1994) Variation in mycorrhizal development and growth stimulation by 20 *Pisolithus* isolates inoculated on *Eucalyptus grandis* W. Hill ex Maiden. New Phytol 127:731–739
Cavagnaro TR, Gao L-L, Smith FA, Smith SE (2001) Morphology of arbuscular mycorrhizas is influenced by fungal identity. New Phytol 151:469–475
Chalot M, Brun A (1998) Physiology of organic nitrogen acquisition by ectomycorrhzal fungi and ectomycorrhizas. FEMS Microbiol Rev 22:21–44
Chapin FS III (1980) The mineral nutrition of wild plants. Annu Rev Ecol Syst 11:233–260
Clarkson DT (1985) Factors affecting mineral nutrient acquisition by plants. Annu Rev Plant Physiol 36:77–115
Clayton JS, Bagyaraj DJ (1984) Vesicular-arbuscular mycorrhizas in submerged aquatic plants in New Zealand. Aquat Bot 19:251–262
Colpaert JV, Vandenkoornhuyse P, Adriaensen K, Vangronsveld J (2001) Genetic variation and heavy metal tolerance in the ectomycorrhizal basidiomycete *Suillus luteus*. New Phytol 157:367–379
Dahlberg A (2001) Community ecology of ectomycorrhizal fungi: an advancing interdisciplinary field. New Phytol 150:555–562.
Dexheimer J, Pargney JC (1991) Comparative anatomy of the host-fungus interface in mycorrhizas. Experientia 47:312–321
Dickson S (1999) Phosphate transfer efficiency of two arbuscular mycorrhizal fungi. PhD Thesis, University of Adelaide, Australia
Dighton J (1991) Aquisition of nutrients from the organic resources by mycorrhizal autotrophic plants. Experientia 47:362–369
Duchesne LC (1994) Role of ectomycorrhizal fungi in biocontrol. In: Pfleger FL, Linderman RG (eds) Mycorrhizae and plant health. APS Press, St Paul, Minnesota, pp 27–45
Duckett JG, Read DJ (1995) Ericoid mycorrhizas and rhizoid-ascomycete associations in liverworts share the same mycobiont: isolation of the partners and resynthesis of the associations in vitro. New Phytol 129:439–447
Duponnois R (1992) Les bactéries auxiliaires de la mycorhization du Douglas fir (*Pseudotsuga menziesii* (Mirb.) Franco) par *Laccaria laccata* souche S238. These de l'Université de Nancy I, France
Duponnois R, Garbaye J (1990) Some mechanisms involved in growth stimulation of ectomycorhizal fungi by bacteria. Can J Bot 68:2148–2152
Dupponois R, Garbaye J, Bouchard D, Churin J-L (1993) The fungus-specificity of mycorrhiza helper bacteria (MHBs) used as an alternative to soil fumigation for ectomycorrhizal inoculation of bare-root Douglas fir planting stocks with *Laccaria laccata*. Plant Soil 157:257–262
Eom A-H, Hartnett DC, Wilson GT (2000) Host plant species effects on arbuscular mycorrhizal fungal communities in tallgrass prairies. Oecologia 122:435–444
Erland S, Finlay R (1992) Effects of temperature and incubation time on the ability of three ectomycorrhizal fungi to colonize *Pinus sylvestris* roots. Mycol Res 96:270–272

Facelli E, Facelli J, McLaughlin MJ, Smith SE (1999) Interactive effects of arbuscular mycorrhizal symbiosis, intraspecific competition and resource availability using *Trifolium subterraneum* L. cv. Mt Barker. New Phytol 141:535–547
Finlay RD (1989) Functional aspects of phosphorus uptake and carbon translocation in incompatible ectomycorrhizal associations between *Pinus sylvestris* and *Suillus grevillei* and *Boletus cavipes*. New Phytol 112:185–192
Finlay RD, Read DJ (1986a) The structure and function of the vegetative mycelium of ectomycorrhizal plants. I. Translocation of ^{14}C-labelled carbon between plants interconnected by a common mycelium. New Phytol 103:143–156
Finlay RD, Read DJ (1986b) The structure and function of the vegetative mycelium of ectomycorrhizal plants. II. The uptake and distribution of phosphorus by mycelial interconnecting host plants. New Phytol 103:157–165
Finlay R, Söderström B (1992) Mycorrhiza and carbon flow to the soil. In: Allen MF (ed) Mycorrhizal functioning: an integrative plant-fungal process. Chapman and Hall, New York, pp 134–160
Fitter AH (1991) Costs and benefits of mycorrhizas: implications for functioning under natural conditions. Experientia 47:350–355
Fitter AH, Heinemeyer A, Staddon PL (2000) The impact of elevated CO_2 and global climate change on arbuscular mycorrhizas: a mycocentric approach. New Phytol 147: 179–187
Foster RC, Marks GC (1966) The fine structure of the mycorrhizas of *Pinus radiata* D. Don. Aust J Biol Sci 19:1027–1038
Francis R, Read DJ (1984) Direct transfer of carbon between plants connected by vesicular-arbuscular mycorrhizal mycelium. Nature 307:53–56
Franken P, Requena N (2001) Analysis of gene expression in arbuscular mycorrhizas: new approaches and challenges. New Phytol 150:517–523
Gallaud I (1905) Études sur les mycorrhizes endotrophes. Rév Gen Bot 17:5–48, 66–83, 123–136, 223–239, 313–325 425, 479–500
Garbaye J (1994) Helper bacteria: a new dimension to the mycorrhizal symbiosis. New Phytol 128:197–210
Gardes M, Dahlberg A (1996) Mycorrhizal diversity in arctic and alpine tundra: an open question. New Phytol 133:147–157
Graham JH (2001) What do root pathogens see in mycorrhizas? New Phytol 149:357–359
Graves JD, Watkins NK, Fitter AH, Robinson D, Scrimgeour C (1997) Intraspecific transfer of carbon between plants linked by a common mycorrhizal network. Plant Soil 192:153–159
Griffiths RP, Caldwell BA, Cromack KC, Morita RY (1990) Douglas-fir forest soils colonized by ectomycorrhizal mats. I. Seasonal variation in nitrogen chemistry and nitrogen cycle transformation rates. Can J Forest Res 20:211–218
Grime JP, Mackey JML, Hillier SH, Read DJ (1987) Floristic diversity in a model system using experimental microcosms. Nature 328:420–422
Hamel C, Furlan V, Smith DL (1991) N_2-fixation and transfer in a field grown mycorrhizal corn and soybean intercrop. Plant Soil 133:177–185
Hartnett DC, Wilson GWT (1999) Mycorrhizae influence plant community structure and diversity in tallgrass prairie. Ecology 80:1187–1195
Hartnett DC, Wilson GWT (2002) The role of mycorrhizas in plant community structure and dynamics. Plant Soil 244:319–331
Hartnett DC, Hetrick BAD, Wilson GWT, Gibson DJ (1993) Mycorrhizal influence on intra- and interspecific neighbour interactions among co-occurring prairie grasses. J Ecol 81:787–795
Haselwandter K, Leyval C, Sanders FE (1994) Impact of arbuscular mycorrhizal fungi on plant uptake of heavy metals and radionuclides from soil. In: Gianinazzi S, Schüepp H

(eds) Impact of arbuscular mycorrhizas on sustainable agriculture and natural ecosystems. Birkhäuser, Berlin, pp 179–189

Helgason T, Fitter AH, Young JPW (1999) Molecular diversity of arbuscular mycorrhizal fungi colonising *Hyacinthoides non-scripta* (bluebell) in a semi-natural woodland. Mol Ecol 8:659–666

Hetrick BAD (1991) Mycorrhizas and root architecture. Experientia 47:355–362

Hetrick BAD, Wilson GWT, Hartnett DC (1989) Relationship between mycorrhizal dependence and competitive ability of two tallgrass prairie grasses. Can J Bot 67:2608–2615

Hetrick BAD, Hartnett DC, Wilson GWT, Gibson DJ (1994) Effects of mycorrhizae, phosphorus availability, and plant density on yield relationships among competing tallgrass prairie grasses. Can J Bot 72:168–176

Hobbie EA, Weber NS, Trappe JM (2001) Mycorrhizal vs saprophytic status of fungi: the isotopic evidence. New Phytol 150:601–610

Hooker JE, Munro M, Atkinson D (1992) Vesicular-arbuscular mycorrhizal fungi induced alteration in poplar root system morphology. Plant Soil 145:207–214

Imhof S (1999) Anatomy and mycotrophy of the achlorophyllous *Afrothismia winkleri.* New Phytol 144:533–540

Jakobsen I, Abbott LK, Robson AD (1992) External hyphae of vesicular-arbuscular mycorrhizal fungi associated with *Trifolium subterraneum* L. I. Spread of hyphae and phosphorus inflow into roots. New Phytol 120:371–380

Jakobsen I, Smith SE, Smith FA (2002) Function and diversity of arbuscular mycorrhizas in carbon and mineral nutrition. In: van der Heijden MGA, Sanders IR (eds) Mycorrhizal ecology. Springer, Berlin Heidelberg New York, pp 75–92

Janos DP (1980) Mycorrhizas influence tropical succession. Biotropia 12:56–64

Janos DP (1995) Mycorrhizas, succession and the rehabilitation of deforested lands in the humid tropics. In: Frankland JC, Magan N, Gadd GM (eds) Fungi and environmental change. Cambridge University Press, Cambridge, pp 129–162

Johnson NC, Graham JH, Smith FA (1997) Functioning of mycorrhizal associations along the mutualism-parasitism continuum. New Phytol 135:575–586

Jumpponen A, Trappe JM (1998) Dark septate endophytes: a review of facultative root-colonizing fungi. New Phytol 140:295–310

Juniper S, Abbott LK (1993) Vesicular-arbuscular mycorrhizas and soil salinity. Mycorrhiza 4:45–57

Khan AG (1974) The occurrence of mycorrhizas in halophytes, hydrophytes and xero phytes, and of *Endogone* spores in adjacent soils. J Gen Microbiol 81:7–14

Khan AG (1993) Occurrence and importance of mycorrhizae in aquatic trees of New South Wales, Australia. Mycorrhiza 3:31–38

Koide RT (1991a) Nutrient supply, nutrient demand and plant response to mycorrhizal infection. New Phytol 117:365–386

Koide RT (1991b) Density-dependent response to mycorrhizal infection in *Abutilon theophrasti* Medic. Oecologia 85:389–395

Koide RT (2000) Functional complementarity in the arbuscular mycorrhizal symbiosis. New Phytol 147:233–235

Koide RT, Kabir Z (2000) Extraradical hyphae of the mycorrhizal fungus *Glomus intraradices* can hydrolyse organic phosphate. New Phytol 148:511–517

Koide RT, Li M, Lewis J, Irby C (1988) Role of mycorrhizal infection in the growth and reproduction of wild vs. cultivated plants. 1. Wild vs. cultivated oats. Oecologia 77:537–542

Koide RT, Dickie IA, Goff MD (1999) Phosphorus deficiency, plant growth and the phosphorus efficiency index. Funct Ecol 13:733–736

Koide RT, Goff MD, Dickie IA (2000) Component growth efficiencies of mycorrhizal and nonmycorrhizal plants. New Phytol 148:163–168

Kope HH, Warcup JH (1986) Synthesised ectomycorrhizal associations of some Australian herbs and shrubs. New Phytol 104:591–599

Laiho O (1965) Further studies on the ectendotrophic mycorrhiza. Acta Forestalia Fennica 79:1–56

Lambers H, Poorter H (1992) Inherent variation in growth rate between higher plants: a search for physiological causes and ecological consequences. Adv Ecol Res 23:188–216

Leake JR (1994) The biology of myco-heterotrophic ('saprophytic') plants. New Phytol 127:171–216

Leake JR, Read DJ (1997). Mycorrhizal fungi in terrestrial habitats. In: Wicklow DT, Söderström BE (eds) The Mycota, vol IV. Environmental and microbial relationships, Springer, Berlin Heidelberg New York, pp 281–301

Leyval C, Berthelin J (1986) Comparison between the utilization of phosphorus from insoluble mineral phosphates by ectomycorrhizal fungi and rhizobacteria. In: Gianinazzi-Pearson V, Gianinazzi S (eds) Physiological and genetical aspects of mycorrhiza. INRA, Dijon, France, pp 345–349

Leyval C, Turnau K, Haselwandter K (1997) Effect of heavy metal pollution on mycorrhizal colonization and function: physiological and applied aspects. Mycorrhiza 7:139–153

Li CY, Massicotte HB, Moore LVH (1992) Nitrogen-fixing *Bacillus* sp. associated with Douglas-fir tuberculate ectomycorrhizae. Plant Soil 140:35–40

Linderman RG (1994) Role of VAM fungi in biocontrol. In: Pfleger FL, Linderman RG (ed) Mycorrhizae and plant health. APS Press, St Paul, Minnesota, pp 1–25

Liu G, Chambers SM, Cairney JWG (1998) Molecular diversity of ericoid mycorrhizal endophytes isolated from *Woollsia pungens* (Cav.) F. Muell. (Epacridaceae). New Phytol 140:145–154

Mäder P, Vierheilig H, Streitwolf-Engel R, Boller T, Frey B, Christie P, Wiemkin A (2001) Transport of ^{15}N from a soil compartment separated by a polytetrafluoroethylene membrane to plant roots via the hyphae of arbuscular mycorrhizal fungi. New Phytol 146:155–161

Manjunath A, Habte M (1991) Root morphological characteristics of host species having distinct mycorrhizal dependency. Can J Bot 69:671–676

Marler MJ, Zabinski CA, Callaway RM (1999) Mycorrhizae indirectly enhance competitive effects of an invasive forb on native bunchgrass. Ecology 80:1180–1186

Marschner H (1995) Mineral nutrition of plants, 2nd edn. Academic Press, London

Martin F (2001) Frontiers in molecular mycorrhizal research – genes, loci, dots and spins. New Phytol 150:499–505

Martin FM, Hilbert JL (1991) Morphological, biochemical and molecular changes during ectomycorrhizal development. Experientia 47:321–331

McGee PA (1996) The Australian zygomycetous mycorrhizal fungi: the genus *Densospora* gen. nov. Aust Syst Bot 9:329–336

McKendrick SL, Leake JR, Read DJ (2000) Symbiotic germination and development of myco-heterotrophic plants in nature: transfer of carbon from ectomycorrhizal *Salix repens* and *Betula pendula* to the orchid *Corallorhiza trifida* through shared hyphal connections. New Phytol 145:539–548

Meyer FH (1973) Distribution of ectomycorrhizae in native and man-made forests. In: Marks GC, Kozlowski TT (ed) Ectomycorrhizae. Academic Press, New York, pp 79–106

Mikola P (1965) Studies in ectendotrophic mycorrhiza of pine. Acta For Fenn 79:1–56

Miller SL, Allen EB (1992) In: Allen MF (ed) Mycorrhizal functioning: an integrative plant-fungal process. Chapman and Hall, London, pp 301–332

Molina R, Massicotte H, Trappe JM (1992) Specificity phenomena in mycorrhizal symbiosis: community-ecological consequences and practical implications. In: Allen MF (ed) Mycorrhizal functioning: an integrative plant-fungal process. Chapman and Hall, London, pp 357–423

Newman EI (1988) Mycorrhizal links between plants: their functioning and ecological significance. Adv Ecol Res 18:243–270

Newman EI, Reddell P (1987) The distribution of mycorrhizas among families of vascular plants. New Phytol 106:745–751

Nurmiaho-Lassila E-L, Timonen S, Haahtela K, Sen R (1997) Bacterial colonisation patterns of intact Scots pine mycorrhizospheres in dry pine forest soil. Can J Microbiol 43:1017–1035

Nye PH, Tinker PBH (1977) Solute movement in the soil-root system. Blackwell Scientific Publications, Oxford

O'Connor PJ, Smith SE, Smith FA (2002) Arbuscular mycorrhizas influence plant diversity and community structure in a semiarid herbland. New Phytologist 154:209–218

Odum EP (1969) The strategy of ecosystem development. Science 146:262:270

Olsson PA, Jakobsen I, Wallander H (2002) Foraging and resource allocation strategies of mycorrhizal fungi in a patchy environment. In: van der Heijden MGA, Sanders IR (eds) Mycorrhizal ecology. Springer, Berlin Heidelberg New York pp 93–115

Pedersen CT, Sylvia DM, Shilling DG (1999) *Pisolithus arhizus* ectomycorrhiza affects plant competition for phosphorus between *Pinus elliottii* and *Panicum chamaelonche*. Mycorrhiza 9:199–204

Perez-Moreno J, Read DJ (2000) Mobilization and transfer of nutrients from litter to tree seedlings via the vegetative mycelium of ectomycorrhizal plants. New Phytol 145:301–309

Perotto S, Bonfante P (1997) Bacterial associations with mycorrhizal fungi: close and distant friends in the rhizosphere. Trends Microbiol 5:496–501

Perry DA, Margolis H, Choquette C, Molina R, Trappe JM (1989) Ectomycorrhizal mediation of competition between coniferous tree species. New Phytol 112:501–511

Peterson RL, Currah RS (1990) Synthesis of mycorrhizae between protocorms of *Goodyera repens* (Orchidaceae) and *Ceratobasidium cereale*. Can J Bot 68:1117–1125

Pfeffer PE, Bago B, Schachar-Hill Y (2001) Exploring mycorrhizal function with NMR spectroscopy. New Phytol 150:543–553

Pfleger FL, Stewart EL, Noyd RK (1994) Role of VAM fungi in mine land revegetation. In: Pfleger FL, Linderman RG (ed) Mycorrhizae and plant health. APS Press, St Paul, Minnesota, pp 47–81

Read DJ (1984) The structure and function of the vegetative mycelium of mycorrhizal roots. In: Jennings DH, Rayner ADM (eds) The ecology and physiology of the fungal mycelium. Cambridge University Press, Cambridge, pp 215–240

Read DJ (1991) Mycorrhizas in ecosystems. Experientia 47:376–391

Read DJ (1992) The mycorrhizal mycelium. In: Allen MF (ed) Mycorrhizal functioning: an integrative plant-fungal process. Chapman and Hall, London, pp 102–133

Read DJ (1993) Mycorrhiza in plant communities. Adv Plant Path 9:1–31

Read DJ (2000) Links between genetic and functional diversity – a bridge too far? New Phytol 145:363–365

Reeves B (1988) Mineral nutrition, mycorrhizal fungi and succession in semiarid environments. In: Ng FSP (ed) Trees and mycorrhiza. The Asian Seminar, Kuala Lumpur, Malaysia, pp 33–50

Robinson D (1994) The responses of plants to non-uniform supplies of nutrients. New Phytol 127:635–674

Robinson D, Fitter AH (1999) The magnitude and control of carbon transfer between plants linked by a common mycorrhizal network. J Exp Bot 50:9–13

Sanders IR, Koide RT, Shumway DL (1999) Diversity and structure in natural communities. In: Varma A, Hock B (eds) Mycorrhiza. Structure, function, molecular biology and biotechnology. Springer, Berlin Heidelberg New York

Sarand L, Timonen S, Rajmaki M, Petola R, Nurmiaho-Lassila E-L, Koivula T, Yrjala K, Haahtela K, Romantschuk M, Sen R (1998) Microbial biofilms and catabolic plasmid harbouring degradative fluorescent pseudomonads in the Scots pine mycorrhizosphere developed on petroleum contaminated soil. FEMS Microbiol Ecol 27:115–126
Scales P, Peterson RL (1991a) Structure and development of *Pinus banksiana-Wilcoxina* ectendomycorrhizae. Can J Bot 69:2135–2148
Scales P, Peterson RL (1991b) Structure of ectomycorrhizae formed by *Wilcoxina mikolae* var. *mikolae* with *Picea mariana* and *Betula alleghaniensis*. Can J Bot 69:2149–2157
Schachtman DP, Reid RJ, Ayling SM (1998) Phosphorus uptake by plants: from soil to cell. Plant Physiol 116:447–453
Schadt CW, Mullen RB, Schmidt SK (2001) Isolation and phylogenetic identification of a dark-septate fungus associated with the alpine plant *Ranunculus adoneus*. New Phytol 150:747–755
Schüßler A, Schwartzott D, Walker C (2001) A new fungal phylum, the *Glomeromycota*: phylogeny and evolution. Mycol Res 105:1413–1421
Sen R, Hietala A, Zelmer A (1999) Common anastomosis and ITS-RFLP groupings among binuculate *Rhizoctonia* isolates representing root endophytes of *Pinus sylvestris*, *Ceratorhiza* spp from orchid mycorrhizas and a phytopathogenic anastomosis group. New Phytol 144:331–341
Sharples J, Chambers SM, Meharg A, Cairney JWG (2000). Genetic diversity of root-associated fungal endophytes of *Calluna vulgaris* at contrasting field sites. New Phytol 148:153–162
Silberbush M, Barber SA (1984) Phosphorus and potassium uptake of field-grown soybean cultivars predicted by a simulation model. Soil Sci Soc Am J 48:592–596
Simard SW, Jones MD, Durall DM, Perry DA, Myrold DD, Molina R (1997a) Reciprocal transfer of carbon isotopes between ectomycorrhizal *Betula papyrifera* and *Pseudotsuga menziesii*. New Phytol 137:529–542
Simard SW, Perry DA, Jones MD, Myrold DD, Durall DM, Molina R (1997b) Net transfer of carbon between ectomycorrhizal tree species in the field. Nature 388:579–582
Smith FA (2000) Measuring the influence of mycorrhizas. New Phytol 148:4–6
Smith SE (1980) Mycorrhizas of autotrophic higher plants. Biol Rev Cambridge Philos Soc 55:475–510
Smith FA, Smith SE (1996a) Mutualism and parasitism: diversity in function and structure in the "arbuscular" (VA) mycorrhizal symbiosis. Adv Bot Res 22:1–43
Smith SE, Smith FA (1996b) Membranes in mycorrhizal interfaces: specialized functions in symbiosis. In: Smallwood M, Knox JP, Bowles DJ (eds) Membranes: specialized functions in plants. Bios Scientific Publishers, Oxford, pp 525–542
Smith FA, Smith SE (1997) Structural diversity in (vesicular)-arbuscular mycorrhizal fungi. New Phytol 137:373–388
Smith SE, Read DJ (1997) Mycorrhizal symbiosis. Academic Press, London
Smith FA, Jakobsen I, Smith SE (2000) Spatial differences in acquisition of soil phosphate between the arbuscular mycorrhizal fungi *Scutellospora calospora* and *Glomus caledonium* in symbiosis with *Medicago truncatula*. New Phytol 147:357–366
Soberon MJ, Martinez del Rio C (1985) Cheating and taking advantage in mutualistic symbioses. In: Boucher D (ed) The biology of mutualism. Croom Helm, London, pp 192–216
Söderström B (1992) Ecological potential of ectomycorrhizal mycelium. In: Read DJ, Lewis DH, Fitter AH, Alexander IJ (eds) Mycorrhizas in ecosystems. Cambridge University Press, Cambridge, pp 77–83
St. John TV, Coleman DC (1983) The role of mycorrhizas in plant ecology. Can J Bot 61:1005–1015

Strzelczyk E, Pokjsa-Burdziej A (1984) Production of auxins and gibberellin-like substances by mycorrhizal fungi, bacteria and actinomycetes isolated from soil and the mycorrhizosphere of pine (*Pinus sylvestris* L.). Plant Soil 81:185–194

Thomson BD, Robson AD, Abbott LK (1986) Effects of phosphorus on the formation of mycorrhizas by *Gigaspora calospora* and *Glomus fasciculatum* in relation to root carbohydrates. New Phytol 103:751–765

Tilman D (1988) Plant strategies and the dynamics and structure of plant communities. Princeton University Press, Princeton, New Jersey

Timonen S, Tammi H, Sen R (1997) Outcome of interactions between two *Suillus* spp. and different *Pinus sylvestris* genotype combinations: identity and distribution of ectomycorrhiza and effects of early seedling growth in N-limited nursery soil. New Phytol 137:691–702

Timonen S, Jørgensen KS, Haahtela K, Sen R (1998) Bacterial community structure of Scots pine-*Suillus bovinus* and *Paxillus involutus* mycorrhizospheres in dry pine forest soil and nursery peat. Can J Microbiol 44:499–513

Tommerup IC, Abbott LK (1981) Prolonged survival and viability of V.A. mycorrhizal hyphae after root death. Soil Biol Biochem 13:431–433

Trappe JM (1987) Phylogenetic and ecologic aspects of mycotrophy in the angiosperms from and evolutionary standpoint. In: Safir GR (ed) Ecophysiology of VA mycorrhizal plants. CRC Press, Boca Raton, pp 5–25

Treseder KK, Allen MF (2000) Mycorrhizal fungi have a potential role in soil carbon storage under elevated CO_2 and nitrogen deposition New Phytol 147:189–200

van der Heijden MGA, Sanders IR (eds) (2002) Mycorrhizal ecology. Springer, Berlin Heidelberg New York

van der Heijden MGA, Boller T, Wiemken A, Sanders IR (1998a) Different arbuscular mycorrhizal fungal species are potential determinants of plant community structure. Ecology 79:2082–2091

van der Heijden MGA, Klironomos JN, Ursic M, Moutoglis P, Streitwolf-Engel R, Boller T, Wiemken A, Sanders IR (1998b) Mycorrhizal fungal diversity determines plant biodiversity, ecosystem variability and productivity. Nature 396:69–72

Varese GC, Portinaro S, Trotta A, Scannerini S, Luppi-Mosca AM, Martinotti G (1996) Bacteria associated with *Suillus grevillei* sporocarps and ectomycorrhizae and their effects on in vitro growth of the mycobiont. Symbiosis 21:129–147

Varma A, Hock B (eds) (1999) Mycorrhiza. Structure, function, molecular biology and biotechnology. Springer, Berlin Heidelberg New York

Varma A, Verma S, Nirmal, Sahay S, Butehorn B, Franken P (1999) *Piriformospora indica*, a cultivable plant-growth-promoting root endophyte. Appl Environ Microbiol 65:2741–2744

Warcup JH (1980) Ectomycorrhizal associations of Australian indigenous plants. New Phytol 85:531–535

Warcup JH (1985a) *Rhizanthella gardneri* (Orchidaceae), its *Rhizoctonia* endophyte and close association with *Melaluca uncinata* (Myrtaceae) in western Australia. New Phytol 99:273–280

Warcup JH (1985b) Ectomycorrhiza formed by *Glomus tubiforme*. New Phytol 99:267–272

Warcup JH (1990) Taxonomy and culture and mycorrhizal associations of some zygosporic Endogonaceae. Mycol Res 94:173–178

Wardle DA (1999) Is 'sampling effect' a problem for experiments investigating biodiversity-ecosystem function relationships? Oikos 87:403–407

Watkins NK, Fitter AH, Graves JD, Robinson D (1996) Carbon transfer between C_3 and C_4 plants linked by a common mycorrhizal network, quantified using stable carbon isotopes. Soil Biol Biochem 28:471–477

West H (1996) Influence of arbuscular mycorrhizal infection on competition between *Holcus lanatus* and *Dactylis glomerata*. J Ecol 84:429–438

Whittingham J, Read DJ (1982) Vesicular-arbuscular mycorrhiza in natural vegetation systems. III. Nutrient transfer between plants with mycorrhizal interconnections. New Phytol 90:277–284

Widden P (1997) Competition and the fungal community. In: Wicklow DT, Soderstrom BE (eds) Environmental and microbial relationships. The Mycota, vol IV. Springer, Berlin Heidelberg New York, pp 135–147

Wilkinson KG, Dixon KW, Sivasithamparam K, Ghisalberti EL (1994) Effects of IAA on symbiotic germination of an Australian orchid and its production by orchid-associated bacteria. Plant Soil 159:291–295

Wong KKY, Montpetit D, Piché Y, Lei J (1990) Root colonization by four closely related genotypes of the ectomycorrhizal basidiomycete *Laccaria bicolor* (Maire) Orton – comparative studies using the electron microscope. New Phytol 116:669–679

Zhu YG, Cavagnaro TR, Smith SE, Dickson S (2001) Backseat driving? Most plants depend on arbuscular mycorrhizal fungi to access phosphate beyond the rhizosphere depletion zone. Trends Plant Sci 6:194–195

12 Signalling in Rhizobacteria–Plant Interactions

L.C. van Loon and P.A.H.M. Bakker

12.1 Introduction

Bacteria are by far the most abundant organisms in soil and they play a key role in nutrient cycling and soil fertility. The rhizosphere – the zone of 1–2 mm around plant roots – is rich in nutrients and provides niches different from those in bulk soil for bacteria to thrive. Microbial diversity in the soil and in the rhizosphere is huge. Multiple interactions occur between the bacteria and between bacteria and other microorganisms, involving competition, antibiosis, parasitism and predation. Various interactions also occur between bacteria and plant roots that can be beneficial, neutral or harmful to the plant. Deleterious effects comprise phytotoxic and pathogenic activities of the bacteria. Conversely, plants profit from bacterially induced growth promotion and protection against pathogens. Growth promotion can be the result of bacterial activities that increase the availability of water and mineral nutrients, as well as of symbiotic relationships such as the formation of root nodules in leguminous plants, in which atmospheric nitrogen is made available in reduced form. Nodulation of roots involves an intricate interplay of molecular signals between the bacterium and its host, and illustrates how such plant–rhizobacteria interactions proceed in an exquisitely controlled manner. In similar ways, suppression of disease-provoking microorganisms can occur through microbial antagonism in the rhizosphere as well as by specific interactions between the protective bacterium and its host. While antagonism involves mostly mechanisms that rhizobacteria likewise use to compete with other microorganisms in the root environment, interactions with plant roots may trigger an induced systemic resistance that enhances the defensive capacity of the plant to subsequent pathogen attack. In this way, the plant becomes better protected not only against soil-borne pathogenic fungi, but also to necrotrophic foliar pathogens. The chapter provides an overview of these beneficial relationships between rhizobacteria and their plant hosts with emphasis on the communicative signals that are involved in regulating the activities of both partners that lead to plant growth promotion and disease suppression.

Ecological Studies, Vol. 168
H. de Kroon, E.J.W. Visser (Eds.)
Root Ecology

12.2 Plant Growth Promotion by Rhizobacteria

Substantially more microorganisms are present near plant root surfaces than in bulk soil. This "rhizosphere effect" is caused by the release of exudates from growing root tissues and the lysis of cells of older root parts (Lynch and Whipps 1991). Bacteria rapidly colonize growing root tips, using simple sugars, organic acids and amino acids as nutrients, whereas saprophytic fungi are more prevalent on older root parts, where cortical cells are being degraded. Numerous strains of bacteria can be isolated from plant roots with, in most cases, little specificity being apparent. However, release of selected nutrients from roots that are preferentially utilizable by specific bacterial strains favors selective colonization by the latter (Bowen 1991; Flores et al. 1999).

Root-colonizing bacteria are commonly referred to as "rhizobacteria". Most rhizobacteria remain confined to the root surface (rhizoplan), but some enter the root interior and behave as endophytes (Sturz et al. 2000). Several rhizobacterial strains have been found to increase plant growth after inoculation on to seeds and are therefore called "plant growth-promoting rhizobacteria" (PGPR*; Kloepper et al. 1980b). Such PGPR improved plant stand and increased yield, e.g. of potato (*Solanum tuberosum*), radish (*Raphanus sativus*), and sugar beet (*Beta vulgaris*; Kloepper et al. 1991), suggesting that plants benefit from rhizobacteria that live on the nutrients lost from the roots. The mechanisms of growth promotion by these PGPR are complex and appear to comprise both changes in the microbial balance in the rhizosphere and alterations in host plant physiology (Glick et al. 1999). By competition and production of antimicrobial compounds, PGPR can reduce populations of plant pathogens and deleterious rhizobacteria, which restrict plant growth. Some of these disease-suppressing activities, such as production of HCN, can reduce plant growth as well, but more often the net effect is improved plant development, resulting in more vigorous growth and increased yield of agricultural crops (Dowling and O'Gara 1994).

Growth promotion through direct stimulation of plant development is more difficult to demonstrate. In radish several rhizobacterial strains strongly increased average plant weight under non-sterile conditions, but failed to do so in a gnotobiotic system in which bacteria were introduced into sterilized soil (Kloepper and Schroth 1981). However, *Pseudomonas flu-*

---

* Abbreviations: ACC: 1-aminocyclopropane-1-carboxylic acid; DAPG: 2,4-diacetylphloroglucinol; ENOD: early nodulin; EPS: extracellular polysaccharide; For: Fusarium oxysporum f. sp. raphani; IAA: indole-3-acetic acid; ISR: induced systemic resistance; JA: jasmonic acid; LPS: lipopolysaccharide; MeJA: methyl jasmonate; NPR1: non-expressor of pathogenesis-related proteins 1; OA: O-antigenic side chain; PCA: phenazine-1-carboxylic acid; PGPR: plant growth-promoting rhizobacteria; PRs: pathogenesis-related proteins; Pst: Pseudomonas syringae pv. tomato; SA: salicylic acid; SAR: systemic acquired resistance; VA: vesicular-arbuscular.

orescens strain WCS374 did increase radish leaf dry weight, but not tuber yield, in gnotobiotic culture, whereas, in non-sterile soil, the effect on leaf weight was non-significant and tuber fresh weight was increased (Table 12.1). During in vitro propagation of statice (*Limonium sinuatum*) the presence of an endophytic *Flavobacterium* sp. promoted growth and rooting (Van Zaayen et al. 1992). Such beneficial effects of microbial inoculants in in vitro cultures of plant tissue explants have been noted also in other species. For instance, in potato, bacterization increased stem length, shoot biomass and root biomass (Bensalim et al. 1998). In vitro culture of tomato (*Lycopersicon esculentum*) seedlings with the PGPR *Pseudomonas* sp. strain PsJN promoted shoot dry weight and increased resistance of transplants to verticillium wilt (Pillay and Nowak 1997; Sharma and Nowak 1998). Prevention of excessive moisture content and water soaking in oregano (*Origanum vulgare*) shoot cultures was sustained through multiple subcultures by selected polysaccharide-producing soil bacteria without re-inoculation (Ueno and Shetty 1998). These in vitro responses caused by the inoculants are referred to as "biotization" (Nowak 1998), and demonstrate that rhizobacteria can directly influence plant growth as well as enhance their tolerance to abiotic and biotic stresses.

The ways in which PGPR directly promote plant growth are not known with any certainty. When plants are grown in culture solution under gnotobiotic conditions, bacteria influence the uptake of ions. Under some conditions ion uptake by plant roots can be stimulated in the presence of bacteria, probably through the PGPR providing chelating agents or compounds promoting active ion transport. However, under other conditions, bacteria can be inhibitory, either by competing for nutrients or by producing phytotoxic compounds (Lynch 1982).

Most microorganisms produce siderophores when iron availability in the environment is low. These are low-molecular-weight metabolites with a high affinity for Fe^{3+} (Höfte 1993). They chelate Fe^{3+} from the environment and transport the iron into the microbial cells after being recognized by a specific siderophore receptor protein (Neilands 1981; De Weger et al. 1986; Leong 1986). Low availability of iron in soil for microorganisms is mainly due to the low solubility of ferric oxyhydroxy polymers. In well-oxidized soils the solubility of iron is largely controlled by $Fe(OH)_3$ (Lindsay and Schwab 1982). The solubility constant of this compound is extremely low ($K_{sol}=10^{-38}$), resulting in a concentration of 1.4×10^{-9} M Fe^{3+} at pH 7 or even lower in the presence of phosphate, whereas a concentration of 10^{-6} M is needed to support microbial growth (Neilands et al. 1987; Chipperfield and Ratledge 2000). Thus, the production of siderophores by microorganisms in slightly acidic, neutral and alkaline soils has to be considered a common phenomenon. The presence of microbially produced siderophores has indeed been demonstrated in a variety of soils (Powell et al. 1980; Akers 1981). Recently, the use of a reporter gene system has allowed monitoring of iron availability for microorganisms in the

Table 12.1. Effect of treatment with a rhizobacterial strain on growth of radish under gnotobiotic and non-sterile conditions. (After M. Leeman, P.A.H.M. Bakker and B. Schippers, unpubl.)

Treatment (gnotobiotic)	Leaf dry weight (mg)	Tuber dry weight (mg)
Control	4.2 a	2.2 a
Pseudomonas fluorescens WCS374	5.7 b	2.3 a
Treatment (non-sterile)	Leaf fresh weight (g)	Tuber fresh weight (g)
Control	10.8 a	8.0 a
Pseudomonas fluorescens WCS374	12.1 a	10.2 b

rhizosphere, identifying situations that are conducive to the production of siderophores in this environment (Loper and Lindow 1994; Loper and Henkels 1997; Duijff et al. 1999).

The influence of microbial siderophores on plant iron nutrition depends on the ferric-chelating properties of the siderophores, as well as on the iron acquisition mechanism of the plant. Becker et al. (1985a) demonstrated negative effects of 10 μM of the pyoverdin siderophore of *Pseudomonas* sp. strain B10 on iron nutrition in pea (*Pisum sativum*). In contrast, the catechol siderophore of *Agrobacterium tumefaciens* stimulated chlorophyll synthesis in pea (Becker et al. 1985b). Pseudobactin 358, the pyoverdin siderophore of *Pseudomonas putida* strain WCS358, increased iron uptake and stimulated chlorophyll synthesis in barley (*Hordeum vulgare*; Duijff et al. 1994b), but had differential effects in the carnation (*Dianthus caryophyllus*) cultivars Lena and Pallas (Duijff et al. 1994a). The latter difference was attributed to iron-deficient plants of cv. Lena producing more and longer root hairs than iron-deficient plants of cv. Pallas, and the ferric-reducing activity of cv. Lena being higher than that of cv. Pallas.

Some bacteria solubilize organic phosphate by secreting phosphatase or inorganic phosphate from soil particles by releasing organic acids, and this could make phosphorus as well as micronutrients more readily available for plant growth in some soils (Kloepper et al. 1989). Free-living nitrogen-fixing bacteria, particularly of the genera *Azobacterium*, *Azospirillum* and *Clostridium*, are present in most soils and in plant rhizospheres. Also, some *Pseudomonas* spp. have the ability to fix nitrogen. However, it has been suggested that the contribution of bacterially fixed nitrogen to plants is minimal and that enhanced growth by an inoculated plant does not necessarily mean that the bacteria associated with the roots do fix nitrogen or pass the products of nitrogen fixation to the plant (James and Olivares 1997). Inoculation of young wheat (*Triticum aestivum*) plants with *Serratia rubidea* increased efflux of carbon compounds from roots and promoted nitrogen uptake and dry matter yield (Merbach and Ruppel 1992). However, it is not clear whether this was due to a direct effect on nitrogen uptake or the result of other physiological changes in the plant caused by root bacterization.

By contrast, there is evidence linking nitrogen-reducing endophytes to biological nitrogen fixation in rice (*Oryza sativa*), sugar cane (*Saccharum officinale*) and sorghum (*Sorghum bicolor*; Reinhold-Hurek and Hurek 1998). Moreover, plant growth can be increased by dual inoculation with *Azospirillum* and phosphate-solubilizing bacteria. Combined inoculation of *A. brasilense* and the phosphate-solubilizing bacteria *Pseudomonas striata* or *Bacillus polymixa* significantly increased nitrogen and phosphorus content as well as grain yield of sorghum (Alagawadi and Gaur 1992). Similar increases in plant growth have been reported as a result of co-inoculation of diazotrophic PGPR with vesicular arbuscular (VA) mycorrhizas (Toro et al. 1998). Interactions between mycorrhizal fungi and rhizosphere bacteria relating to plant growth promotion are discussed more fully by Smith et al. elsewhere in this volume (Chap. 11).

Many bacteria have the ability to produce auxins, gibberellins, cytokinins and ethylene (Frankenberger and Arshad 1995). It has often been inferred that rhizobacterially produced auxins are responsible for growth promotion. However, compared with stem elongation, root growth is only slightly stimulated by auxin, and generally only at concentrations of 10^{-9} to 10^{-11} M, higher concentrations being strongly inhibitory (Thimann 1937). Indoleacetic acid (IAA) promotes ethylene production by stimulating the rate-limiting enzyme in the ethylene biosynthetic pathway, 1-aminocyclopropane-1-carboxylic acid (ACC) synthase (Kende 1993), and the ethylene thus formed inhibits root elongation. If the auxin concentration reached in the root after uptake of bacterially produced IAA does not fall within the limits indicated above, no root growth promotion can be expected. Auxin is transported basipetally towards the root tip, but some might enter the phloem and be transported to the shoot. However, concentrations required for shoot growth are unlikely to be reached. On the other hand, auxin at 10^{-4} to 10^{-6} M promotes lateral root formation, and it cannot be excluded that locally bacterial microcolonies can produce auxin in amounts that would stimulate this process and, thereby, contribute to enhanced uptake of water and nutrients. For example, mutant strains of the PGPR *Azospirillum brasilense* that synthesized very low amounts of IAA compared with the wild-type strain no longer promoted the formation of lateral roots of wheat seedlings (Barbieri et al. 1986; Barbieri and Galli 1993). A mutant strain of the PGPR *P. fluorescens* strain BsP53a, that overproduced IAA, stimulated root development of blackcurrant cv. Shirjaevskaja softwood cuttings, but inhibited that of sour cherry cv. Vladimirskaja (Dubeikovsky et al. 1993). In potato plantlets grown in vitro, strain PsJN increased cytokinin content by inducing synthesis in the early stages of plant growth and development (Lazarovits and Nowak 1997). Thus, it appears that rhizobacteria also affect hormone metabolism and reactivity within the plant itself.

Interestingly, growth promotion was linked recently not to the production of stimulatory hormones, but to reduction of the inhibitory hormone ethylene. Ethylene has been identified as a common component of the soil atmosphere and under certain conditions has been shown to reach concentrations high enough to influence plant growth and development (Smith 1976; Frankenberger and Arshad 1995). The PGPR *P. putida* strain GR12-2 was mutagenized to select for variants that were unable to utilize the ethylene precursor ACC as a sole nitrogen source. These mutants proved to be devoid of the ACC-deaminase activity that is present in wild-type GR12-2 cells. They had also lost the ability to promote root elongation of developing canola (*Brassica campestris*) seedlings under gnotobiotic conditions (Glick et al. 1994), and no longer promoted shoot growth of seedlings planted in soil (Glick et al. 1997).

Conversely, transforming *Escherichia coli* or *Pseudomonas* spp. strains with a cloned ACC deaminase gene enabled the bacteria to grow on ACC as a sole source of nitrogen and to promote the elongation of seedling roots (Shah et al. 1998). These results were interpreted in terms of a model in which the bacterial strains promote root elongation by binding to germinating seeds or developing roots and hydrolyzing ACC leaking from the plant tissues through deamination to ammonia and α-ketobutyrate (Fig. 12.1). By

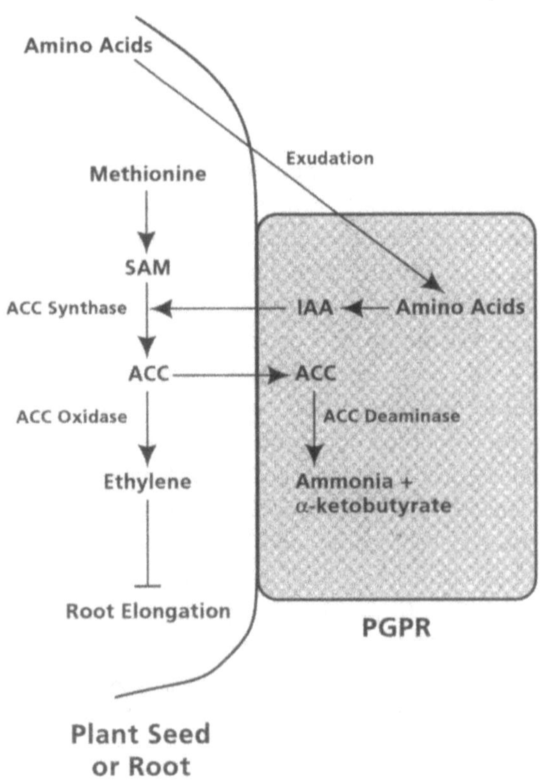

Fig. 12.1. Mechanism of the promotion of plant root elongation by rhizobacteria that possess ACC deaminase. *ACC* 1-Aminocyclopropane-1-carboxylate; *IAA* indole-3-acetic acid; *SAM* S-adenosylmethionine (adapted from Glick et al. 1999)

reducing the level of unbound ACC, re-uptake would be lowered, less ethylene would be produced and, consequently, roots grow longer (Glick et al. 1994, 1998). The ability to utilize ACC as a sole nitrogen source appeared to be limited to soil bacteria that are capable of stimulating plant growth (Glick et al. 1995), linking ACC-deaminating activity to growth promotion. Canola, lettuce (*Lactuca sativa*), tomato and wheat all responded with increased root length when seeds were treated with wild-type GR12-2 or with the chemical inhibitor of ethylene synthesis aminoethoxyvinylglycine, but not with the mutant strain. However, barley and oats (*Avena sativa*) did not respond to wild-type GR12-2, suggesting that promotion of plant growth by mechanisms that include hydrolyzing ACC could be limited largely to dicots (Hall et al. 1996).

12.3 Rhizobium–Plant Interactions

Whether bacteria promote growth through production of plant hormones or through modulating hormone metabolism of the plant, it is clear that signalling between the bacteria and plant roots is of central importance. Apparently, upon colonization of the roots, bacteria start to produce signal molecules that are perceived and transduced within the cells of the plant, leading to a response resulting in increased growth of the whole plant. The *Rhizobium*-legume symbiosis is a special case in which the specific interaction between a rhizobacterium and a leguminous host leads to the formation of nitrogen-fixing root nodules as a result of a "two-way molecular conversation". This interaction can serve as a paradigm of how rhizobacteria and root cells can influence each other's activities.

The interaction of soil bacteria of the genera *Azorhizobium*, *Bradyrhizobium*, *Mesorhizobium* and *Rhizobium* (collectively referred to as "rhizobia") and legumes starts with attachment of the bacteria to developing root hairs through lectin binding (Gray et al. 1992; Kijne et al. 1992; Noel 1992). The root hairs then deform and curl at the tip, and the bacteria invade the root by a newly formed infection thread, which grows through the root hairs into the cortex. Simultaneously, cortical cells are mitotically activated, giving rise to the nodule primordium. Infection threads grow towards the primordium and the bacteria, surrounded by a plant-derived, enclosing peribacteroid membrane, are released into the cytoplasm of the host cells. The nodule primordium then develops into the nodule, while the bacteria differentiate into their endosymbiontic form, the bacteroids, able to fix gaseous nitrogen into ammonia through the action of the nitrogenase enzyme complex. Carbon skeletons provided by the plant are converted into amino acids and amides, which can be utilized by the plant for growth on nitrogen-poor soils (Heidstra and Bisseling 1996).

Extensive signalling occurs in each of the steps leading to effective nodule formation. The initial interaction between the bacterium and its host is triggered by the perception by the bacterium of certain (iso)flavonoids that are secreted from the plant roots. Historically, the presence of flavonoid compounds in legumes has been associated most closely with pathogenic attack. Stimulation of isoflavonoid biosynthesis in plants is a common feature of their response to pathogens, irrespective of whether those are bacteria, fungi, viruses or nematodes. The general antimicrobial activity of these isoflavonoids appears to contribute to resistance to various infecting microorganisms and parasites (Dakora and Phillips 1996). Rhizobia are mutualistic symbionts, but the early events, such as root hair deformation and curling, infection thread formation and cortical cell division, suggest that these organisms evolved from a pathogenic ancestor (Djordjevic et al. 1987). In fact, plant defense reactions are evident in cases where the final stage of the symbiosis (e.g. nitrogen fixation) is genetically blocked in the microsymbiont (Parniske et al. 1991). Symbiotic VA mycorrhizas and rhizobia all enhance isoflavonoid production or exudation in their host legume by producing β-glucan elicitors of the types that induce defense reactions in plant cells in response to e.g. pathogenic fungi (Dakora and Phillips 1996).

Host specificity is a prominent aspect of root nodule formation (Long 1996; Broughton and Perret 1999). Most rhizobia have a narrow host range and only form nodules on a very limited set of legume species. (Iso)flavonoids are recognized in a rhizobacterial strain-specific manner by binding to the constitutively expressed bacterial nodulation protein NodD (Fig. 12.2), a product specified by the large *Sym* plasmid on which the genes for bacterial nodulation (*nod* genes) and nitrogen fixation (*nif* genes) are located. Binding to specific flavonoids turns NodD into a transcriptional activator of other *nod* genes that encode proteins involved in the synthesis of specific lipo-oligosaccharides (Nod factors) that, in turn, are recognized by host legumes (Fig. 12.2). Nod factors all have a chitin β-1,4-linked *N*-acetyl-D-glucosamine backbone, varying in length between three and six sugar units, and a fatty acyl chain on the C-2 position of the non-reducing sugar (Dénarié and Roche 1992). Host-specific *nod* genes in different rhizobial species are involved in the modification of the fatty acyl chain or the addition of strain-specific substitutions that confer host specificity. For instance, *R. meliloti* produces Nod factors that contain a sulfate group at the C-6 position of the glucosamine residue at the reducing end (Fig. 12.2). When this sulfate group is absent due to a mutation in the bacterium, the modified factors no longer allow nodulation of the normal host, alfalfa (*Medicago sativa*). Instead, the mutant bacteria have acquired the ability to nodulate common vetch (*Vicia sativa*) that is host to *R. leguminosarum* bv. *viciae*, a strain that itself produces Nod factors without a sulfate group substitution (Geurts and Franssen 1996; Heidstra and Bisseling 1996). It has been postulated that another level of recognition may exist, because Nod factors can be hydrolyzed by specific chitinases of plant or microbial ori-

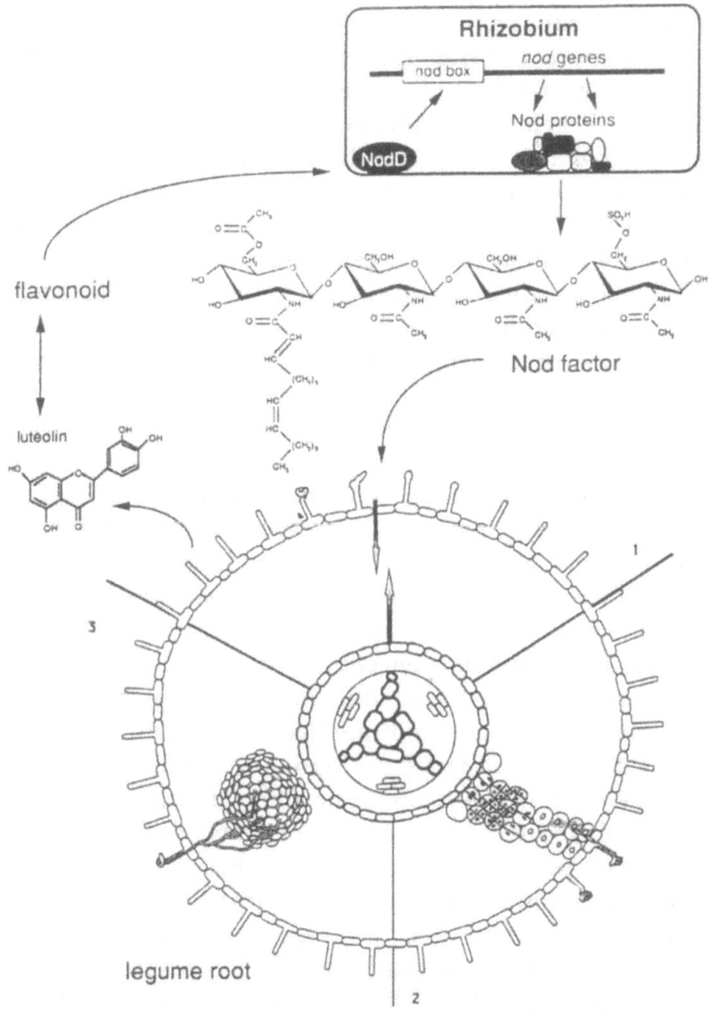

Fig. 12.2. Symbiotic interaction between rhizobia and leguminous plants. The plant exudes flavonoids, such as luteolin, that activate the NodD protein in the bacterium, leading to the production and secretion of Nod factors. The Nod factors induce root hair deformation (stage 1); rhizobia initiate infection and cortical cells start dividing (stage 2); the infection thread grows towards the developing nodule primordium where cells get infected (stage 3). Reproduced with permission from K. van de Sande and T. Bisseling, 1997. Essays in Biochemistry, vol. 32. Copyright The Biochemical Society

gin (Staehelin et al. 1994; Krishnan et al. 1999). Thus, effective nodulation could be prevented by the destruction of the rhizobial Nod factor by the plant itself or its rhizosphere microflora (Mellor and Collinge 1995).

Upon perception by epidermalroot cells, Nod factors induce the typical responses of root hair deformation, induction of plant "nodulin gene" expression, and formation of nodule primordia. Moreover, Nod factors promote the (iso)flavonoid biosynthesis that is stimulated by the bacterial elicitors. The earliest responses observed after application of Nod factors to legume roots are depolarization of root hair plasma membrane potential, spiking of cytoplasmic calcium levels in the root hairs, alkalinization of the root hair cytoplasm, rearrangement of the actin filaments and increased protoplasmic streaming, which all occur within minutes and prior to root hair deformation (Heidstra et al. 1994). After 3 h the root hairs are fully deformed and expression of plant early nodulin (ENOD) genes starts in anticipation of bacterial invasion. However, purified Nod factors alone are not sufficient to induce infection thread formation. Interaction of root hair cells with bacterial surface components, such as exopolysaccharide (EPS) or lipopolysaccharide (LPS), plays a further important role in the infection process (Gray et al. 1992; Noel 1992). These compounds are likely to act as additional signal molecules for eliciting infection thread formation (Hirsch 1992). Moreover, a concentration of Nod factor three orders of magnitude greater than for root hair deformation is required, suggesting that a different signalling pathway is involved.

Only root hairs just bulging out from the epidermal cells are sensitive to Nod factors: neither epidermal cells that have not yet formed root hairs, nor the old root hairs respond to Nod factors (Kurkdjian 1995). Root hair deformation is induced by picomolar concentrations of Nod factors, indicative of a hormone-like nature of the latter. Different Nod factor structural requirements of the various responses indicate that more than a single receptor is likely to be involved in Nod factor perception (e.g. Felle et al. 1996). Recent data suggest that lectin-like nucleotide phosphohydrolases (apyrases) possess the ability to bind Nod factors. Treatment of roots of *Dolichos biflorus* with antibodies against apyrase prevented nodulation, suggesting that apyrase is involved in the initiation of the root hair deformation (Etzler et al. 1999). In both soybean (*Glycine max*) and *Medicago trunculata*, apyrase mRNA is induced within 3 h after inoculation with rhizobia. Several mutants that are defective in nodulation lack the ability to express apyrase mRNA or to induce apyrase expression in response to rhizobial inoculation (Stacey 1999). Whether the apyrase protein really functions as a Nod signal receptor and how its enzymatic action may be coupled to signal transduction are questions that are currently attracting attention.

Purified Nod factors applied to the root surface not only induce responses in epidermal cells, but also in tissue inside the root, the pericycle and the cortex, that are not in direct physical contact with the medium containing the Nod factors. About 3 h after Nod factor addition and preceding the induction of cell divisions, a gene encoding a 10–13 amino acid long peptide (ENOD40) is induced in the pericycle (Vijn et al. 1995; Albrecht et al. 1999). After 16 h a substantial and prolonged inhibition of polar auxin transport was observed in vetch roots (Boot et al. 1999), preceding the first root cortical cell divisions that lead to the formation of nodule primordia or even nodule-like structures in the absence of bacteria. It seems unlikely that Nod factors themselves are transported to the innermost layers. Rather, secondary signal molecules seem to be generated that are transported from the epidermis and interact with cortical and pericycle cells. However, no such signal has been identified so far.

The position in the root cortex where the nodule primordia are formed is almost exclusively opposite the protoxylem poles of the root vascular bundle. Positional information is specified by stele-derived uridine (Smit et al. 1995), which diffuses into the cortex in the protoxylem zones and positively stimulates cortical cell division. Localized production of ethylene in the phloem sectors of the root appears to act as a negative regulator by inhibiting cortical cell division (Heidstra et al. 1997). ACC oxidase, the enzyme catalyzing the last step in ethylene biosynthesis, is expressed specifically in the cell layers opposite the phloem in that part of the root where nodule primordia are induced upon inoculation with *Rhizobium*. This expression pattern, together with the inhibitory effect of ethylene on cell division, suggests that ethylene can locally suppress the formation of nodule primordia. That ethylene can act as a negative regulator of nodule formation is clearly seen in the ethylene-insensitive *M. truncatula* mutant *sickle*, which forms many more nodules than wild-type plants (Penmetsa and Cook 1997). Also, nodulation by low-nodulating pea *sym5* mutants is fully restored by application of Ag^+, a competitive inhibitor of ethylene perception (Fearn and LaRue 1991; Guinel and LaRue 1991). However, in soybean, varying effects of ethylene perception in the regulation of the numbers of nodules formed have been reported (Caba et al. 1999; Ligero et al. 1999; Schmidt et al. 1999). Nodule formation itself is taken to result from an increased cytokinin level in the root, which triggers cell division in conjunction with an increase in auxin resulting from inhibition of auxin transport due to flavonoids inhibiting this process (cf. Long 1996).

Although an overall picture of rhizobial root nodule formation is emerging, many details are still unknown of this complex but fascinating interaction between a rhizobacterium and its host. Thus, the ENOD40 peptide has homologs in non-leguminous plant species and appears to play a general role in regulating plant development by modulating sensitivity to auxin. It has been postulated that ENOD40 produced in the pericycle acts as a peptide hormone by diffusing into the inner cortex, changing the auxin/cytokinin balance and, thereby, triggering the onset of cell division (Van de Sande et al. 1996).

Root-exuded flavonoids that activate rhizobial *nod* genes can also stimulate growth of mycorrhizal fungi prior to infection, and plant genes encoding early nodulins are likewise activated in root tissues upon infection by VA mycorrhizal fungi (Van Rhijn et al. 1997). Genetic studies have demonstrated that some plant genes that regulate initial steps in the nodulation response of pea also control early stages of VA mycorrhizal development (Gianinazzi-Pearson 1996; Harrison 1997; Albrecht et al. 1999). In some mycorrhiza-resistant mutants of pea the mutant phenotype can be partially reverted by treatment with the auxin transport inhibitor triiodobenzoic acid (Muller 1999), indicative of a common regulation by alteration of the hormone balance of the root. Since both legumes and non-legumes are able to establish VA mycorrhizas with the same fungal species, one must assume that most vascular plants possess common symbiosis genes. In both bacterial and fungal symbiosis, the microsymbionts do not colonize root meristems or the central cylinder, but instead infect plants through the epidermis and multiply within the cortical parenchyma without triggering obvious defense reactions by the plant. Such findings suggest that the genetic program for nodulation may have arisen by adaptation of an ancestral mechanism regulating VA mycorrhizal symbiosis.

12.4 Disease Suppression by Rhizobacteria

Disease-suppressive properties are displayed by epiphytic, endophytic and symbiotic rhizobacteria. Extensive colonization of plant surfaces can prevent pathogens from establishing themselves on or in the plant. However, in addition both direct and indirect interactions between rhizobacteria and pathogens can reduce disease development or severity (Whipps 2001).

Soil-borne plant pathogens cause significant damage to crop production worldwide. Disease symptoms caused by these plant pathogens include damping-off, root rots, foot rots and wilting. For several soil-borne plant pathogens, including *Gaeumannomyces graminis* var. *tritici*, *Fusarium oxysporum*, *Fusarium solani*, *Phytophthora cinnamomi*, *Rhizoctonia solani*, and *Sclerotium cepivorum*, disease-suppressive soils have been described (Cook and Baker 1983). In these soils expression of disease is limited despite the presence of a virulent pathogen, a susceptible crop and environmental conditions favorable for disease development. In several of these suppressive soils microbial populations that are antagonistic towards the pathogen play a key role in disease suppression. Selected strains from many genera of bacteria isolated from these suppressive soils have the potential to reduce plant diseases when applied to the plant root environment (Weller 1988). Using transposon mutagenesis, complementation studies, and reporter gene systems, the fluorescent pseudomonads in particular have received much attention with respect to the mechanisms involved in biocontrol. Mechanisms have been studied in detail not only to satisfy the curiosity of scientists, but also notably to improve the performance of biological control, either through selection of more effective strains, or through genetic modification of strains with traits desired. The modes of action that deal with a direct interference of the biological control agent with the pathogen include competition for substrates, siderophore-mediated competition for iron, antibiosis, and lytic activity.

12.4.1 Competition for Substrate

Competition between pathogenic and saprophytic microorganisms for organic materials released from the roots can reduce growth and/or pathogenic activity of the pathogens. For this mode of action the classical approach of comparing biocontrol activities of specific catabolic mutants with wild-type strains is not simple, since any of a number of substrates could be utilized (Loper et al. 1997). The involvement of competition for nutrients in biological control by fluorescent *Pseudomonas* spp. was suggested in several studies. It was found that in vitro antagonistic activity is based on competition, and correlated with disease suppression. Moreover, addition of specific substrates to the plant pathogen system reduced biological control (Elad and Baker 1985; Elad and Chet 1987). For non-pathogenic *Fusarium oxysporum* isolate Fo47 the involvement of competition for carbon in the effective suppression of fusarium wilt on different crops has been studied in detail (Alabouvette et al. 1998).

Enterobacter cloacae can effectively suppress damping-off and root rot diseases caused by *Pythium* species (Nelson and Maloney 1992). It was demonstrated for *P. ultimum* that germination of sporangia is stimulated specifically by seed exudates (Nelson and Hsu 1994). *E. cloacae* is able to catabolize long-chain fatty acids, such as linoleic acid, a predominant stimulant of germination of *Pythium* sporangia in cotton (*Gossypium hirsutum*) seed exudate (Van Dijk and Nelson 1998). A mutant of *E. cloacae* not able to utilize linoleic acid showed a reduced suppression of *Pythium* seed rots, and restoration of linoleic acid utilization by complementation of this mutant also restored suppression of seed rot (Van Dijk and Nelson 1997). Thus, the importance of competition for this specific stimulatory compound in disease suppression was elegantly demonstrated.

12.4.2 Competition for Iron by Siderophores

As discussed above, rhizobacteria produce various types of siderophores to chelate the scarcely available Fe and, thereby, can deprive pathogens from acquiring iron (Fig. 12.3). Using Tn5 transposon mutagenesis in plant growth-promoting *Pseudomonas putida* WCS358, mutants defective in siderophore biosynthesis were obtained (Marugg et al. 1985). Whereas the wild-type strain WCS358 increased potato root growth and tuber yield significantly in pot and field experiments, respectively, the mutants defective in siderophore biosynthesis had no such effect (Bakker et al. 1986, 1987). In these experiments with potato the increased plant growth was due to suppression of deleterious rhizosphere microorganisms (Schippers et al. 1987). The involvement of siderophore production in disease suppression by

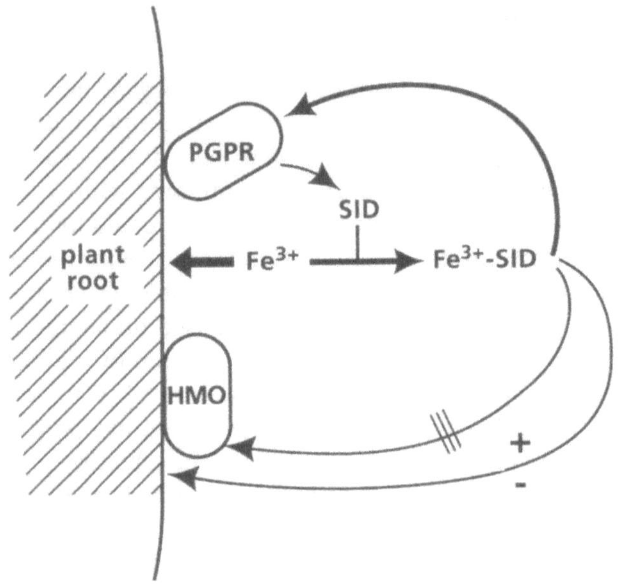

Fig. 12.3. Competition for iron between microorganisms in the rhizosphere: a plant growth-promoting rhizobacterium (*PGPR*) deprives a harmful microorganism (*HMO*) of iron by secreting siderophores (*SID*), which can (+) or cannot (-) also be used by plant roots. (After Bakker 1989)

WCS358 was further studied on carnation, radish, and flax (*Linum usitatissimum*) using, respectively, *Fusarium oxysporum* f.sp. *dianthi*, *F. oxysporum* f.sp. *raphani* and *F. oxysporum* f.sp. *lini* as the pathogen. In all cases the siderophore mutant was less effective than the wild-type strain in suppression of disease (Duijff et al. 1993; Raaijmakers et al. 1995). Also in the combined effects of non-pathogenic *F. oxysporum* Fo47 and WCS358, the siderophore produced by the *Pseudomonas* strain plays a key role in suppression of fusarium wilt (Lemanceau et al. 1992, 1993; Leeman et al. 1996a; Duijff et al. 1999). Whereas the combination of Fo47 with the parental strain WCS358 suppressed fusarium wilt of carnation significantly better compared with the single treatments, a siderophore mutant of WCS358 had no such effect (Fig. 12.4). In this case the combined effects of siderophore-mediated competition for iron by WCS358 and effective competition for carbon by the non-pathogenic Fo47 explain the effective suppression of disease (Lemanceau et al. 1993). Siderophore production by fluorescent *Pseudomonas* spp. has been suggested or demonstrated to be similarly involved in the suppression of *Pythium* spp. (Becker and Cook 1988; Loper 1988), *G. graminis* var. *tritici* (Kloepper et al. 1980a), and *F. oxysporum* (Sneh et al. 1984; Elad and Baker 1985; Baker et al. 1986).

Siderophore mutants of fluorescent *Pseudomonas* spp. strains are comparable to the parental strains with regard to their abilities to colonize the rhizosphere (Bakker et al. 1990), in spite of their reduced competitiveness in acquiring iron. However, the mutants do produce a functional siderophore receptor and, thus, are still able to utilize the parental siderophore (Bitter et al. 1991). For different strains of fluorescent pseudomonads, including *P. putida*

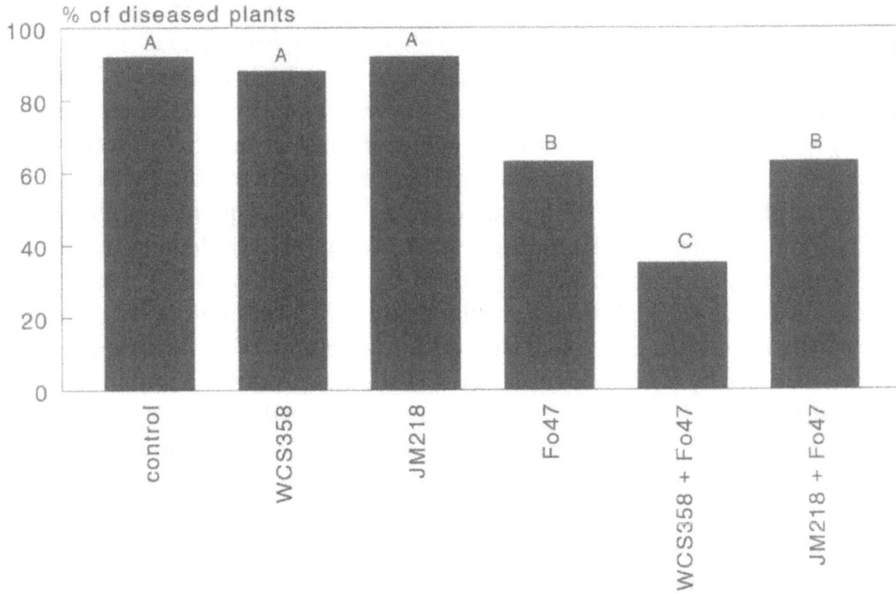

Fig. 12.4. Suppression of fusarium wilt of carnation by *Pseudomonas putida* WCS358 and a siderophore minus mutant of this strain (JM218), both applied either singly or in combination with non-pathogenic *Fusarium oxysporum* Fo47. *Different letterings* indicate significant differences. (Adapted from Lemanceau et al. 1992)

WCS358, utilization of siderophores is not restricted to the siderophore produced by the strain itself, but they can use those produced by many heterologous strains for their iron acquisition (Bakker et al. 1990; Raaijmakers et al. 1994; Koster et al. 1995). The latter observation can explain why siderophore mutants reach similar population densities as the wild type in rhizosphere colonization. Uptake of heterologous siderophores in WCS358 is regulated effectively. For instance, the *pupB* gene, encoding an outer membrane protein that recognizes the siderophores pseudobactin BN7 and pseudobactin BN8, is only expressed in the presence of the heterologous siderophores that are recognized (Koster et al. 1993).

12.4.3 Antibiosis

It has been questioned for many years whether antibiotics are produced by soil microorganisms in quantities large enough to play a significant role in microbial interactions (Williams and Vickers 1986). The introduction of genetic techniques and methods has provided clear evidence for the involvement of antibiotics in suppression of plant diseases by biological control agents (Fravel 1988; Loper et al. 1994). Antibiosis is now often implicated as an important mechanism of biological control, resulting from the fact that it is an attractive mechanism to study and can provide a highly effective mode of action (Handelsman and Stabb 1996). *P. fluorescens* strain 2-79 is suppressive to *G. graminis* var. *tritici*, the causal agent of take-all in wheat (Weller and Cook 1983). Using Tn5 transposon mutants defective in the production of the antibiotic phenazine-1-carboxylic acid (PCA) and subsequent complementation of these mutants, Thomashow and Weller (1988) demonstrated the involvement of this antibiotic in control of take-all disease by strain 2-79 coated on wheat seeds. Using a similar approach, Keel et al. (1992) demonstrated the importance of 2,4-diacetylphloroglucinol (DAPG) in suppression of root diseases by *P. fluorescens* strain CHA0, that produces a wide variety of antifungal metabolites (Table 12.2). Other antibiotics described recently to be involved in disease suppression by fluorescent pseudomonads are phenazine-1-carboxamide (Chin-A-Woeng et al. 1998) and anthranilate (Anjaiah et al. 1998). For the biocontrol agent *Bacillus cereus* strain UW85 production of kanosamine and zwittermicin A was suggested to be important for its biocontrol activity (Silo-Suh et al. 1994; Milner et al. 1996). Another class of bacterial metabolites with antibiotic properties are rhamnolipid biosurfactants (Stanghellini and Miller 1997). Fungal zoospores lack a protective cell wall, leaving the plasma membrane exposed and vulnerable to influences from the environment. Strains of *Pseudomonas* spp. can rapidly kill zoospores by disrupting the plasma membrane by the rhamnolipid biosurfactant (Stanghellini and Miller 1997).

Table 12.2. Production of 2,4-diacetylphloroglucinol (DAPG) by *Pseudomonas fluorescens* strain CHA0 and its derivatives in the rhizosphere of wheat grown under gnotobiotic conditions and relationship between DAPG production and suppression of *Gaeumannomyces graminis* var. *tritici* induced take-all disease by the bacteria. (Data from Keel et al. 1992)

P. fluorescens	*G. graminis*	µg DAPG per g root	Root fresh weight (mg)	Disease rating
None	–	<0.01	320^a	0^e
	+	<0.01	156^c	3.1^a
CHA0 (DAPG$^+$)	–	0.94±0.48	332^a	0^e
	+	1.36±0.16	323^a	0.7^d
CHA625 (DAPG–)	–	<0.01	320^a	0^e
	+	<0.01	249^b	1.9^b
CHA625/pME3128 (DAPG$^+$)	–	0.26±0.14	335^a	0^e
	+	0.19±0.05	294^a	1.3^c

CHA625 is a mutant of CHA0 that lacks DAPG production; CHA625/pME3128 is a transformant of mutant CHA625 in which DAPG production is restored by complementation. Disease severity was rated on a 0–4 scale (0=no disease; 4=plants dead).
Different letterings indicate significant differences at the 5 % level.

Antibiosis as a highly effective means of control of a soil-borne pathogen in natural soils was described recently by Raaijmakers and Weller (1998). They demonstrated that in take-all decline the build-up of populations of DAPG-producing *Pseudomonas* spp. plays a key role in the development of disease suppressiveness.

Production of antibiotics by rhizosphere bacteria is controlled by complex regulatory networks, in which plant, bacterial and environmental signals are involved. Cell density influences production of antibiotics by fluorescent pseudomonads in the rhizosphere (Pierson III et al. 1998). Diffusible *N*-acylhomoserine lactones are produced and utilized by *P. aureofaciens* strain 30-84 and they control the production of PCA (Pierson et al. 1998). These so-called autoinducers can diffuse freely across bacterial membranes. They accumulate in the environment, but also in the producing cells, as the density of the cells increases. When the autoinducer reaches a certain concentration within the cell, the transcription of specific genes is activated. It was recently demonstrated that *N*-acylhomoserine lactone mediated communication occurs between bacterial populations in complex consortia (Pierson et al. 1998; Eberl 1999), as well as between plants and bacteria (Teplitski et al. 2000). Thus, the production of antibiotics by introduced microorganisms can be influenced by the indigenous microflora. There is also evidence for signalling between the pathogen and the introduced biocontrol agent. A promoterless *lacZ* reporter gene was used to generate a transcriptional gene fusion library in *P. fluorescens* strain F113 in order to detect promotors whose activities are altered under specific environmental conditions. Using this library five gene clusters in F113 were identified that are repressed by the presence of *Pythium ultimum* and, interestingly, these gene clusters are important in colonization of the rhizosphere by this strain of *P. fluorescens* (Fedi et al. 1997).

Crown and root rot of tomato caused by *F. oxysporum* f.sp. *radicis-lycopersici* can be controlled by *P. fluorescens* CHA0. Duffy and Défago (1997) report that a metabolite of this fungal pathogen, fusaric acid, represses the production of DAPG and pyoluteorin, both metabolites of CHA0 with antifungal activity. Also, the plant can regulate promoter activity in pseudomonads. By using a library of transcriptional fusion mutants, Van Overbeek and Van Elsas (1995) identified a gene responding to the presence of wheat root exudate. This reporter gene was also induced in the presence of maize and grass roots, but not by roots of clover, suggesting crop-specific interactions. Finally, environmental factors have a significant influence on the production of specific metabolites by fluorescent pseudomonads. In the model strain *P. fluorescens* CHA0, it was recently demonstrated that production of antifungal metabolites can be differentially influenced by specific environmental factors (Duffy and Défago 1999).

12.4.4 Lytic Activity

Certain biological control agents have been demonstrated to suppress disease by parasitizing the plant pathogen. In most cases the biocontrol agent is a fungus that parasitizes on a plant pathogenic fungus. Lytic activity has been demonstrated to be involved in this phenomenon and to comprise degradation of the chitin and glucans in the fungal cell wall and osmotic disruption of the cellular membrane. Transformants of *Trichoderma harzianum* that overexpress a chitinase are more effective in inhibition of growth of *Rhizoctonia solani* (Limon et al. 1999). More interestingly, transformants of *Trichoderma longibrachiatum* that overexpress the β-1,4-endoglucanase gene *egl1* were more effective in controlling effects of *P. ultimum* on cucumber plant emergence and health (Migheli et al. 1998). Woo et al. (1999) describe the genetic modification of *T. harzianum* strain P1, resulting in disruption of a single copy gene that encodes a 42-kDa endochitinase. The endochitinase mutant was compared with the wild-type strain with regard to its biocontrol activity against *Botrytis cinerea*, *R. solani* and *P. ultimum*. Whereas the mutant was as effective as the parental strain in controlling *P. ultimum* – an oomycete lacking chitin – it was less effective against the chitin-containing fungus *B. cinerea* and, surprisingly, it was more effective against *R. solani* (Woo et al. 1999). Thus, endochitinase activity is important in biocontrol of *B. cinerea* by *T. harzianum*, but for control of *Pythium* and *Rhizoctonia* other mechanisms appear to play a role.

For bacteria the role of lytic activity in biological control of plant pathogens is less clear. Many chitin-degrading soil bacteria have the ability to inhibit fungal growth. However, in many cases, bacterial antagonism was not associated with chitinase production (De Boer et al. 1998). On the other hand, it has been suggested that lysis of fungal cell walls of *F. oxysporum* f. sp. *cucumerinum* by *Paenibacillus* sp. 300 and *Streptomyces* sp. 385 is involved in biological control of fusarium wilt by these bacteria (Singh et al. 1999).

12.5 Rhizobacteria-Mediated Induced Systemic Resistance

Induced resistance results from perception of rhizobacteria by plant roots giving rise to an increased level of resistance that is expressed upon subsequent infection by a pathogen. Localized induction of resistance at the site where eliciting bacteria are present on the roots is difficult to demonstrate, because a challenging pathogen will also be subject to bacterial antagonism at this same location. In contrast, no direct interaction between inducing bacteria and a challenging pathogen is possible when each is present at spatially separated sites and no contact between the two is established. Under such conditions, it was demonstrated that various non-pathogenic rhizobacterial strains can induce systemic resistance against fungi, bacteria, and viruses in Arabidopsis (*Arabidopsis thaliana*), bean (*Phaseolus vulgaris*), carnation, cucumber (*Cucumis sativus*), radish, tobacco (*Nicotiana tabacum*) and tomato (Van Loon et al. 1998). Relatively little is known about the "molecular conversation" between these bacteria and the plant compared with the *Rhizobium*-legume symbiosis. However, critical steps are being defined, in large part as a result of the analysis of Arabidopsis mutants that are impaired in resistance signalling pathways.

Rhizobacterially mediated induced systemic resistance (ISR) is phenotypically similar to the better-known systemic acquired resistance (SAR), the induced state that develops when plants successfully activate their defense mechanism in response to primary infection by a pathogen, notably when the latter induces a hypersensitive reaction through which it becomes limited in a local necrotic lesion of brown, desiccated tissue (Ryals et al. 1996; Sticher et al. 1997; Fig. 12.5). SAR is a generally occurring phenomenon that confers an enhanced defensive capacity against all types of pathogens. Under the influence of the primary infection, a signal – the nature of which is still unclear – is generated and transported throughout the plant, thereby establishing the induced state. SAR is characterized by a requirement for salicylic acid (SA) as a signal and by the SA-mediated accumulation of several families of pathogenesis-related proteins (PRs), among which are chitinases and glucanases with potential antifungal activity (Durner et al. 1997; Kombrink and Somssich 1997).

ISR resembles SAR in that it is effective against different types of pathogens, but it differs from SAR in that the inducing rhizobacterium does

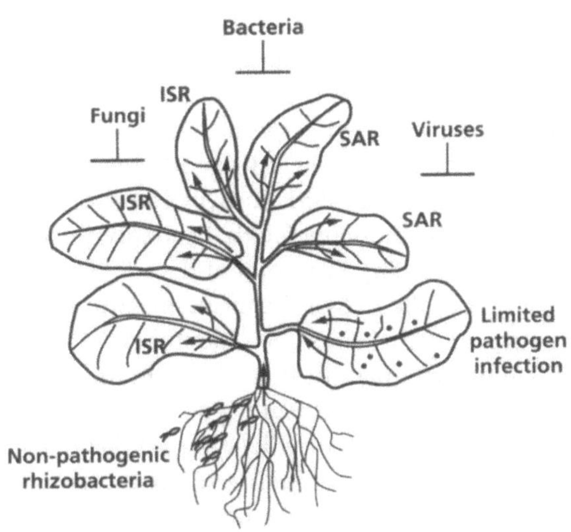

Fig. 12.5. Diagrammatic representation of systemic resistance against fungi, bacteria and viruses induced locally by root colonization by non-pathogenic rhizobacteria (*left* ISR) or by limited pathogen infection (*right* SAR). *ISR* Induced systemic resistance; *SAR* systemic acquired resistance

not cause any visible symptoms in the host. At least in *Arabidopsis*, SA is not involved as a signal and no PRs accumulate (Pieterse et al. 1996). ISR thus constitutes a mechanistically different type of induced resistance and has so far been found to be triggered only by selected rhizobacterial strains (Van Loon 1997). Several *Pseudomonas* spp. isolates have the ability to elicit ISR, but do so differentially in different plant species. For instance, when using *F. oxysporum* as the challenging pathogen, strain WCS358 elicits ISR in *Arabidopsis* but not in radish, strain WCS374 elicits ISR in radish but not in *Arabidopsis*, and *P. fluorescens* strain WCS417 elicits ISR in both *Arabidopsis* and radish. Such species specificity implies that *Arabidopsis* and radish – although both crucifers – either do not offer an environment in which bacterial determinants for resistance induction are expressed in a similar manner, or the same bacterial signals are differentially perceived or transduced.

A prerequisite for resistance induction in general is that the rhizobacteria are able to colonize the roots to a sufficient level. In radish the minimal number of bacteria required was determined to be 10^5 colony-forming units (cfu) per g root (Raaijmakers et al. 1995). Upon isolation of fluorescent pseudomonads from roots of different crop plants growing in a silty loam soil, a high diversity of isolates was recovered (Glandorf et al. 1993), suggesting that in nature no single strain is likely to exceed the threshold level for eliciting ISR. This can explain why plants with levels of up to 10^6 cfu of pseudomonads and 10^9 cfu of total bacteria per g root are not usually found to be induced already. Species specificity cannot be readily explained by differential root colonization. Growing plants in autoclaved soil mixed with individual bacterial strains always led to similar levels of root colonization well above the threshold concentration (Van Wees et al. 1997).

The question of which bacterial determinants are involved in the elicitation of ISR has been addressed by investigating effects of the purified factors and comparing levels of resistance induced by wild-type strains and by selected mutants. It was established that in carnation and radish the O-antigenic side chain of the bacterial outer membrane lipopolysaccharide (LPS) acts as the main determinant (Van Peer and Schippers 1992; Leeman et al. 1995). Treatment of roots with purified bacterial LPS was as effective as living bacteria in eliciting ISR. In radish, bacterial mutants lacking the O-antigenic side chain of the LPS (OA–) did not trigger ISR (Leeman et al. 1995). Thus, cell surface components present in the LPS appear to be the inducing factor. Probably, the carbohydrate side chain of the LPS is recognized by a receptor at the root surface. However, neither the detailed structure of the LPS of the inducing strains, nor a binding entity on roots of carnation or radish, has been identified.

The situation in *Arabidopsis* is more complex in that LPS-containing cell wall preparations of strain WCS417 elicit ISR in this plant species, but an OA– mutant still induced levels of protection similar to wild-type WCS417. This indicates that ISR-inducing bacteria produce more than a single factor triggering ISR in Arabidopsis (Van Wees et al. 1997). Siderophores have also been implicated in the induction of resistance in Arabidopsis (Van Loon et al. 1998), as well as in tobacco (Maurhofer et al. 1994) and radish (Leeman et al. 1996b). However, their contribution to the elicitation of ISR by bacteria in the rhizosphere is uncertain.

The picture is further complicated by the capacity of certain bacterial strains to produce SA under the iron-limiting conditions that are likely to occur in the rhizosphere. For *Pseudomonas aeruginosa* strain 7NSK2 it has been demonstrated that induction of resistance in tobacco against tobacco mosaic virus (De Meyer and Höfte 1998) and in bean against gray mold caused by *Botrytis cinerea* (De Meyer and Höfte 1997) is dependent on the production of SA, because bacterial mutants impaired in SA biosynthesis were no longer inducive. Such experiments have yet to be performed for WCS374 and WCS417. Both these strains can produce SA and induce a resistance in radish under iron-limiting conditions that is not abolished in OA– mutants (Leeman et al. 1996b). Bacterially produced SA can be readily taken up by plant roots and be transported to distant plant parts. The type of induced resistance resulting resembles SAR in its requirement for SA and differs from that induced by non-SA-producing strains. Thus, it is clear that different bacterial components can act as determinants and that these are differentially recognized by different plant species.

The systemic resistance induced in Arabidopsis by WCS358 or WCS417 is equally effective against the fungal root pathogen *F. oxysporum* f. sp. *raphani* (For) and the bacterial leaf pathogen *Pseudomonas syringae* pv. *tomato* (Pst). When using Pst as the challenging pathogen, it was observed that most Arabidopsis ecotypes reacted to induction by the rhizobacteria with a reduction in the proportion of leaves with symptoms of bacterial speck disease. However, ecotypes RLD and Ws-O were non-responsive to the rhizobacteria and became as diseased as non-bacterized control plants infected with Pst (Ton et al. 1999). Non-responsiveness was not caused by poor root colonization or inability to perceive inducing determinants. Rather, the ecotypes appeared to be impaired in a step in the signal-transduction pathway leading to ISR. Different crossings established that responsiveness was inherited as a monogenic, dominant trait, and was correlated with basal resistance against Pst, i.e. RLD and Ws-O are more sensitive to Pst than other ecotypes. These observations suggest that ISR makes use of a signal-transduction pathway that is likewise involved in plant defense against primary infection (Ton et al. 1999).

Testing of known resistance-signalling mutants in Arabidopsis revealed that development of ISR does indeed require components that are also implicated in genetically determined primary resistance to pathogens. Using the jasmonate (JA) response mutant *jar1*, the ethylene response mutant *etr1*, and the SAR signalling mutant *npr1*, it became clear that ISR requires responsiveness to both plant hormones, and shares with SAR a dependency on the regulatory protein NPR1. NPR1 is an ankyrin repeat-containing protein with homology to the mammalian transcription inhibitory regulatory factor IκBα, which plays a role in disease resistance responses in a wide range of higher organisms (Cao et al. 1997; Ryals et al. 1997). NPR1 has been shown to interact with transcription factors involved in the expression of PR-mRNAs (Zhang et al. 1999). Because PRs are not induced during ISR, it is surprising that NPR1 is required not only for SAR, but also for ISR. This dual requirement demonstrates that both signal-transduction pathways share at least one component (Fig. 12.6), pointing to various pathogens and non-pathogenic rhizobacteria stimulating partly overlapping signalling pathways leading to the induced resistant state.

JA and ethylene are produced together with SA during pathogen-induced necrotizing reactions giving rise to SAR, but, in contrast to SA, they are not involved in the establishment of SAR (Pieterse et al. 1998). Application of either methyl jasmonate (MeJA) or the ethylene precursor ACC induces a resistance in *Arabidopsis* against Pst that fully mimics the effect of root colonization with WCS417. However, no increases in endogenous JA or ethylene production were apparent in *Arabidopsis* treated with WCS417 (Pieterse et al. 2000). Yet, JA and ethylene must be involved, because, if responsiveness to either is lost, no ISR develops. Treatment with MeJA still induced ISR in the ethylene non-responsive mutant *etr1*, whereas treatment with ACC did not elicit ISR in the JA response-mutant *jar1*. These data demonstrate that the JA and ethylene responses are engaged in this order in triggering ISR. Because JA can increase sensitivity to ethylene (Tsai et al. 1996), it may be envisaged that

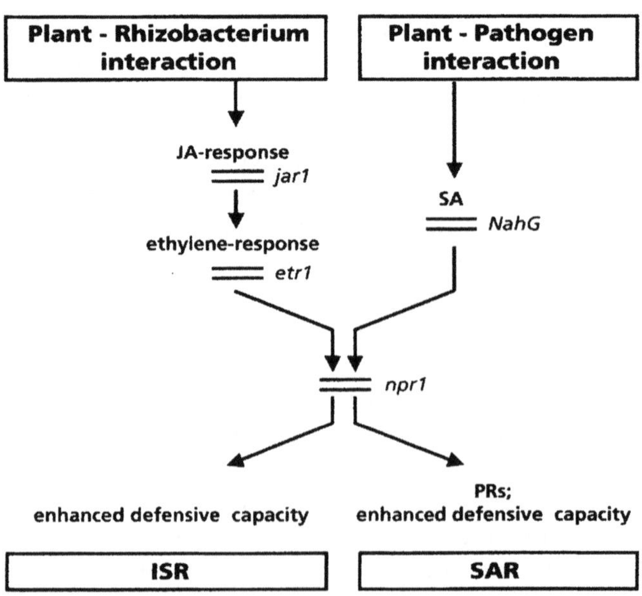

Fig. 12.6. Signalling in *Arabidopsis thaliana* leading to rhizobacteria-mediated induced systemic resistance (*ISR*) or to pathogen-induced systemic acquired resistance (*SAR*). *JA* Jasmonate; *PRs* pathogenesis-related proteins; *SA* salicylate. (Van Loon et al. 1998)

the JA response leads to an enhanced sensitivity to ethylene. Whether the requirement of the JA response in the absence of an increase in endogenous JA also comprises an increase in sensitivity to JA remains an open question. Nevertheless, it must be concluded that, in the induction of ISR, recognition of specific bacterial determinants at the root surface modulates JA and ethylene signalling in the plant and, through the action of the protein NPR1, results in systemically enhanced resistance (Fig. 12.6).

By taking advantage of an *Arabidopsis* mutant that is impaired in ethylene perception in the roots but not in the shoots, it was investigated whether the requirement for ethylene signalling occurs during induction of ISR in the roots or during expression of ISR in the leaves. This *eir1* mutant did not express ISR upon application of WCS417 to the roots, but did exhibit ISR when the inducing bacteria were infiltrated into the leaves. These results demonstrate that, for the induction of ISR, ethylene responsiveness is required at the site of application of the inducing rhizobacteria (Knoester et al. 1999). Because JA perception is required prior to ethylene perception, by inference it can be concluded that also JA-dependent signalling must occur in the root cells that are contacted by the inducing bacteria. How bacterial determinants activate the JA signalling pathway in root cells is totally unclear at present. Equally unclear are the nature of the signal that is transported from the root to other plant parts, and how the induced state becomes established. So far no consistent changes in antimicrobial compounds, enzyme activities, protein patterns or mRNA abundance have been observed upon induction; only after challenge inoculation is an enhanced defensive capacity expressed in the plant.

The same rhizobacterial strains may suppress disease by both microbial antagonism and eliciting ISR, as well as promote plant growth. Thus, WCS417 not only induces resistance in *Arabidopsis*, but was also found to increase fresh weight by 32% (Pieterse and Van Loon 1999). Such results bear testimony to the intricate interactions between root-colonizing bacteria and their hosts and indicate that not only rhizobia, but also free-living rhizobacteria have intimate relationships with plant roots. Signals are continuously being exchanged, which influence the physiology of both the plant and the microbial partner. Recently, it has become clear that plants have a sensitive perception system for the most conserved domain of bacterial flagellin and react by activating an early defense response. However, *Rhizobium* and some plant pathogenic bacteria exhibit divergence in the N-terminal conserved domain of flagellin, suggesting that this difference enables them to evade plant defenses and to invade the host (Felix et al. 1999). The presence of substantial numbers of a wide variety of microorganisms on plant roots also allows extensive signalling between microbes (Pierson III et al. 1998), resulting in antagonistic or synergistic effects on root colonization, plant growth, and suppression of disease. For example, certain strains of PGPR increased growth and development, nodulation and nitrogen fixation by *Rhizobium* in bean (Srinivasan et al. 1996, 1997) and soybean (Shabayev et al. 1996; Dashti et al. 1997). Similarly, ectomycorrhizal formation on eucalypt (*Eucalyptus diversicolor*) seedlings was significantly increased upon inoculation with specific PGPR (Dunstan et al. 1998). A combination of two *Pseudomonas* strains, antagonising For by competition for iron and inducing resistance, respectively, reduced fusarium wilt in radish more than each strain by itself (De Boer et al. 1999).

12.6 Summary and Prospects

In view of the multiple and dynamic interactions between microorganisms and plant roots, the rhizosphere must be considered a signalling network between many partners, the details of which form a challenge for scientific research, as well as a promise for environmentally friendly agronomic practices. Although some of the mechanisms involved are being elucidated in increasing detail at the molecular level, the complexities appear far greater than anticipated. Growth promotion by rhizobacteria has been associated with deamination of root-derived ACC, but many PGPR with plant growth-promoting properties do not possess ACC deaminase. Clearly, their stimulatory activity must be based on other mechanisms. None of these have been unequivocally established, nor is it evident to what extent increases in nutrient availability as a result of bacterial action in the rhizosphere contribute quantitatively to plant nutrition. Root nodule formation in legumes by symbiotic rhizobia is far better understood and can serve as a model for unraveling the complex interactions between bacteria and plant roots. However, it is still unknown how rhizobial Nod factors are perceived by the roots and how nodule initiation and differentiation are controlled. There are similarities between these symbiotic and mycorrhizal and pathogenic host–microbe interactions, and a more profound understanding of the mechanisms involved may lead to the development of new agricultural applications that can increase sustainability by making use of already evolved and functioning mechanisms for nutrient supply and biological disease control.

Many rhizobacteria can antagonize pathogens either directly through competition for nutrients, production of antimicrobial compounds or secretion of lytic enzymes, or indirectly by stimulating plant host defenses. Many details about what factors are involved in the expression of these mechanisms in the rhizosphere are still lacking, and the molecular basis of rhizobacterially mediated induced systemic resistance has yet to be clarified. However, bacteria are easily transformable and the use of well-defined mutants together with complementation analysis allows the significance of individual genes to be assessed. To provide direct evidence of the functioning of relevant mechanisms in the rhizosphere, various reporter genes are available to monitor gene expression in situ in time and space by using suitable transformants carrying promoter-reporter constructs. Gene expression profiling through the use of microarrays will be the method of choice to identify novel genes that are expressed in both rhizobacteria and plant hosts during their interactions. These approaches will lead to a far fuller understanding of the many physiological processes involved, their regulation at the molecular level, and their ecological and evolutionary implications for the functioning of plant roots.

References

Akers HA (1981) The effect of waterlogging on the quantity of microbial iron chelators (siderophores) in soil. Soil Sci 132:150–152

Alabouvette C, Schippers B, Lemanceau P, Bakker PAHM (1998) Biological control of fusarium wilts; toward development of commercial products. In: Boland GJ, Kuykendall LD (eds) Plant-microbe interactions and biological control. Marcel Dekker, New York, pp 15–36

Alagawadi AR, Gaur AC (1992) Inoculation of *Azospirillum brasilense* and phosphate-solubilizing bacteria on yield of sorghum [*Sorghum bicolor* (L.) Moench] in dry land. Trop Agric 69:347–350

Albrecht C, Geurts R, Bisseling T (1999) Legume nodulation and mycorrhizae formation; two extremes in host specificity meet. EMBO J 18:281–288

Anjaiah V, Koedam N, Nowak-Thompson B, Loper JE, Höfte M, Tambong JT, Cornelis P (1998) Involvement of phenazines and anthranilate in the antagonism of *Pseudomonas aeruginosa* PNA1 and Tn5 derivatives toward *Fusarium* spp. and *Pythium* spp. Mol Plant-Microbe Interact 11:847–854

Baker R, Elad Y, Sneh B (1986) Physical, biological and host factors in iron competition in soils. In: Swinburne TR (ed) Iron, siderophores and plant diseases. Plenum Press, New York, pp 77–84

Bakker PAHM (1989) Siderophore-mediated plant growth promotion and colonization of roots by strains of *Pseudomonas* spp. PhD Thesis, Utrecht University, Utrecht

Bakker PAHM, Lamers JG, Bakker AW, Marugg JD, Weisbeek PJ, Schippers B (1986) The role of siderophores in potato tuber yield increase by *Pseudomonas putida* in a short rotation of potato. Neth J Plant Pathol 92:249–256

Bakker PAHM, Bakker AW, Marugg JD, Weisbeek PJ, Schippers B (1987) Bioassay for studying the role of siderophores in potato growth stimulation by *Pseudomonas* spp. in short potato rotations. Soil Biol Biochem 19:443–449

Bakker PAHM, Van Peer R, Schippers B (1990) Specificity of siderophores and siderophore receptors and biocontrol by *Pseudomonas* spp. In: Hornby D (ed) Biological control of soil-borne plant pathogens. CAB International, Wallingford, pp 131–142

Barbieri P, Galli E (1993) Effect on wheat root development of inoculation with an *Azospirillum brasilense* mutant with altered indole-3-acetic acid production. Res Microbiol 144:69–75

Barbieri P, Zanelli T, Galli E, Zanetti G (1986) Wheat inoculation with *Azospirillum brasilense* Sp6 and some mutants altered in nitrogen fixation and indole-3-acetic acid production. FEMS Microbiol Lett 36:87–90

Becker JO, Hedges RW, Messens E (1985a) Inhibitory effect of pseudobactin on the uptake of iron by higher plants. Appl Environ Microbiol 49:1090–1093

Becker JO, Messens E, Hedges RW (1985b) The influence of agrobactin on the uptake of ferric iron by plants. FEMS Microbiol Ecol 31:171–175

Becker O, Cook RJ (1988) Role of siderophores in suppression of *Pythium* species and production of increased-growth response of wheat by fluorescent pseudomonads. Phytopathology 78:778–782

Bensalim S, Nowak J, Asiedu SK (1998) A plant growth promoting rhizobacterium and temperature effects on performance of 18 clones of potato. Am J Potato Res 75:145–152

Bitter W, Marugg JD, De Weger LA, Tommassen J, Weisbeek PJ (1991) The ferric-pseudobactin receptor PupA of *Pseudomonas putida* WCS358: homology to TonB dependent *Escherichia coli* receptors and specificity of the protein. Mol Microbiol 5:647–655

Boot KJM, Van Brussel AAN, Tak T, Spaink HP, Kijne JW (1999) Lipochitin oligosaccharides from *Rhizobium leguminosarum* bv. *viciae* reduce auxin transport capacity in *Vicia sativa* subsp. *nigra* roots. Mol Plant-Microbe Interact 12:839–844

Bowen GD (1991) Microbial dynamics in the rhizosphere: possible strategies in managing rhizosphere populations. In: Keister DL, Cregan PB (eds) The rhizosphere and plant growth. Kluwer, Dordrecht, pp 25–32

Broughton WJ, Perret X (1999) Genealogy of legume-*Rhizobium* symbioses. Curr Opin Plant Biol 2:305–311

Caba JM, Poveda JL, Gresshoff PM, Ligero F (1999) Differential sensitivity of nodulation to ethylene in soybean cv. Bragg and a supernodulating mutant. New Phytol 142:233–242

Cao H, Glazebrook J, Clarke JD, Volko S, Dong X (1997) The *Arabidopsis NPR1* gene that controls systemic acquired resistance encodes a novel protein containing ankyrin repeats. Cell 88:57–63

Chin-A-Woeng TFC, Bloemberg GV, Van der Bij AJ, Van der Drift KMGM, Schripsema J, Kroon B, Scheffer RJ, Keel C, Bakker PAHM, Tichy HV, De Bruijn FJ, Thomas-Oates JE, Lugtenberg BJJ (1998) Biocontrol by phenazine-1-carboxamide-producing *Pseudo monas chlororaphis* PCL1391 of tomato root rot caused by *Fusarium oxysporum* f. sp. *radicis-lycopersici*. Mol Plant-Microbe Interact 11:1069–1077

Chipperfield JR, Ratledge C (2000) Salicylic acid is not a bacterial siderophore: a theoretical study. BioMetals 13:165–168

Cook RJ, Baker KF (1983) The nature and practice of biological control of plant pathogens. APS Press, St. Paul

Dakora FD, Phillips DA (1996) Diverse functions of isoflavonoids in legumes transcend anti-microbial definitions of phytoalexins. Physiol Mol Plant Pathol 49:1–20

Dashti N, Zhang F, Hynes R, Smith DL (1997) Application of plant growth-promoting rhizobacteria to soybean (*Glycine max* [L.] Merr.) increases protein and dry matter yield under short-season conditions. Plant Soil 188:33–41

De Boer M, Van der Sluis I, Van Loon LC, Bakker PAHM (1999) Combining fluorescent *Pseudomonas* spp. strains to enhance suppression of fusarium wilt of radish. Eur J Plant Pathol 105:201–219

De Boer W, Klein Gunnewiek JA, Lafeber P, Janse JD, Spit BE, Woldendorp JW (1998) Anti-fungal properties of chitinolytic dune soil bacteria. Soil Biol Biochem 30:193–203

De Meyer G, Höfte M (1997) Salicylic acid produced by the rhizobacterium *Pseudomonas aeruginosa* 7NSK2 induces resistance to leaf infection by *Botrytis cinerea* on bean. Phytopathology 87:588–593

De Meyer G, Höfte M (1998) Induction of systemic resistance by the rhizobacterium *Pseudomonas aeruginosa* 7NSK2 is a salicylic acid dependent phenomenon in tobacco. In: Duffy B, Rosenberger U, Défago G (eds) Molecular approaches in biological control, IOBC wprs Bull 21:117–121

De Weger LA, Van Boxtel R, Van der Burg B, Gruters RA, Geels FP, Schippers B, Lugtenberg B (1986) Siderophores and outer membrane proteins of antagonistic, plant growth-stimulating, root-colonizing *Pseudomonas* spp. J Bacteriol 165:585–594

Dénarié J, Roche P (1992) *Rhizobium* nodulation signals. In: Verma DPS (ed) Molecular signals in plant-microbe communications. CRC Press, Boca Raton, pp 295–324

Djordjevic MA, Gabriel DW, Rolfe BG (1987) *Rhizobium* – the refined parasite of legumes. Annu Rev Phytopathol 25:145–168

Dowling DN, O'Gara F (1994) Metabolites of *Pseudomonas* involved in the biocontrol of plant disease. Trends Biotechnol 12:133–141

Dubeikovsky AN, Mordukhova EA, Kochetkov VV, Polikarpova FY, Boronin AM (1993) Growth promotion of blackcurrant softwood cuttings by recombinant strain *Pseudomonas fluorescens* BSP53a synthesizing an increased amount of indole-3-acetic acid. Soil Biol Biochem 25:1277–1281

Duffy BK, Défago G (1997) Zinc improves biocontrol of fusarium crown and root rot of tomato by *Pseudomonas fluorescens* and represses the production of pathogen metabolites inhibitory to bacterial antibiotic biosynthesis. Phytopathology 87:1250–1257

Duffy BK, Défago G (1999) Environmental factors modulating antibiotic and siderophore biosynthesis by *Pseudomonas fluorescens* biocontrol strains. Appl Environ Microbiol 65:2429–2438

Duijff BJ, Meijer JW, Bakker PAHM, Schippers B (1993) Siderophore-mediated competition for iron and induced resistance in the suppression of fusarium wilt of carnation by fluorescent *Pseudomonas* spp. Neth J Plant Pathol 99:277–289

Duijff BJ, Bakker PAHM, Schippers B (1994a) Ferric pseudobactin 358 as an iron source for carnation. J Plant Nutr 17:2069–2078

Duijff BJ, De Kogel WJ, Bakker PAHM, Schippers B (1994b) Influence of pseudobactin 358 on the iron nutrition of barley. Soil Biol Biochem 26:1681–1688

Duijff BJ, Recorbet G, Bakker PAHM, Loper JE, Lemanceau P (1999) Microbial antagonism at the root level is involved in suppression of fusarium wilt by the combination of nonpathogenic *Fusarium oxysporum* Fo47 and *Pseudomonas putida* WCS358. Phytopathology 89:1073–1079

Dunstan WA, Malajczuk N, Dell B (1998) Effects of bacteria on mycorrhizal development and growth of container grown *Eucalyptus diversicolor* F. Muell. seedlings. Plant Soil 201:241–249

Durner J, Shah J, Klessig DF (1997) Salicylic acid and disease resistance in plants. Trends Plant Sci 2:266–274

Eberl L (1999) *N*-Acylhomoserinelactone-mediated gene regulation in gram-negative bacteria. System Appl Microbiol 22:493–506

Elad Y, Baker R (1985) The role of competition for iron and carbon in suppression of chlamydospore germination of *Fusarium* spp. by *Pseudomonas* spp. Phytopathology 75:1053–1059

Elad Y, Chet I (1987) Possible role of competition for nutrients in biocontrol of *Pythium* damping-off by bacteria. Phytopathology 77:190–195

Etzler ME, Kalsi G, Ewing NN, Roberts NJ, Day RB, Murphy JB (1999) A nod factor binding lectin with apyrase activity from legume roots. Proc Natl Acad Sci USA 96:5856–5861

Fearn JC, LaRue TA (1991) Ethylene inhibitors restore nodulation to *sym5* mutants of *Pisum sativum* L. cv. Sparkle. Plant Physiol 96:239–244

Fedi S, Tola E, Moenne-Loccoz Y, Dowling DN, Smith LM, O'Gara F (1997) Evidence for signaling between the phytopathogenic fungus *Pythium ultimum* and *Pseudomonas fluorescens* F113: *P. ultimum* represses the expression of genes in *P. fluorescens* F113, resulting in altered ecological fitness. Appl Environ Microbiol 63:4261–4266

Felix G, Duran JD, Volko S, Boller T (1999) Plants have a sensitive perception system for the most conserved domain of bacterial flagellin. Plant J 18:265–276

Felle HH, Kondorosi E, Kondorosi A, Schultze M (1996) Rapid alkalinization in alfalfa root hairs in response to rhizobial lipochitooligosaccharide signals. Plant J 10:295–301

Flores HE, Vivanco JM, Loyola-Vargas VM (1999) 'Radicle' biochemistry: the biology of root-specific metabolism. Trends Plant Sci 4:220–226

Frankenberger WT, Arshad M (1995) Phytohormones in soils; microbial production and function. Marcel Dekker, New York

Fravel DR (1988) Role of antibiosis in the biocontrol of plant diseases. Annu Rev Phytopathol 26:75–91

Geurts R, Franssen H (1996) Signal transduction in *Rhizobium*-induced nodule formation. Plant Physiol 112:447–453

Gianinazzi-Pearson V (1996) Plant cell responses to arbuscular mycorrhizal fungi: getting to the roots of the symbiosis. Plant Cell 8:1871–1883

Glandorf DCM, Peters LGL, Van der Sluis I, Bakker PAHM, Schippers B (1993) Crop specificity of rhizosphere pseudomonads and the involvement of root agglutinins. Soil Biol Biochem 25:981–989

Glick BR, Jacobson CB, Schwarze MMK, Pasternak JJ (1994) 1-Aminocyclopropane-1-carboxylic acid deaminase mutants of the plant growth promoting rhizobacterium *Pseudomonas putida* GR12-2 do not stimulate canola root elongation. Can J Microbiol 40:911–915

Glick BR, Karaturovíc DM, Newell PC (1995) A novel procedure for rapid isolation of plant growth promoting pseudomonads. Can J Microbiol 41:533–536

Glick BR, Liu C, Ghosh S, Dumbroff EB (1997) Early development of canola seedlings in the presence of the plant growth-promoting rhizobacterium *Pseudomonas putida* GR12-2. Soil Biol Biochem 29:1233–1239

Glick BR, Penrose DM, Li, J (1998) A model for the lowering of plant ethylene concentrations by plant growth-promoting bacteria. J Theor Biol 190:63–68

Glick BR, Patten CL, Holguin G, Penrose DM (1999) Biochemical and genetic mechanisms used by plant growth promoting bacteria. Imperial College Press, London

Gray JX, De Maagd RA, Rolfe BG, Johnston AWB, Lugtenberg BJJ (1992) The role of the Rhizobium cell surface during symbiosis. In: Verma DPS (ed) Molecular signals in plant-microbe communications. CRC Press, Boca Raton, pp 359–376

Guinel FC, LaRue TA (1991) Light microscopy study of nodule initiation in *Pisum sativum* L cv. Sparkle and in its low-nodulating mutant E2 (*sym 5*). Plant Physiol 97:1206–1211

Hall JA, Peirson D, Ghosh S, Glick BR (1996) Root elongation in various agronomic crops by the plant growth promoting rhizobacterium *Pseudomonas putida* GR12-2. Israel J Plant Sci 44:37–42

Handelsman J, Stabb EV (1996) Biocontrol of soilborne plant pathogens. Plant Cell 8:1855–1869

Harrison MJ (1997) The arbuscular mycorrhizal symbiosis: an underground association. Trends Plant Sci 2:54–60

Heidstra R, Bisseling T (1996) Nod factor induced host responses and mechanisms of Nod factor perception. New Phytol 133:25–43

Heidstra R, Geurts R, Franssen H, Spaink HP, Van Kammen A, Bisseling T (1994) Root hair deformation activity of nodulation factors and their fate on *Vicia sativa*. Plant Physiol 105:787–797

Heidstra R, Yang WC, Yalcin Y, Peck S, Emons AM, Van Kammen A, Bisseling T (1997) Ethylene provides positional information on cortical cell division but is not involved in Nod factor-induced root hair tip growth in *Rhizobium*-legume interaction. Development 124:1781–1787

Hirsch AM (1992) Developmental biology of legume nodulation. New Phytol 122:211–237

Höfte M (1993) Classes of microbial siderophores. In: Barton LL, Hemming BC (eds) Iron chelation in plants and soil microorganisms. Academic Press, San Diego, pp 3–26

James EK, Olivares FL (1997) Infection of sugar cane and other graminaceous plants by endophytic diazotrophs. Crit Rev Plant Sci 17:77–119

Keel C, Schnider U, Maurhofer M, Voisard C, Laville J, Burger U, Wirthner P, Haas D, Défago G (1992) Suppression of root diseases by *Pseudomonas fluorescens* CHA0: importance of the bacterial metabolite 2,4-diacetylphloroglucinol. Mol Plant-Microbe Interact 5:4–13

Kende H (1993) Ethylene biosynthesis. Annu Rev Plant Physiol Plant Mol Biol 44:283–307

Kijne JW, Lugtenberg BJJ, Smit G (1992) Attachment, lectin and initiation of infection in *(Brady)rhizobium*-legume interactions. In: Verma DPS (ed) Molecular signals in plant-microbe communications. CRC Press, Boca Raton, pp 281–294

Kloepper JW, Schroth MN (1981) Plant growth-promoting rhizobacteria and plant growth under gnotobiotic conditions. Phytopathology 71:642–644

Kloepper JW, Leong J, Teintze M, Schroth MN (1980a) *Pseudomonas* siderophores: a mechanism explaining disease suppressive soils. Curr Microbiol 4:317–320

Kloepper JW, Leong J, Teintze M, Schroth MN (1980b) Enhanced plant growth by siderophores produced by plant growth-promoting rhizobacteria. Nature 286:885–886

Kloepper JW, Lifshitz R, Zablotowicz RM (1989) Free living bacterial inocula for enhancing crop productivity. Trends Biotechnol 7:39–43

Kloepper JW, Zablotowicz RM, Tipping EM, Lifshitz R (1991) Plant growth promotion mediated by bacterial rhizosphere colonizers. In: Keister DL, Cregan PB (eds) The rhizosphere and plant growth. Kluwer, Dordrecht, pp 315–326

Knoester M, Pieterse CMJ, Bol JF, Van Loon LC (1999) Systemic resistance in Arabidopsis induced by rhizobacteria requires ethylene-dependent signaling at the site of application. Mol Plant-Microbe Interact 12:720–727

Kombrink E, Somssich IE (1997) Pathogenesis-related proteins and plant defense. In: Carroll G, Tudzynski P (eds) The mycota, vol V, part A: Plant relationships. Springer, Berlin Heidelberg New York, pp 107–128

Koster M, Van de Vossenberg J, Leong J, Weisbeek PJ (1993) Identification and characterization of the *pupB* gene encoding an inducible ferric-pseudobactin receptor of *Pseudomonas putida* WCS358. Mol Microbiol 8:591–601

Koster M, Ovaa W, Bitter W, Weisbeek PJ (1995) Multiple outer membrane receptors for uptake of ferric-pseudobactins in *Pseudomonas putida* WCS358. Mol Gen Genet 248:735–743

Krishnan HB, Kim KY, Krishnan AH (1999) Expression of a *Serratia marcescens* chitinase gene in *Sinorhizobium fredii* USDA191 and *Sinorhizobium meliloti* RCR2011 impedes soybean and alfalfa nodulation. Mol Plant-Microbe Interact 12:748–751

Kurkdjian AC (1995) Role of the differentiation of root epidermal cells in Nod factor (from *Rhizobium meliloti*)-induced root-hair depolarization of *Medicago sativa*. Plant Physiol 107:783–790

Lazarovits G, Nowak J (1997) Rhizobacteria for improvement of plant growth and establishment. Hortscience 32:188–192

Leeman M, Van Pelt JA, Den Ouden FM, Heinsbroek M, Bakker PAHM, Schippers B (1995) Induction of systemic resistance against fusarium wilt of radish by lipopolysaccharides of *Pseudomonas fluorescens*. Phytopathology 85:1021–1027

Leeman M, Den Ouden FM, Van Pelt JA, Cornelissen C, Matamala-Garros A, Bakker PAHM, Schippers B (1996a) Suppression of fusarium wilt of radish by co-inoculation of fluorescent *Pseudomonas* spp. and root-colonizing fungi. Eur J Plant Pathol 102:21–31

Leeman M, Den Ouden FM, Van Pelt JA, Dirkx FPM, Steijl H, Bakker PAHM, Schippers B (1996b) Iron availability affects induction of systemic resistance against fusarium wilt of radish by *Pseudomonas fluorescens*. Phytopathology 86:149–155

Lemanceau P, Bakker PAHM, De Kogel WJ, Alabouvette C, Schippers B (1992) Effect of pseudobactin 358 production by *Pseudomonas putida* WCS358 on suppression of fusarium wilt of carnations by nonpathogenic *Fusarium oxysporum* Fo47. Appl Environ Microbiol 58:2978–2982

Lemanceau P, Bakker PAHM, De Kogel WJ, Alabouvette C, Schippers B (1993) Antagonistic effect of nonpathogenic *Fusarium oxysporum* Fo47 and pseudobactin 358 upon pathogenic *Fusarium oxysporum* f. sp. *dianthi*. Appl Environ Microbiol 59:74–82

Leong J (1986) Siderophores: their biochemistry and possible role in the biocontrol of plant pathogens. Annu Rev Phytopathol 24:187–209

Ligero F, Poveda JL, Gresshoff PM, Caba JM (1999) Nitrate- and inoculation-enhanced ethylene biosynthesis in soybean roots as a possible mediator of nodulation control. J Plant Physiol 154:482–488

Limon MC, Pintor-Toro JA, Benitez T (1999) Increased antifungal activity of *Trichoderma harzianum* transformants that overexpress a 33-kDa chitinase. Phytopathology 89:254–261

Lindsay WL, Schwab AP (1982) The chemistry of iron in soils and its availability to plants. J Plant Nutr 5:821–840

Long SR (1996) *Rhizobium* symbiosis: Nod factors in perspective. Plant Cell 8:1855–1898

Loper JE (1988) Role of fluorescent siderophore production in biological control of *Pythium ultimum* by a *Pseudomonas fluorescens* strain. Phytopathology 78:166–172

Loper JE, Henkels MD (1997) Availability of iron to *Pseudomonas fluorescens* in rhizosphere and bulk soil evaluated with an ice nucleation reporter gene. Appl Environ Microbiol 63:99–105

Loper JE, Lindow SE (1994) A biological sensor for iron available to bacteria in their habitats on plant surfaces. Appl Environ Microbiol 60:1934–1941

Loper JE, Corbell N, Kraus J, Nowak-Thomson B, Henkels MD, Carnegie S (1994) Contributions of molecular biology towards understanding mechanisms by which rhizosphere pseudomonads effect biological control. In: Ryder MH, Stephens PM, Bowen GD (eds) Improving plant productivity with rhizosphere bacteria. CSIRO, Glen Osmond, pp 89–96

Loper JE, Nowak-Thomson B, Whistler CA, Hagen MJ, Corbell NA, Henkels MD, Stockwell VO (1997) Biological control mediated by antifungal metabolite production and resource competition: an overview. In: Ogoshi A, Kobayashi K, Homma Y, Kodama F, Kondo N, Akino S (eds) Plant growth-promoting rhizobacteria – present status and future prospects. Fac Agric Hokkaido Univ, Sapporo, pp 73–79

Lynch JM (1982) Interactions between bacteria and plants in the root environment. In: Rhodes-Roberts ME, Skinner FA (eds) Bacteria and plants. Academic Press, London, pp 1–23

Lynch JM, Whipps JM (1991) Substrate flow in the rhizosphere. In: Keister DL, Cregan PB (eds) The rhizosphere and plant growth. Kluwer, Dordrecht, pp 15–24

Marugg JD, Van Spanje M, Hoekstra WPM, Schippers B, Weisbeek PJ (1985) Isolation and analysis of genes involved in siderophore biosynthesis in plant-growth-stimulating *Pseudomonas putida* WCS358. J Bacteriol 164:563–570

Maurhofer M, Hase C, Meuwly P, Métraux J-P, Défago G (1994) Induction of systemic resistance of tobacco to tobacco necrosis virus by the root-colonizing *Pseudomonas fluorescens* strain CHA0: influence of the *gacA* gene and of pyoverdine production. Phytopathology 84:139–146

Mellor RB, Collinge DB (1995) A simple model based on known plant defence reactions is sufficient to explain most aspects of nodulation. J Exp Bot 46:1–18

Merbach W, Ruppel S (1992) Influence of microbial colonization on $^{14}CO_2$ assimilation and amounts of root-borne ^{14}C compounds in soil. Photosynthetica 26:551–554

Migheli Q, Gonzalez-Candelas L, Dealessi L, Camponogara A, Ramon-Vidal D (1998) Transformants of *Trichoderma lomgibrachiatum* overexpressing the β-1,4-endoglucanase gene *egl1* show enhanced biocontrol of *Pythium ultimum* on cucumber. Phytopathology 88:673–677

Milner JL, Silo-Suh LA, Lee JC, He H, Clardy J, Handelsman J (1996) Production of kanosamine by *Bacillus cereus* UW85. Appl Environ Microbiol 62:3061–3065

Muller J (1999) Mycorrhizal fungal structures are stimulated in wildtype peas and in isogenic mycorrhiza-resistant mutants by triiodobenzoic acid (TIBA), an auxin-transport-inhibitor. Symbiosis 26:379–389

Neilands JB (1981) Microbial iron compounds. Annu Rev Biochem 50:715–731

Neilands JB, Konopka K, Schwyn B, Coy M, Francis RT, Paw BH, Bagg A (1987) Comparative biochemistry of microbial iron assimilation. In: Winkelmann G, Van der Helm D, Neilands JB (eds) Iron transport in microbes, plants, and animals. VCH, Weinheim, pp 3–33

Nelson EB, Hsu JST (1994) Nutritional factors affecting responses of sporangia of *Pythium ultimum* to germination stimulants. Phytopathology 84:677–683

Nelson EB, Maloney AP (1992) Molecular approaches for understanding biological control mechanisms in bacteria: studies of the interaction of *Enterobacter cloacae* with *Pythium ultimum*. Can J Plant Pathol 14:106–114

Noel KD (1992) Rhizobial polysaccharides required in symbioses with legumes. In: Verma DPS (ed) Molecular signals in plant-microbe communications. CRC Press, Boca Raton, pp 341–357

Nowak J (1998) Benefits of in vitro "biotization" of plant tissue cultures with microbial inoculants. In Vitro Cell Dev Biol Plant 34:122–130

Parniske M, Fischer H-M, Hennecke H, Werner D (1991) Accumulation of the phytoalexin glyceollin I in soybean nodules infected by a *Bradyrhizobium japonicum nifA* mutant. Z. Naturforsch 46:318–320

Penmetsa RV, Cook DR (1997) A legume ethylene-insensitive mutant hyperinfected by its rhizobial symbiont. Science 275:527–530

Pierson EA, Wood DW, Cannon JA, Blachere FM, Pierson LS III (1998) Interpopulation signaling via *N*-acylhomoserine lactones among bacteria in the wheat rhizosphere. Mol Plant-Microbe Interact 11:1078–1084

Pierson LS III, Wood DW, Pierson EA (1998) Homoserine lactone-mediated gene regulation in plant-associated bacteria. Annu Rev Phytopathol 36:207–225

Pieterse CMJ, Van Loon LC (1999) Salicylic acid-independent plant defence pathways. Trends Plant Sci 4:52–58

Pieterse CMJ, Van Wees SCM, Hoffland E, Van Pelt JA, Van Loon LC (1996) Systemic resistance in *Arabidopsis* induced by biocontrol bacteria is independent of salicylic acid accumulation and pathogenesis-related gene expression. Plant Cell 8:1225–1237

Pieterse CMJ, Van Wees SCM, Van Pelt JA, Knoester M, Laan R, Gerrits H, Weisbeek PJ, Van Loon LC (1998) A novel signaling pathway controlling induced systemic resistance in *Arabidopsis*. Plant Cell 10:1571–1580

Pieterse CMJ, Van Pelt JA, Ton J, Parchmann S, Mueller MJ, Buchala AJ, Métraux J-P, Van Loon LC (2000) Rhizobacteria-mediated induced systemic resistance (ISR) in *Arabidopsis* requires sensitivity to jasmonate and ethylene but is not accompanied by an increase in their production. Physiol Mol Plant Pathol 57:123–134

Pillay VK, Nowak J (1997) Inoculum density, temperature, and genotype effects on in vitro growth promotion and epiphytic and endophytic colonization of tomato

(*Lycopersicon esculentum* L.) seedlings inoculated with a pseudomonad bacterium. Can J Microbiol 43:354–361

Powell PE, Cline GR, Reid CPP, Szaniszlo PJ (1980) Occurrence of hydroxamate siderophore iron chelators in soils. Nature 287:833–834

Raaijmakers JM, Weller DM (1998) Natural plant protection by 2,4-diacetylphloroglucinol producing *Pseudomonas* spp. in take-all decline soils. Mol Plant-Microbe Interact 11:144–152

Raaijmakers JM, Bitter W, Punte HLM, Bakker PAHM, Weisbeek PJ, Schippers B (1994) Siderophore-receptor PupA as a marker to monitor wild-type *Pseudomonas putida* WCS358 in natural environments. Appl Environ Microbiol 60:1184–1190

Raaijmakers JM, Leeman M, Van Oorschot MPM, Van der Sluis I, Schippers B, Bakker PAHM (1995) Dose-response relationships in biological control of fusarium wilt of radish by *Pseudomonas* spp. Phytopathology 85:1075–1081

Reinhold-Hurek B, Hurek T (1998) Interactions of graminaceous plants with *Azoarcus* spp. and other diazotrophs: identification, localization and perspectives to study their function. Crit Rev Plant Sci 17:29–54

Ryals JA, Neuenschwander UH, Willits MG, Molina A, Steiner H-Y, Hunt MD (1996) Systemic acquired resistance. Plant Cell 8:1809–1819

Ryals JA, Weymann K, Lawton K, Friedrich L, Ellis D, Steiner H-Y, Johnson J, Delaney TP, Jesse T, Vos P, Uknes S (1997) The *Arabidopsis* NIM1 protein shows homology to the mammalian transcription factor inhibitor IκB. Plant Cell 9:425–439

Schippers B, Bakker AW, Bakker PAHM (1987) Interactions of deleterious and beneficial rhizosphere microorganisms and the effect of cropping practices. Annu Rev Phytopathol 25:339–358

Schmidt JS, Harper JE, Hoffman TK, Bent AF (1999) Regulation of soybean nodulation independent of ethylene signaling. Plant Physiol 119:951–959

Shabayev VP, Smolin VY, Mudrik VA (1996) Nitrogen fixation and CO_2 exchange in soybeans (*Glycine max* L.) inoculated with mixed cultures of different microorganisms. Biol Fertil Soils 23:425–430

Shah S, Li J, Moffatt BA, Glick BR (1998) Isolation and characterization of ACC deaminase genes from two different plant growth-promoting rhizobacteria. Can J Microbiol 44:833–843

Sharma VK, Nowak J (1998) Enhancement of verticillium wilt resistance in tomato transplants by in vitro co-culture of seedlings with a plant growth promoting rhizobacterium (*Pseudomonas* sp. strain PsJN). Can J Microbiol 44:528–536

Silo-Suh LA, Lethbridge BJ, Raffel SJ, He H, Clardy J, Handelsman J (1994) Biological activities of two fungistatic antibiotics produced by *Bacillus cereus* UW85. Appl Environ Microbiol 60:2023–2030

Singh PP, Shin YC, Park CS, Chung YR (1999) Biological control of fusarium wilt of cucumber by chitinolytic bacteria. Phytopathology 89:92–99

Smit G, De Koster CC, Schripsema J, Spaink HP, Van Brussel AAN, Kijne JW (1995) Uridine, a cell division factor in pea roots. Plant Mol Biol 29:869–873

Smith AM (1976) Ethylene in soil biology. Annu Rev Phytopathol 14:53–73

Sneh B, Dupler M, Elad Y, Baker R (1984) Chlamydospore germination of *Fusarium oxysporum* f.sp. *cucumerinum* as affected by fluorescent and lytic bacteria from *Fusarium*-suppressive soil. Phytopathology 74:1115–1124

Srinivasan M, Petersen DJ, Holl FB (1996) Influence of indoleacetic-acid-producing *Bacillus* isolates on the nodulation of *Phaseolus vulgaris* by *Rhizobium etli* under gnotobiotic conditions. Can J Microbiol 42:1006–1014

Srinivasan M, Petersen DJ, Holl FB (1997) Nodulation of *Phaseolus vulgaris* by *Rhizobium etli* is enhanced by the presence of *Bacillus*. Can J Microbiol 43:1–8

Stacey G (1999) Nod signal perception. Abstr 9th Int Congr Mol Plant-Microbe Interact, Amsterdam, SP24, p 16

Stanghellini ME, Miller RM (1997) Biosurfactants: their identity and potential efficacy in the biological control of zoosporic plant pathogens. Plant Dis 81:4–12

Staehelin C, Granado J, Müller J, Wiemken A, Mellor RB, Felix G, Regenass M, Broughton WJ, Boller T (1994) Perception of *Rhizobium* nodulation factors by tomato cells and inactivation by root chitinases. Proc Natl Acad Sci USA 91:2196–2200

Sticher L, Mauch-Mani B, Métraux J-P (1997) Systemic acquired resistance. Annu Rev Phytopathol 35:235–270

Sturz AV, Christie BR, Nowak J (2000) Bacterial endophytes: potential role in developing sustainable systems of crop production. Crit Rev Plant Sci 19:1–30

Teplitski M, Robinson JB, Bauer WD (2000) Plants secrete substances that mimic bacterial *N*-acyl homoserine lactone signal activities and affect population density-dependent behaviors in associated bacteria. Mol Plant-Microbe Interact 13:637–648

Thimann KV (1937) On the nature of inhibitions caused by auxin. Am J Bot 24:407–412

Thomashow LS, Weller DM (1988) Role of a phenazine antibiotic from *Pseudomonas fluorescens* in biological control of *Gaeumannomyces graminis* var. *tritici*. J Bacteriol 170:3499–3508

Ton J, Pieterse CMJ, Van Loon LC (1999) Identification of a locus in Arabidopsis controlling both the expression of rhizobacteria-mediated induced systemic resistance (ISR) and basal resistance against *Pseudomonas syringae* pv. *tomato*. Mol Plant-Microbe Interact 12:911–918

Toro M, Azcón R, Barea JM (1998) The use of isotopic dilution techniques to evaluate the interactive effects of *Rhizobium* genotype, mycorrhizal fungi, phosphate-solubilizing rhizobacteria and rock phosphate on nitrogen and phosphorus acquisition by *Medicago sativa*. New Phytol 138:265–273

Tsai FY, Hung KT, Kao CH (1996) An increase in ethylene sensitivity is associated with jasmonate-promoted senescence of detached rice leaves. J Plant Growth Regul 154:197–200

Ueno K, Shetty K (1998) Prevention of hyperhydricity in oregano shoot cultures is sustained through multiple subcultures by selected polysaccharide-producing soil bacteria without re-inoculation. Appl Microbiol Biotechnol 50:119–124

Van de Sande K, Pawlowski K, Czaja I, Wieneke U, Schell J, Schmidt J, Walden R, Matvienko M, Wellink J, Van Kammen A, Franssen H, Bisseling T (1996) Modification of phytohormone response by a peptide encoded by *ENOD40* of legumes and a nonlegume. Science 273:370–373

Van Dijk K, Nelson EB (1997) Fatty acid uptake and beta-oxidation by *Enterobacter cloacae* is necessary for seed rot suppression of *Pythium ultimum*. Phytopathology 87:S100

Van Dijk K, Nelson EB (1998) Inactivation of seed exudate stimulants of *Pythium ultimum* sporangium germination by biocontrol strains of *Enterobacter cloacae* and other seed-associated bacteria. Soil Biol Biochem 30:183–192

Van Loon LC (1997) Induced resistance in plants and the role of pathogenesis-related proteins. Eur J Plant Pathol 103:753–765

Van Loon LC, Bakker PAHM, Pieterse CMJ (1998) Systemic resistance induced by rhizosphere bacteria. Annu Rev Phytopathol 36:453–483

Van Overbeek LS, Van Elsas JD (1995) Root exudate-induced promoter activity in *Pseudomonas fluorescens* mutants in the wheat rhizosphere. Appl Environ Microbiol 61:890–898

Van Peer R, Schippers B (1992) Lipopolysaccharides of plant-growth promoting *Pseudomonas* sp. strain WCS417r induce resistance in carnation to fusarium wilt. Neth J Plant Pathol 98:129–139

Van Rhijn P, Fang Y, Galili S, Shaul O, Atzmon N, Wininger S, Eshed Y, Lum M, Li Y, To V, Fusishige N, Kapulnik Y, Hirsch AM (1997) Expression of early nodulin genes in alfalfa mycorrhizae indicates that signal transduction pathways used in forming arbuscular mycorrhizae and *Rhizobium*-induced nodules may be conserved. Proc Natl Acad Sci USA 94:5467–5472

Van Wees SCM, Pieterse CMJ, Trijssenaar A, Van 't Westende Y, Hartog F, Van Loon LC (1997) Differential induction of systemic resistance in *Arabidopsis* by biocontrol bacteria. Mol Plant-Microbe Interact 10:716–724

Van Zaayen A, Van Eijk C, Versluijs JMA (1992) Production of high quality, healthy ornamental crops through meristem culture. Acta Bot Neerl 41:425–433

Vijn I, Martinez-Abarca F, Yang W-C, Das Neves L, Van Brussel A, Van Kammen A, Bisseling T (1995) Early nodulin gene expression during Nod factor-induced processes in *Vicia sativa*. Plant J 8:111–119

Weller DM (1988) Biological control of soil-borne plant pathogens in the rhizosphere with bacteria. Annu Rev Phytopathol 26:379–407

Weller DM, Cook RJ (1983) Suppression of take-all of wheat by seed treatments with fluorescent pseudomonads. Phytopathology 73:463–469

Whipps JM (2001) Microbial interactions and biocontrol in the rhizosphere. J Exp Bot 52:487–511

Williams ST, Vickers JC (1986) The ecology of antibiotic production. Microbial Ecol 12:43–52

Woo SL, Donzelli B, Scala F, Mach R, Harman GE, Kubicek CP, Del Sorbo G, Lorito M (1999) Disruption of the *ech*42 (endochitinase-encoding) gene affects biocontrol activity in *Trichoderma harzianum* P1. Mol Plant-Microbe Interact 12:419–429

Zhang Y, Fan W, Kinkema M, Li X, Dong X (1999) Interaction of NPR1 with basic leucine zipper protein transcription factors that bind sequences required for salicylic acid induction of the PR-1 gene. Proc Natl Acad Sci USA 96:6523–6528

13 Interactions Between Oxygen-Releasing Roots and Microbial Processes in Flooded Soils and Sediments

P.L.E. Bodelier

13.1 Introduction

It is hard to imagine that wetland soils and sediments are adverse habitats for plant growth, considering the diverse vegetation that flourishes in fresh- and salt-water marshes (tidal as well as non-tidal), bogs, fens, forested swamps, and floodplains. However, the permanent or periodic flooding of soils or sediments has a number of physical, chemical and biological consequences for the soil/sediment environment (e.g. Ponnamperuma 1984; Gambrell et al. 1991). The most important ramification of the water layer on top of flooded soils and sediments is the restricted entry of atmospheric oxygen. The diffusion of oxygen in water is 10,000 times slower than in air. Oxygen is rapidly depleted in the upper layers of inundated soil due to chemical and biological oxidation processes, resulting in a soil profile where the presence of oxygen is limited to the upper mm's (e.g. Frenzel et al. 1992; Lorenzen et al. 1998). Plants have developed various mechanisms to equip them for colonisation and growth in anoxic flooded soils and sediments, as reviewed elsewhere (Drew 1983; Armstrong et al. 1994; Blom and Voesenek 1996; Blom 1999). Changes in anatomy, morphology and metabolism are of paramount importance for surviving in anoxic root environments. The most important adaptation is the development of aerenchyma, which creates a gas-space continuum that stretches from the stomata to the root-root cap junction, referred to as aerenchyma. Aerenchymatous tissue is present in a vast array of plant species (e.g. Justin and Armstrong 1987) and reduces the longitudinal diffusive resistance of gases, thereby facilitating the diffusion of gases from the atmosphere to the roots. In this way, the roots are supplied with oxygen for respiration. In addition to their own consumption, the roots also lose substantial amounts of oxygen to the surrounding anoxic soil, as has been demonstrated in numerous studies (cited in Armstrong et al. 2000). By means of this process of radial oxygen loss (ROL), plants directly affect soil microbiological processes, which, in turn, may have a feedback on plant growth. By creating oxic/anoxic

Ecological Studies, Vol. 168
H. de Kroon, E.J.W. Visser (Eds.)
Root Ecology

interfaces, wetland plants facilitate the coupling of diverse metabolic processes, as schematically depicted in Fig. 13.1A,B.

This chapter will focus on the interactions of wetland plants with aerobic as well as anaerobic microbes involved in C- and N-cycling in flooded soils and sediments. These interactions are mediated through the release of oxygen by aerenchymatous roots. In the oxic rhizosphere, the oxygen released from the

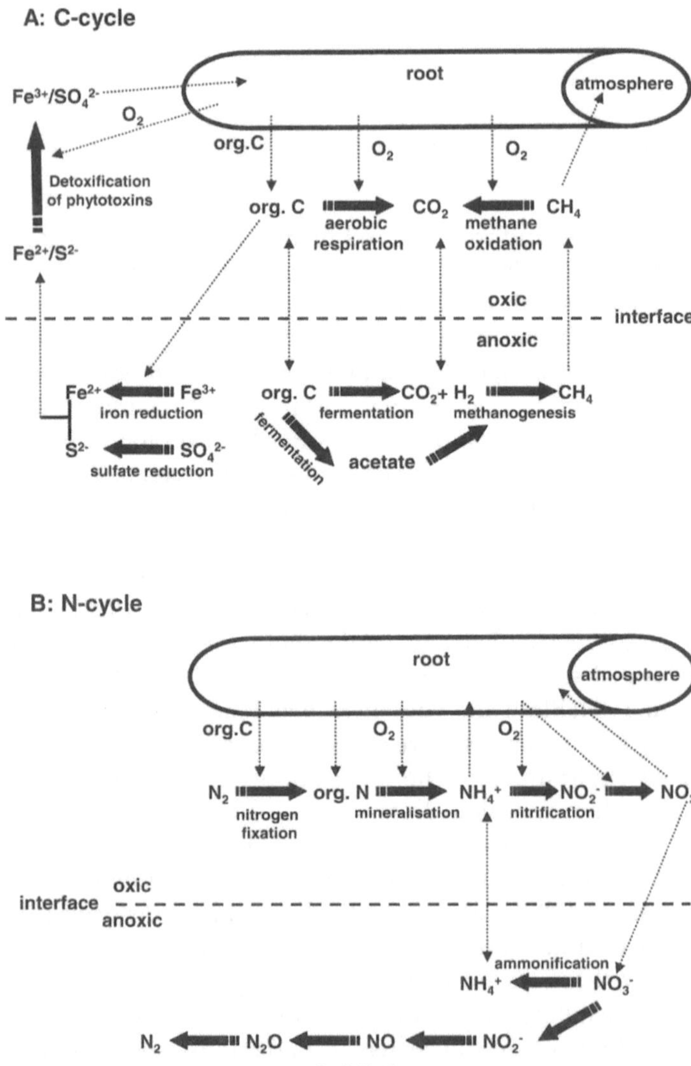

Fig. 13.1A, B. Schematic presentation of the interactions between oxygen-releasing roots and relevant aerobic and anaerobic microbial processes involved in **A** carbon and **B** nitrogen cycling in flooded soils and sediments. *Solid arrows* indicate microbial processes, while *dotted arrows* depict diffusion processes

roots may stimulate aerobic microbial processes. Heterotropic, methane-oxidising and nitrifying bacteria all use molecular oxygen as electron acceptor for energy generation. Heterotrophic bacteria can use plant-derived organic carbon compounds (e.g. exudates, decaying root material) as electron donor (Fig. 13.1A, Sect. 13.4.1) to generate energy and to form cell mass. Roots can profit from such enhanced aerobic respiration because the accumulation of potential phytotoxic (e.g. acetic acid) compounds in the rhizosphere is prevented.

With oxygen derived from the roots, obligate methane-oxidising bacteria (Fig. 13.1A; Sect. 13.4.2) can convert CH_4 to CO_2, which is simultaneously incorporated into cell carbon. The methane carbon, which would otherwise be non-degradable or escape to the atmosphere via the aerenchyma of the plant roots, is therefore conserved for the ecosystem in this way. Oxygen-releasing roots can also stimulate the conversion of ammonia to nitrite by ammonia-oxidising bacteria and the oxidation of nitrite to nitrate by nitrite-oxidising bacteria (Fig. 13.1B; Sect. 13.4.3). Enhanced nitrification may be profitable for plant growth due to the presence of nitrate as a nitrogen source.

Outside the oxic rhizosphere, where no oxygen is available, other electron acceptors are used by microorganisms to respire plant-derived carbon for energy generation. Denitrifying (Fig. 13.1B, Sect. 13.5.1), iron-reducing and sulphate-reducing (Fig. 13.1A, Sect. 13.5.2) bacteria use nitrate, ferric iron and sulphate, respectively, as alternate electron acceptors. The order in which electron acceptors are utilised is based on thermodynamic theory (Zehnder and Stumm 1988) and on the characteristics of the bacteria involved (Laanbroek 1990; Peters and Conrad 1996). The electron acceptor with the highest capacity to take up electrons (i.e. highest redox potential), and thus with the highest energy gain for microbial metabolism, will be utilised first. Thus, inorganic electron acceptors will be reduced in the following sequence: nitrate, manganese, ferric iron, sulphate, and carbon dioxide (Patrick and Jugsujinda 1992; Peters and Conrad 1996; Ratering and Schnell 2000). This may lead to a negative feedback due to the formation of phytotoxic, reduced products (NH_4^+, Mn^{2+}, Fe^{2+}, S^{2-}). However, these compounds can be reoxidised upon diffusion to the oxic rhizosphere and thereby transformed back their non-toxic counterparts (NO_3^-, Mn^{4+}, Fe^{3+}, SO_4^{2-}). Upon depletion of inorganic electron acceptors, fermentation (Fig. 13.1A) and methanogenesis (Fig. 13.1A; Sect. 13.5.3) are the dominating energy-generating microbial processes in soil (Conrad 1996). Both processes can be inhibited by oxygen released from the root, but can also be stimulated by the provision of organic matter. Fermentation may affect plant roots through the formation of a vast range of compounds (Ponnamperuma 1984), including fatty acids like acetic acid, butyric acid, formic acid and proprionic acid (Rothfuss and Conrad 1993), all of which have phytotoxic potential (Armstrong et al. 1996).

13.2 Methodology in Rhizosphere Microbiology

The difficulties associated with the study of microbes in soil and in particular in the rhizosphere can be summed up as: "Trying to characterise organisms which are invisible, unidentified and living in a habitat we cannot define." The habitat definition is already a major problem in rhizosphere microbiology. The rhizosphere is typically defined as the volume of soil that is influenced by the root. Since this influence is not a visible or measurable entity, it is a difficult task to locate and sample soil that is actually influenced by the roots. A frequently used approach is the forced growth of roots in a physically separated compartment that becomes densely rooted and is regarded as a "macro-rhizosphere" (e.g. Klemedtsson et al. 1987; Stienstra et al. 1993; Bodelier et al. 1997). It is assumed that all the soil from these compartments is under the influence of the roots. To actually see the soil and roots, model systems are in use with transparent walls that make it possible to sample or analyse near roots, whose age and growth can be monitored (e.g. Højberg and Sørensen 1993; Jaeger et al. 1999; Arth and Frenzel 2000).

Once one has defined the rhizosphere, whatever the assumptions or compromises involved, the next hurdle is to locate, identify, and determine the activities of the bacteria present in this habitat. Numerous methods have been developed to address these issues. The wide variety of strategies applied to study the fundamental issues in plant–microbe interactions gives some indication of the complexity of the problem. Most methods in microbial ecology rely on the generation of pure cultures as a reference for detection and quantification of specific bacterial populations in the environment. However, the available isolation and cultivation techniques are estimated to yield only 1 % of the total bacterial diversity present in soils (Amann et al. 1995). Culture-independent nucleic acid-based techniques have been developed to circumvent the cultivation problem. However, from non-cultivated strains, only sequence information can be obtained, which gives little information regarding activity and physiology. A number of newly developed techniques are starting to link in situ activity to microbial identity. These include the incorporation of radiotracers or stable isotopes into taxonomically relevant phospholipids (PLFA; Boschker et al. 1998; Bodelier et al. 2000b), the combination of fluorescent in situ hybridisation (FISH) with microautoradiography (MAR; Lee et al. 1999), and the incorporation of stable isotopes into DNA (SIP; Radajewski et al. 2000); all of these techniques offer great opportunities. In addition, the application of genomics and proteomics in microbial ecology should also advance our understanding of the in situ "behaviour" of bacteria. However, these methods have yet to be applied in the rhizosphere of wetland plants. Combining available techniques, classical as well as new, and searching for new strategies of isolation of yet unknown species can only assess this highly complex habitat.

13.3 Quantitative and Qualitative Aspects of Root Oxygen Release

The phenomena of rhizosphere oxidation and the release of oxygen by roots have been topics of investigation for well over a century (Molisch 1887). Past and present research has aimed to assess the qualitative and quantitative aspects of radial oxygen loss (ROL). This research has been driven by the impact of ROL on the biogeochemistry of ecosystems and its importance for the survival and fitness of plants in wetland habitats. The actual release of oxygen from roots has been visualised using dye techniques (Armstrong and Armstrong 1988; Laan et al. 1989; Chabbi et al. 2000) and measured and quantified using polarographic, cylindrical oxygen electrodes (e.g. Armstrong 1964; Laan et al. 1989; Colmer et al. 1998), monitoring oxygen accumulation of complete root systems in deoxygenated solutions (Bedford et al. 1991; Howes and Teal 1994; Sorrell and Armstrong 1994), and colorimetrically using the oxidation of Ti^{3+}-citrate solution as a measure of root oxygen release (Kludze et al. 1993; Sorrell and Armstrong 1994; Sorrell 1999). However, none of these studies has assessed oxygen release of roots in soils or sediments in situ. The oxygen sink strength of natural soils or sediments is a critical factor (Armstrong and Beckett 1987; Kludze et al. 1993; Sorrell and Armstrong 1994), which is very difficult to mimic in solution- or agar-based systems. However, the information gathered with these techniques demonstrates the complexity of the process of ROL. Numerous variables have been shown to influence this process. In addition to plant species, also type and age of the root appear to be important factors. Rates of oxygen release in wetland plants are generally highest in the sub-apical region of roots and decrease with distance from the apex (Armstrong 1971; Connell et al. 1999; Chabbi et al. 2000; Visser et al. 2000). Oxygen release from fine laterals at the base of roots can be significant, but no release of oxygen is generally detected from old roots or rhizomes (Armstrong and Armstrong 1988), except for a number of *Rumex* species (Laan et al. 1989). Root walls of the non-leaking root areas are impermeable due to suberised and lignified layers in the hypodermis and outer cortex (Armstrong and Armstrong 1988). These barriers to ROL will divert more oxygen to the apical meristem to sustain root growth. Furthermore, it has been hypothesised that these permeability barriers would also prevent the entry of phytotoxins (Armstrong 1979). However, an elegant study by Armstrong et al. (2000) has recently demonstrated that enhanced respiration of the sub-apical epidermal-hypodermal cylinder region can account for the local lack of oxygen release. This implies that the entry of gasses, nutrients and phytotoxins is still possible within these barrier areas.

The mechanism of oxygen transport can also have a major impact on the amount of oxygen released. Supplementing diffusion, many emergent macrophytes have an additional pressurised gas flow that accelerates the oxygen flux through their tissues (Brix et al. 1992; Grosse et al. 1996). These convective flows are driven by gas-pressure differences induced by temperature, humidity and wind-velocity gradients between the atmosphere and the internal gas spaces. The oxygen flow into rhizomes and roots can be increased directly by these mechanisms, as was demonstrated for *Phragmites australis* (Armstrong and Armstrong 1990; Armstrong et al. 1992). Since these physical parameters are subject to diurnal and seasonal changes, the ROL associated with convection processes will also fluctuate accordingly.

Radial oxygen loss from roots of submerged macrophytes depends strongly on the photosynthetic oxygen production of the leaves. Daily and seasonal changes in oxygen released by the roots of such plants have been demonstrated (Caffrey and Kemp 1991; Christensen et al 1994; Pedersen et al 1995; Risgaardpedersen and Jensen 1997).

Soil physicochemical factors can also influence oxygen loss by wetland plants. Several authors have shown that the redox potential of the soil or sediment influences oxygen release. Lowering the redox potential of the root environment to values close to naturally flooded soils or sediments (–200 to –300 mV) resulted in enhanced aerenchyma formation in the roots and subsequent oxygen loss (Kludze et al. 1993; Sorrell and Armstrong 1994; Kim et al. 1999). Engelaar and co-workers (1993) demonstrated that soil compaction can lead to a reduction in aerenchyma formation and ROL from *Rumex* species.

Given the multiple variables that affect ROL, it is difficult to make generalisations regarding quantitative data on ROL, and the complexity of this process is illustrated by the disparate results presented in Table 13.1. The range of values found is as variable as the number of controlling factors involved. Furthermore, the methodology involved and the way of expressing ROL are not consistent. Relating ROL to total plant or root biomass can easily overestimate ROL, considering that only parts of the root surface are actually releasing oxygen. Electrode data are probably more suitable for comparative analysis, because this method is fairly standard across different laboratories. Data are also given in reference to root surface area that is actually releasing oxygen. However, all these data still say little about the in situ oxygenation of the rhizosphere. This can only be demonstrated using microelectrodes that are robust and resistant to the mechanical stress, inflicted upon penetration into soil and roots. Arth and Frenzel (2000) demonstrated the presence of oxygen in a densely rooted soil compartment of a microcosm planted with rice (Fig. 13.2). Within this profile, oxygen concentrations increased between 15 and 25 mm depth to values of 70 μM. The authors assumed that in this region the electrode passed a rice root releasing oxygen. Similar in situ approaches have also demonstrated the presence of oxygen in the vicinity of oxygen-releasing roots (Pedersen et al 1995; Risgaardpedersen and Jensen 1997; Gilbert and Frenzel 1998). In sediments inhabited with the submerged macrophyte *Lobelia dortmanna* L., O_2 concentrations of up to 300 μM were

Table 13.1. Collection of radial oxygen loss (ROL) values of roots from wetland plants

Dimension	ROL value	Method	Plant species	Reference
μmol O_2 plant^{-1} day^{-1}<?3>	4–58	Deoxygenated solution	Various emergent macrophytes	Bedford et al. (1991)
	20–35	Ti(III) citrate solution	*Oryza sativa* L.	Kludze et al. (1993)
	25–45	Ti(III) citrate solution	*Oryza sativa* L.	Kim et al. (1999)
	144; 385	Ti(III) citrate solution	*Juncus effusus*; *Juncus inflexus*	Sorrell (1999)
μmol O_2 g root^{-1} h^{-1}<?6>	0.25–5.14	Deoxygenated solution	Various macrophytes	Moorhead and Reddy (1988)
	5.2–19.5	Deoxygenated solution	Various macrophytes	Reddy et al. (1989a)
	0–99	Deoxygenated solution	*Potamogeton perfoliatus* L. and *Zostera marina* L.	Caffrey and Kemp (1991)
	126	Ti(III) citrate solution	*Juncus ingens*	Sorrell and Armstrong (1994)
	120–200	Ti(III) citrate solution	*Typha latifolia*	Jespersen et al. (1998)
	8.1; 14.5	Ti(III) citrate solution	*Juncus effusus*; *Juncus inflexus*	Sorrell (1999)
	1.75; 2	Ti(III) citrate solution	*Cladium jamaicense*; *Typha domigensis*	Chabbi et al. (2000)
ng O_2 cm^{-2} root min^{-1}<?5>	14–289	Cylindrical electrode	Various bog plants	Armstrong (1964)
	0–80	Cylindrical electrode	*Rumex* species	Laan et al. (1989)
	35–60	Cylindrical electrode	*Litorella uniflora* (L.) Ascheron	Christensen et al. (1994)
	76–200	Cylindrical electrode	*Oryza sativa* L.	Gilbert and Frenzel (1998)
	4–72	Cylindrical electrode	*Halophila ovalis*	Connell et al. (1999)
	0–90	Cylindrical electrode	Various wetland species	Visser et al. (2000)

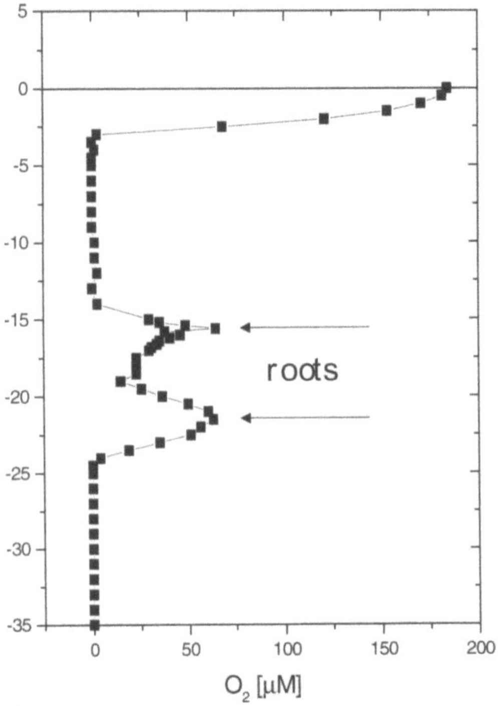

Fig. 13.2. Oxygen profile in the root compartment of a microcosm planted with rice. The profiles were recorded using a microelectrode of the Clark-type with a tip diameter of ca. 30 µm (from Arth and Frenzel 2000, with permission). The *arrows* indicate the soil layers with increased oxygen concentrations. The authors interpreted this as the presence of an oxygen-releasing root

found (Pedersen et al. 1995; Risgaardpedersen and Jensen 1997). However, although these profiles prove that oxygen is finding its way to the rhizosphere at a given point in time and space, they do not provide any indication about the total impact of such oxygen on biogeochemical cycles and bacteria in the rhizosphere on a relevant time scale. Several authors have developed models to estimate the soil volume oxygenated by ROL. For rice roots it was estimated that oxygen diffusing from the roots would penetrate 1–3 mm into the soil (Armstrong 1970; VanBodegom et al. 2001a). Such models are useful when assessing the potential effects of environmental change or disturbance on rhizosphere oxygenation, and they indicate potential ranges of microenvironmental conditions for bacteria in the rhizosphere. However, the actual effect of ROL on bacteria in flooded soils and sediments can better be assessed by directly studying microbial population dynamics (i.e. activity, population size and composition, localisation).

13.4 Interactions Between Oxygen-Releasing Roots and Aerobic Microbial Processes Involved in C- and N-Cycling

13.4.1 Heterotrophic Bacteria

Plants translocate significant amounts (30–60 %) of photosynthetically fixed carbon to their roots. As much as 40 % of this carbon can end up in the soil or sediment through rhizodeposition (Whipps 1990; Marschner 1995). Rhizodeposited materials include lysates liberated by autolysis of sloughed cells and tissues, as well as root exudates released passively or actively from intact cells. Predominant compounds in root exudates are sugars and organic acids (Marschner 1995). The quantity and quality of root exudation materials are influenced by light intensity, temperature, nutrient status of the plant, plant age and species, mechanical impedance, soil type, and the presence of bacteria (Curl and Truelove 1986; Whipps 1990; Marschner 1995). Just like oxygen release, carbon exudation is confined to specific zones of the root system. Root tips and sites where lateral roots emerge are the main sites of exudation (see references cited in Semenov et al. 1999). It is assumed that the exudation of low molecular weight carbon compounds stimulates mineralisation, thereby enhancing nutrient availability (Marschner 1995). However, in the case of flooded soils or sediments, carbon input by plants may have deleterious effects as root-derived carbon can be converted anaerobically into phytotoxic compounds (e.g. acetic acid; Cížková et al. 1999). The obligate and facultative aerobic bacteria in the rhizosphere have the potential to prevent such accumulation of harmful compounds. By respiring root exudates or fermentation products diffusing into the rhizosphere, these microorganisms may play an important role in wetland plant functioning. However, detailed studies on the activity and population dynamics of aerobically respiring heterotrophs in the rhizosphere of oxygen-releasing plants are extremely scarce. Plant–microorganism interaction studies have mainly been focused on microbial associations with agricultural plants in well-aerated soils. Furthermore, the number of microbial species is large and range of potential growth substrates is wide, complicating the detailed study of population dynamics.

To determine the potential of aerobic bacteria to utilise root-derived carbon and oxygen, Bodelier and coworkers (1997, 1998) developed a gnotobiotic (i.e. defined number of species present) microcosm system with a physical separation between the rhizosphere and bulk soil, suitable for experiments under flooded conditions. Growth of a facultative, aerobic heterotroph *Pseudomonas chlororaphis* was significantly enhanced in the rhizosphere of the emergent macrophyte *Glyceria maxima* (Fig. 13.3). The average rhizosphere effect (i.e. number of cells in the rhizosphere/number of cells in the bulk soil) was 20. Elongation of the photoperiod of the plant by 6 h resulted in higher growth rates of the *Pseudomonas chlororaphis*. The authors concluded

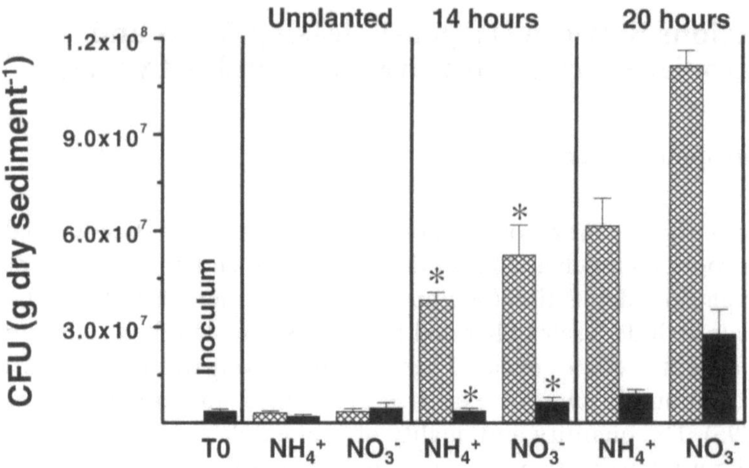

Fig. 13.3. Growth of the facultative heterotroph *Pseudomonas chlororaphis* on exudates of the emergent macrophyte *Glyceria maxima*. Displayed is the number of colony-forming units of *P. chlororaphis* in the root (*hatched bars*) and non-root compartment (*black bars*) of gnotobiotic (i.e. defined number of species present) microcosms planted with or without the emergent macrophyte *G. maxima*. The plants were grown under two different photoperiods (14 vs. 20 h) and with NH_4^+ or NO_3^- as nitrogen source. T0 is the initial inoculum density of *P. chlororaphis* determined 3 days after inoculation. *Bars* are means (±1SE) of four replicate microcosms. *Asterisks* indicate significant differences between 14- and 20-h photoperiod ($p<0.05$, one-way ANOVA; Bodelier et al. 1997)

that the absolute amount of photosynthetically derived carbon released into the rhizosphere was enhanced due to the prolonged day length. Upon addition of nitrate to these systems, growth was even enhanced further due to the ability of this bacterium to use nitrate as electron acceptor to oxidise organic carbon by denitrification. This demonstrates that the amount of oxygen released by the plant was not sufficient for complete respiration of the exuded carbon compounds. Some carbon apparently remained available for anaerobic bacterial metabolism, and this amount increased with the photosynthetic period of the plant.

Studies on the association between aerobic heterotrophs and roots of wetland plants are rare. Gilbert and Frenzel (1998) detected higher numbers of aerobic heterotrophs in the rhizosphere and on the rhizoplane of rice as compared with bulk and unplanted soil. Other related studies report higher numbers of heterotrophic bacteria in the root zone of seagrasses (Donnelly and Herbert 1999) compared with the bulk soil, and higher sediment oxygen consumption rates were found in the presence of the macrophyte *Lobelia dortmanna* (Pedersen et al. 1995). However, taking the spatial and temporal heterogeneity of oxygen availability in the rhizosphere into account, it is difficult to determine whether these bacteria grew aerobically in the rhizosphere or if part of the oxygen consumption is due to aerobic carbon respiration.

The level of aerobic carbon respiration most certainly depends on the outcome of competition for oxygen between heterotrophs and other microbiological and chemical oxygen-consuming processes, but experimental data on this matter are not available. However, VanBodegom and co-workers (2001a) designed a mechanistic model describing the kinetics of aerobic oxidation of the most important electron donors and the diffusion of these donors from a rice root surface. The model predicted that, after 2 days of ROL, 80 % of the oxygen in the rice rhizosphere would be consumed by chemical oxidation of ferric iron, while 15 % would be consumed by aerobic heterotrophs. Apparently, the kinetic properties of these organisms allow them to compete successfully with other oxygen-consuming microbial processes like methane and ammonia oxidation. These model predictions, however, have been made assuming a non-growing root. Taking into account that oxygen and carbon are released preferentially near the root tip, which is moving through soil, additional spatial and temporal aspects have to be considered. In the case of wheat roots, it has been demonstrated that bacterial populations show temporal oscillations upon passage of a carbon-releasing root through the soil (Semenov et al. 1999). However, nothing is known about these dynamic aspects in relation to oxygen-releasing roots and aerobic heterotrophs.

13.4.2 Methane-Consuming Bacteria

Methane is the end point of carbon degradation and accumulates in wetland soils or sediments in the absence of oxygen. However, when oxygen-releasing roots are present, methane-oxidising bacteria close this gap in the carbon cycle. These obligatory aerobic bacteria oxidise methane with oxygen as electron acceptor to carbon dioxide, while incorporating part of the methane into cell material (Hanson and Hanson 1996). These bacteria not only ensure recycling of carbon, but also prevent the diffusion of large quantities of soil methane to the atmosphere. Of the methane potentially emitted by emergent wetland plants, 10–90 % can be oxidised by methane-oxidising bacteria (Denier van der Gon and Neue 1996; Schipper and Reddy 1996; Bosse and Frenzel 1998). Thus, these bacteria reduce the potential effects of this greenhouse gas on atmospheric chemistry and global warming (Conrad 1996). This important attribute has sparked intense study regarding the associations of methane-oxidising bacteria with the roots of wetlands plants in recent years, as reviewed by Frenzel (2000). It has been demonstrated that methanotrophs are able to grow in the rhizosphere of oxygen-releasing rice roots, where higher numbers and activities are observed compared with non-rooted soil (Gilbert and Frenzel 1998; Bodelier et al. 2000a,b). Also, on the rhizoplane of wetland plants, substantial numbers of methanotrophs are present and active (Bosse and Frenzel 1997; Calhoun and King 1998; Gilbert and Frenzel 1998). By means of most probable number counts performed on surface-sterilised rice roots and basal culms, Bosse and Frenzel (1997) demonstrated that methanotrophs are also able to colonise the interior tissues of wetland plants. This was also demonstrated using polyclonal antisera and 16S rRNA probes (Gilbert et al. 1998).

Despite these tight associations, wetland plants do not exclusively promote the activity of methane-oxidising bacteria. As discussed in the previous section, roots promote heterotrophic respiration by exuding organic carbon. Methanotrophs will have to compete with these bacteria for the available oxygen. Van-Bodegom (2001a,b) isolated dominant methanotrophs and heterotrophs from the rhizosphere of rice and performed competition experiments. Using the kinetic properties of these organisms, a mechanistic model was created which describes the competition around a rice root. It appeared that methanotrophs were only able to outcompete heterotrophs where conditions of low oxygen (<5 μM) and organic carbon concentrations prevail, which is not the case directly at the root surface. The outcome of these model simulations holds true only if oxygen and carbon are excreted from the same part of the root. Scherer and Frenzel (pers. comm.) found methanotrophs and methane oxidation activity exclusively on older parts of rice roots using quadrupole mass spectrometry and molecular detection techniques. The root tips were free of methanotrophs, which may indicate that at this location they are outcompeted by other bacteria or that root colonisation by these bacteria was slow.

Recently, it has been shown that plant ammonium uptake can lead to severe suppression of methane consumption in wetland environments (Bodelier et al. 2000a,b). N-Limitation in the rhizosphere compartment of unfertilised microcosms planted with rice (Fig. 13.4A) resulted in complete inactivation of methane-consuming bacteria. Fertilisation with urea or $(NH_4)_2HPO_4$ leads to enhanced activity of methanotrophs in these systems. This was also true for methanotrophic biomass, as measured by the abundance of specific phospholipid ester-linked fatty acids (PLFA) in the soil (Fig. 13.4B). These components are structural parts of the bacterial cell membrane and are correlated with cell biomass (Zelles 1999). On the basis of PLFA composition, cell morphology, physiology and phylogeny, methane-oxidising bacteria have been divided into types I and II. As can be gleaned from Fig. 13.4B, the repression of methanotrophic growth by plant ammonium uptake is more severe for type I methanotrophs than type II methanotrophs, the former having the highest growth rates and methane-consuming potentials. This effect was consistent with the results of radiotracer experiments and molecular biological community composition analyses (Bodelier et al. 2000b). Hence, by selectively affecting microbial community composition, wetland plants can affect biogeochemistry. The repression of methane consumption by plant uptake of mineral nitrogen, as demonstrated in these laboratory experiments, has recently been confirmed by field surveys (Dan et al. 2001; Krüger et al. 2001).

Methanotrophic bacteria themselves may also affect the functioning of wetland plants. These bacteria are able to utilise a broad range of substrates

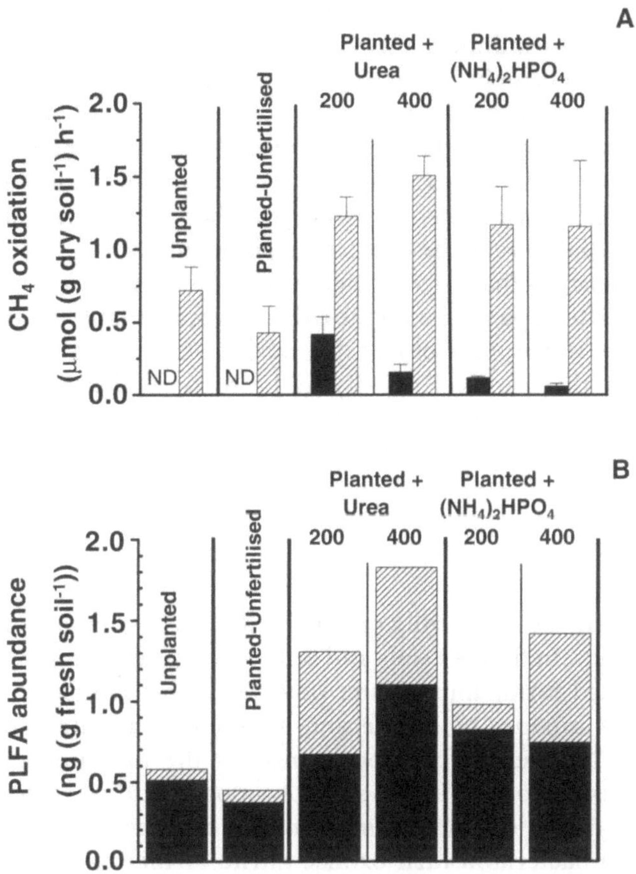

Fig. 13.4A, B. Effect of ammonium-based fertilisation of rice plants on A activity and B phospholipid ester-linked fatty acid (PLFA) abundance of methanotrophs in compartmented microcosms which were either unfertilised or supplemented with urea (200 or 400 kg N ha^{-1}) or $(NH_4)_2HPO_4$ (200 or 400 kg N ha^{-1}). **A** Arithmetic means of the initial (*black bars*) and induced methane-oxidising activities (*hatched bars*) of four replicate microcosms (±1 SD). Depletion plots of methane in oxidation assays often have a sigmoidal shape. The oxidation rate calculated over the initial decline (initial methane oxidation rate) is regarded as a measure for activity in situ. The activity calculated over the subsequent steep decline in the oxidation curve is called induced methane oxidation activity and is regarded as a measure of total oxidation potential. *ND* Assays without a detectable initial oxidation rate. **B** Abundance of PLFA which are specific for type I (16:1ω8 c; *hatched bars*) and type II (18:1ω8 c; *black bars*) methanotrophs. The PLFAs are structural components of the bacterial cell membrane and therefore are directly related to biomass. (Bodelier et al. 2000b)

coupled to the oxidation of methane, including certain xenobiotics (i.e. non-naturally occurring chemicals) and the plant hormone ethylene (Bèdard and Knowles 1989). Since methanotrophs are active and present inside roots of wetland plants, they could reduce ethylene concentrations in the roots. Since ethylene is a regulating factor in the adaptive response of plants to flooded conditions (e.g. Blom 1999), the methanotrophic community may influence plant growth in wetland environments. However, these hypothetical plant–microbe interactions have yet to be explored experimentally.

As in the case of ethylene, methane-oxidising bacteria have the ability to convert ammonium to nitrite in parallel with the oxidation of methane. It has been demonstrated for the rhizosphere of rice that methanotrophs are more important nitrifiers than the "classical" nitrifying bacteria (Bodelier and Frenzel 1999). Hence, these bacteria may also be of relevance for nitrogen cycling and thus for plant nutrition in wetland environments.

13.4.3 Nitrifying Bacteria

Nitrogen is an essential element in all living organisms as a basic component of proteins and nucleic acids. In flooded soils and sediments, nitrogen will be present in its most reduced form, ammonium. However, a vast number of plants prefer nitrate or a mixture of nitrate and ammonium as nitrogen source (Ellenberg 1977). Nitrate can only be formed by the oxidation of ammonium with oxygen, which occurs in two steps. Ammonia-oxidising bacteria that catalyse the oxidation of ammonium to nitrite carry out the first reaction. Nitrite-oxidising bacteria in turn oxidise nitrite further to nitrate (see Kowalchuk and Stephen 2001 for the most recent review on the ecology of these bacteria). Ammonia-oxidising bacteria are not able to grow in the absence of oxygen. Hence, as is the case for methanotrophs, oxygen-releasing roots may create a niche for these bacteria in flooded soils and sediments. Enhanced nitrification activity in the rhizosphere of oxygen-releasing plants has been demonstrated using various methods (e.g. Reddy et al. 1989b; Bodelier et al. 1996; Risgaardpedersen and Jensen 1997; Arth et al. 1998). Recently, Arth and Frenzel (2000) have demonstrated the stimulatory effect of oxygen release on ammonia oxidation in situ. Using a newly developed multi-channel microelectrode, these authors detected an increase in nitrate, as well as a corresponding decrease in ammonium, in the vicinity of rice roots growing in a rhizotron system (Fig. 13.5).

However, ammonia oxidisers have to compete for oxygen and ammonium with other bacteria and plants. Uptake of ammonium by oxygen-releasing plants has also been shown to repress nitrification in the rhizosphere (Engelaar et al. 1991; Bodelier et al. 1998; Bodelier et al. 2000a). By excreting organic carbon compounds, wetland plants can also affect ammonia-oxidising bacteria indirectly by increasing competition for oxygen between nitri-

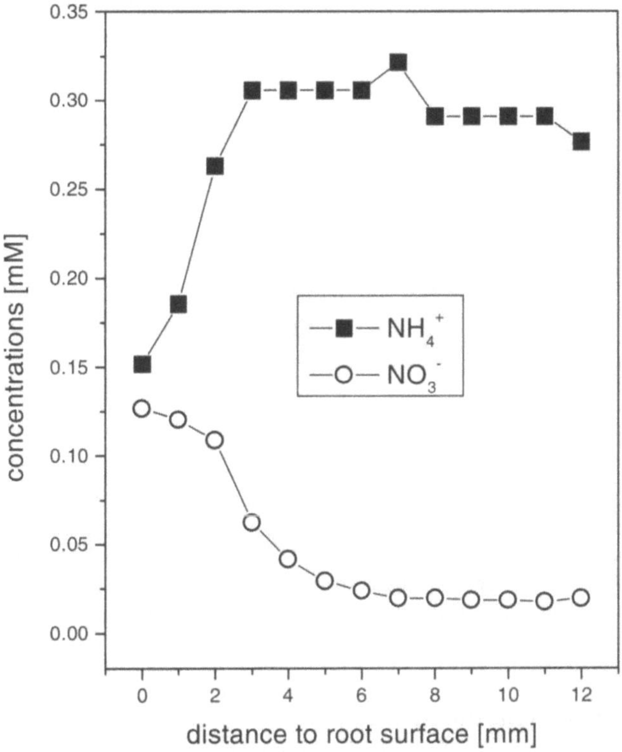

Fig. 13.5. Ammonium and nitrate concentration as a function of the distance from an isolated root of a rice plant grown in a rhizotron. Ammonium and nitrate concentrations were measured with a multi-channel microelectrode recorded in a range of 0–12 mm perpendicular to the root surface, at a distance behind the root apex of 2 mm. (Arth and Frenzel 2000, with permission)

fiers and heterotrophs. Due to their low biomass yield and slow growth rate, nitrifiers are generally poor competitors compared with heterotrophic bacteria for limiting amounts of oxygen (Bodelier and Laanbroek 1997). Nevertheless, ammonia oxidisers can be active and survive in the rhizosphere of oxygen-releasing plants. When comparing ammonia oxidisers of the rhizosphere of *Glyceria maxima* with those of oxic soils, it appeared that these organisms were adapted to functioning at low oxygen concentrations and fluctuating oxic-anoxic periods (Bodelier et al. 1996). After anoxic incubation, ammonia oxidisers originating from the *G. maxima* rhizosphere started to nitrify immediately and restored their nitrifying capacity, while nitrification in the oxic soils started after a lag phase with a loss of 90 % of the initial nitrifying capacity (Fig. 13.6). The ammonia oxidisers from the rhizosphere of *G. maxima* also displayed higher affinities (lower apparent K_m) for oxygen, enabling them to be active at lower oxygen availability. Community analysis using molecular biological techniques did not reveal any consistent differences in ammonia oxidiser community that could be related to the observed adaptations (Kowalchuk et al. 1998). Thus, it appeared that these differences in physiological properties of ammonia-oxidising bacteria in these habitats were due to adaptation of organisms within the same phylogenetic group, rather than selection of specific bacterial lineages for the different habitats.

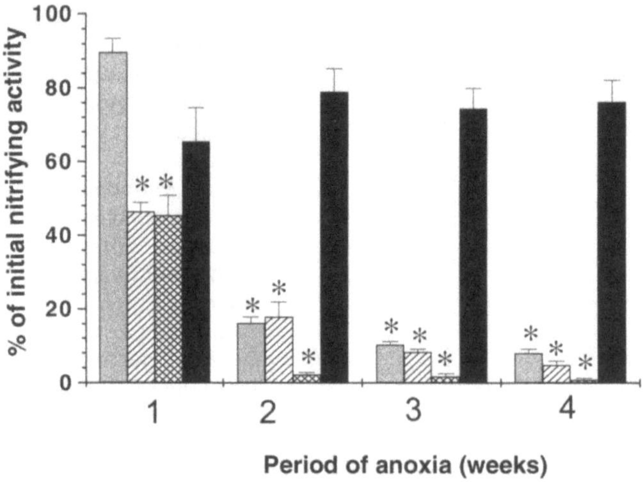

Fig. 13.6. Effect of prolonged anoxic incubation on ammonium-oxidising bacteria in sediment/soil cores from a calcareous grassland (*grey bars*), a chalk grassland (*striped bars*), a dune top (*hatched bars*), and from the root zone of *Glyceria maxima* (*black bars*). *Bars* represent the means (±1 SE) of five replicate cores. *Asterisks* indicate significant effects of the anoxic treatments on the initial ammonium-oxidising activities. (Bodelier et al. 1996)

Almost nothing is known about the interactions between nitrite-oxidising bacteria and oxygen-releasing plants. Two studies have shown higher population sizes of these bacteria in the rhizosphere of G. *maxima* (Both et al. 1992; Bodelier et al. 1996) as compared with non-vegetated sediment. However, the growth and activity of these bacteria cannot always be related to nitrite oxidation, as they are also able to utilise organic carbon compounds and may even grow anaerobically by denitrification. These aspects of nitrite oxidiser ecology are discussed elsewhere in this chapter (Sect. 13.4.1, heterotrophic bacteria and Sect. 13.5.1, denitrifying bacteria). Nevertheless, these bacteria are critical to the production of the important plant nutrient nitrate, and, despite the complications addressed above, it is remarkable how few studies have focused on their interactions with wetland plants.

13.5 Interactions Between Oxygen-Releasing Roots and Anaerobic Microbial Processes Involved in C- and N-Cycling

13.5.1 Denitrifying Bacteria

Reduction of nitrate in flooded soils and sediments is dominated by two dissimilatory processes (i.e. energy-generating) that use organic carbon compounds as electron donor, denitrification and ammonification (see Fig.

13.1B). Denitrification is a bacterial respiratory process that couples electron transport phosphorylation to the stepwise, sequential reduction of nitrogenous oxides (Zumft 1997). The end product, N_2, and reaction intermediates, NO and N_2O, can be emitted via aerenchymatous tissue of wetland plants to the atmosphere, leading to loss of nitrogen from ecosystems. Moreover, NO and N_2O contribute to the greenhouse effect and breakdown of the ozone layer (Crutzen 1981).

Dissimilatory nitrate reduction to ammonium is used by a number of (facultative) fermentative bacteria (Zumft 1997). The capacity to denitrify is present in almost all phylogenetic lineages (Zumft 1997). Both nitrate reduction pathways are inhibited by oxygen and therefore generally occur in anoxic habitats, although exceptions have been demonstrated in pure culture studies (Robertson et al. 1995).

Oxygen-releasing plants can influence the overall denitrification process by introducing oxygen, releasing organic carbon and taking up nitrate. Enhanced nitrification in the rhizosphere, which results in diffusion of nitrate to anaerobic soil parts, and carbon released by the plant both lead to elevated nitrate production. This has been demonstrated for flooded soils supporting rice (Reddy et al. 1989b; Arth et al. 1998; Engelaar et al. 2000), *Glyceria maxima* (Bodelier et al 1996), *Lobelia dortmanna* L. (Risgaardpedersen and Jensen 1997) and seagrasses (Donnelly and Herbert 1999). In contrast, the suppression of nitrate reduction has also been observed as a result of ammonium or nitrate uptake by wetland plants (Bodelier et al. 1996; Welsh et al. 2000) or due to oxygen release (Prade and Trolldenier 1990). Plant carbon release is strongly influenced by photosynthetic activity and nutritional status, and diurnal as well as seasonal effects of oxygen-releasing plants have been coupled to photosynthesis (Christensen and Sørensen 1986; Bodelier et al. 1996; Bodelier et al. 1997; Risgaardpedersen and Jensen 1997). Nutrient deficiency of potassium and/or phosphorus can lead to higher carbon loss by plants and subsequently stimulate denitrification (Prade and Trolldenier 1990).

Wetland plants not only influence denitrification by modifying environmental conditions, but can also induce changes in the composition of the microbial community involved in this process. For instance, Nijburg et al. (1997) found that the rhizosphere of *Glyceria maxima* was dominated by *Bacillus* species, whereas the bulk soil mainly consisted of *Pseudomonas* and *Acinetobacter* strains (Nijburg et al. 1997). Brunel et al. (1992) observed Enterobacteriaceae and *Aeromonas/Vibrio* as the dominant denitrifying bacteria in the rhizosphere of *Typha angustifolia*.

Since denitrification depends on the presence of oxygen, nitrate and carbon, root uptake and release activities will also regulate this process. As was discussed in Sections 13.2 and 13.3.1, these activities will vary with root age, type and part. However, there has only been a single study to date that reports on denitrification linked to root physiology and anatomy. Arth and

Frenzel (2000) used microsensors to measure nitrate profiles in the direct vicinity of rice roots growing in rhizotrons (see also Sect. 13.4.3). From these nitrate profiles, production and consumption activities could be calculated taking diffusion coefficients and sediment porosity into account. These calculations showed that denitrification activity occurred at a distance of 1–5 mm from the root surface and was only detected at the root apex and the maturation zone (50–90 mm behind the root tip). The elongation zone (20–35 mm behind the root tip) did not show any denitrification activity, leading the authors to conclude that the hypothetical barrier to ROL in this part of the root prevented nitrate production and subsequent denitrification in the rhizosphere.

The interactions between wetland plants and denitrification have traditionally only been viewed from the plant perspective, and information on the effects of fluctuations in denitrification activity on wetland plants is lacking. Denitrifiers could potentially increase or reduce nitrate supply to plants. Also, the oxidation of phytotoxic compounds (e.g. acetic acid), mediated by denitrification with nitrate, may be beneficial to the plant. High denitrification activities could also prevent the formation of phytotoxic fermentation products. By respiring organic carbon compounds with nitrate, these compounds remain unavailable for fermentative organisms. Unfortunately, no information is available concerning these hypothetical effects of denitrifying bacteria on wetland plants.

13.5.2 Iron- and Sulphate-Reducing Bacteria

Iron- and sulphate-reducing bacteria can contribute to or even dominate the mineralisation of organic carbon in freshwater and marine wetland environments. Ferric iron (Fe^{3+}) reducing bacteria use ferric iron as electron acceptor to oxidise organic carbon compounds (Lovley 1995). In this process, the phytotoxic ferrous (Fe^{2+}) iron is produced, which has to be re-oxidised for subsequent microbial iron reduction. This can be accomplished in the rhizosphere of wetland plants by chemical as well as biological oxidation, facilitated by root-derived oxygen. The plant root forms a "redox shuttle" providing carbon to reduce ferric iron and oxygen to oxidise the produced ferrous iron. During this process, Fe(III) (hydroxy)oxide precipitates often are formed that cover the root surface (Snowden and Wheeler 1995). These plaques serve as adhering surfaces for ferric iron dependent carbon oxidation, which makes roots of wetland plants ideal habitats for these bacteria, as demonstrated for roots of freshwater and marine macrophytes (King and Garey 1999). Furthermore, bacteria responsible for the biological ferrous iron oxidation can also be present within iron oxide plaques on roots of wetland plants (Emerson et al. 1999). Physical and chemical adsorption processes may allow such iron plaques to scavenge nutrients, trace elements and dissolved organic matter

(Sposito 1984), resulting in the immobilisation of nutrients. Active microbial iron reduction will serve to remobilise these nutrients.

The action of iron reducers can not only affect plants directly by the formation of the potentially phytotoxic ferrous iron, but may also prevent the accumulation of toxic sulphide, which precipitates as FeS. In extreme cases, such precipitation can lead to iron-limiting conditions for plant growth (Smolders and Roelofs 1995).

The process of iron reduction can have a major impact on wetland biogeochemistry. In the rhizosphere of the emergent macrophyte *Juncus effuses*, the root-mediated regeneration of ferric iron and subsequent microbial reduction accounted for 65 % of the total carbon metabolism in vegetated freshwater sediments (Roden and Wetzel 1996). The iron-reducing bacteria in this sediment were apparently successful in the competition for carbon with sulphate reducers and methanogens. The iron reducers can utilise acetate as well as hydrogen thereby reducing the levels of these substrates to values that are below the threshold for utilisation by sulphate reducers and methanogens (Lovley and Phillips 1987). The suppression of the latter two processes will lead to lower levels of sulphide and reduced emission of methane. Next to effects on methane and sulphate reduction, the production of ferrous iron will lead to oxygen stress for aerobic bacteria, which have to compete for oxygen with ferrous iron oxidation processes. This can lead to reduced rhizospheric nitrate production (see Sect. 13.4.3) and methane oxidation (see Sect. 13.4.2).

However, the impact of microbial iron reduction on wetland systems will strongly depend on iron availability. When iron concentrations are low, and sufficient sulphate is present, sulphate-reducing bacteria will dominate the anaerobic carbon metabolism. Because of the high sulphate availability, marine wetland habitats have been most intensively studied with respect to sulphate reduction. The roots and rhizomes of seagrasses and salt marsh plants are environments of intense sulphate reduction (e.g. Hines et al. 1999; Nielsen et al. 2001). The obligatory anaerobic sulphate reducers use the carbon released by plants to generate energy while reducing sulphate to sulphide. Hines and co-workers (1999) demonstrated a seasonal trend for activities as well as numbers of sulphate reducers in the rhizosphere of *Spartina alterniflora*. They observed low numbers and activities during flowering of the plants and the opposite during the vegetative growth periods. These authors related this effect to the lower exudate release of the plants during the flowering stage. Diurnal cycles of sulphate reduction rates were recorded in sediments inhabited by the seagrass *Zostera marina* (Blaabjerg et al. 1998). Rates increased by 30 % during the day due to higher exudation of photosynthates.

In addition to being found in the vicinity of wetland plant roots, sulphate reducers have also been detected on the rhizoplane of rice roots (Scheid and Stubner 2001) and on the rhizoplane, as well as inside epidermis and cortex cells, of the seagrass *Halodule wrightii* (Küsel et al. 1999). The high and instantaneous sulphate reduction rates upon anaerobic incubation of roots of wetland plants (Nielsen et al. 2001) also suggest that these anaerobes are well adapted to function within and in the vicinity of roots releasing oxygen. The presence and activity of bacteria recycling sulphide to sulphate in the oxic rhizosphere (Stubner et al. 1998) may facilitate lifestyles of sulphate reduction in wetland environments.

The presence of plant roots also promotes sulphate reduction in freshwater wetlands. Sulphate reduction rates were found to be higher in planted versus unplanted microcosms and decreased with distance from the rhizoplane of rice roots (Wind and Conrad 1995, 1997). However, the typically low sulphate concentrations in freshwater systems, combined with plant sulphate uptake, will suppress sulphate reduction in these systems, unless there is an input of external sulphate. The intense use of water for agricultural, industrial and domestic purposes often requires the inlet of river water into wetland systems to balance water losses. However, such water sources are often polluted with sulphate, which has led to an increase in sulphate concentrations in many European freshwater wetlands. Enhanced sulphate reduction can lead to toxic sulphide levels and iron limitation, which can affect plant growth (Smolders and Roelofs 1993). In addition, heightened sulphide levels lead to chemical reactions within the soil or sediment that mobilise nutrients like phosphate and ammonium in a process known as internal eutrophication (Smolders and Roelofs 1995). These effects may change the composition of the vegetation in freshwater wetlands in favour of fast growing, sulphide-tolerant species (Lamers et al. 1998). Also, the die-back of the common reed, *Phragmites australis,* in Europe has been connected to enhanced sulphate reduction due to the toxic levels of sulphide that have accumulated upon eutrophication (Armstrong and Armstrong 2001).

13.5.3 Methanogenic Bacteria

As already mentioned in Section 13.4.2, wetlands, and the plants that inhabit them, contribute significantly (15–45 %) to global emission of the greenhouse gas CH_4 to the atmosphere (Segers 1998). A major source of this methane comes from obligatory anaerobic methanogenic microorganisms. Methanogens grow only on a limited number of substrates (H_2/CO_2, formate, methanol, methylamines and acetate; Whitman et al. 1992). Methane is produced mainly from H_2 and CO_2 (hydrogenotrophic methanogenesis, see also Fig. 13.1A) or using acetate (acetoclastic methanogenesis, see also Fig. 13.1A), the latter being the predominant pathway in freshwater wetland environments (Krüger et al. 2001). Methanogenic substrates are the end product of the degradation of organic matter by fermenting bacteria or compounds directly released by roots. It has been demonstrated that 3–52 % of carbon fixed photosynthetically by rice plants can be converted to methane after being released by the roots (Minoda et al. 1996; Dannenberg and Conrad

1999). In densely rooted rice soil, methane production potentials were 15 times higher than in unplanted soil (Bodelier et al. 2000a), demonstrating the distinct interaction between wetland plants and methanogenesis. As outlined in Section 13.4.1, carbon release by plants, and thereby methanogenesis, depends on numerous factors. P-deficiency, for instance, led to higher root exudation and subsequent methane emission from rice plants (Lu et al. 1999). Wang and Adachi (2000) demonstrated differences in the numbers of methanogens between soils planted with different rice varieties. The number of methanogens positively correlated with the carbon exudation levels of the rice varieties examined.

The roots of wetland plants can also influence methanogenesis indirectly. It appears that roots of wetland plants harbour a range of fermentative microorganisms including acetogens (Conrad and Klose 1999; Küsel et al. 1999). Acetate-producing bacteria have been shown to be present and active inside roots of the seagrass *Halodule wrightii* (Küsel et al. 1999). It may be advantageous for methanogens to be present in the close vicinity of the source of substrate. However, methanogens are strictly anaerobic and oxygen is generally regarded to be toxic. Thus, aerenchymatous plants may inhibit methanogenesis directly, by releasing oxygen, or indirectly by facilitating the formation of nitrate, ferric iron and sulphate. The latter ions may serve as electron acceptor for nitrate-reducing, iron-reducing and sulphate-reducing bacteria that compete with methanogens for substrate. In the rhizosphere of rice (Frenzel et al. 1999) and *Juncus effusus* (Roden and Wetzel 1996), inhibition of methanogenesis by the presence of ferric iron has been demonstrated. Grünfeld and Brix (1999) detected lower methane production rates in sediments planted with *Phragmites australis*, and Schipper and Reddy (1996) demonstrated an inhibiting effect of *Sagitarria lancifolia* L. Per. Both studies attributed these effects to oxygen toxicity. However, despite the released oxygen, methanogens associated with roots of macrophytes produce methane immediately upon anoxic incubation (King 1994; Lehmann-Richter et al. 1999). Using specific fluorescent oligonucleotide probes in combination with confocal-laser scanning microscopy, known and novel methanogenic lineages were detected on washed rice roots (Grosskopf et al 1998). Methanogens also constituted a significant part of the total microbial biomass on rice roots from Philippine rice paddies as demonstrated using specific lipid biomarkers (Reichardt et al. 1997). Hence, these root-associated methanogens appear to be tolerant to the presence of oxygen and can profit immediately from exudates when anoxic conditions occur. It is also possible that the methanogens on, e.g., rice roots colonise the root parts that are not releasing oxygen. Unfortunately, no experiments have been reported to test this hypothesis.

Practically nothing is known concerning the possible effect of methanogens on wetland plants. Methane has no known effect on plants. Hypothetically, methanogens can contribute to reduce levels of phytotoxic fermentation products. However, the majority of the research to date has been performed on rice agriculture and little information exists concerning the interaction of methanogens and plants in natural wetlands.

13.5.4 Nitrogen-Fixing bacteria

Molecular nitrogen can only become available for ecosystems through microbiological nitrogen fixation (diazotrophy). N_2 is converted to NH_3 by the enzyme nitrogenase. This reaction, however, demands large amounts of energy. The bacteria that can perform this reaction, aerobes as well as anaerobes, derive the energy for this reaction by respiring organic carbon released by plants. However, the enzyme nitrogenase is extremely oxygen-sensitive and nitrogen fixation by the enzyme can only proceed under microaerobic conditions. Symbiotic nitrogen fixers are well protected against high oxygen in the root nodules of leguminous plants, where the oxygen concentration is maintained low by respiration. Oxygen-releasing wetland plants also support a great variety of free-living heterotrophic nitrogen fixers associated with their roots and rhizomes (Stoltzfus et al. 1997; Bagwell et al. 1998; Lovell et al. 2000). Apparently, oxygen-releasing plants may create ideal conditions for aerobic free-living nitrogen fixers by providing them with microaerobic conditions and carbon. The activity of these bacteria is generally thought to be of great importance for the primary productivity in *Spartina* marshes (Bagwell et al. 1998) and seagrass beds (Donnelly and Herbert 1999; Welsh 2000). For seagrasses it has been shown that nitrogen-fixing bacteria can supply up to 65 % of the nitrogen used for plant growth (Hansen et al. 2000). This study also showed that nitrogen fixation rates were much higher in zones directly associated with roots and rhizomes as compared to the surrounding sediment. A similar strong association between nitrogen fixation and roots and rhizomes of wetland plants has also been reported recently by Nielsen et al. (2001). These authors reported that 31 % and 91 % of the total rhizosphere nitrogen fixation was associated with roots and rhizomes of the coastal marine wetland plans *Zostera noltii* and *Spartina maritima*, respectively. The nitrogen fixation rates were even up to 650-fold higher as compared with bulk sediments as calculated on a dry weight basis. It was proposed that diazotrophic sulphate-reducing bacteria were mainly responsible for the observed nitrogen fixation, as was previously suggested by others (see Welsh 2000).

The interior of wetland plant roots is also colonised by nitrogen-fixing bacteria. Research in this area has been driven by the financial implications of finding alternatives to costly chemical N-fertiliser input in the rice cultivation industry. In the case of rice (Stoltzfus et al. 1997; Egener et al. 1998) and Kallar grass (*Leptochloa fusca* L. Kunth; Reinhold-Hurek and Hurek 1998), nitrogen-fixing bacteria have been detected inside roots using various molecular biological techniques. However, these techniques only demonstrate the presence of nitrogen fixers, not the actual fixation of nitrogen or its transfer to the plants. The nature of the associations of these endophytes with wetland plants, and thus their contribution to nitrogen cycling in plant-inhabited flooded soils and sediments, remains to be studied.

13.6 Summary and Prospects

Flooded soils and sediments are adverse habitats for plant growth. The absence of oxygen hampers root growth, leading to low mineralisation rates and the accumulation of phytotoxic compounds. Wetland plants counteract these adverse conditions by supplying the roots with oxygen via aerenchymatous tissue. Parts of the roots lose oxygen to the surrounding soil or sediment, thereby creating oxic/anoxic interfaces within this otherwise anoxic habitat. These interfaces facilitate aerobic microbial processes (e.g. nitrification, iron oxidation, methane oxidation) that oxidise compounds that have been reduced in the anaerobic zones during organic carbon degradation processes (e.g. denitrification, iron and sulphate reduction, methanogenesis). The latter processes are fuelled by root-derived carbon and use electron acceptors produced in the oxic rhizosphere. During the course of these processes, nutrients are mobilised and become available for plant growth. The phytotoxins liberated by the anaerobic conversions can be detoxified upon diffusion into the rhizosphere. Hence, ROL by wetland plants, and the subsequent interactions with soil/sediment microbes, seems to be relatively straightforward and in most cases mutualistic.

However, the interactions between wetland plants and soil/sediment microbes can be more complex, as we have illustrated in this chapter. Primary interactions between wetland plants and microbial processes are the consequence of direct effects on soil/sediment bacteria following the release of carbon and oxygen and the uptake of nutrients. Roots can also take up oxygen from the rhizosphere to respire, as is the case during darkness. These primary interactions can stimulate or inhibit aerobes as well as anaerobes by providing or depriving bacteria of substrates. It is evident that the nature of these primary interactions will depend on variables influencing the "root function" (e.g. plant species, root physiology/morphology, time in the growing season, soil physics/chemistry). Secondary interactions occur when aerobes as well as anaerobes influence each other as a consequence of the primary plant influence. Secondary interactions that can be envisaged include competition for plant exudates, or for substrates that become limiting due to plant uptake, as well as stimulation or inhibition due to product formation or consumption. The microbial community composition and characteristics of the species present will help to determine the nature of these secondary interactions. The results of primary and secondary interactions can lead to a positive or negative feedback to the plant roots, which can be defined as a tertiary interaction. Such tertiary interactions continue the feedback cycle by reinforcing the primary interactions of the plant system. Although the spatial and temporal variability of these highly interactive systems leads to a high degree of complexity, these complicated interactions seem to be in balance in natural, undisturbed wetlands. Anthropogenic influences may however lead to uncoupling and imbalance of plant–microbe interactions in wetlands. Excessive external input of nutrients or organic carbon has been shown to lead to die-back of common reed in European freshwater wetlands, to loss of seagrass beds all over the world, and to reduced species diversity in wetlands receiving polluted river water. Due to the uncoupling of interactions, the "uncontrolled" negative feedback of soil/sediment microbes results in plant death, exacerbating these negative effects.

The realisation that wetland plants and soil/sediment bacteria are involved in the emission of greenhouse gasses has stimulated research in the interactions of these organisms, especially with reference to the eutrophication in wetland habitats. Wetland rice agriculture, which is responsible for the world's most important food staple, has also been the focus of much attention. However, despite the large amount of attention to this area of research, the majority of the studies to date have mainly addressed the biogeochemical fluxes facilitated by wetland plants and microbes. The approach employed has typically been of comparative nature, comparing microbial processes or population dynamics between vegetated and non-vegetated soils/sediments. Studies using this approach often have been performed in microcosms with confined compartments, or pot experiments with non-representative root/sediment ratios and young plants. A positive development with the advent of molecular biological techniques is that we can now explore the in situ microbial diversity inside and outside oxygen-releasing roots.

Despite the insight gained in recent years, we still have only a "top-down" view of the wetland system. We have an idea of how plants affect soil/sediment microbes, but we have no information whatsoever regarding the microbial feedback of these systems. We need to open the "rhizosphere black box" and start studying natural systems in a relevant time scale with respect to the life history of the plants and bacteria. In the field situation the use of stable isotopes (^{13}C, ^{15}N, ^{18}O) or radiotracers offers opportunities to investigate the feedback of bacteria in more detail. However, the rhizosphere in the field situation will remain a black box if we cannot manipulate the system. Therefore, the best way to understand wetland plant–microbe interactions will be to use systems in which plants can grow under natural conditions, in a relevant time scale, with the possibility to localise and monitor root growth. Rhizolabs as well as rhizotrons (e.g. VandeGeijn et al. 1994; Grierson and Comerford 2000) have been developed to study growth and development of roots in soil. Such systems can be adapted to allow for minimally invasive micro-sampling techniques (Gregory and Hinsinger 1999) and manipulations (e.g. addition of labelled substrates) in the vicinity of visible roots, thus allowing spatial and temporal root–microbe interactions to be studied. Methods have been developed to study the root colonisation, distribution, activity and physiological state of plant pathogens and symbionts (Killham and Yeomans 2001), and such techniques can be adopted in wetland research. To this end it is crucial to have isolates of important functional groups of the representative study systems. The isolation of key community members will also facilitate the use of new techniques to study individual microbial cells in situ, such as microautoradiography combined with FISH (fluorescent in situ hybridisation; Lee et al. 1999). Clearly, efforts to obtain and cultivate root and rhizosphere bacteria have to be continued. Also, the introduction of genomics and proteomics will yield new information with respect to in situ expression of genes of both roots and bacteria as a consequence of their interaction. However, it is highly questionable whether these techniques will take us further with respect to the feedback of bacteria on plant roots.

Another field that is open for study is whether there is some sort of communication between wetland plant roots and bacteria, as has been found for symbiotic and pathogenic bacteria (Teplitski et al. 2000). Taking the tight associations between oxygen-releasing roots and soil/sediment bacteria into account, it is hard to imagine that these interactions are random processes. Nevertheless, of all the opportunities offered by new techniques, only few have been applied to natural wetland systems. The challenge for the future will be to employ newly developed techniques in a minimally invaded rhizosphere of plants grown under natural conditions on a "life-history" relevant time scale.

Acknowledgements. The author would like to thank Dr. H.J. Laanbroek and Dr. G.A. Kowalchuk for their critical comments on the manuscript.

References

Amann RI, Ludwig W, Schleifer KH (1995) Phylogenetic identification and in situ detection of individual microbial cells without cultivation. Microbiol Rev 59:143–169

Armstrong W (1964) Oxygen diffusion from the roots of some British bog plants. Nature 204:801–802

Armstrong W (1970) Rhizosphere oxidation in rice and other species: a mathematical model based on the oxygen flux component. Physiol Plant 23:296–303

Armstrong W (1971) Radial oxygen losses from intact rice roots as affected distance from the apex, respiration and waterlogging. Physiol Plant 25:192–197

Armstrong W (1979) Aeration in higher plants. In: Woolhouse HWW (ed) Advances in botanical research, vol 7. Academic Press, London, pp 225–332

Armstrong J, Armstrong W (1988) *Phragmites australis* – a preliminary study of soil oxidising sites and internal gas transport pathways. New Phytol 108:373–382

Armstrong J, Armstrong W (1990) Light enhanced convective throughflow increases oxygenation in rhizomes and rhizosphere of *Phragmites australis* (Cav.) Trin. Ex Steud. New Phytol 114:121–128

Armstrong J, Armstrong W (2001) An overview of the effects of phytotoxins on *Phragmites australis* in relation to die-back. Aquat Bot 69:251–268

Armstrong W, Beckett PM (1987) Internal aeration and the development of stelar anoxia in submerged roots. A multishelled mathematical model combining axial diffusion of oxygen in the cortex with radial losses to the stele, the wall layers and the rhizosphere. New Phytol 105:221–245

Armstrong J, Armstrong W, Beckett PM (1992) *Phragmites australis* – Venturi-induced and humidity-induced pressure flows enhance rhizome aeration and rhizosphere oxidation. New Phytol 120:197–207

Armstrong W, Brändle R, Jackson MB (1994) Mechanisms of flood tolerance in plants. Acta Bot Neer 43:307–358

Armstrong J, Afreen-Zobayed F, Armstrong W (1996) *Phragmites* die-back: sulphide- and acetic acid-induced bud and root death, lignifications, and blockages within aeration vascular systems. New Phytol 134:601–614

Armstrong W, Cousins D, Armstrong J, Turner DW, Beckett PM (2000) Oxygen distribution in wetland plant roots and permeability barriers to gas-exchange with the rhizosphere: a microelectrode and modelling study with *Phragmites australis*. Ann Bot 86:687–703

Arth I, Frenzel P (2000) Nitrification and denitrification in the rhizosphere of rice: the detection of processes by a new multi-channel electrode. Biol Fertil Soils 31:427–435

Arth I, Frenzel P, Conrad R (1998) Denitrification coupled to nitrification in the rhizosphere of rice. Soil Biol Biochem 30:509–515

Bagwell CE, Piceno YM, Ashburne-Lucas A, Lovell CR (1998) Physiological diversity of the rhizosphere diazotroph assemblages of selected salt marsh grasses. Appl Environ Microbiol 64: 4276–4282

Bèdard C, Knowles R (1989) Physiology, biochemistry and specific inhibitors of CH_4, NH_4^+ and CO oxidation by methanotrophs and nitrifiers. Microbiol Rev 53:68–84

Bedford BL, Bouldin RR, Bethany BD (1991) Net oxygen and carbon dioxide balances in solutions bathing roots of wetland plants. J Ecol 79:943–959

Blaabjerg V, Mouritsen KM, Finster K (1998) Diel cycles of sulphate reduction rates in sediments of a *Zostera marina* bed (Denmark). Aquat Microb Ecol 15:97–102

Blom CWPM (1999) Adaptations to flooding stress: from plant to molecule. Plant Biol 1:261–273

Blom CWPM, Voesenek LACJ (1996) Flooding: the survival strategies of plants Tree 11:290–295

Bodelier PLE, Frenzel P (1999) Contribution of methanotrophic and nitrifying bacteria to CH_4 and NH_4^+ oxidation in the rhizosphere of rice plants as determined by new methods of discrimination. Appl Environ Microbiol 65:1826–1833

Bodelier PLE, Laanbroek HJ (1997) Oxygen uptake kinetics of *Pseudomonas chlororaphis* growing in glucose- or glutamate-limited continuous cultures. Arch Microbiol 167:392–395

Bodelier PLE, Libochant JA, Blom CWPM, Laanbroek HJ (1996) Dynamics of nitrification and denitrification in root-oxygenated sediments and adaptations of ammonia-oxidising bacteria to low oxygen or anoxic habitats. Appl Environ Microbiol 62:4100–4107

Bodelier PLE, Wijlhuizen AG, Blom CWPM, Laanbroek HJ (1997) Effects of photoperiod on growth of and denitrification by *Pseudomonas chlororaphis* in the root zone of *Glyceria maxima*, studied in a gnotobiotic microcosm. Plant Soil 190:91–103

Bodelier PLE, Duyts H, Blom CWPM, Laanbroek HJ (1998) Interactions between nitrifying and denitrifying bacteria in gnotobiotic microcosms planted with the emergent macrophyte *Glyceria maxima*. FEMS Microbiol Ecol 25:63–78

Bodelier PLE, Hahn AP, Arth I, Frenzel P (2000a) Effects of ammonium-based fertilisation on microbial processes involved in methane emission from soils planted with rice. Biogeochemistry 51: 225–257

Bodelier PLE, Roslev P, Henckel T, Frenzel P (2000b) Stimulation of ammonium-based fertilisers of methane oxidation in soil around rice roots. Nature 403:421–424

Boschker HTS, Nold SC, Wellsbury P, Bos D, DeGraaf W, Pel R, Parkes RJ, Cappenberg TE (1998) Direct linking of microbial populations to specific biogeochemical processes by ^{13}C-labelling of biomarkers. Nature 392:801–805

Bosse U, Frenzel P (1997) Activity and distribution of methane-oxidising bacteria in flooded rice soil microcosms and in rice plants. Appl Environ Microbiol 63:1199–1207

Bosse U, Frenzel P (1998) Methane emission from rice microcosms: the balance of production, accumulation and oxidation. Biogeochemistry 41:199–214

Both GJ, Gerards S, Laanbroek HJ (1992) The occurrence of chemolithoautotrophic nitrifiers in water-saturated soils. Microb Ecol 23:15–26

Brix H, Sorrell BK, Orr PhT (1992) Internal pressurisation and convective gas-flow in some emergent freshwater macrophytes. Limnol Oceanogr 37:1420–1433

Brunel B, Janse JD, Laanbroek HJ, Woldendorp JW (1992) Effect of transient oxic conditions on the composition of the nitrate-reducing community from the rhizosphere of *Typha angustifolia*. Microb Ecol 24:51–61

Caffrey JM, Kemp WM (1991) Seasonal and spatial pattern of oxygen production, respiration and root-rhizome release in *Potamogeton perfoliatus* L. and *Zostera marina* L. Aquat Bot 40:109–128

Calhoun A, King GM (1998) Characterisation of root-associated methanotrophs from three freshwater macrophytes: *Pontederia cordata*, *Sparganium eurycarpum*, and *Sagittaria latifolia*. Appl Environ Microbiol 64:1099–1105

Chabbi A, McKee KL, Mendelssohn IA (2000) Fate of oxygen losses from *Typha domingensis* (Typhaceae) and *Cladium jamacense* (Cyperaceae) and consequence for root metabolism. Am J Bot 87:1081–1090

Christensen PB, Sørensen J (1986) Temporal variation of denitrification activity in plant-covered, littoral sediment from Lake Hampen, Denmark. Appl Environ Microbiol 51:1174–1179

Christensen PB, Revsbech NP, Sand-Jensen K (1994) Microsensor analysis of oxygen in the rhizosphere of the aquatic macrophyte *Littorella uniflora* (L.) Ascherson. Plant Physiol 105:847–852

Čížková H, Brix H, Kopecký J, Lukavská J (1999) Organic acids in the sediments of wetlands dominated by *Phragmites australis*: evidence of phytotoxic concentrations. Aquat Bot 64:303–315

Colmer TD, Gibberd MR, Wiengweera A, Tinh TK (1998) The barrier to radial oxygen loss from roots of rice (*Oryza sativa* L.) is induced by growth in stagnant solutions. J Exp Bot 49:1431–1436

Connell EL, Colmer TD, Walker DI (1999) Radial oxygen loss from intact roots of *Halophila ovalis* as a function of distance behind the root tip and shoot illumination. Aquat Bot 63:219–228

Conrad R (1996) Soil micro-organisms as controllers of atmospheric trace gases (H_2, CO, CH_4, OCS, N_2O, and NO). Microbiol Rev 60:609–640

Conrad R, Klose M (1999) Anaerobic conversion of carbon dioxide to methane, acetate and proprionate on washed rice roots. FEMS Microbiol Ecol 30:147–155

Crutzen PJ (1981) Atmospherical chemical processes of the oxides of nitrogen, including nitrous oxide. In: Delwiche CC (ed) Denitrification, nitrification and atmospheric nitrous oxide. Wiley, New York, pp 17–74

Curl A, Truelove B (1986) The rhizosphere. Springer, Berlin Heidelberg New York

Dan J, Krüger M, Frenzel P, Conrad R (2001) Effect of late season urea fertilisation on methane emission from a rice field in Italy. Agric Ecosyst Environ 83:191–199

Dannenberg S, Conrad R (1999) Effect of rice plants on methane production and rhizospheric metabolism in paddy soil. Biogeochemistry 45:53–71

Denier van der Gon HAC, Neue HU (1996) Oxidation of methane in the rhizosphere of rice plants. Biol Fertil Soils 22:359–366

Donnelly AP, Herbert RA (1999) Bacterial interactions in the rhizosphere of seagrass communities in shallow coastal lagoons. J Appl Microbiol Symp Suppl 85: 151S-160S

Drew MC (1983) Plant injury and adaptation to oxygen deficiency in the root environment: a review. Plant Soil 75:179–199

Egener T, Hurek T, Reinhold-Hurek B (1998) Use of green fluorescent protein to detect expression of *nif* genes of *Azoarcus* sp. BH72, a grass-associated diazotroph, on rice roots. MPMI 11:71–75

Ellenberg H (1977) Stickstoff als Standortfactor, insbesondere für mitteleuropäische Pflanzengesellschaften. Oecol Plant 12:1–22

Emerson D, Weiss JV, Megonical JP (1999) Iron-oxidising bacteria associated with ferric hydroxide precipitates (Fe-plaque) on roots of wetland plants. Appl Environ Microbiol 65:2758–2761

Engelaar WMHG, Bodelier PLE, Laanbroek HJ, Blom CWPM (1991) Nitrification in the rhizosphere of a flooding-resistant and flooding-non-resistant *Rumex* species under drained and waterlogged conditions. FEMS Microbiol Ecol 86:33–42

Engelaar WMHG, VanBruggen MW, VandenHoek WPM, Huyser MAH, Blom CWPM (1993) Root porosities and radial oxygen losses of *Rumex* and *Plantago* species as influenced by soil pore diameter and soil aeration. New Phytol 125:565–574

Engelaar WMHG, Matsumaru T, Yoneyama T (2000) Combined effects of soil waterlogging and compaction on rice (*Oryza sativa* L.) growth, soil aeration, soil N transformations and ^{15}N discrimination. Biol Fertil Soils 32:484–493

Frenzel P (2000) Plant-associated methane oxidation in rice field and wetlands. In: Schink B (ed) Advances in microbial ecology, vol 16. Kluwer, New York, pp 85–114

Frenzel P, Bosse U, Janssen PH (1999) Rice roots and methanogenesis in a paddy soil: ferric iron as an alternative electron acceptor in the rooted soil. Soil Biol Biochem 31:421–430

Frenzel P, Rothfuss F, Conrad R (1992) Oxygen profiles and methane turnover in a flooded rice microcosm. Biol Fertil Soils 14:84–89

Gambrell RP, Delaune RD, Patrick WH Jr (1991) Redox processes is soils following oxygen depletion. In: Jackson MB, Davies DD, Lambers H (eds) Plant life under oxygen deprivation. SPB Academic Publishing bv, The Hague, The Netherlands

Gilbert B, Frenzel P (1998) Rice roots and CH_4 oxidation: the activity of bacteria, their distribution and the microenvironment. Soil Biol Biochem 30:1903–1916

Gilbert B, Aßmuss B, Hartmann A, Frenzel P (1998) In situ localisation of two methanotrophic strains in the rhizosphere of rice plants. FEMS Microbiol Ecol 25:117–128

Gregory PJ, Hinsinger P (1999) New approaches to studying chemical and physical changes in the rhizosphere: an overview. Plant Soil 211:1–9

Grierson PF, Comerford NB (2000) Non-destructive measurements of acid phosphatase activity in the rhizosphere using nitrocellulose membranes and image analysis. Plant Soil 218–49–57

Grosse W, Armstrong J, Armstrong W (1996) A history of pressurised gas-flow studies in plants. Aquat Bot 54:87–100

Grosskopf R, Stubner S, Liesack W (1998) Novel euryarchaeotal lineages detected on rice roots and in the anoxic bulk soil of flooded rice microcosms. Appl Environ Microbiol 64:4983–4989

Grünfeld S, Brix H (1999) Methanogenesis and methane emissions: effects of water table, substrate type and the presence of *Phragmites australis*. Aquat Bot 64:63–75

Hansen JW, Udy JW, Perry CJ, Dennison WC, Lomstein BA (2000) Effects of the seagrass *Zostera capricorni* on sediment microbial processes. Mar Ecol Prog Ser 199:83–96

Hanson RS, Hanson TE (1996) Methanotrophic bacteria. Microbiol Rev 60:439–471

Hines ME, Evans RS, Sharak Genthner BR, Willis SG, Friedman S, Rooney-Varga JN, Devereux R (1999) Molecular phylogenetic and biogeochemical studies of sulphate-reducing bacteria in the rhizosphere of *Spartina alterniflora*. Appl Environ Microbiol 65:2209–2216

Højberg O, Sørensen J (1993) Microgradients of microbial oxygen consumption in a barley rhizosphere model system. Appl Environ Microbiol 59:431–437

Howes BL, Teal JM (1994) Oxygen loss from *Spartina alterniflora* and its relationship to salt marsh oxygen balance. Oecologia 97:431–438

Jaeger CH, Lindow SE, Miller W, Clark E, Firestone MK (1999) Mapping of sugar and amino acid availability in soil around roots with bacterial sensors of sucrose and tryptophan. Appl Environ Microbiol 65:2685–2690

Jespersen DN, Sorrell BK, Brix H (1998) Growth and root oxygen release by *Typha latifolia* and its effects on sediment methanogenesis. Aquat Bot 61:165–180

Justin SHFW, Armstrong W (1987) The anatomical characteristics of roots and plant response to soil flooding. New Phytol 106:465–495

Killham K, Yeomans C (2001) Rhizosphere carbon flow measurements and implications: from isotopes to reporter genes. Plant Soil 232:91–96

Kim JD, Jugsujinda A, Carbonell-Barrachina AA, Delaune RD, Patrick WH Jr (1999) Physiological functions and methane and oxygen exchange in Korean rice cultivars under controlled soil redox potential. Bot Bull Acad Sin 40:185–191

King GM (1994) Associations of methanotrophs with the roots and rhizomes of aquatic vegetation. Appl Environ Microbiol 60:3220–3227

King GM, Garey MA (1999) Ferric iron reduction by bacteria associated with the roots of freshwater and marine macrophytes. Appl Environ Microbiol 65:4393–4398

Klemedtsson L, Berg P, Clarholm M, Schürer J (1987) Microbial nitrogen transformations in the root environment of barley. Soil Biol Biochem 19:551–558

Kludze HK, Delaune RD, Patrick WH Jr (1993) Aerenchyma formation and methane and oxygen exchange in rice. Soil Sci Soc Am J 57:386–391

Kowalchuk GA, Stephen JR (2001) Ammonia-oxidising bacteria: a model for molecular microbial ecology. Annu Rev Microbiol 55:485–529

Kowalchuk GA, Bodelier PLE, Heilig GHJ, Stephen JR, Laanbroek HJ (1998) Community analysis of ammonia-oxidising bacteria, in relation to oxygen availability in soils and root-oxygenated sediments, using PCR, DGGE and oligonucleotide probe hybridisation. FEMS Microbiol Ecol 27:339–350

Krüger M, Frenzel P, Conrad R (2001) Microbial processes influencing methane emission from rice fields. Glob Change Biol 7:49–63

Küsel K, Pinkart H, Drake HL, Devereux R (1999) Acetogenic and sulfate reducing bacteria inhabiting the rhizoplane and the deep cortex cells of the sea grass *Halodule wrightii*. Appl Environ Microbiol 65:5117–5123

Laan P, Smolders A, Blom CWPM, Armstrong W (1989) The relative roles of internal aeration, radial oxygen losses, iron exclusion and nutrient balances in flood-tolerance of *Rumex* species. Acta Bot Neerl 38:131–145

Laanbroek HJ (1990) Bacterial cycling of minerals that affect plant growth in waterlogged soils: a review. Aquat Bot 38:109–125

Lamers PM, Tomassen HBM, Roelofs (1998) Sulphate-induced eutrophication and phytotoxicity in freshwater wetlands. Environ Sci Technol 32:199–205

Lee N, Nielsen PH, Andreasen KH, Juretschenko S, Nielsen JP, Schleifer KH, d Wagner M (1999) Combination of fluorescent in situ hybridisation and microautoradiography – a new tool for structure-function analyses in microbial ecology. Appl Environ Microbiol 65:1289–1297

Lehmann-Richter S, Grosskopf R, Liesack W, Frenzel P, Conrad R (1999) Methanogenic archaea and CO_2-dependent methanogenesis on washed rice roots. Environ Microbiol 1:159–166

Lorenzen J, Hauer Larsen L, Kjaer T, Revsbech NP (1998) Biosensor determination of the microscale distribution of nitrate, nitrate assimilation, nitrification, and denitrification in a diatom inhabited freshwater sediment. Appl Environ Microbiol 64:3264–3269

Lovell CR, Piceno YM, Quattro JM, Bagwell CE (2000) Molecular analysis of diazotroph diversity in the rhizosphere in the rhizosphere of the smooth cordgrass, *Spartina alterniflora*. Appl Environ Microbiol 66:3814–3822

Lovley DR (1995) Microbial reduction of iron, manganese and other metals. Adv Agron 54:175–231

Lovley DR, Phillips EJP (1987) Competitive mechanisms for inhibition of sulphate reduction and methane production in the zone of ferric iron reduction in sediments. Appl Environ Microbiol 53:2636–2641

Lu Y, Wassmann R, Neue HU, Huang C (1999) Impact of phosphorus supply on root exudation, aerenchyma formation and methane emission of rice plants. Biogeochemistry 47:203–218

Marschner H (1995) Mineral nutrition of higher plants. Academic Press, London, 889 pp

Minoda T, Kimura M, Wada E (1996) Photosynthates as dominant source of CH_4 and CO_2 in soil water and CH_4 emitted to the atmosphere from paddy fields. J Geophys Res 101:21091–21097

Molisch H (1887) Über Wurzelausscheidungen und deren Einwirkung auf organische Substanzen. Sitzungsber Akad Wiss Wien Math Nat Kl 96:84–109

Moorhead KK, Reddy KR (1988) Oxygen transport through selected aquatic macrophytes. J Environ Qual 17:138–142

Nielsen LB, Finster K, Welsh DT, Donnely A, Herbert RA, DeWit R, Lomstein BA (2001) Sulphate reduction and nitrogen fixation rates with roots, rhizomes and sediments from *Zostera noltii* and *Spartina maritima* meadows. Environ Microbiol 3:63–71

Nijburg JW, Coolen JL, Gerards S, Klein Gunnewiek PJA, Laanbroek HJ (1997) Effects of nitrate availability and the presence of *Glyceria maxima* on the composition and activity of the dissimilatory nitrate-reducing bacterial community. Appl Environ Microbiol 63:931–937

Patrick WH Jr, Jugsujinda A (1992) Sequential reduction and oxidation of inorganic nitrogen, manganese, and iron in flooded soil. Soil Sci Soc Am J 56:1071–1073

Pedersen O, Sand-Jensen K, Revsbech NP (1995) Diel pulses of O_2 and CO_2 in sandy lake sediments inhabited by *Lobelia dortmanna*. Ecology 76:1536–1545

Peters V, Conrad R (1996) Sequential reduction processes and initiation of CH_4 production upon flooding of oxic upland soils. Soil Biol Biochem 28:371–382

Ponnaperuma FN (1984) Effects of flooding on soils. In: Kozlowski TT (ed) Flooding and plant growth. Academic Press, New York, pp 9–45

Prade K, Trolldenier G (1990) Denitrification in the rhizosphere of rice and wheat seedlings as influenced by plant K-status, air filled porosity and substrate organic matter. Soil Biol Biochem 22:769–773

Radajewski S, Ineson P, Parekh NR, Murrell JC (2000) Stable-isotope probing as a tool in microbial ecology. Nature 403:646–649.

Ratering S, Schnell S (2000) Localisation of iron-reducing activity in paddy soil by profile studies. Biogeochemistry 48:341–365

Reddy KR, D'Angelo EM, Debusk TA (1989a) Oxygen transport through aquatic macrophytes: the role in wastewater treatment. J Environ Qual 19:261–267

Reddy KR, Patrick WH Jr, Lindau CW (1989b) Nitrification and denitrification at the plant root-sediment interface. Limnol Oceanogr 34:1004–1013

Reichardt W, Mascariña G, Padre B, Doll J (1997) Microbial communities of continuously cropped, irrigated rice fields. Appl Environ Microbiol 63:233–238

Reinhold-Hurek B, Hurek T (1998) Life in grasses: diazotrophic endophytes. Tree 6:139–143

Risgaardpedersen N, Jensen K (1997) Nitrification and denitrification in the rhizosphere of the aquatic macrophyte *Lobelia dortmanna* L. Limnol Oceanogr 42:529–537

Robertson LA, Dalsgaard T, Revsbech NP, Kuenen JG (1995) Confirmation of 'aerobic' denitrification in batch cultures, using gas chromatography and ^{15}N mass spectrometry. FEMS Microbiol Ecol 18:113–120

Roden EE, Wetzel RG (1996) Organic carbon oxidation and suppression of methane production by microbial Fe(III) oxide reduction in vegetated and unvegetated freshwater wetland sediments. Limnol Oceanogr 41:1733–1748

Rothfuss F, Conrad R (1993). Thermodynamics of methanogenic intermediary metabolism in littoral sediment of Lake Constance. FEMS Microbiol Ecol 12:265–276

Schipper LA, Reddy KR (1996) Determination of methane oxidation in the rhizosphere of *Sagittaria lancifolia* using methyl fluoride. Soil Sci Soc Am J 60:611–616

Scheid D, Stubner S (2001) Structure and diversity of gram-negative sulphate-reducing bacteria on rice roots. FEMS Microbiol Ecol 36:175–183

Segers R (1998) Methane production and methane consumption: a review of processes underlying wetland methane fluxes. Biogeochemistry 41:23–51

Semenov AM, VanBruggen AHC, Zelenev VV (1999) Moving waves of bacterial populations and total organic carbon along roots of wheat. Microb Ecol 37:116–128

Smolders A, Roelofs JGM (1993) Sulphate-mediated iron limitation and eutrophication in aquatic ecosystems. Aquat Bot 46:247–253

Smolders A, Roelofs JGM (1995) Internal eutrophication, iron limitation and sulphide accumulation due to the inlet of River Rhine water in peaty shallow waters in the Netherlands. Arch Hydrobiol 133:349–365

Snowden RED, Wheeler (1995) Chemical changes in selected wetland plant species with increasing Fe supply, with specific reference to root precipitates and Fe tolerance. New Phytol 131:503–520

Sorrell BK (1999) Effect of external oxygen demand on radial oxygen loss by *Juncus* roots in titanium citrate solutions. Plant Cell Environ 22:1587–1593

Sorrell BK, Armstrong W (1994) On the difficulties of measuring oxygen release by root systems of wetland plants. J Ecol 82:177–183

Sposito G (1984) The surface chemistry of soils. Oxford University Press, Oxford, 234 pp

Stienstra AW, Both GJ, Gerards S, Laanbroek HJ (1993) Numbers of nitrite-oxidising bacteria in the root zone of grassland plants. FEMS Microbiol Ecol 12:207–214

Stoltzfus JR, So R, Malarvithi PP, Ladha JK, de Bruijn FJ (1997) Isolation of endophytic bacteria from rice and assessment of their potential for supplying rice with biologically fixed nitrogen. Plant Soil 194:25–36

Stubner S, Wind T, Conrad R (1998) Sulphur oxidation in rice field soil: activity, enumeration, isolation and characterisation of thiosulphate-oxidising bacteria. Syst Appl Microbiol 21:569–578

Teplitski M, Robinson JB, Bauer WD (2000) Plants secrete substances that mimic bacterial *N*-acyl homoserine lactone signal activities and affect population density-dependent behaviours in associated bacteria. MPMI 13:637–648

VanBodegom P, Goudriaan J, Leffelaar P (2001a) A mechanistic model on methane oxidation in a rice rhizosphere. Biogeochemistry 55:145–177

VanBodegom P, Stams F, Mollema L, Boeke S, Leffelaar P (2001b) Methane oxidation and competition for oxygen in the rice rhizosphere. Appl Environ Microbiol 67:3586–3597

VandeGeijn SC, Vos J, Groenwold J, Goudriaan J, Leffelaar PA (1994) The Wageningen rhizolab-facility to study soil-root-shoot-atmosphere interactions in crops. Plant Soil 161:275–287

Visser EJW, Colmer TD, Blom CWPM, Voesenek LACJ (2000) Changes in growth, porosity, and radial oxygen loss from adventitious rots of selected mono-and dicotyledonous wetland species with contrasting types of aerenchyma. Plant Cell Environ 23:1237–1245

Wang B, Adachi K (2000) Differences among rice cultivars in root exudation, methane oxidation, and populations of methanogenic and methanotrophic bacteria in relation to methane emission. Nutr Cycl Agroecosyst 58:349–356

Welsh DT (2000) Nitrogen fixation in seagrass meadows: regulation, plant-bacteria interactions and significance to primary productivity. Ecol Lett 3:58–71

Welsh DT, Bartoli M, Nizzoli D, Castaldelli G, Riou SA, Viaroli P (2000) Denitrification, nitrogen fixation, community primary productivity and inorganic-N and oxygen fluxes in an intertidal *Zostera noltii* meadow. Mar Ecol Prog Ser 208:65–77

Whipps JM (1990) Carbon economy. In: Lynch JM (ed) The rhizosphere. Wiley, Chichester, pp 59–97

Whitman WB, Bowen TL, Boone DR (1992) The methanogenic bacteria. In: Balows A, Trüper HG, Dworkin M, Harder W, Schleifer KH (eds) The prokaryotes. Springer, Berlin Heidelberg New York, pp 719–767

Wind T, Conrad R (1995) Sulphur compounds, potential turnover of sulphate and thiosulphate, and numbers of sulphate-reducing bacteria in planted and unplanted paddy soil. FEMS Microb Ecol 18:257–266

Wind T, Conrad R (1997) Localisation of sulphate reduction in planted and unplanted rice field soil. Biogeochemistry 37:253–278

Zehnder AJB, Stumm W (1988) Geochemistry and biogeochemistry of anaerobic habitats. In: Zehnder AJB (ed) Biology of anaerobic microorganisms. Wiley, New York, pp 1–38

Zelles L (1999) Fatty acid patterns of phospholipids and lipopolysaccharides in the characterisation of microbial communities. Biol Fertil Soils 29:111–129

Zumft WG (1997) Cell biology and molecular basis of denitrification. Microbiol Rev 61:533–616

14 Root–Animal Interactions

J.B. Whittaker

14.1 Introduction

In view of the relative inaccessibility and invisibility of belowground organisms, and the prevalence of invertebrates among root herbivores, it is not surprising that understanding of belowground ecological processes has proceeded at a much slower rate than aboveground studies. When Cragg (1961) was writing about soil animals, almost all of the information available to him was descriptive, or, if quantitative, was unable to go much beyond population counts and biomass estimates.

Thirty years later, in an excellent review, Brown and Gange (1990) brought together some 300 references of work on belowground herbivory by insects up to that date, though it was still the case that relatively little experimental work had been attempted. Their findings and conclusions still stand:

Most belowground feeders are different species or different life-cycle stages from foliar feeders and they often have long life cycles, perhaps because of their access only to low quality food sources; an extreme example being *Magicicada septendecim* feeding below ground on roots of trees for up to 17 years.

Plant damage is not necessarily proportional to herbivore density, much of it occurring at lower densities before intraspecific competition between the herbivores trips in. In general, root-feeding can be very destructive and significantly reduce plant growth and species richness often by modifying the competitive balance between species of plants.

Up to 70 % reduction in plant biomass can occur as a result of belowground herbivory, which could translate in agricultural situations to 50 % loss in yield. Some plant compensation may take place, for example, through stimulation of lateral roots, and even overcompensation leading to some short-term benefits for the plant.

Here we will largely focus on work published since that review. In particular, the call of Brown and Gange (1990) for experimental and interactive studies has begun to be met and consideration of these will form the core of this chapter. Experimental studies of herbivory had largely concentrated on

Ecological Studies, Vol. 168
H. de Kroon, E.J.W. Visser (Eds.)
Root Ecology

aboveground processes, but, in recent years, these have extended to root–animal interactions. It was already clear that belowground herbivory is an important phenomenon which cannot be neglected in any attempt to understand community structure and functioning, plant population dynamics and the links between processes taking place above- and belowground.

After briefly considering the organisms involved in root herbivory, indirect effects on roots of aboveground grazing are discussed followed by sections on direct herbivory on roots and the interactions between above- and belowground herbivory. Physiological responses of individual plants are then considered and the whole placed in the context of community responses.

14.2 The Organisms Involved

It will be shown later that a wide range of herbivorous animals grazing, browsing or sucking on plant tissues aboveground may influence root morphology, productivity and biomass, and functioning. Those specialising in root herbivory, however, are the plant-feeding nematodes (Stanton 1988; Mortimer et al. 1999), representatives of a restricted number of families in nine insect orders, predominantly, Collembola, Hemiptera, Isoptera, Orthoptera, Coleoptera, Diptera and Lepidoptera (Brown and Gange 1990), and rodents which display fossorial behaviour in order to access plant roots as food material (Andersen 1987).

Rarely has the root fauna of a plant species been so thoroughly investigated as that for *Centaura* spp. in Europe (Müller 1989; Fig. 14.1). A total of 21 insect species of root herbivores was found on *C. maculosa*, 12 on *C. diffusa* and 11 on *C. vallesiaca*.

The range of invertebrates feeding on living roots of clover (*Trifolium repens*) in sheep-grazed pasture in North Wales was investigated by Bayliss et al. (1986) using ^{32}P as a radiotracer. Those found to have taken up ^{32}P, and therefore presumed to have fed on roots or root exudates, included some earthworms (*Lumbricus rubellus*, *Aporrectodea caliginosa*, *A. longa*, but not *Allolobophora chlorotica* and *A. longa*), Enchytraeidae, mites (Acari), Collembola, Nematoda, larvae of Diptera and Coleoptera, adult beetles and spiders. The surprising inclusion of earthworms in this list echoes the observations by Piearce (1978) that some earthworms may feed on plant roots or exudates, a phenomenon also observed by Gunn and Cherrett (1993) in their rhizotron studies allowing direct observation of root herbivory. Their findings are summarised in Table 14.1. Whilst these are qualitative records of invertebrates feeding on roots, quantitative studies made in the field suggest that they can be very abundant, though often with highly aggregated distributions. There can be, for instance, 1–6 × 10^6 nematodes m^{-2} of grassland soil of which between 50 and 80 % may be plant feeders (Cragg 1961; Yeates et al. 1997).

Root structures infested by phytophagous insect species	Centaurea Mac.	diff	vall.
1. Central meristem of rosette			
Stenodes straminea Haw.	+	+	+
Pegohylemyia centaureae Hennig	+	+	+
2. Root collar			
Apion pentrans Germer	+	+	-
Apion onopordi Kirby	+	+	+
Apion orientale Gerst.	+	-	-
Apion alliariae Herbst	+	-	-
Cheilosia sp.	+	+	-
3. Central vascular tissue			
Cyphocleonus achates Faber	+	+	-
Cleonus piger Scop.	+	+	-
Sphenoptera jugoslavica Ob.	+	+	-
Pterolonchа inspersa Stg.	+	+	-
4. Root cortex			
Agapeta zoegana L.	+	-	-
Pelochrista medullana Stgr.	+	+	-
5. External on root			
Trama centaureae CB	?	+	?
Sminthurodes betae Westw.	?	+	?

Feeding niches

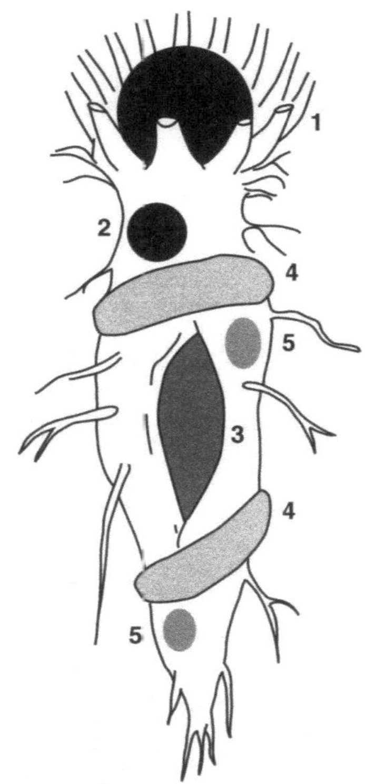

Fig. 14.1. Common insect species associated with the roots of *Centaurea maculosa* (*Mac.*), *C. diffusa* (*diff.*) and *C. vallesiaca* (*vall.*) in Europe. (Müller 1989)

Table 14.1. Groups of root herbivores observed in a rhizotron in grassland. (Simplified from Gunn and Cherrett 1993)

	Main lateral roots	Root tips	Root hairs	Trans-parent rootlets	Decaying roots	Mycor-rhizae
Earthworms	x		x	x	x	x
Enchytraeidae	x		x		x	
Mollusca	x	x	x		x	x
Isopoda					x	x
Acari	x					
Diplopoda	x		x		x	
Symphyla	x					
Insecta						
Diplura	x	x	x	x	x	x
Collembola	x	x	x	x	x	x
Coleoptera						
Elateridae larvae	x					
Diptera larvae	x					
Homoptera Aphidoidea	x					

Leather-jackets (*Tipula paludosa*; Tipulidae) may reach densities of 250 m^{-2} (French 1969) whilst Coulson and Whittaker (1978) calculated that productivity of root-feeding tipulid larvae on a range of habitats on the Moor House National Nature Reserve in northern England was comparable to, or exceeded that, of the sheep. Salt et al. (1996b) found mean densities of root-feeding aphids on Sitka spruce (*Picea sitchensis*) of 1625 m^{-2}, although population turnover rates were much slower than in comparable aboveground aphid populations. These densities compare with 1950–17,700 m^{-2} recorded by Macfadyen (1952) in grasslands. Strong et al. (1995) found that 197 out of 250 sampled roots of bush lupines (*Lupinus arboreus*) in an area of bush die-back contained evidence of ghost moth larval damage.

Although Gunn and Cherrett (1993) were not able to detect much selective feeding by root herbivores in grassland, Crutchfield and Potter (1994) found that the Japanese beetle (*Popillia japonica*) consistently showed preference for perennial ryegrass (*Lolium perenne*) over all other turf grasses tested, whilst the southern masked chafer beetle (*Cyclocephala lurida*) showed no such preference, nor did it discriminate between endophyte-infected and endophyte-free *Festuca arundinacea*. Even among the plant-feeding nematodes, there are different feeding strategies (Yeates et al. 1999).

14.3 Indirect Effects of Aboveground Grazing on Roots

It has long been argued that the effects of aboveground grazing on growing plant tissue are as likely to be seen in changes in root growth and morphology as in changes aboveground (Jameson 1963). This might be expected because it is known that damage to one part of a plant by herbivores is frequently seen in a response elsewhere. Phloem-feeding insects, for example, may act as physiological sinks (Llewellyn 1982). Re-allocation of carbon from roots to re-growing shoots is a well-documented response to aboveground herbivory (Ryle and Powell 1975; Briske and Richards 1994).

Many of these findings cited above are derived, however, from short-term pot experiments (McNaughton et al. 1998), and some more recent field experiments have thrown some doubt on this generalisation and tend to suggest, that at least in some grasslands, grazing has little effect on belowground production (Matter 2001). McNaughton et al. (1998) sampled grazed and ungrazed (22–25 year exclosure from large wild mammal grazing) roots in a range of Serengeti grasslands. They found no evidence that grazing affected root productivity in the long or short term. On the other hand, Bortolus et al. (1998) found that rhizome and root biomass of seagrass (*Ruppia maritima*) was reduced by water fowl grazing and Ford and Grace (1998) identified a decrease in root zone expansion in a coastal marsh when grazed by muskrat, rabbit, racoon, and particularly nutria (coypu) and wild boar.

Herbivory experiments, especially those concerned with effects of large grazers, sometimes use artificial defoliation to simulate grazing. In maize, defoliation had no effect on root biomass (Collantes et al. 1998), whereas artificial defoliation of sand bluestem grass (*Andropogon hallii*) in containers in the natural habitat of the plant resulted in reduced root biomass (Engels et al. 1998). However, experiments using artificial defoliation should be interpreted with caution since plants often respond differently to clipping and real herbivory (e.g. Bentley and Whittaker 1979).

By contrast, some observations used small rhizotrons (Hendrick and Pregitzer 1992) into which video cameras were inserted to permit monthly images of roots to be taken in Alaskan taiga plots. These were browsed by snowshoe hare (*Lepus americanus*) and moose (*Alces alces*) or protected from browsing by exclosures (Ruess et al. 1998). Aboveground herbivory was found to significantly reduce fine root production (annual unbrowsed: 453.8±49.8; browsed 311.4±31.7 mm tube^{-1} year^{-1}) over a 3-year period (Fig. 14.2) and turnover of fine roots was higher in the browsed plots.

Combined aboveground cattle grazing and belowground invertebrate herbivory by larvae of June beetles was studied by Coffin et al. (1998) who found that the combined effects led to greater plant mortality on short-grass steppes than in ungrazed conditions.

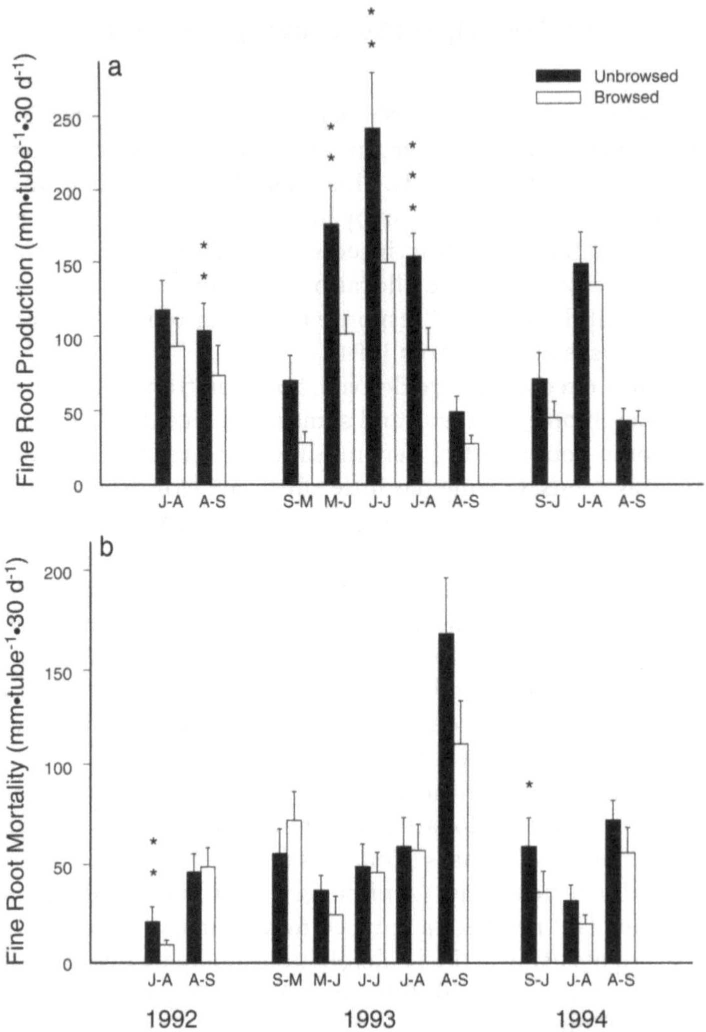

Fig. 14.2. **a** Fine root production and **b** mortality for Alaskan taiga forest stands browsed and unbrowsed by moose and snowshoe hares, measured over ten time intervals from 1992 to 1994 (*J*=June in both 1992 and 1994). Measurements are of mm root length per mini-rhizotron tube per 30 days±1 SE. (Ruess et al. 1998)

There are fewer studies of effects of aboveground grazing by invertebrates on root performance, but Bentley and Whittaker (1979) found that a chrysomelid beetle, *Gastrophysa viridula*, grazing on *Rumex crispus* significantly reduced root dry weight and resulted in a decrease in root/shoot (R/S) ratio, whereas, on *Rumex obtusifolius*, the same level of grazing resulted in increased R/S ratio. These differences in the responses of two closely related plant species explained their relative abundance on a frequently flooded river shingle bank (Whittaker 1982) because the larger root stock of *R. obtusifolius* provided adequate anchorage in floods whereas the reduced root stock of the *R. crispus* did not. A combination of environmental stresses was also found in a study by Bach (1998) who showed that it was the interactive effects of leaf and apical meristem herbivory by the tortoise beetle (*Aspidomorpha deusta*) and the noctuid caterpillar (*Aedia leucomelas*) and sand burial which led to reduction of root producing nodes of the tropical dune plant, *Ipomoea pes-caprae*.

Even established trees may show responses belowground to repeated insect defoliations. When Kosola et al. (2001) manipulated defoliation of hybrid poplar by the fire tussock moth (*Lymantria dispar*) they found effects on whole-tree physiology including depression of nitrate and ammonium uptake and transient declines in carbon allocation to starch in the fine roots. Root production and mortality, however, were not affected.

Scale insects which feed aboveground as phloem suckers on seedlings of the eucalypt *Eucalyptus blakelyi* consistently affect roots more than shoots (Vranjic and Ash 1997). This was especially noticeable at relatively low densities of the insects: at higher densities, all parts of the seedlings were affected (Fig. 14.3). Decreases in volume and dry weight of roots also occurred in

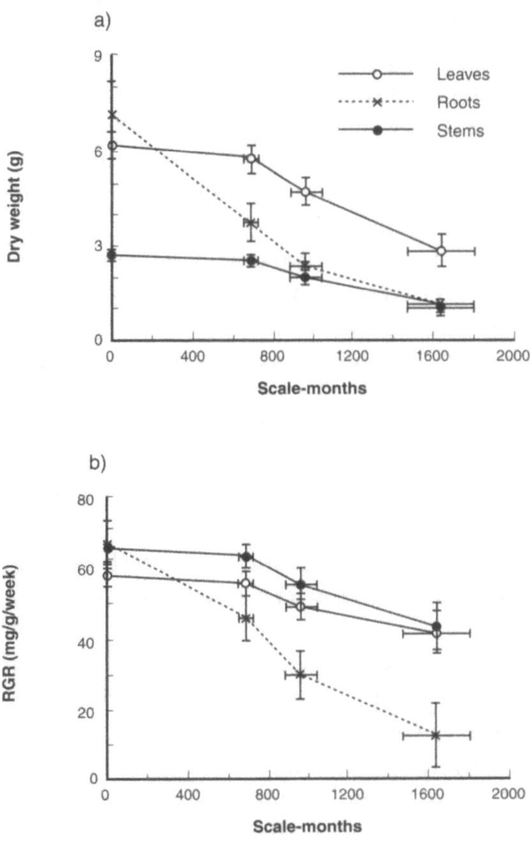

Fig. 14.3. **a** Dry weight and **b** relative growth rate (mg g^{-1} $week^{-1}$); means (±SE) of roots, stems and leaves of seedlings of *Eucalyptus blakelyi* subjected to different infestation loads (summed monthly population over the period of the experiment) of the scale insect *Eriococcus coriaceus*. (Vranjic and Ash 1997)

aphid infestations of Douglas-fir (*Pseudotsuga menziesii*) seedlings (Smith and Schowalter 2001) and willow trees (Collins et al. 2001).

Despite such consequences of aboveground herbivory to root growth, Morón-Ríos et al. (1997) found no effect of simulated aboveground defoliation of the grass *Muhlenbergia quadridentata* on survival of the root-feeding beetle *Phyllophaga* sp.

These examples underline the difficulty of making meaningful generalisations about the effects of aboveground grazing on roots. The response of some plants to aboveground herbivory will be to re-allocate to new shoot growth at the expense of roots whereas others will accumulate root material or produce more root exudates as aboveground plant parts are damaged, some of which may act as attractants to root feeders of the same or neighbouring plants (Hibbard and Bjostad 1988; Dicke and Dijkman 2001).

14.4 Direct Herbivory on Roots

Belowground herbivory by insects has been thoroughly reviewed by Brown and Gange (1990). There is also an extensive literature on nematode herbivory of roots (Ingham and Detling 1986), whilst Anderson (1987) and Stanton (1988) review belowground herbivory in more general terms. Scott et al. (1979) calculated that whereas some 2–7 % of total primary production in grasslands was consumed aboveground, 7–20 % was lost belowground. They concluded that plant-feeding nematodes were the major consumers in grassland, accounting for 46–67 % of total plant consumption.

Prominent among root-feeding insects are the Scarabaeidae (Coleoptera). Their feeding illustrates an important consequence of root herbivory, which is that the impact may be seen in the aboveground foliage as well as belowground. Ridsdill-Smith (1977) showed this to be the case in a study of *Sericesthis nigrolineata*, a pest of pastures in New South Wales. In greenhouse pots of perennial rye grass containing different densities of the beetle larvae, there was a reduction in new root and foliage growth accompanied by water stress in the foliage. These effects were seen at low insect densities, and did not increase linearly with insect numbers, a phenomenon that has been observed in some other studies of root grazing. The explanation seems to be that at higher densities individual larvae eat less and grow at a slower rate (Ridsdill-Smith and Roberts 1976).

Similar aboveground consequences were seen when potted *Carduus nutans* plants were exposed to attack by *Hadroplontus trimaculatus* (weevil) larvae and *Cheilosia corydon* (syrphid fly) larvae feeding on the root crown and root vascular tissue and meristem, respectively (Sheppard et al. 1995). Weevil damage resulted in significantly more stems being produced whilst fly damage caused significantly fewer seeds to be produced. Diverse consequences of root-feeding by different herbivore species was also observed by Steinger and Müller-Schärer (1992) when a lepidopteran caterpillar, *Agapeta zoegama*, and a weevil, *Cyphocleonus achates*, were added to pots containing *Centaurea maculosa* seedlings. The moth caterpillar had no significant effect on shoot or root biomass whilst the weevil reduced shoot mass, but not root mass. This was not explained by differential amounts of tissue consumption, but more probably by the consumption of different tissues, i.e. they occupied different feeding niches. In particular, the weevil may have destroyed the xylem vessels preventing compensatory root growth above the site of damage.

Experimental weevil (*Sitona flavescens*) infestations (foliar and root-feeding) also resulted in reduction in foliar biomass in white clover (*Trifolium repens*) seedlings (Murray et al. 1996). This was accompanied by impaired nitrogen fixation, a significant increase in C/N ratios and a higher proportion of amino-N (mainly aspartic acid and serine) in the root exudate of infested plants.

Simulated belowground grazing by root-clipping also resulted in reduction of tiller production in *Bouteloua gracilis* (Detling et al. 1980) and 35 % lower net photosynthesis and a reduction in the proportion of photosynthetically fixed ^{14}C allocated to the roots in the first few days after treatment. This, however, gradually switched to an increase in allocation to roots compared with controls, by 3 weeks later. Gange and Brown (1989) used the leaf-feeding aphid *Aphis fabae* as a bioassay to investigate the effect of root herbivory by a scarabaeid, *Phyllopertha horticola*, on growth of *Capsella bursa-pastoris*. A combination of low watering and root herbivory caused a reduction in vegetative biomass and an increase in the proportion of material allocated to reproduction in the plant, coinciding with an increase in mean aphid weight and growth rate. Ample watering reduced, but did not eliminate, these effects.

Little is known about aphid damage to roots, but Brown et al. (1991) reported that *Eriosoma lanigerum*, the apple woolly aphid, caused galls and reduced growth of roots of apple which led to reduced yield in orchards (Brown et al. 1995). There was a proliferation of non-functional xylem tissue in the galls which was highly resistant to water flow and may have contributed to the observed decrease in plant growth. *Pachypappa vesicalis* feeding on roots of Sitka spruce is known to reduce root extension and cause root-bending close to the tip (Fig. 14.4; Cook 1996).

The effects of aboveground herbivory are often seen in changes in competitive advantage of the affected plants (Whittaker 1979). Evidence is emerging that this is similarly true for belowground herbivory. An early example of this was the study by Sibma et al. (1964) of the effects of root nematode (*Heterodera avenae*) feeding on the outcome of competition between barley and oat plants in a pot experiment. In the presence of 500–600 nematodes per 200 ml of soil, the competitive advantage enjoyed by oats was almost completely lost. Recently, Verschoor (2001) has been shown that plant-feeding

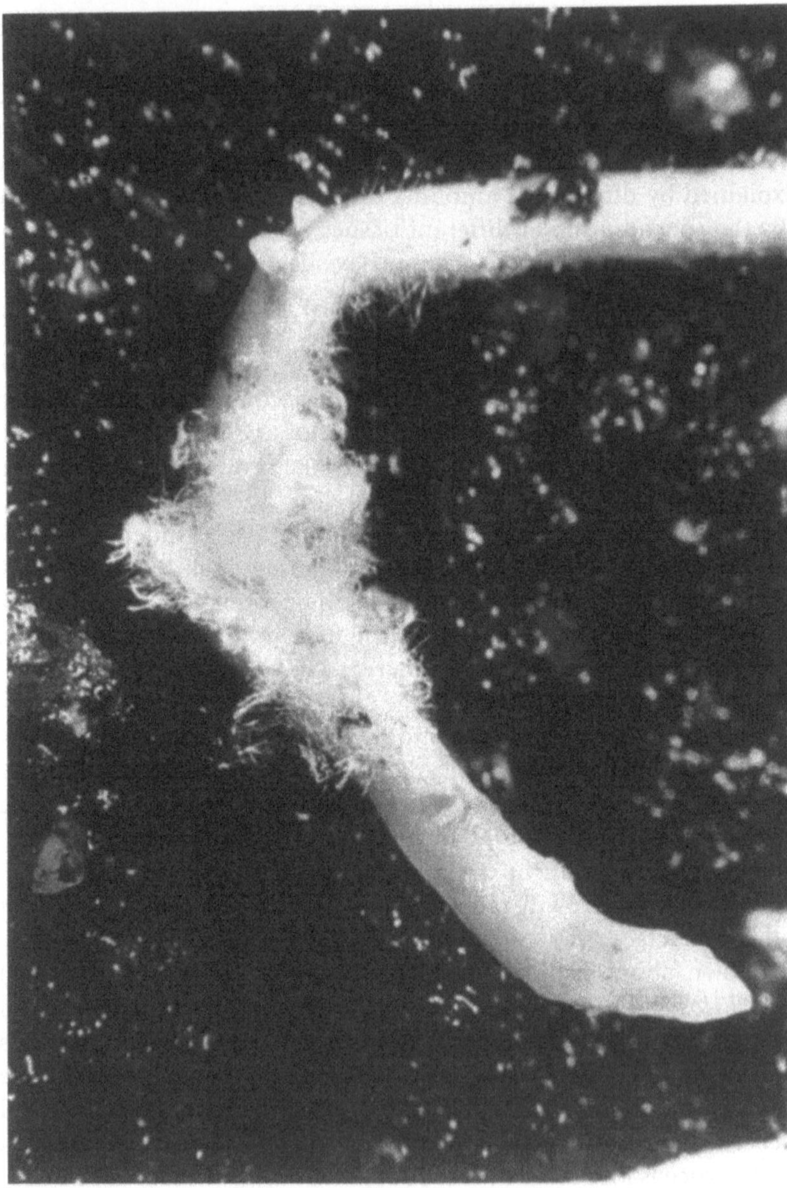

Fig. 14.4. Root bending associated with phloem feeding by *Pachypappa versicalis* (Aphididae) on roots of Sitka spruce. (Photo by P.A. Cook)

nematodes can reduce the competitive suppression of *Anthoxanthum odoratum* by *Holcus lanatus* under conditions of low nutrient availability, thus affecting the outcome of reverse succession. A similar interaction between root herbivory and competition was found by Steinger and Müller-Schärer (1992) in their study of root herbivory on *Centaurea maculosa*. Competition of the *Centaurea* with grass (*Festuca pratensis*) was intense and significantly affected all plant growth parameters. Root feeding by the weevil *C. achates* mainly reduced shoot growth and hence shoot/root ratios. However, production of new leaves in the following year of plants attacked by the weevil depended on whether the plants had been in competition or not, and on nitrogen availability. The dependence on nitrogen for plant compensation for root herbivory might be expected to be reduced when the *Centaura* were competing with grass.

When purple loosestrife (*Lythrum salicaria*) was exposed to larvae of the root-feeding weevil *Hylobius transversovittatus*, or separately to competition with the grass *Phleum pratense*, there was no effect of either on the purple loosestrife in year one (Notzold et al. 1998). However, if the plants were simultaneously exposed to root herbivory and competition, there was a significant reduction in plant height and shoot biomass. Both herbivory and competition separately affected plant growth parameters in the second year of growth, but the interaction of herbivory and competition added an extra response by significantly delaying flowering. Müller-Shärer (1991) and McEvoy et al. (1993) also found evidence of interaction between competition and root herbivory in *Centaurea maculosa* and a root-feeding caterpillar, *Agapeta*, and in ragwort and the rag-wort flea beetle, respectively. It is difficult to predict the outcome of these interactions, as Wardle and Barker (1997) found in their experimental study of aboveground and root herbivory and competition in grassland plant species. Different species responded quite differently to competition and herbivory, including root herbivory.

It is not yet known whether these examples are exceptional or whether interactions between root herbivory and competition are commonplace. Sometimes the effect of root herbivory on a plant is contrary to expectation. When Ramsell et al. (1993) exposed *Lolium perenne*, which was in competition with *Rumex obtusifolius*, to root grazing by the leather-jacket *Tipula paludosa*, the grass became a more effective competitor despite the clear preference of the insect for the grass. It appeared that a response of the grass to grazing was a relative increase in its shoot production leading to increased competitive ability.

Evidence of mortality of plants subjected to root herbivory is hard to obtain in the field. Strong et al. (1995) reported deaths of individual lupine bushes (*Lupinus arboreus*) and, in some cases, large stands caused by the caterpillars of the ghost moth *Hepialus californicus* (Fig. 14.5). The highest densities (mean 37.5, max. 62 caterpillars per root) resulted in 41 % mortality in a stand of large, old *L. arboreus*. This was much more dramatic than damage by the folivorous caterpillar, the western tussock moth (*Orgyia vetusta*), which rarely caused the death of plants. Maron (1998) also recorded 18 % mortality of *L. arboreus* when subjected to root herbivory by *H. californicus*.

It is a characteristic of many of these studies that the measured impact of root herbivory on aboveground production is greater than would be pre-

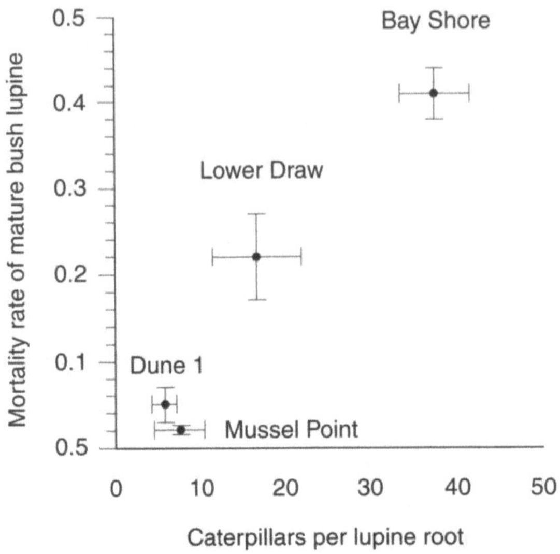

Fig. 14.5. Mortality (mean ± SE) in four stands of *Lupinus arboreus* versus number of ghost moth caterpillars per lupine root in each stand. (Strong et al. 1995)

dicted from calculations based on the insect's requirements for food (Detling et al. 1980). Compensation by the affected plant is also reported (e.g. Quinn and Hall 1992; Steinger and Müller-Schärer 1992) just as it is in aboveground herbivory.

14.5 Interactions Between Above- and Belowground Herbivory

The known frequency of herbivore attack above- and belowground makes it unlikely that plants often suffer one without the other. Unlike in most of the experiments summarised above, where they have been deliberately separated, they may often be in a position to interact. Although it has been argued (Lawton and Strong 1981) that there is little competition between phytophagous insects in the same community, there is increasing evidence that, on the same plant, interactions do occur. Morón-Ríos et al. (1997) examined the nature of this interaction by either simultaneously or separately exposing the grass *Muhlenbergia quadridentata* to artificial defoliation aboveground and beetle (*Phyllophaga* sp.) herbivory belowground. The interaction between the two was significant for all components of plant biomass. This seemed to be because in this case the plants did not compensate for the carbon drain from the grazed roots when leaves were not removed, whereas plants under heavy defoliation showed no extra loss in the presence of root feeders. Root herbivory led to decreased N content in lightly defoliated live tillers, but not when it occurred in the presence of heavy defoliation (Fig. 14.6). In fact, in the presence of heavy defoliation (removal of 70 % of plant height), above- and below-

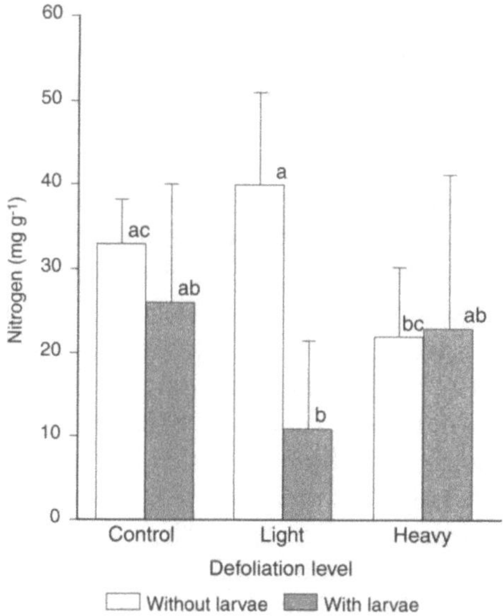

Fig. 14.6. Nitrogen content of live tillers of *Muhlenbergia* in response to defoliation and belowground herbivory (means ±1 SD; from Morón-Ríos et al. 1997)

ground herbivory appear to compensate each other in terms of effect on nitrogen content of live tillers. This seems to be a consequence of the very high defoliation which allows more young re-growth and greater photosynthetic capacity thus increasing carbon for both above- and belowground production, enabling compensation for N lost to the root feeders. Some of these consequences are the opposite of those predicted by Masters et al. (1993) and remind us that the exact conditions of herbivory (e.g. relative levels and timing) may be very important, making synthesis difficult from a range of experiments conducted under different conditions (e.g. pots, field, artificial defoliation, etc.).

In their study, Morón Ríos et al. (1997) could find no effect of simulated aboveground defoliation of the grass *Muhlenbergia quadridentata* on survival of the root-feeding beetle *Phyllophaga* sp. However, root aphid populations (*Pemphigus populitransversus*) were found by Salt et al. (1996a) to be significantly reduced in size on *Cardamine pratensis* when in the presence of *Aphis fabae* feeding on the shoots of the same plants, compared with populations on plants without shoot-feeding aphids.

There is increasing evidence that foliar herbivory, whether by vertebrates or invertebrates, results in allocation of more plant biomass to the roots, if those roots are themselves being grazed (Bardgett et al. 1998). Ingham and Detling (1986) showed in a laboratory experiment that, when the grass *Bouteloua curtipendula* was defoliated, more plant biomass was allocated to the roots in plants which were infected by root-feeding nematodes than when nematodes were absent. Presumably this is because the root damage causes more carbon to be allocated to the roots and they speculate that this may result in increased root exudation. However, the consequences of this are not clear since foliar herbivory generally reduces allocation of carbon to ungrazed roots (Bardgett et al. 1998) which may lead to reduction in the synthesis of carbon-based defence chemicals in the roots, thus encouraging root herbivory which may in turn lead to restoration of the carbon allocation to roots.

Root exudates may have feed-back consequences to root herbivory. As Kilham (1994) remarks, dormant cysts of organisms like the soil nematode *Heterodera* may persist in soils for long periods and be reactivated by exudates of a new potato crop and then attracted to feed on the roots. However, when root exudates of *Thuja occidentalis* were released after attack by a weevil (*Otiorhynchus sulcatus*), they in turn attracted a nematode parasite of the weevil larvae (Van Tol et al. 2001). In addition, Yeates et al. (1998) have shown using $^{14}CO_2$ labelling that, when white clover (*Trifolium repens*) is infested by clover cyst nematodes (*Heterodera trifolii*), there is increased leakage of carbon from the roots with consequences for the soil microflora. Thus root feeding may indirectly modify the soil microbial community in the vicinity of the plant, for example, increasing the biomass of bacteria at the expense of fungi (Denton et al. 1996).

14.6 Physiological Responses

Measurement of belowground carbon fluxes following exposure to ^{13}C and subsequent aboveground artificial defoliation have been made by Briske et al. (1996) in C-4 perennial grasses. "Herbivore-sensitive" grasses such as *Andropogon gerardii* held most of the ^{13}C which was in the roots and did not reallocate it to re-growth, whereas "herbivore-tolerant" grasses such as *Bouteloua rigidiseta* showed greatest quantities of ^{13}C in the re-growth and little in the roots. In a similar experiment which exposed maize to ^{14}C and then herbivory by an aboveground generalist herbivore, the lubber grasshopper (*Romalea guttata*), Holland et al. (1996) found an increase in fluxes of carbon belowground in the form of root biomass, root exudates and also increased rhizosphere respiration. There is also some evidence (Murray and Hatch 1994) that ryegrass in the presence of clover infested with *Sitona* (weevil) larvae took up more nitrogen than ryegrass in the presence of uninfested clover. Consequences of the aboveground herbivory of Rumex plants by beetles reported by Bentley and Whittaker (1979), and which affected root biomass, were the reduction of up to 26 % in concentrations of fructans, starch and total non-structural carbohydrates in the roots (Hatcher 1996) and the fact that, after defoliation, it may take 3 weeks for root carbohydrate concentrations to return to previous levels (Hidaka 1973).

Measurement of ^{11}C in tomato seedlings subjected to nematode root herbivory (Freckman et al.1991) also showed that nematode grazing led to decreased net photosynthetic rates and reduction in the pool of labile C in the roots, and in *Panicum coloratum*, a C-4 grass, plants with nematode root feeders had a lower stem sink activity, root activity and root transport speed. These results are in line with previous findings by Loveys and Bird (1973) and Wallace (1974).

Steinger and Müller-Schärer (1992) found that root herbivory of *Centaurea* resulted in a significant increase in N concentration in the roots, accompanied in the case of attack by a root-boring weevil in reduction of leaf nitrogen as well. Soluble carbohydrates were substantially reduced, a response also reported by Dintenfass and Brown (1988) in alfalfa (*Medicago sativa*) grazed by weevils. No effect was found on xylem water potential. The increase in nitrogen concentration is consistent with a plant that is under stress, in this case herbivory (Rabe 1990). Is this a case of the feeding herbivore modifying the plant to its own nutritional advantage, but also, of course, to the advantage of other herbivores? A consequence of water stress may be greater levels of root herbivory to add to the existing stress (Barber and Martin 1976). One of the consequences of root-feeding by aphids is the production by the aphids of large quantities of secreted wax which is extremely hydrophobic and may reduce water and nutrient acquisition by the infested roots. It is also possible that damage to roots may induce root dormancy because of interference with root ABA (Philipson and Coutts 1979).

Just as aboveground herbivores show a wide range of feeding habits, belowground there are root grazers destroying or damaging fine roots, root borers, sap feeders, etc. However, the majority of plants seem to respond to root damage from any source by mobilising amino acids and carbohydrates from roots to foliage (Masters and Brown 1996). Root herbivory also tends to decrease foliar water content and increase soluble nitrogen and carbohydrates in leaves (Masters 1995). The effect of these changes may be to make the plant more susceptible to attack by aboveground herbivores. Thus, Gange and Brown (1989) found that root damage to *Capsella bursa-pastoris* by the scarab *Phyllopertha horticola* resulted in increased weight, growth rate, fecundity and adult longevity of the aphid *Aphis fabae* feeding aboveground. They attributed this to a sequence of events in which the root feeding led to water stress and then increases in soluble nitrogen, especially proline, which benefited the aphid.

Sometimes these interactions may be reversed. When leaves of *Chenopodium album* were galled by *Hayhurstia atriplicis*, a root aphid on the same plant, *Pemphigus betae* decreased in population by up to 91 % on some plant genotypes, whereas *P. betae* had no measurable effect on *H. atriplicis* (Moran and Whitham 1990). Leaf-galling aphids are known to be a strong sink for plant nutrients (Llewellyn 1982), perhaps diverting nutrients from the root feeders.

14.7 Community Responses

Interactions like those discussed in Section 14.5 make it difficult to predict the consequences of root herbivory in stands of vegetation. Pioneer work to answer this question has been carried out at Silwood Park by V.K. Brown and her associates, by treating experimental plots of vegetation in various stages of succession with foliar systemic insecticide, soil insecticide, or a mixture of the two. The results are summarised in Fig. 14.7 (from Brown 1990) in which cover abundance of annual forbs, perennial forbs and perennial grasses, and plant species richness are used as a measure of plant performance and shifts in vegetation structure. Application of both insecticides greatly increased cover particularly of annual forbs and perennial grasses, especially in the first season. For our present purposes, however, it is particularly interesting to note that, by years two and three, it was the removal of belowground herbivores which had a bigger effect on annual forbs than removing foliar insects. The reverse was true in the case of perennial grasses such as *Holcus lanatus*. The biggest effect on species richness of the plant community was by removing soil fauna, especially in years two and three. Three annual and twelve perennial species of forbs were found only in the soil-insecticide treated plots. In fact, there was a striking difference in the effects of below- and aboveground herbivory on species richness. Foliar insecticide (removal of aboveground herbivory) depressed richness below the controls whereas removal of belowground herbivores led to a much larger number of forb (particularly perennial) species.

Denton et al. (1999) present evidence that low levels of herbivory by root-feeding nematodes, below the threshold level of damage to the food plant, *Trifolium repens*, led to significant increases in total microbial biomass in the soil. Higher levels of herbivory, beyond the damage threshold to *T. repens*, had no such effect.

Just as effects of root herbivory cannot be readily separated from consequences of aboveground herbivory on the same plants (Sect. 14.5), it is becoming increasingly clear that it has to be viewed in the context of the wider ecological community. For instance, Murray and Clements (1998) found that the root-feeding weevil *Sitona lepidus* on clover resulted in the transfer of nitrogen from the clover to adjacent wheat plants. Bardgett et al. (1998) link root herbivory with aboveground activity and discuss its consequences to other soil organisms. These in turn may be expected to interact with root herbivores. For example, the harmful effects of root herbivory may be moderated by the presence of mycorrhizas, either directly (Eissenstat et al. 2000) or by mitigating the effects of one life stage of a herbivore on another (Gange 2001). Eissenstat et al. (2000) also suggest that root life span may be shortened by herbivory and that it may indeed be a function of herbivore pressure and the extent to which the roots produce defence mechanisms

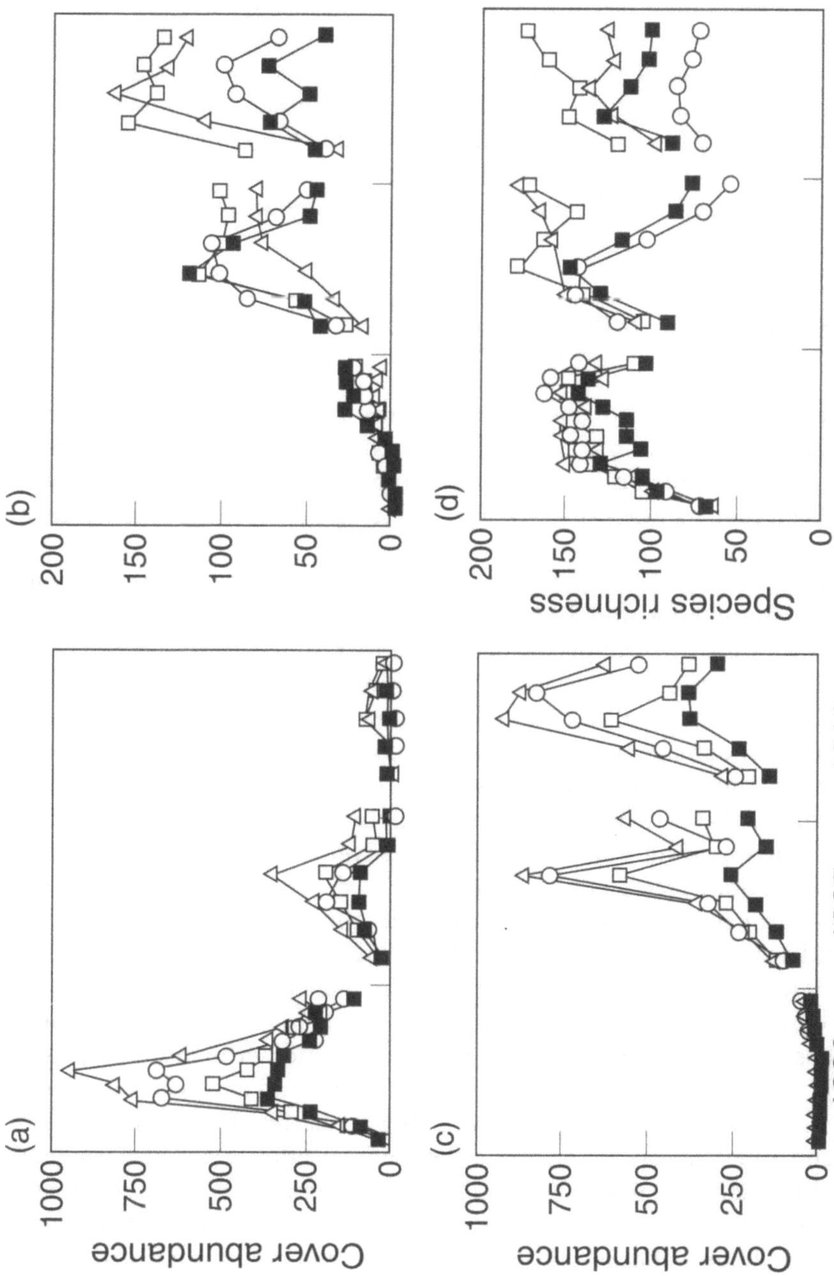

Fig. 14.7a–d. Effects of foliar and root herbivory on cover abundance and species richness of a plant community during 3 years of succession (1986–1988) at Silwood Park, Berkshire, UK. ■ Control; ○ addition of foliar insecticide; □ addition of soil insecticide, Δ addition of foliar and soil insecticide. Cover abundance of **a** annual forbs; **b** perennial forbs; **c** perennial grasses; and **d** species richness. (Brown 1990)

against herbivory. They support this argument by pointing out that, whereas a cost/benefit analysis of root retention and shedding would predict low mortality in young roots and peak mortality at optimal life span, actually survivorship curves of roots tend to show fairly constant mortality at all root ages (Kosola et al.1995) or high mortality of young roots (e.g. Hendrick and Pregitzer 1993). It is expected that young roots will be more susceptible to herbivory than older ones (see distribution of aphids in Fig. 14.4).

Root herbivory can even be seen to have effects on components of the community far removed from the root system. When Masters et al. (2001) manipulated root herbivory by applying a belowground insecticide Dursban to *Cirsium palustre* they found that not only did it lead to larger flower heads in the host plant compared with controls, but the abundance of tephritid seed predators increased, and the parasitoids of these flies also increased.

14.8 Summary and Prospects

Because it is difficult to observe and measure, root herbivory has been somewhat neglected in the ecological literature, though less so in the disciplines of agriculture and forestry where emphasis is usually placed on yield losses. Even many studies of aboveground herbivory have ignored effects on roots.

There are clear parallels between the responses of plants to root herbivores and foliage herbivores, and there are significant effects of many foliage herbivores on root growth and performance. Plants respond to root herbivory by increasing allocation of biomass to the root at the expense of shoot growth. More nitrogen may be allocated to roots allowing increased metabolic activity of roots for damage repair and possibly compensatory root growth. Sometimes there may be severe distortion of the root. There is often an increase in root exudates that can have a very important influence on the microbial communities surrounding roots and may also adversely influence plant growth and distribution. Yeates et al. (1999) attributed the microbial responses to enhanced leakage of carbon from nematode-infested plants which was then observed as increased ^{14}C activity in the surrounding microbial biomass.

Where it has been possible to separate the effects of foliar and root herbivory on the same plants in the field, there is evidence that root herbivory is at least as important a phenomenon as aboveground herbivory, though Masters et al. (1993) consider that above- and belowground herbivores interact in a plus-minus fashion (root herbivory having a positive effect on foliar herbivores and foliar herbivory having a detrimental effect on root herbivores) which is mediated by the host plant. This may be especially so in those cases where low levels of root herbivory were found to have a bigger effect on the community than higher levels. Its effectiveness has even been tested in integrated biological control such as that of *Senecio jacobaea* by the root-feeding flea-beetle *Longitarsus jacobaeae* (McEvoy et al. 1993).

Inhibition of root elongation by root herbivores may lead to reduction in root "foraging" and hence to a decrease in the availability of soil for absorption of nutrients by the plant. On the other hand, there is some evidence that some root herbivores, especially aphids, may increase the quality of food available to themselves or other root feeders by manipulation of host physiology. Indeed the whole question of how root-feeders influence physiological processes in their food plants, and in turn respond to such changes, is a fruitful area for future research.

We need to know more about how plants that are subjected to normal levels of root herbivory perform, in contrast to the "perfect" plants more usually studied in laboratory physiological studies, and what the consequences of root herbivory are to community processes and dynamics. Sometimes pot experiments may be appropriate, but it is clear from recent work at the community level that understanding responses of plants and their herbivores to events in natural field conditions is the goal and must take into account the complex links between above- and belowground processes, the role of plant competition as modified by herbivory and the significance of other changes, especially microbial, in the soil community. There is still very considerable scope for experimental studies of root herbivory, especially concerning physiological responses of the affected plants.

All these topics are ripe for future investigation and are clearly very important if we are to develop our understanding of relationships between plants and their herbivores, especially those concealed belowground.

References

Andersen DC (1987) Below-ground herbivory in natural communities: A review emphasizing fossorial animals. Q Rev Biol 62:261–286

Bach CE (1998) Interactive effects of herbivory and sand burial on growth of a tropical dune plant, *Ipomoea pes-caprae*. Ecol Entomol 23:238–245

Barber DA, Martin JK (1976) The release of organic substances by cereal roots into soil. New Phytol 76:69–80

Bardgett RD, Wardle DA, Yeates GW (1998) Linking above-ground and below-ground interactions: how plant responses to foliar herbivory influence soil organisms. Soil Biol Biochem 30:1867–1878

Bayliss JP, Cherrett JM Ford JB (1986) A survey of the invertebrates feeding on living clover roots (*Trifolium repens* L.) using 32P as a radiotracer. Pedobiologia 29:201–208

Bentley S, Whittaker JB (1979) Effects of grazing by a chrysomelid beetle, *Gastrophysa viridula*, on competition between *Rumex obtusifolius* and *Rumex crispus*. J Ecol 67:79–90

Bortolus A, Iribarne OO, Martinez MM (1998) Relationship between water fowl and the seagrass *Ruppia maritima* in a southwestern Atlantic coastal lagoon. Estuaries 21:718–717

Briske DD, Boulton TW, Wang Z (1996) Contribution of flexible allocation priorities to herbivory tolerance in C-4 perennial grasses: an evaluation with ^{13}C labeling. Oecologia 105:151–159
Briske DD, Richards JH (1994) Physiological responses of individual plants to grazing: current status and ecological significance. In: Vavra M, Laycock WA, Pieper RD (eds) Ecological implications of livestock herbivory in the west. Soc Range Manage, Denver, pp 147–176
Brown VK (1990) Insect herbivory and its effect on plant succession. In: Burdon JJ, Leather SR (eds) Pests, pathogens and plant communities. Blackwell, Oxford
Brown VK, Gange AC (1990) Insect herbivory below ground. Adv Ecol Res 20:1–58
Brown MW, Glenn DM, Wisniewski ME (1991) Functional and anatomical disruption of apple roots by the woolly apple aphid (Homoptera: Aphididae). J Econ Entomol 84:1823–1826
Brown MW, Schmitt JJ, Ranger S, Hogmire HW (1995) Yield reduction in apple by edaphic woolly apple aphid (Homoptera: Aphididae) populations. J Econ Entomol 88:127–133
Coffin DP, Laycock WA, Lauenroth WK (1998) Disturbance intensity and above- and below-ground herbivory effects on long-term (14 y) recovery of semi-arid grassland. Plant Ecol 139:221–233
Collantes HG, Gianoli E, Niemeyer HM (1998) Changes in growth and chemical defences upon defoliation in maize. Phytochemistry 49:1921–1923
Collins CM, Rosado RG, Leather SR (2001) The impact of the aphids *Tuberolachnus salignus* and *Pteocomma salicis* on willow trees. Ann Appl Biol 138:133–140
Cook PA (1996) Effects of the spruce root aphid, *Pachypappa vesicali*, on the growth and physiology of Sitka spruce seedlings. PhD Thesis, Lancaster University
Coulson JC, Whittaker JB (1978) Ecology of moorland animals. In: Heal OW, Perkins DF (eds) Production ecology of British moors and montane grasslands. Ecological Studies 27. Springer, Berlin Heidelberg New York
Cragg JB (1961) Some aspects of the ecology of moorland animals. J Ecol 49:477–506
Crutchfield BA, Potter DA (1994) Preferences of Japanese beetle and southern masked chafer (Coleoptera: Scarabaeidae) grubs among cool-season turfgrasses. J Entomol Sci 29:398–406
Denton CS, Bardgett RD, Cook R (1996) Nematode-microbial interactions in the rhizosphere of grassland plants. Abstracts of the Annual Biotechnology and Biological Sciences Research Council Plant Microbe Interactions Initiative Meeting, March 1996, University of Manchester
Denton CS, Bardgett RD, Cook R, Hobbs PJ (1999) Low amounts of root herbivory positively influence the rhizosphere microbial community in a temperate grassland soil. Soil Biol Biochem 31:155–165
Detling JK, Winn DT, Proctor-Gregg C, Painter KL (1980) Effects of simulated grazing by below-ground herbivores on growth, CO_2 exchange, and carbon allocation patterns of *Bouteloua gracilis*. J Appl Ecol 17:771–778
Dicke M, Dijkman H (2001) Within-plant circulation of systemic elicitor of induced defence and release from roots of elicitor that affects neighbouring plants. Biochem Syst Ecol 29:1075–1087
Dintenfass LP, Brown GC (1988) Quantifying effects of clover root curculio (Coleoptera: Curculionidae) larval feeding on biomass and root reserves of alfalfa. J Econ Entomol 81:641–648
Eissenstat DM, Wells CE, Yanai RD, Whitbeck JL (2000). Building roots in a changing environment: implications for root longevity. New Phytol 147:33–42
Engels RK, Nichols JT, Dodd JL, Brummer JE (1998) Root and shoot responses of sand bluestem to defoliation. J Range Manage 51:42–46

Ford MA, Grace JB (1998) Effects of vertebrate herbivores on soil processes, plant biomass, litter accumulation and soil elevation changes in a coastal marsh. J Ecol 86:974–982

Freckman DW, Barker KR, Coleman DC, Acras M, Dyer MI, Strain BR, McNaughton SJ (1991) The use of the ^{11}C technique to measure plant responses to herbivorous soil nematodes. Funct Ecol 5:810–818

French N (1969) Assessment of leather-jacket damage to grassland and economic aspects of control. Proceedings of the 5th British Insecticide and Fungicide Conference, vol 2, pp 511–521

Gange AC (2001) Species-specific responses of a root- and shoot-feeding insect to arbuscular mycorrhizal colonization of its host plant. New Phytol 150:611–618

Gange AC, Brown VK (1989) Effects of root herbivory by an insect on a foliar-feeding species, mediated through changes in the host plant. Oecologia 81:38–42

Gunn A, Cherrett JM (1993) The exploitation of food resources by soil meso- and macroinvertebrates. Pedobiologia 37:303–320

Hatcher PE (1996) The effect of insect-fungus interactions on the autumn growth and over-wintering of *Rumex crispus* and *R. obtusifolius* seedlings. J Ecol 84:101–109

Hendrick RL, Pregitzer KS (1992) The demography of fine roots in a northern hardwood forest. Ecology 73:1094–1104

Hendrick RL, Pregitzer KS (1993) Patterns of fine root mortality in two sugar maple forests. Nature 361:59–61

Hibbard BF, Bjostad LB (1988) Behavioral responses of western corn rootworm larvae to volatile semiochemicals from corn seedlings. J Chem Ecol 14:1523–1539

Hidaka M (1973) Effect of cutting on the total non-structural carbohydrates (TNC) contents in the roots and crowns of *Rumex obtusifolius* L. J JPN Grassl Sci 19: 313–317 (in Japanese with English summary)

Holland JN, Cheng WX, Crossley DA (1996) Herbivore-induced changes in plant carbon allocation: assessment of below-ground C fluxes using ^{14}C. Oecologia 107:87–94

Ingham RE, Detling JK (1986) Effects of defoliation and nematode consumption on growth and leaf gas exchange in *Bouteloua curtipendula*. Oikos 46:23–28

Jameson DA (1963) Responses of individual plants to harvesting. Bot Rev 29:532–594

Kilham K (1994) Soil ecology. Cambridge University Press, Cambridge

Kosola KR, Eissenstat DM, Graham JH (1995) Root demography of mature citrus trees: the influence of *Phytophthora nicotianae*. Plant Soil 171:283–288

Kosola KR, Dickman DI, Paul EA et al. (2001) Repeated insect defoliation effects on growth, nitrogen acquisition, carbohydrates, and root demography of poplars. Oecologia 129:65–74

Lawton JH, Strong DR (1981) Community patterns and competition in folivorous insects. Am Nat 118:317–338

Llewellyn M (1982) The energy economy of fluid-feeding herbivorous insects. In: Visser JH, Minks AK (eds) Proceedings of the 5th International Symposium on Insect-Plant Relationships , Wageningen, Netherlands, pp 243–252

Loveys BR, Bird AF (1973) The influence of nematodes on photosynthesis in tomato plants. Physiol Plant Path 3:525–52

Macfadyen A (1952) The small arthropods of a Mollinia fen at Cothill. J Anim Ecol 21:87–117

Maron JL (1998) Insect herbivory above- and belowground: Individual and joint effects on plant fitness. Ecology 79:1281–1293

Masters GJ (1995) The effect of herbivore density on host plant mediated interactions between two insects. Ecol Res 10:125–133

Masters GJ, Brown VK (1996) Host-plant mediated interactions between spatially separated herbivores: effects on community structure. In: Gange AC, Brown VK (eds) Multitrophic interactions in terrestrial ecosystems. 36th Symposium of the British Ecological Society. Blackwell Science, Oxford

Masters GJ, Brown VK, Gange AC (1993) Plant mediated interactions between above- and below-ground insect herbivores. Oikos 66:148–151

Masters GJ, Jones TH, Rogers M (2001) Host-plant mediated effects of root herbivory on insect seed predators and their parasitoids. Oecologia 127:246–250

Matter SF (2001) Effects of above and below ground herbivory by *Tetraopes tetraophthalmus* (Coleoptera: Cerambycidae) on the growth and reproduction of *Asclepias syriaca* (Asclepidacae). Env Entomol 30:333–338

McEvoy PB, Rudd NT, Cox CS, Huso M (1993) Disturbance, competition, and herbivory effects on ragwort *Senecio jacobaeae* populations. Ecol Monogr 63:55–75

McNaughton SJ, Banyikua FF, McNaughton MM (1998) Root biomass and productivity in a grazing ecosystem: The Serengeti. Ecology 79:587–592

Moran NA, Whitham TG (1990). Interspecific competition between root-feeding and leaf galling aphids mediated by host plant resistance. Ecology 71:1050–1058

Morón-Ríos A, Dirzo R, Jaramillo VJ (1997) Defoliation and below-ground herbivory in the grass *Muhlenbergia quadridentata*: effects on plant performance and on the root-feeder *Phyllophaga* sp. (Coleoptera, Melolonthidae). Oecologia 110:237–241

Mortimer S, Van der Putten WH, Brown VK (1999) Insect and nematode herbivory below-ground: interactions and role in vegetation development. In: Olff H, Brown VK, Drent RH (eds) Herbivores: between plants and predators. Blackwell, Oxford, pp 205–238

Müller H (1989) Structural analysis of the phytophagous insect guilds associated with the roots of *Centaura maculosa* Lam., *C. diffusa* Lam., and *C. vallesiaca* Jordan in Europe. Oecologia 78:41–52

Müller-Shärer H (1991) The impact of root herbivory as a function of plant density and competition – survival, growth and fecundity of *Centaurea maculosa* in field plots. J Appl Ecol 28:759–766

Murray PJ, Clements RO (1998) Transfer of nitrogen between clover and wheat: effect of root herbivory. Eur J Soil Biol 34:25–30

Murray PJ, Hatch DJ (1994) *Sitona* weevils (Coleoptera; Curculionidae) as agents for rapid transfer of nitrogen from white clover (*Trifolium repens* L.) to perennial rye grass (*Lolium perenne* L.). Ann Appl Biol 125:29–33

Murray PJ, Hatch DJ, Cliquet JB (1996) Impact of insect root herbivory on the growth and nitrogen and carbon contents of white clover (*Trifolium repens*) seedlings. Can J Bot 74:1591–1595

Notzold R, Blossey B, Newton E (1998) The influence of below ground herbivory and plant competition on growth and biomass allocation of purple loosestrife. Oecologia 113:82–93

Piearce TG (1978) Gut contents of some lumbricid earthworms. Pedobiologia 18:153–157

Philipson JJ, Coutts MP (1979) The induction of root dormancy in *Picea sitchensis* (Bong.) Carr. by abscisic acid. J Exp Bot 30:371–380

Quinn MA, Hall MH (1992) Compensatory response of a legume root-nodule system to nodule herbivory by *Sitona hispidulus*. Entomol Exp Appl 64:167–176

Rabe E (1990) Stress physiology: the functional significance of the accumulation of nitrogen-containing compounds. J Hortic Sci 65:231–243

Ramsell J, Malloch AJC, Whittaker JB (1993) When grazed by *Tipula paludosa*, *Lolium perenne* is a stronger competitor of *Rumex obtusifolius*. J Ecol 81:777–786

Ridsdill-Smith TJ (1977) Effects of root-feeding by Scarabaeid larvae on growth of perennial ryegrass plants. J Appl Ecol 14:73–80

Ridsdill-Smith TJ, Roberts RJ (1976) Insect density effects in root feeding by larvae of *Sericeothis nigrolineata* (Coleoptera: Scarabaeidae). J Appl Ecol 13:423–428

Ruess RW, Hendrick RL, Bryant JP (1998) Regulation of fine root dynamics by mammalian browsers in early successional Alaskan taiga forests. Ecology 79:2706–2720
Ryle GJA, Powell CE (1975) Defoliation and regrowth in the graminaceous plant: the role of current assimilate. Ann Bot 39:297–310
Salt DT, Fenwick P, Whittaker JB (1996a) Interspecific herbivore interactions in a high CO_2 environment: root and shoot aphids feeding on *Cardamine*. Oikos 77:326–330
Salt DT, Major E, Whittaker JB (1996b) Population dynamics of root aphids feeding on Sitka spruce in two commercial plantations. Pedobiologia 40:1–11
Scott, JA, French NR, Leethan JW (1979) Patterns of consumption in grasslands. In: French NR (ed) Perspectives in grassland ecology. Ecological Studies 32. Springer, Berlin Heidelberg New York, pp 89–105
Sheppard AW, Aeschlimann J-P, Sagliocco J-L, Vitou J (1995) Below-ground herbivory in *Carduus nutans* (Asteraceae) and the potential for biological control. Biocontrol Sci Technol 5:261–270
Sibma L, Kort J, de Wit CT (1964) Experiments on competition as a means of detecting possible damage by nematodes. Jaarboek, Instituut voor biologischen scheikundig onderzoek van Landbouwgewassen 1964:119–124
Smith JP, Schowalter TD (2001) Aphid-induced reduction of shoot and root growth in Douglas-fir seedlings. Ecol Entomol 26:411–416
Stanton NL (1988) The underground in grasslands. Annu Rev Ecol Syst 19:573–589
Steinger T, Müller-Schärer H (1992) Physiological and growth responses of *Centaurea maculosa* (Asteraceae) to root herbivory under varying levels of interspecific plant competition and soil nitrogen availability. Oecologia 91:141–149
Strong DR, Maron JL, Connors PG, Whipple A, Harrison S, Jefferies RL (1995) High mortality, fluctuation in numbers, and heavy subterranean insect herbivory in bush lupine, *Lupinus arboreus*. Oecologia 104:85–92
Van Tol RWHM, Van der Sommen ATC, Boff MIC, et al. (2001) Plants protect their roots by alerting the enemies of grubs. Ecol Lett 4:292–294
Verschoor BC (2001) Nematode-plant interactions in grasslands under restoration management. PhD thesis, Wageningen University, The Netherlands
Vranjic JA, Ash JE (1997) Scale insects consistently affect roots more than shoots: the impact of infestation size on growth of eucalypt seedlings. J Ecol 85:143–149
Wallace HR (1974) The influence of root-knot nematode, *Meloidogyne javanica*, on photosynthesis and nutrient demand by roots of tomato plants. Nematologica 17:154–166
Wardle DA, Barker GM (1997) Competition and herbivory in establishing grassland communities: implications for plant biomass, species diversity and soil microbial activity. Oikos 80:470–480
Whittaker JB (1979) Invertebrate grazing, competition and plant dynamics. In: Anderson RM, Turner BD, Taylor LR (eds) Population dynamics. Symposium of the British Ecological Society 20:207–222. Blackwell, Oxford
Whittaker JB (1982) The effect of grazing by a chrysomelid beetle, *Gastrophysa viridula*, on growth and survival of *Rumex crispus* on a shingle bank. J Ecol 70:291–296
Yeates GW, Bardgett RD, Cook R, Hobbs PJ, Bowling PJ, Potter JF (1997) Faunal and microbial diversity in three Welsh grassland soils under conventional and organic management regimes. J Appl Ecol 34:453–70
Yeates GW, Saggar S, Denton CS, Mercer CF (1998) Impact of clover cyst nematode (*Heterodera trifolii*) infection on soil microbial activity in the rhizosphere of white clover (*Trifolium repens*) – a pulse labelling experiment. Nematologica 44:81–90
Yeates GW, Bardgett RD, Mercer CF, Saggar S, Feltham CW (1999) The impact of feeding by five nematodes on ^{14}C distribution in soil microbial biomass and nematodes: initial observations. N Z J Zool 26:87

Subject Index

S

Ecological Studies

Volumes published since 1997

Volume 129
Pelagic Nutrient Cycles: Herbivores as Sources and Sinks (1997)
T. Andersen

Volume 130
Vertical Food Web Interactions: Evolutionary Patterns and Driving Forces (1997)
K. Dettner, G. Bauer, and W. Völkl (Eds.)

Volume 131
The Structuring Role of Submerged Macrophytes in Lakes (1998)
E. Jeppesen et al. (Eds.)

Volume 132
Vegetation of the Tropical Pacific Islands (1998)
D. Mueller-Dombois and F.R. Fosberg

Volume 133
Aquatic Humic Substances: Ecology and Biogeochemistry (1998)
D.O. Hessen and L.J. Tranvik (Eds.)

Volume 134
Oxidant Air Pollution Impacts in the Montane Forests of Southern California (1999)
P.R. Miller and J.R. McBride (Eds.)

Volume 135
Predation in Vertebrate Communities: The Białowieża Primeval Forest as a Case Study (1998)
B. Jędrzejewska and W. Jędrzejewski

Volume 136
Landscape Disturbance and Biodiversity in Mediterranean-Type Ecosystems (1998)
P.W. Rundel, G. Montenegro, and F.M. Jaksic (Eds.)

Volume 137
Ecology of Mediterranean Evergreen Oak Forests (1999)
F. Rodà et al. (Eds.)

Volume 138
Fire, Climate Change and Carbon Cycling in the North American Boreal Forest (2000)
E.S. Kasischke and B. Stocks (Eds.)

Volume 139
Responses of Northern U.S. Forests to Environmental Change (2000)
R. Mickler, R.A. Birdsey, and J. Hom (Eds.)

Volume 140
Rainforest Ecosystems of East Kalimantan: El Niño, Drought, Fire and Human Impacts (2000)
E. Guhardja et al. (Eds.)

Volume 141
Activity Patterns in Small Mammals: An Ecological Approach (2000)
S. Halle and N.C. Stenseth (Eds.)

Volume 142
Carbon and Nitrogen Cycling in European Forest Ecosystems (2000)
E.-D. Schulze (Ed.)

Volume 143
Global Climate Change and Human Impacts on Forest Ecosystems: Postglacial Development, Present Situation and Future Trends in Central Europe (2001)
J. Puhe and B. Ulrich

Volume 144
Coastal Marine Ecosystems of Latin America (2001)
U. Seeliger and B. Kjerfve (Eds.)

Volume 145
Ecology and Evolution of the Freshwater Mussels Unionoida (2001)
G. Bauer and K. Wächtler (Eds.)

Volume 146
Inselbergs: Biotic Diversity of Isolated Rock Outcrops in Tropical and Temperate Regions (2000)
S. Porembski and W. Barthlott (Eds.)

Volume 147
Ecosystem Approaches to Landscape Management in Central Europe (2001)
J.D. Tenhunen, R. Lenz, and R. Hantschel (Eds.)

Volume 148
A Systems Analysis of the Baltic Sea (2001)
F.V. Wulff, L.A. Rahm, and P. Larsson (Eds.)

Volume 149
Banded Vegetation Patterning in Arid and Semiarid Environments (2001)
D. Tongway and J. Seghieri (Eds.)

Volume 150
Biological Soil Crusts: Structure, Function, and Management (2001)
J. Belnap and O.L. Lange (Eds.)

Volume 151
Ecological Comparisons of Sedimentary Shores (2001)
K. Reise (Ed.)

Volume 152
Global Biodiversity in a Changing Environment: Scenarios for the 21st Century (2001)
F.S. Chapin, O. Sala, and E. Huber-Sannwald (Eds.)

Volume 153
UV Radiation and Arctic Ecosystems (2002)
D.O. Hessen (Ed.)

Volume 154
Geoecology of Antarctic Ice-Free Coastal Landscapes (2002)
L. Beyer and M. Bölter (Eds.)

Volume 155
Conserving Biological Diversity in East African Forests: A Study of the Eastern Arc Mountains (2002)
W.D. Newmark

Volume 156
Urban Air Pollution and Forests: Resources at Risk in the Mexico City Air Basin (2002)
M.E. Fenn, L. I. de Bauer, and T. Hernández-Tejeda (Eds.)

Volume 157
Mycorrhizal Ecology (2002)
M.G.A. van der Heijden and I.R. Sanders (Eds.)

Volume 158
Diversity and Interaction in a Temperate Forest Community: Ogawa Forest Reserve of Japan (2002)
T. Nakashizuka and Y. Matsumoto (Eds.)

Volume 159
Big-Leaf Mahogany: Genetic Resources, Ecology and Management (2003)
A. E. Lugo, J. C. Figueroa Colón, and M. Alayón (Eds.)

Volume 160
Fire and Climatic Change in Temperate Ecosystems of the Western Americas (2003)
T. T. Veblen et al. (Eds.)

Volume 161
Competition and Coexistence (2002)
U. Sommer and B. Worm (Eds.)

Volume 162
How Landscapes Change: Human Disturbance and Ecosystem Fragmentation in the Americas (2003)
G.A. Bradshaw and P.A. Marquet (Eds.)

Volume 163
Fluxes of Carbon, Water and Energy of European Forests (2003)
R. Valentini (Ed.)

Volume 164
Herbivory of Leaf-Cutting Ants: A Case Study on *Atta colombica* in the Tropical Rainforest of Panama (2003)
R. Wirth, H. Herz, R.J. Ryel, W. Beyschlag, B. Hölldobler

Volume 165
Population Viability in Plants: Conservation, Management, and Modeling of Rare Plants (2003)
C.A Brigham, M.W. Schwartz (Eds.)

Volume 166
North American Temperate Deciduous Forest Responses to Changing Precipitation Regimes (2003)
P. Hanson and S.D. Wullschleger (Eds.)

Volume 167
Alpine Biodiversity in Europe (2003)
L. Nagy et al. (Eds.)

Volume 168
Root Ecology (2003)
H. de Kroon and E.J.W. Visser (Eds.)

MIX
Papier aus verantwortungsvollen Quellen
Paper from responsible sources
FSC® C105338

If you have any concerns about our products,
you can contact us on
ProductSafety@springernature.com

In case Publisher is established outside the EU,
the EU authorized representative is:
Springer Nature Customer Service Center GmbH
Europaplatz 3, 69115 Heidelberg, Germany

Printed by Libri Plureos GmbH
in Hamburg, Germany